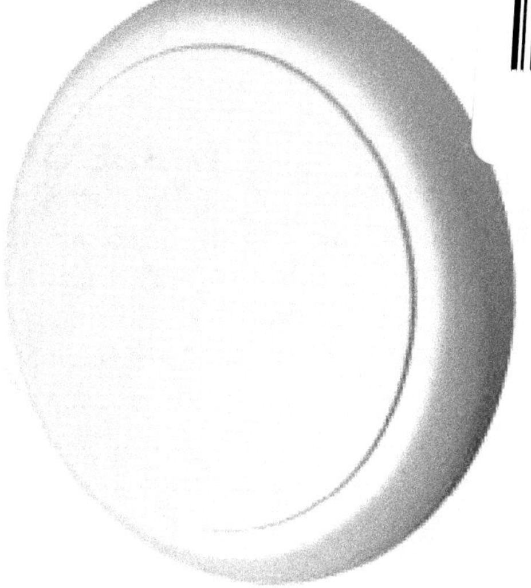

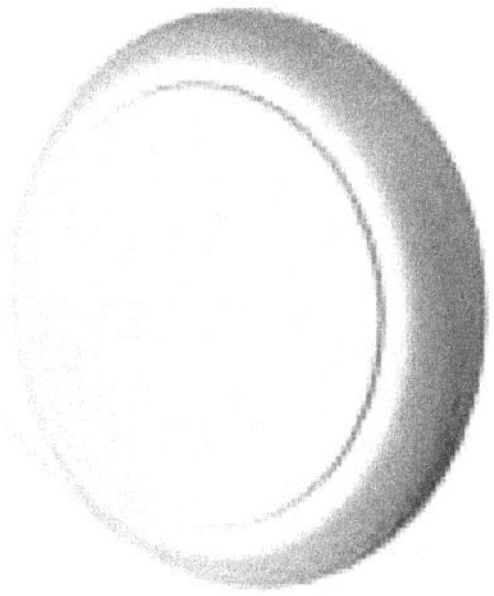

Everything is a Circle:

A New Model For Orbits of Bodies In The Universe

ASLI PINAR TAN

" Allah alone is the God of the two Easts and of the two Wests."
– Quran, Surah Ar-Rahman / 17

TABLE OF CONTENTS

INTRODUCTION

Based on measured astronomical position data of heavenly objects in the Solar System and other planetary systems, all bodies in space seem to move in some kind of elliptical motion with respect to each other. According to Kepler's 1st Law, "orbit of a planet with respect to the Sun is an ellipse, with the Sun at one of the two foci." Orbit of the Moon with respect to Earth is also distinctly elliptical, but this ellipse has a varying eccentricity as the Moon comes closer to and goes farther away from Earth in a harmonic style along a full cycle of this ellipse.

In this book, our research results are presented in detail in a set of scientific papers as Articles, where in **ARTICLE 1** it is first mathematically demonstrated and proven that the "distance between points around any two different circles in three dimensional space" is equivalent to the "distance of points around a vector ellipse to another fixed or moving point, as in two dimensional space". What is done is equivalent to showing that bodies moving on two different circular orbits in space vector wise behave as if moving on an elliptical path with respect to each other, and virtually seeing each other as positioned at an instantaneously stationary point in space on their relative ecliptic plane, whether they are moving with the same angular velocity, or different but fixed angular velocities, or even with different and changing angular velocities with respect to their own centers of revolution. This mathematical revelation has the potential to lead to far reaching discoveries in physics, enabling more insight into forces of nature, with a formulation of a new fundamental model regarding the motions of bodies in the Universe, including the Sun, Planets, and Satellites in the Solar System and elsewhere, as well as at particle and subatomic level. Based on the demonstrated mathematical analysis, as they exhibit almost fixed elliptic orbits relative to one another over time, the assertion is made that the Sun, the Earth, and the Moon must each be revolving in their individual circular orbits of revolution in space. With this expectation, individual orbital parameters of Sun and Earth within their relative motion are calculated in **ARTICLE 3** based on observed Earth to Sun distance data, and individual orbital parameters of Earth and Moon within their relative motion are calculated in **ARTICLE 4** based on observed Earth to Moon distance data, also using analytical methods developed as part of this research in **ARTICLE 2**, **ARTICLE 3**, and **ARTICLE 4** to an approximation. This calculation and analysis process have revealed additional results aligned with observation, and these also support our assertion that the Sun, the Earth, and the Moon must actually be revolving in individual circular orbits. An overall description of these research results as a whole is presented in **ARTICLE 5**. An overview summary of these original research results are presented in Turkish in **ARTICLE 6 (TÜRKÇE)**, which is also under Notary registration in Istanbul, Turkey where the Author lives, and this section has been included in this book as a reference for Turkish readers.

ABOUT THE AUTHOR

Aslı Pınar Tan is the only author contributing to this Book, and her findings in this Article are part of all her findings as a result of her personal theoretical studies and research over the years independent from any institution or university, which she has published in this book and will further publish in a set of books and other articles.

She is an Electrical & Electronics Engineer, with Bsc. and Msc. from Bilkent University, Ankara, Turkey, and she also holds a degree in Multi-Disciplinary Space Studies program from International Space University, Strasbourg, France. She has been working professionally as a Senior Consultant & Director in the global ICT / Telecom industry, in addition to her other academic career in Electromagnetics and Space Applications, all independent from this research. She has been granted a Young Scientist Award by URSI (Union Radio Science Internationale) based on her Master Thesis Research results presenting a systematic methodology for designing High-Frequency Radar Antennas using the Genetic Algorithm and Equivalent Edge Currents, to generate plain waves in the near field, which is utilized in the industry since then.

(For more info: http://www.linkedin.com/in/apinartan)

ARTICLE 1

POINTS ON TWO CIRCLES IN SPACE MIMIC AN ELLIPSE WITH RESPECT TO A POINT

Author: Aslı Pınar Tan[I]

SUMMARY

Based on measured astronomical position data of heavenly objects in the Solar System and other planetary systems, all bodies in space seem to move in some kind of elliptical motion with respect to each other. In this article, it is mathematically demonstrated and proven that the "distance between points on any two different circles in three-dimensional space" is equivalent to the "distance of points on a vector ellipse from another fixed or moving point, like in two-dimensional space". Equivalently, it amounts to showing that moving points on two circles in space vector-wise mimic an elliptical path with respect to each other, even when they are moving with *different* and *changing* angular velocities with respect to their own centers of revolution. This mathematical revelation leads to far-reaching discoveries in physics, including formulation of a fundamental new model of motion for bodies in the Universe, enabling more insight into forces in nature.

ARTICLE

According to Kepler's 1st Law[1,2], "the orbit of a planet with respect to the Sun is an ellipse[5], with the Sun at one of the two foci", which is an empirical rule concluded through near estimation based on the observation of astronomical position data measured over time in the Solar System. The orbit of the Moon[3] with respect to the Earth is also distinctly elliptical, but with a varying eccentricity in a harmonic style as the Moon comes closer to and goes farther away from the Earth harmonically along a full cycle of this ellipse, again based on the observation of astronomical position data measured over time. In this article, it is mathematically demonstrated and proven that the "distance between points on any two different circles in three-dimensional space" is equivalent to the "distance of points on a vector ellipse from another fixed or moving point, similar to two-dimensional space". In other words, it is shown that moving points on two circles in space vector-wise mimic an elliptical path with respect to each other, whether they are moving with the *same* angular velocity, or *different* but *fixed* angular velocities, or even with *different* and *changing* angular velocities with respect to their own centers of revolution, virtually seeing each other as positioned at a instantaneously stationary point in space on their respective virtual ecliptic plane. The observation result by Very Large Baseline Array (VLBA) of antennas published[20] by National Radio Astronomy Observatory on August 4, 2020 about the orbits of a Saturn-sized planet and its small, cool star 35 light years from Earth is a proof of this assertion.

Consider a system of two circles in three-dimensional space, with the geometry of the system demonstrated as in **Figure 1** in Cartesian $(\hat{x},\hat{y},\hat{z})$ coordinates. The two circles have vector radii $\bar{r}_1$ (1) - (3) and $\bar{r}_2$ (4) - (5), with constant magnitudes r_1 (6) and r_2 (6), respectively, where the unit vector is $\hat{r}_1(\phi)$ (2) in the direction of $\bar{r}_1$ (1) - (3) and the unit vector is $\hat{r}_2(\phi)$ (5) in the

---

[I] *ASLI PINAR TAN*
Linkedin Website: *https://www.linkedin.com./in/apinartan*

direction of $\vec{r}_2$ (4) - (5). The scalar magnitude r_1 (6) of the radius vector $\vec{r}_1$ (1) - (3) is calculated as the square root of r_1^2 (6), which in turn is calculated in terms of the Dot Product[4] $[\vec{r}_1 \cdot \vec{r}_1]$ (6) of the vector $\vec{r}_1$ (1) - (3) with itself. The scalar magnitude r_2 (6) of the radius vector $\vec{r}_2$ (4) - (5) is calculated as the square root of r_2^2 (6), which in turn is calculated in terms of the Dot Product[4] $[\vec{r}_2 \cdot \vec{r}_2]$ (6) of the vector $\vec{r}_2$ (4) - (5) with itself. The centers of these two circles are displaced by a constant or variable vector $\vec{\ell}(\phi)$ (7) with magnitude $\ell(\phi)$ (8) at each phase ϕ. The scalar distance between the centers of the two circles at every phase ϕ, namely magnitude $\ell(\phi)$ (8) of $\vec{\ell}(\phi)$ (7), is calculated as the square root of $\ell^2(\phi)$ (8), which in turn is calculated in terms of the Dot Product[4] $\left[\vec{\ell}(\phi) \cdot \vec{\ell}(\phi)\right]$ (8) of the vector $\vec{\ell}(\phi)$ (7) with itself.

Points $\mathbf{P}_1$ and $\mathbf{P}_2$ are defined on these two circles, respectively, phased apart by a constant or time(t)-dependent angle ϕ_0 (1), demonstrated as in **Figure 1** spatially in the most generic form, and expressed in the form of generic vector equations below in (1) - (8). More explicitly, in the Cartesian $(\hat{x}, \hat{y}, \hat{z})$ coordinate configuration of **Figure 1**, at each ϕ, the location of $\mathbf{P}_1$ phased at $(\phi+\phi_0)$ is defined by the vector $\left[\vec{r}_1(\phi)+\vec{\ell}(\phi)\right]$ based on (1) - (3) and (7), and the location of $\mathbf{P}_2$ phased at (ϕ) is defined by the vector $\vec{r}_2(\phi)$ (4) - (5). Note that the inclination angle β (1) between the planes of these two circles is also taken to be constant.

$$\vec{r}_1 = \vec{r}_1(\phi+\phi_0) = \hat{x}\, r_1 Cos(\phi+\phi_0) Cos\,\beta + \hat{y}\, r_1 Sin(\phi+\phi_0) + \hat{z}\, r_1 Cos(\phi+\phi_0) Sin\,\beta \tag{1}$$

$$\hat{r}_1(\phi) = \hat{x}\, Cos(\phi+\phi_0) Cos\,\beta + \hat{y}\,{}_1 Sin(\phi+\phi_0) + \hat{z}\, Cos(\phi+\phi_0) Sin\,\beta \;\Rightarrow\; \vec{r}_1 = \hat{r}_1(\phi) r_1 \tag{2}$$

$$\begin{aligned} \Rightarrow\; \vec{r}_1 = \vec{r}_1(\phi) = &\left(\hat{x}\, r_1 Cos\,\beta\, Cos\,\phi_0 + \hat{y}\, r_1 Sin\,\phi_0 + \hat{z}\, r_1 Sin\,\beta\, Cos\,\phi_0\right)\; Cos\,\phi\, + \\ &\left(-\hat{x}\, r_1 Cos\,\beta\, Sin\,\phi_0 + \hat{y}\, r_1 Cos\,\phi_0 - \hat{z}\, r_1 Sin\,\beta\, Sin\,\phi_0\right) Sin\,\phi \end{aligned} \tag{3}$$

$$\vec{r}_2 = \vec{r}_2(\phi) = \hat{x}\, r_2\, Cos\,\phi + \hat{y}\, r_2\, Sin\,\phi \tag{4}$$

$$\hat{r}_2(\phi) = \hat{x}\, Cos\,\phi + \hat{y}\, Sin\,\phi \;\Rightarrow\; \vec{r}_2 = \hat{r}_2(\phi) r_2 \tag{5}$$

$$\vec{r}_1 \cdot \vec{r}_1 = r_1^2 \;\;;\;\; |\vec{r}_1| = r_1 = \sqrt{\vec{r}_1 \cdot \vec{r}_1} \;\;;\;\; \vec{r}_2 \cdot \vec{r}_2 = r_2^2 \;\;;\;\; |\vec{r}_2| = r_2 = \sqrt{\vec{r}_2 \cdot \vec{r}_2} \tag{6}$$

$$\vec{\ell} = \vec{\ell}(\phi) = \hat{x}\,\ell_x(\phi) + \hat{y}\,\ell_y(\phi) + \hat{z}\,\ell_z(\phi) \tag{7}$$

$$\left|\vec{\ell}(\phi)\right| = \ell(\phi) = \sqrt{\ell^2(\phi)} = \sqrt{\vec{\ell}(\phi)\cdot\vec{\ell}(\phi)} \;\;;\;\; \ell^2(\phi) = \vec{\ell}(\phi)\cdot\vec{\ell}(\phi) = \ell_x^{\,2}(\phi) + \ell_y^{\,2}(\phi) + \ell_z^{\,2}(\phi) \tag{8}$$

The vector distance $\vec{d}(\phi)$ (9) - (12) between any of these two points $\mathbf{P}_1$ phased at $(\phi+\phi_0)$ and $\mathbf{P}_2$ phased at (ϕ) on the two respective circles, and its magnitude $d(\phi)$ (13) which is also the scalar distance between $\mathbf{P}_1$ and $\mathbf{P}_2$, is as demonstrated in **Figure 2** and expressed in the following vector equations in (9) - (17), based on (1) - (8). It is important to note that the scalar distance $d(\phi)$ (13) between $\mathbf{P}_1$ and $\mathbf{P}_2$ is calculated as the square root of $d^2(\phi)$ (14) - (16), which in turn is calculated in terms of the Dot Product[4] $\left[\vec{d}(\phi)\cdot\vec{d}(\phi)\right]$ (14) - (16) of the vector distance $\vec{d}(\phi)$ (9) - (12) with itself. Unit vector in the direction of $\vec{d}(\phi)$ (9) - (12) is $\hat{d}(\phi)$ (17).

$$\vec{d}(\phi)=\vec{r}_1(\phi+\phi_0)-\vec{r}_2(\phi)+\vec{\ell}(\phi)=\vec{r}_1-\vec{r}_2+\vec{\ell} \tag{9}$$

$$\begin{aligned}\Rightarrow\ \vec{d}(\phi)=&\hat{x}\ [r_1Cos(\phi+\phi_0)Cos\,\beta-r_2\,Cos\,\phi+\ell_x(\phi)]+\hat{y}\ [r_1Sin(\phi+\phi_0)-r_2Sin\phi+\ell_y(\phi)]\\&+\hat{z}\ [r_1Cos(\phi+\phi_0)Sin\,\beta+\ell_z(\phi)]\end{aligned} \tag{10}$$

$$\begin{aligned}\Rightarrow\ \vec{d}(\phi)=&\left[\hat{x}(r_1Cos\,\beta\,Cos\,\phi_0-r_2)+\hat{y}\,r_1Sin\phi_0+\hat{z}\,r_1Sin\,\beta\,Cos\,\phi_0\right]\ Cos\,\phi+\\&\left[-\hat{x}\,r_1Cos\,\beta\,Sin\,\phi_0+\hat{y}(r_1Cos\,\phi_0-r_2)-\hat{z}\,r_1Sin\,\beta\,Sin\,\phi_0\right]Sin\ \phi+\\&\left[\hat{x}\ell_x(\phi)+\hat{y}\ell_y(\phi)+\hat{z}\ell_z(\phi)\right]\end{aligned} \tag{11}$$

$$\vec{d}(\phi)=\hat{x}d_x(\phi)+\hat{y}d_y(\phi)+\hat{z}d_z(\phi)\ \Rightarrow\ \begin{cases}d_x(\phi)=r_1Cos(\phi+\phi_0)Cos\,\beta-r_2\,Cos\phi+\ell_x(\phi)\\d_y(\phi)=r_1Sin(\phi+\phi_0)-r_2Sin\phi+\ell_y(\phi)\\d_z(\phi)=r_1Cos(\phi+\phi_0)Sin\,\beta+\ell_z(\phi)\end{cases} \tag{12}$$

$$d(\phi)=\left|\vec{d}(\phi)\right|=\sqrt{d^2(\phi)}=\sqrt{\vec{d}(\phi)\cdot\vec{d}(\phi)}=\sqrt{\left[d_x(\phi)\right]^2+\left[d_y(\phi)\right]^2+\left[d_z(\phi)\right]^2} \tag{13}$$

$$d^2(\phi)=\vec{d}(\phi)\cdot\vec{d}(\phi)=(\vec{r}_1-\vec{r}_2+\vec{\ell})\cdot(\vec{r}_1-\vec{r}_2+\vec{\ell})=\vec{r}_1\cdot\vec{r}_1-2\,\vec{r}_1\cdot\vec{r}_2+\vec{r}_2\cdot\vec{r}_2+2(\vec{r}_1-\vec{r}_2)\cdot\vec{\ell}+\vec{\ell}\cdot\vec{\ell} \tag{14}$$

$$\begin{aligned}d^2(\phi)=&\vec{d}(\phi)\cdot\vec{d}(\phi)\\=&[r_1Cos(\phi+\phi_0)Cos\,\beta-r_2\,Cos\,\phi+\ell_x(\phi)]^2+[r_1Sin(\phi+\phi_0)-r_2\,Sin\phi+\ell_y(\phi)]^2+\\&[r_1Cos(\phi+\phi_0)Sin\,\beta+\ell_z(\phi)]^2\end{aligned} \tag{15}$$

$$d^2(\phi)=\vec{d}(\phi)\cdot\vec{d}(\phi)=\left[d_x(\phi)\right]^2+\left[d_y(\phi)\right]^2+\left[d_z(\phi)\right]^2 \tag{16}$$

$$\hat{d}(\phi)=\frac{\vec{d}(\phi)}{\left|\vec{d}(\phi)\right|}=\frac{\vec{d}(\phi)}{d(\phi)}=\hat{x}\frac{d_x(\phi)}{d(\phi)}+\hat{y}\frac{d_y(\phi)}{d(\phi)}+\hat{z}\frac{d_z(\phi)}{d(\phi)}\ \Rightarrow\ \vec{d}(\phi)=\hat{d}(\phi)\left|\vec{d}(\phi)\right|=\hat{d}(\phi)d(\phi) \tag{17}$$

The vector distance $\vec{d}(\phi)$ (11) at any value of the phase ϕ, between points $\mathbf{P}_1$ phased at $(\phi+\phi_0)$ and $\mathbf{P}_2$ phased at (ϕ) on two respective circles, can be equivalently expressed as $\vec{d}(\phi)$ (18) in terms of virtual vectors $\vec{\mathbf{X}}(\phi)$ (19) and $\vec{\mathbf{Y}}(\phi)$ (20), defined utilizing $\vec{r}_1$ (1) - (3) and $\vec{r}_2$ (4) - (5). Based on the definition of $\vec{\mathbf{X}}(\phi)$ (19) from (18), (7) and (11), the virtual vector $\vec{a}$ (21) is also defined, and based on the definition of $\vec{\mathbf{Y}}(\phi)$ (20) from (18), (7) and (11), another virtual vector $\vec{b}$ (22) is also defined.

The magnitude $X(\phi)$ (19) of the virtual vector $\vec{\mathbf{X}}(\phi)$ (19) is calculated as the square root of $X^2(\phi)$ (19), which in turn is calculated in terms of the Dot Product[4] $\left[\vec{\mathbf{X}}(\phi)\cdot\vec{\mathbf{X}}(\phi)\right]$ (19) of the vector $\bar{\mathbf{X}}(\phi)$ (19) with itself. The magnitude $Y(\phi)$ (20) of the virtual vector $\bar{\mathbf{Y}}(\phi)$ (20) is calculated as the square root of $Y^2(\phi)$ (20), which in turn is calculated in terms of the Dot Product[4] $\left[\bar{\mathbf{Y}}(\phi)\cdot\bar{\mathbf{Y}}(\phi)\right]$ (20) of the vector $\bar{\mathbf{Y}}(\phi)$ (20) with itself. The magnitude a (23) of the virtual vector $\vec{a}$ (21) is calculated as the square root of a^2 (23), which in turn is calculated in terms of the Dot Product[4] $(\vec{a}\cdot\vec{a})$ (23) of the vector $\vec{a}$ (21) with itself. The magnitude b (24) of the virtual vector $\vec{b}$ (22) is calculated as the square root of b^2 (24), which in turn is calculated in terms of the Dot Product[4] $(\vec{b}\cdot\vec{b})$ (24) of the vector $\vec{b}$ (22) with itself. When the phase difference ϕ_0 (1) is constant for all ϕ, the magnitudes a (23) and b (24) are constant for all ϕ, and when the phase difference ϕ_0 (1) is time(t)-dependent, i.e. if $\phi_0 = \phi_0(t)$, a (23) and b (24) are also time(t)-dependent for different ϕ. Based on the definitions of $\vec{a}$ (21) and $\vec{b}$ (22), their Dot Product[4] $(\vec{a}\cdot\vec{b})$ (25) is also defined.

$$\vec{d}(\phi)=\bar{\mathbf{X}}(\phi)+\bar{\mathbf{Y}}(\phi)+\vec{\ell}(\phi) \qquad where \qquad \vec{r}_1(\phi)-\vec{r}_2(\phi)=\bar{\mathbf{X}}(\phi)+\bar{\mathbf{Y}}(\phi)=\vec{a}\,Cos\,\phi+\vec{b}\,Sin\phi \quad (18)$$

$$\bar{\mathbf{X}}(\phi)=\vec{a}\,Cos\,\phi \quad ; \quad \bar{\mathbf{X}}(\phi)\cdot\bar{\mathbf{X}}(\phi)=X^2(\phi)=\vec{a}\cdot\vec{a}\;Cos^2\phi=a^2Cos^2\phi \quad ; \quad \left|\bar{\mathbf{X}}(\phi)\right|=X(\phi) \quad (19)$$

$$\bar{\mathbf{Y}}(\phi)=\vec{b}\,Sin\phi \quad ; \quad \bar{\mathbf{Y}}(\phi)\cdot\bar{\mathbf{Y}}(\phi)=Y^2(\phi)=\vec{b}\cdot\vec{b}\;Sin^2\phi=b^2Sin^2\phi \quad ; \quad \left|\bar{\mathbf{Y}}(\phi)\right|=Y(\phi) \quad (20)$$

$$\vec{a}=\hat{x}\left(r_1Cos\,\beta\,Cos\,\phi_0-r_2\right)+\hat{y}\,r_1Sin\phi_0+\hat{z}\,r_1Sin\,\beta\,Cos\,\phi_0 \;\Rightarrow\; \vec{a}=\left[\vec{r}_1(\phi)-\vec{r}_2(\phi)\right](\phi=0) \quad (21)$$

$$\vec{b}=-\hat{x}\,r_1Cos\,\beta\,Sin\phi_0+\hat{y}\left(r_1Cos\,\phi_0-r_2\right)-\hat{z}\,r_1Sin\,\beta\,Sin\phi_0 \;\Rightarrow\; \vec{b}=\left[\vec{r}_1(\phi)-\vec{r}_2(\phi)\right]\left(\phi=\frac{\pi}{2}\right) \quad (22)$$

$$\vec{a}\cdot\vec{a}=a^2=r_1^{\,2}-2\,r_1\,r_2\,Cos\beta\,Cos\,\phi_0+r_2^{\,2} \quad ; \quad \left|\vec{a}\right|=a \quad (23)$$

$$\vec{b}\cdot\vec{b}=b^2=r_1^{\,2}-2\,r_1\,r_2\,Cos\,\phi_0+r_2^{\,2} \quad ; \quad \left|\vec{b}\right|=b \quad (24)$$

$$\vec{a}\cdot\vec{b}=r_1\,r_2\left(Cos\beta-1\right)Sin\,\phi_0 \quad (25)$$

According to the definitions of vectors $\bar{\mathbf{X}}(\phi)$ (19), $\bar{\mathbf{Y}}(\phi)$ (20), $\vec{a}$ (21), and $\vec{b}$ (22), as described in (18) - (24), the relation in (26) is valid and holds for all ϕ.

$$\frac{\bar{\mathbf{X}}(\phi)\cdot\bar{\mathbf{X}}(\phi)}{\vec{a}\cdot\vec{a}}+\frac{\bar{\mathbf{Y}}(\phi)\cdot\bar{\mathbf{Y}}(\phi)}{\vec{b}\cdot\vec{b}}=\frac{X^2(\phi)}{a^2}+\frac{Y^2(\phi)}{b^2}=Cos^2\phi+Sin^2\phi=1 \quad (26)$$

$$\Rightarrow \boxed{\frac{\bar{\mathbf{X}}(\phi)\cdot\bar{\mathbf{X}}(\phi)}{\vec{a}\cdot\vec{a}}+\frac{\bar{\mathbf{Y}}(\phi)\cdot\bar{\mathbf{Y}}(\phi)}{\vec{b}\cdot\vec{b}}=1} \quad \left(Definition\ of\ Vector\ Ellipse\ in\ 3-Dimensions\right) \quad (27)$$

$$\Rightarrow \boxed{\frac{X^2(\phi)}{a^2}+\frac{Y^2(\phi)}{b^2}=1} \quad \left(Definition\ of\ Scalar\ Ellipse\ in\ 2-Dimensions\right) \quad (28)$$

Therefore, the relation in (26) reveals the validity of (27) and (28) for the vector pair $\left[\bar{\mathbf{X}}(\phi),\bar{\mathbf{Y}}(\phi)\right]$ (19) - (20) and its magnitude pair $\left[X(\phi),Y(\phi)\right]$ (19) - (20), respectively. As

(28) is the defining equation of an ellipse[5] in two dimensions, where a (23) is the semi-major[6] axis and b (24) is the semi-minor[6] axis of the ellipse[5] when $(a>b)$[5,6] holds, and vice versa, with (28) reducing to the special case of a circle when $(a=b)$[5,6] holds, we can claim that (27) indicates that the vector pair $\left[\vec{\mathbf{X}}(\phi),\vec{\mathbf{Y}}(\phi)\right]$ (19) - (20) defines points on a vector ellipse in three dimensions in the most general case. *In other words, the vector distance $\vec{\boldsymbol{d}}(\phi)$ (11) between two points $\mathbf{P}_1$ phased at $(\phi+\phi_0)$ and $\mathbf{P}_2$ phased at (ϕ) on two respective circles, whose centers are displaced by a constant or variable vector $\vec{\ell}(\phi)$ (7), can equivalently be mathematically expressed and interpreted as the distance $\vec{\boldsymbol{d}}(\phi)$ (18) of points on a virtual vector ellipse, whose locations with respect to a virtual origin at each ϕ are determined by the sum of vector pair $\left[\vec{\mathbf{X}}(\phi),\vec{\mathbf{Y}}(\phi)\right]$ (19) - (20), from another fixed or moving point displaced from the same virtual origin of the ellipse by a constant or variable vector $\left[-\vec{\ell}(\phi)\right]$ (7), where $\vec{\boldsymbol{a}}$* (21) *and $\vec{\boldsymbol{b}}$* (22) *are the fixed or variable semi-major[6] and semi-minor[6] axis vectors of the vector ellipse. This result is mathematically valid even when the phase difference ϕ_0 (1) is a variable function of time (t), i.e. even if $\phi_0=\phi_0(t)$.* This revelation is the core and most significant finding of this Article. Moreover, we have also mathematically introduced the concept of a "vector ellipse" (27) in three-dimensional space based on our analysis introduced along the lines of (1) - (28).

Square of the focal[5,7] distance c^2 (29) of the vector ellipse can be determined using (23) - (24), and the eccentricity[5] e (30) of the vector ellipse can be found using (23) - (24) and (29).

$$c^2=\left|a^2-b^2\right|=2\,r_1 r_2\left(1-Cos\beta\right)\left|Cos\,\phi_0\right| \qquad (Focal\ Distance\ Squared) \tag{29}$$

$$\overset{(a>b)}{e}=\frac{c}{a}=\sqrt{\frac{2\,r_1 r_2\left(1-Cos\beta\right)\left|Cos\,\phi_0\right|}{r_1^2-2\,r_1 r_2\,Cos\beta\,Cos\,\phi_0+r_2^2}} \quad or \quad \overset{(a<b)}{e}=\frac{c}{b}=\sqrt{\frac{2\,r_1 r_2\left(1-Cos\beta\right)\left|Cos\,\phi_0\right|}{r_1^2-2\,r_1 r_2 Cos\,\phi_0+r_2^2}} \tag{30}$$

Let us continue to analyze the features and potential physical meaning and implications of this virtual vector ellipse in three-dimensional space.

Note that $\vec{\boldsymbol{a}}$ (21) is the vector value of $\left[\vec{r}_1(\phi)-\vec{r}_2(\phi)\right]$ (18) when $(\phi=0)$, and $\vec{\boldsymbol{b}}$ (22) is the vector value of $\left[\vec{r}_1(\phi)-\vec{r}_2(\phi)\right]$ (18) when $\left(\phi=\frac{\pi}{2}\right)$. Throughout a respective cycle of the phased points $\mathbf{P}_1$ and $\mathbf{P}_2$ moving around their own circles, namely $\mathbf{P}_1$ phased at $(\phi+\phi_0)$ and $\mathbf{P}_2$ phased at (ϕ) on two respective circles, the $\left[\vec{r}_1(\phi)-\vec{r}_2(\phi)\right]$ (18) vector has a vector value of $\left(\vec{\boldsymbol{a}}\,Cos\,\phi+\vec{\boldsymbol{b}}\,Sin\,\phi\right)$ (18) at each phase value of ϕ, and moves in the plane formed by the $\vec{\boldsymbol{a}}$ (21) and $\vec{\boldsymbol{b}}$ (22) vectors, namely the $\vec{\boldsymbol{a}}-\vec{\boldsymbol{b}}$ plane, which is a <u>variable</u> moving plane in three dimensions if ϕ_0 is a variable function of time t, i.e. if $\phi_0=\phi_0(t)$, and a <u>fixed</u> plane otherwise.

Based on (18) - (25), square of the distance $d^2(\phi)$ between the points $\mathbf{P}_1$ and $\mathbf{P}_2$ can be expressed for all ϕ also as in (31) - (34).

$$\left|\vec{d}(\phi)\right|^2 = d^2(\phi) = \vec{d}(\phi)\cdot\vec{d}(\phi) = \left[\vec{\mathbf{X}}(\phi)+\vec{\mathbf{Y}}(\phi)+\vec{\ell}\right]\cdot\left[\vec{\mathbf{X}}(\phi)+\vec{\mathbf{Y}}(\phi)+\vec{\ell}\right] \;\; ; \;\; \left|\vec{d}(\phi)\right| = d(\phi) \quad (31)$$

$$\Rightarrow \;\; d^2(\phi) = \vec{\mathbf{X}}(\phi)\cdot\vec{\mathbf{X}}(\phi) + 2\vec{\mathbf{X}}(\phi)\cdot\vec{\mathbf{Y}}(\phi) + \vec{\mathbf{Y}}(\phi)\cdot\vec{\mathbf{Y}}(\phi) + 2\left[\vec{\mathbf{X}}(\phi)+\vec{\mathbf{Y}}(\phi)\right]\cdot\vec{\ell} + \vec{\ell}\cdot\vec{\ell} \quad (32)$$

$$\Rightarrow \;\; d^2(\phi) = a^2 Cos^2\phi + 2\vec{a}\cdot\vec{b}\, Sin\phi\, Cos\phi + b^2 Sin^2\phi + 2\vec{a}\cdot\vec{\ell}\, Cos\phi + 2\vec{b}\cdot\vec{\ell}\, Sin\phi + \ell^2 \quad (33)$$

$$d^2(\phi) = b^2 + 2r_1 r_2 (1 - Cos\beta) Cos\phi\, Cos(\phi+\phi_0) + 2(\vec{a} Cos\phi + \vec{b} Sin\phi)\cdot\vec{\ell}(\phi) + \ell^2(\phi) \quad (34)$$

The phases $\phi_1(t)$ (35) and $\phi_2(t)$ (36), respectively, of points $\mathbf{P}_1$ and $\mathbf{P}_2$ moving as a function of time t with respect to the centers of their own circles, and their phase difference $\phi_0(t)$ (37), can be stated as in (35) - (37) in the most general case.

$$\phi_1(t) = \phi + \phi_0 = \phi(t) + \phi_0(t) = \phi_2(t) + \phi_0(t) \qquad (Phase\ of\ \mathbf{P}_1) \qquad (35)$$

$$\phi_2(t) = \phi = \phi(t) \qquad (Phase\ of\ \mathbf{P}_2) \qquad (36)$$

$$\phi_0(t) = \phi_1(t) - \phi_2(t) = \phi_0 \qquad (Phase\ difference\ of\ \mathbf{P}_1\ and\ \mathbf{P}_2) \qquad (37)$$

In this most general case, at each $\left[\phi = \phi(t)\right]$ (36), vector radii $\vec{r}_1$ (3) and $\vec{r}_2$ (4) of the two circles, as well as the vector distance $\vec{d}(\phi)$ (11) between the moving points $\mathbf{P_1}$ and $\mathbf{P_2}$ on two respective circles, and vector distance $\vec{\ell}(\phi)$ (7) with magnitude $\ell(\phi)$ (8) between the centers of these two circles, become as in (38) - (42).

$$\begin{aligned}\vec{r}_1 = \vec{r}_1\left[\phi(t)\right] = &\left(\hat{x} r_1 Cos\beta\, Cos\left[\phi_0(t)\right] + \hat{y} r_1 Sin\left[\phi_0(t)\right] + \hat{z} r_1 Sin\beta\, Cos\left[\phi_0(t)\right]\right) Cos\left[\phi(t)\right] + \\ &\left(-\hat{x} r_1 Cos\beta\, Sin\left[\phi_0(t)\right] + \hat{y} r_1 Cos\left[\phi_0(t)\right] - \hat{z} r_1 Sin\beta\, Sin\left[\phi_0(t)\right]\right) Sin\left[\phi(t)\right]\end{aligned} \quad (38)$$

$$\vec{r}_2 = \vec{r}_2\left[\phi(t)\right] = \hat{x} r_2\, Cos\left[\phi(t)\right] + \hat{y} r_2\, Sin\left[\phi(t)\right] \quad (39)$$

$$\vec{\ell} = \vec{\ell}\left[\phi(t)\right] = \hat{x}\ell_x\left[\phi(t)\right] + \hat{y}\ell_y\left[\phi(t)\right] + \hat{z}\ell_z\left[\phi(t)\right] \quad (40)$$

$$\ell\left[\phi(t)\right] = \left|\vec{\ell}\left[\phi(t)\right]\right| = \sqrt{\ell^2\left[\phi(t)\right]} = \sqrt{\vec{\ell}\left[\phi(t)\right]\cdot\vec{\ell}\left[\phi(t)\right]} = \sqrt{\ell_x^2\left[\phi(t)\right] + \ell_y^2\left[\phi(t)\right] + \ell_z^2\left[\phi(t)\right]} \quad (41)$$

$$\begin{aligned}\vec{d}\left[\phi(t)\right] = &\left[\hat{x}\left(r_1 Cos\beta\, Cos\left[\phi_0(t)\right] - r_2\right) + \hat{y} r_1 Sin\left[\phi_0(t)\right] + \hat{z} r_1 Sin\beta\, Cos\left[\phi_0(t)\right]\right] Cos\left[\phi(t)\right] + \\ &\left[-\hat{x} r_1 Cos\beta\, Sin\left[\phi_0(t)\right] + \hat{y}\left(r_1 Cos\left[\phi_0(t)\right] - r_2\right) - \hat{z} r_1 Sin\beta\, Sin\left[\phi_0(t)\right]\right] Sin\left[\phi(t)\right] + \\ &\left[\hat{x}\ell_x\left[\phi(t)\right] + \hat{y}\ell_y\left[\phi(t)\right] + \hat{z}\ell_z\left[\phi(t)\right]\right]\end{aligned} \quad (42)$$

Here, when the phase difference $\phi_0(t)$ (37) of the moving points $\mathbf{P}_1$ and $\mathbf{P}_2$ on the two respective circles is a <u>variable</u> function of time t, the semi-major[6] and semi-minor[6] axis vectors $\vec{a}(t)$ (43) and $\vec{b}(t)$ (44) are also variable functions of time t, but if $\phi_0(t) = \phi_0$ (37) is constant over time t, $\vec{a}$ (21) and $\vec{b}$ (22) are also both constant over time t, in which case (43) - (44) reduce to (21) - (22).

$$\vec{a}(t)=\hat{x}\{r_1 Cos\,\beta\, Cos[\phi_0(t)]-r_2\}+\hat{y}\,r_1 Sin[\phi_0(t)]+\hat{z}\,r_1 Sin\,\beta\, Cos[\phi_0(t)] \quad (43)$$

$$\vec{b}(t)=-\hat{x}\,r_1 Cos\,\beta\, Sin[\phi_0(t)]+\hat{y}\{r_1 Cos[\phi_0(t)]-r_2\}-\hat{z}\,r_1 Sin\,\beta\, Sin[\phi_0(t)] \quad (44)$$

Along the same lines, when $\phi_0(t)$ (37) is a <u>variable</u> function of time t, the magnitudes of the time(t)-dependent $\vec{a}(t)$ (43) and $\vec{b}(t)$ (44) are $|\vec{a}(t)|=a(t)$ (45) and $|\vec{b}(t)|=b(t)$ (46), respectively, and their Dot Product[4] is $[\vec{a}(t)\cdot\vec{b}(t)]$ (25). When $\phi_0(t)=\phi_0$ (37) is constant over time t, $\vec{a}$ (21) and $\vec{b}$ (22) are also constant over time t, and (45) - (47) reduce to (23) - (25).

$$\vec{a}(t)\cdot\vec{a}(t)=a^2(t)=r_1^2-2r_1r_2\,Cos\beta\, Cos[\phi_0(t)]+r_2^2 \quad ; \quad |\vec{a}(t)|=a(t)=\sqrt{\vec{a}(t)\cdot\vec{a}(t)} \quad (45)$$

$$\vec{b}(t)\cdot\vec{b}(t)=b^2(t)=r_1^2-2r_1r_2\,Cos[\phi_0(t)]+r_2^2 \quad ; \quad |\vec{b}(t)|=b(t)=\sqrt{\vec{b}(t)\cdot\vec{b}(t)} \quad (46)$$

$$\vec{a}(t)\cdot\vec{b}(t)=r_1r_2(Cos\beta-1)Sin[\phi_0(t)] \quad (47)$$

Based on our analysis in (48), $[(1-Cos\beta)\geq 0]$ is always true. As such, the sign of the difference (49) of $[a^2(t)]$ (45) and $[b^2(t)]$ (46) is based on the sign of $Cos[\phi_0(t)]$ (49), and thus the value of $\phi_0(t)$ (49). Thus, depending on the values of the parameters (r_1,r_2,β) (1) - (6), and the value of $\phi_0(t)$ (37) at any time t, the expression in (50) holds for the instantaneous semi-major[6] and semi-minor[6] axes of the ellipse[5], or the instantaneous radii of the vector circle whenever $a(t)=b(t)$.

$$-1\leq Cos\beta\leq 1 \quad \Rightarrow \quad (1-Cos\beta)\geq 0 \quad (48)$$

$$a^2(t)-b^2(t)=2\,r_1r_2(1-Cos\beta)Cos[\phi_0(t)]\begin{cases}\geq 0 \;\; if \;\; Cos[\phi_0(t)]\geq 0 \Rightarrow -\frac{\pi}{2}\leq\phi_0(t)\leq\frac{\pi}{2}\\ \leq 0 \;\; if \;\; Cos[\phi_0(t)]\leq 0 \Rightarrow \frac{\pi}{2}\leq\phi_0(t)\leq\frac{3\pi}{2}\end{cases} \quad (49)$$

$$\begin{cases}a(t)>b(t) \;\Rightarrow\; \vec{a}(t) \textit{ is semi-major axis } \& \; \vec{b}(t) \textit{ is semi-minor axis of vector ellipse}\\ a(t)<b(t) \;\Rightarrow\; \vec{b}(t) \textit{ is semi-major axis } \& \; \vec{a}(t) \textit{ is semi-minor axis of vector ellipse}\\ a(t)=b(t) \;\Rightarrow\; \vec{a}(t) \textit{ and } \vec{b}(t) \textit{ are radii of vector circle}\end{cases} \quad (50)$$

Furthermore, when $\phi_0(t)$ (37) is a <u>variable</u> function of time t, the square of the instantaneous focal[5,7] distance $[c^2(t)]$ (51) of the vector ellipse can be determined using (45) - (46) and (49), and the instantaneous eccentricity[5] $e(t)$ (52) of the vector ellipse can be found using (45) - (46) and (50) - (51).

$$c^2(t)=|a^2(t)-b^2(t)|=2\,r_1r_2(1-Cos\beta)|Cos[\phi_0(t)]| \quad (\textit{Focal Distance Squared}) \quad (51)$$

$$e(t)=\begin{cases}\dfrac{c(t)}{a(t)}=\sqrt{\dfrac{2\,r_1 r_2\left(1-Cos\beta\right)\left|Cos\left[\phi_0(t)\right]\right|}{r_1^2-2r_1 r_2\,Cos\beta\,Cos\left[\phi_0(t)\right]+r_2^2}} & if\quad a(t)>b(t)\\[2ex] \dfrac{c(t)}{b(t)}=\sqrt{\dfrac{2\,r_1 r_2\left(1-Cos\beta\right)\left|Cos\left[\phi_0(t)\right]\right|}{r_1^2-2r_1 r_2 Cos\left[\phi_0(t)\right]+r_2^2}} & if\quad a(t)<b(t)\end{cases}\qquad (Eccentricity)\quad (52)$$

When $\phi_0(t)=\phi_0$ (37) is constant over time t, as (43) - (44) reduce to (21) - (22), (51) - (52) also reduce to (29) - (30), and both focal distance c (29) and eccentricity e (30) become constant over all time t.

In this general case, at each $\left[\phi=\phi(t)\right]$ (36), $\vec{\boldsymbol{d}}(\phi)$ (18), $\vec{\mathbf{X}}(\phi)$ (19), $\vec{\mathbf{Y}}(\phi)$ (20), as well as $d^2(\phi)$ (34) can more explicitly be expressed as in $\vec{\boldsymbol{d}}\left[\phi(t)\right]$ (53) - (54), $\vec{\mathbf{X}}\left[\phi(t)\right]$ (55) - (56), $\vec{\mathbf{Y}}\left[\phi(t)\right]$ (57) - (58), and $d^2\left[\phi(t)\right]$ (59), respectively.

$$\vec{\boldsymbol{d}}\left[\phi(t)\right]=\vec{\mathbf{X}}\left[\phi(t)\right]+\vec{\mathbf{Y}}\left[\phi(t)\right]+\vec{\ell}\left[\phi(t)\right]=\vec{\boldsymbol{r}}_1\left[\phi(t)\right]-\vec{\boldsymbol{r}}_2\left[\phi(t)\right]+\vec{\ell}\left[\phi(t)\right] \qquad (53)$$

$$\vec{\boldsymbol{r}}_1\left[\phi(t)\right]-\vec{\boldsymbol{r}}_2\left[\phi(t)\right]=\vec{\mathbf{X}}\left[\phi(t)\right]+\vec{\mathbf{Y}}\left[\phi(t)\right]=\vec{\boldsymbol{a}}(t)Cos\left[\phi(t)\right]+\vec{\boldsymbol{b}}(t)Sin\left[\phi(t)\right] \qquad (54)$$

$$\vec{\mathbf{X}}\left[\phi(t)\right]=\vec{\boldsymbol{a}}(t)Cos\left[\phi(t)\right]\quad ;\quad \left|\vec{\mathbf{X}}\left[\phi(t)\right]\right|=X\left[\phi(t)\right] \qquad (55)$$

$$\vec{\mathbf{X}}\left[\phi(t)\right]\cdot\vec{\mathbf{X}}\left[\phi(t)\right]=X^2\left[\phi(t)\right]=\vec{\boldsymbol{a}}(t)\cdot\vec{\boldsymbol{a}}(t)\,Cos^2\left[\phi(t)\right]=a^2(t)Cos^2\left[\phi(t)\right] \qquad (56)$$

$$\vec{\mathbf{Y}}\left[\phi(t)\right]=\vec{\boldsymbol{b}}(t)Sin\left[\phi(t)\right]\quad ;\quad \left|\vec{\mathbf{Y}}\left[\phi(t)\right]\right|=Y\left[\phi(t)\right] \qquad (57)$$

$$\vec{\mathbf{Y}}\left[\phi(t)\right]\cdot\vec{\mathbf{Y}}\left[\phi(t)\right]=Y^2\left[\phi(t)\right]=\vec{\boldsymbol{b}}(t)\cdot\vec{\boldsymbol{b}}(t)\,Sin^2\left[\phi(t)\right]=b^2(t)Sin^2\left[\phi(t)\right] \qquad (58)$$

$$\begin{aligned}d^2\left[\phi(t)\right]&=b^2(t)+2r_1 r_2\left(1-Cos\beta\right)Cos\left[\phi(t)\right]Cos\left[\phi(t)+\phi_0(t)\right]\\&+2\left\{\vec{\boldsymbol{a}}(t)Cos\left[\phi(t)\right]+\vec{\boldsymbol{b}}(t)Sin\left[\phi(t)\right]\right\}\cdot\vec{\ell}\left[\phi(t)\right]+\ell^2\left[\phi(t)\right]\end{aligned} \qquad (59)$$

The major consequence in physics, of this analysis and mathematical results in (18) - (59) for moving points $\mathbf{P}_1$ and $\mathbf{P}_2$ on the two respective circles, is that "Particles or bodies moving around different circular orbits in space see themselves positioned on an elliptical path with respect to each other, with fixed or variable semi-major[6] and semi-minor[6] axes over time depending on the time dependencies of the angular velocities of the particles or bodies, where the particles or bodies instantaneously observe their counterpart virtually positioned at a fixed point in space, whose position is determined by the distance vector between the centers of the circles around which they revolve."

Observation also tells us that, apart from their relative motions with respect to each other, spatial bodies such as the Sun[9], the Earth[10,13], and the Moon[11] also rotate about their own axis[12,13], which determines their geographical poles[14], namely North Poles[15] and South Poles[16], along the normal to their equatorial[17] planes. Further, from a vantage point above the North Pole[15] of the Earth[10], the Earth appears to rotate[18] in a counterclockwise direction about its axis, which is expected based on the "right hand rule"[19]. As a result, based on our finding above and these observations, in this Article we make the broader assertion that "Spatial bodies must be revolving around individual circular orbits of revolution in space, which determine their individual equatorial[17] planes, whose normal and the direction of motion of the body around its individual circular orbit determine the axis of self-revolution according to 'right hand rule'[19]. This self-revolution axis and the self-rotation axis[12,13] of each spatial body must be aligned along the normal of its equatorial[17] plane. The direction of self-rotation axis[12,13] of each spatial body is also determined according to 'right hand rule'[19] based on the direction of self-rotation of the body." Based on this assertion, we can also infer that the axial tilt[13] of a body in its motion relative to another body must topologically be based on the angle between their planes of individual circular orbit of revolution in space. These are all fundamental findings and assertions of this Article.

A special case of (35) - (37) is expressed in (61) - (63), when the moving points $\mathbf{P}_1$ and $\mathbf{P}_2$ are phased with ***fixed*** but ***different*** (60) angular velocities ω_1 (61) and ω_2 (62), respectively, with respect to the centers of their own circles of revolution, their phase difference being $\left[\phi_0(t_0)=\varphi_0\right]$ (63) at time $(t=t_0=0)$, the reference timestamp t_0 taken to be the point in time when $\mathbf{P_2}$ has a phase of $\left[\phi_2(t_0)=0\right]$ (62), in the configuration described in **Figure 2**.

$$\omega_1 \neq \omega_2 \qquad\qquad (\omega_1 \;\&\; \omega_2 \; constant) \tag{60}$$

$$\phi_1(t)=\omega_1 t+\varphi_0=\phi_2(t)+\phi_0(t)=\phi(t)+\phi_0(t)=\phi+\phi_0 \quad ; \quad \phi_1(t_0=0)=\varphi_0 \qquad (Phase\ of\ \mathbf{P}_1) \tag{61}$$

$$\phi_2(t)=\omega_2 t=\phi(t)=\phi \quad ; \quad \phi_2(t_0=0)=0 \qquad (Phase\ of\ \mathbf{P}_2) \tag{62}$$

$$\phi_0(t)=\phi_1(t)-\phi_2(t)=(\omega_1-\omega_2)t+\varphi_0 \quad ; \quad \phi_0(t_0=0)=\varphi_0 \quad ; \quad (Phase\ difference\ of\ \mathbf{P}_1\ and\ \mathbf{P}_2) \tag{63}$$

In this special case, at each $\phi(t)$ (62), vector radii $\vec{r}_1\left[\phi(t)\right]$ (38) and $\vec{r}_2\left[\phi(t)\right]$ (39) of the two circles, and the vector distance $\vec{\ell}\left[\phi(t)\right]$ (40) with magnitude $\ell\left[\phi(t)\right]$ (41) between the centers of these two circles, as well as the vector distance $\vec{\boldsymbol{d}}\left[\phi(t)\right]$ (42) between the moving points $\mathbf{P}_1$ and $\mathbf{P}_2$ on two respective circles, the semi-major[6] and semi-minor[6] axis vectors $\vec{\boldsymbol{a}}(t)$ (43) and $\vec{\boldsymbol{b}}(t)$ (44) of the vector ellipse[5], their Dot Product[4] $\left[\vec{\boldsymbol{a}}(t)\cdot\vec{\boldsymbol{b}}(t)\right]$ (47), their magnitudes squared $\left[a^2(t)\right]$ (45) and $\left[b^2(t)\right]$ (46), elliptic representation of vector distance $\vec{\boldsymbol{d}}\left[\phi(t)\right]$ (53) - (54), the elliptic coordinate vectors $\vec{\mathbf{X}}\left[\phi(t)\right]$ (55) - (56) and $\vec{\mathbf{Y}}\left[\phi(t)\right]$ (57) - (58), square of the instantaneous focal[5,7] distance $\left[c^2(t)\right]$ (51) of the vector ellipse[5], and instantaneous eccentricity[5] $e(t)$ (52) of the vector ellipse[5] become as in (64) - (81).

$$\begin{aligned}\vec{r}_1=\vec{r}_1(\omega_2 t)=&\left\{\begin{array}{l}\hat{\boldsymbol{x}}\, r_1 Cos\,\beta\, Cos\left[(\omega_1-\omega_2)t+\varphi_0\right]+\hat{\boldsymbol{y}}\, r_1 Sin\left[(\omega_1-\omega_2)t+\varphi_0\right]+\\ \hat{\boldsymbol{z}}\, r_1 Sin\,\beta\, Cos\left[(\omega_1-\omega_2)t+\varphi_0\right]\end{array}\right\}Cos(\omega_2 t)+\\ &\left\{\begin{array}{l}-\hat{\boldsymbol{x}}\, r_1 Cos\,\beta\, Sin\left[(\omega_1-\omega_2)t+\varphi_0\right]+\hat{\boldsymbol{y}}\, r_1 Cos\left[(\omega_1-\omega_2)t+\varphi_0\right]\\ -\hat{\boldsymbol{z}}\, r_1 Sin\,\beta\, Sin\left[(\omega_1-\omega_2)t+\varphi_0\right]\end{array}\right\}Sin(\omega_2 t)\end{aligned} \tag{64}$$

$$\vec{r}_2=\vec{r}_2(\omega_2 t)=\hat{\boldsymbol{x}}\, r_2\, Cos(\omega_2 t)+\hat{\boldsymbol{y}}\, r_2\, Sin(\omega_2 t) \tag{65}$$

$$\vec{\ell} = \vec{\ell}(\omega_2 t) = \hat{\boldsymbol{x}}\,\ell_x(\omega_2 t) + \hat{\boldsymbol{y}}\,\ell_y(\omega_2 t) + \hat{\boldsymbol{z}}\,\ell_z(\omega_2 t) \tag{66}$$

$$\left|\vec{\ell}(\omega_2 t)\right| = \ell(\omega_2 t) = \sqrt{\ell^2(\omega_2 t)} = \sqrt{\vec{\ell}(\omega_2 t)\cdot\vec{\ell}(\omega_2 t)} = \sqrt{\ell_x^{\,2}(\omega_2 t) + \ell_y^{\,2}(\omega_2 t) + \ell_z^{\,2}(\omega_2 t)} \tag{67}$$

$$\begin{aligned}\vec{\boldsymbol{d}}(\omega_2 t) = &\left\{\begin{array}{l}\hat{\boldsymbol{x}}\left(r_1 Cos\,\beta\, Cos\left[(\omega_1-\omega_2)t+\varphi_0\right]-r_2\right)+\hat{\boldsymbol{y}}\, r_1 Sin\left[(\omega_1-\omega_2)t+\varphi_0\right]+\\ \hat{\boldsymbol{z}}\, r_1 Sin\,\beta\, Cos\left[(\omega_1-\omega_2)t+\varphi_0\right]\end{array}\right\} Cos(\omega_2 t)\\ &+\left\{\begin{array}{l}-\hat{\boldsymbol{x}}\, r_1 Cos\,\beta\, Sin\left[(\omega_1-\omega_2)t+\varphi_0\right]+\hat{\boldsymbol{y}}\left(r_1 Cos\left[(\omega_1-\omega_2)t+\varphi_0\right]-r_2\right)\\ -\hat{\boldsymbol{z}}\, r_1 Sin\,\beta\, Sin\left[(\omega_1-\omega_2)t+\varphi_0\right]\end{array}\right\} Sin(\omega_2 t)\\ &+\left\{\hat{\boldsymbol{x}}\,\ell_x(\omega_2 t)+\hat{\boldsymbol{y}}\,\ell_y(\omega_2 t)+\hat{\boldsymbol{z}}\,\ell_z(\omega_2 t)\right\}\end{aligned} \tag{68}$$

$$\begin{aligned}\vec{\boldsymbol{a}}(t) = &\,\hat{\boldsymbol{x}}\left\{r_1 Cos\,\beta\, Cos\left[(\omega_1-\omega_2)t+\varphi_0\right]-r_2\right\}+\hat{\boldsymbol{y}}\, r_1 Sin\left[(\omega_1-\omega_2)t+\varphi_0\right]+\\ &\hat{\boldsymbol{z}}\, r_1 Sin\,\beta\, Cos\left[(\omega_1-\omega_2)t+\varphi_0\right]\end{aligned} \tag{69}$$

$$\begin{aligned}\vec{\boldsymbol{b}}(t) = &-\hat{\boldsymbol{x}}\, r_1 Cos\,\beta\, Sin\left[(\omega_1-\omega_2)t+\varphi_0\right]+\hat{\boldsymbol{y}}\left\{r_1 Cos\left[(\omega_1-\omega_2)t+\varphi_0\right]-r_2\right\}\\ &-\hat{\boldsymbol{z}}\, r_1 Sin\,\beta\, Sin\left[(\omega_1-\omega_2)t+\varphi_0\right]\end{aligned} \tag{70}$$

$$a^2(t) = \vec{\boldsymbol{a}}(t)\cdot\vec{\boldsymbol{a}}(t) = r_1^2 - 2 r_1 r_2\, Cos\beta\, Cos\left[(\omega_1-\omega_2)t+\varphi_0\right]+r_2^2 \tag{71}$$

$$b^2(t) = \vec{\boldsymbol{b}}(t)\cdot\vec{\boldsymbol{b}}(t) = r_1^2 - 2 r_1 r_2\, Cos\left[(\omega_1-\omega_2)t+\varphi_0\right]+r_2^2 \tag{72}$$

$$\vec{\boldsymbol{a}}(t)\cdot\vec{\boldsymbol{b}}(t) = r_1 r_2 (Cos\beta - 1)\, Sin\left[(\omega_1-\omega_2)t+\varphi_0\right] \tag{73}$$

$$\vec{\boldsymbol{d}}(\omega_2 t) = \vec{\mathbf{X}}(\omega_2 t)+\vec{\mathbf{Y}}(\omega_2 t)+\vec{\ell}(\omega_2 t) = \vec{\boldsymbol{r}}_1(\omega_2 t)-\vec{\boldsymbol{r}}_2(\omega_2 t)+\vec{\ell}(\omega_2 t) \tag{74}$$

$$\vec{\boldsymbol{r}}_1-\vec{\boldsymbol{r}}_2 = \vec{\boldsymbol{r}}_1(\omega_2 t)-\vec{\boldsymbol{r}}_2(\omega_2 t) = \vec{\mathbf{X}}(\omega_2 t)+\vec{\mathbf{Y}}(\omega_2 t) = \vec{\boldsymbol{a}}(t)\, Cos(\omega_2 t)+\vec{\boldsymbol{b}}(t)\, Sin(\omega_2 t) \tag{75}$$

$$\vec{\mathbf{X}}(\omega_2 t) = \vec{\boldsymbol{a}}(t)\, Cos(\omega_2 t) \quad ; \quad \left|\vec{\mathbf{X}}(\omega_2 t)\right| = X(\omega_2 t) \tag{76}$$

$$\vec{\mathbf{X}}(\omega_2 t)\cdot\vec{\mathbf{X}}(\omega_2 t) = X^2(\omega_2 t) = \vec{\boldsymbol{a}}(t)\cdot\vec{\boldsymbol{a}}(t)\, Cos^2(\omega_2 t) = a^2(t)\, Cos^2(\omega_2 t) \tag{77}$$

$$\vec{\mathbf{Y}}(\omega_2 t) = \vec{\boldsymbol{b}}(t)\, Sin(\omega_2 t) \quad ; \quad \left|\vec{\mathbf{Y}}(\omega_2 t)\right| = Y(\omega_2 t) \tag{78}$$

$$\vec{\mathbf{Y}}(\omega_2 t)\cdot\vec{\mathbf{Y}}(\omega_2 t) = Y^2(\omega_2 t) = \vec{\boldsymbol{b}}(t)\cdot\vec{\boldsymbol{b}}(t)\, Sin^2(\omega_2 t) = b^2(t)\, Sin^2(\omega_2 t) \tag{79}$$

$$c^2(t) = \left|a^2(t)-b^2(t)\right| = 2\, r_1 r_2\,(1-Cos\beta)\left|Cos\left[(\omega_1-\omega_2)t+\varphi_0\right]\right| \qquad (Focal\ Distance\ Squared) \tag{80}$$

$$e(t)=\begin{cases}\dfrac{c(t)}{a(t)}=\sqrt{\dfrac{2\,r_1 r_2\,(1-Cos\beta)\left|Cos\left[(\omega_1-\omega_2)t+\varphi_0\right]\right|}{r_1^2-2r_1 r_2\,Cos\beta\,Cos\left[(\omega_1-\omega_2)t+\varphi_0\right]+r_2^2}} & if \quad a(t)>b(t)\\[2ex] \dfrac{c(t)}{b(t)}=\sqrt{\dfrac{2\,r_1 r_2\,(1-Cos\beta)\left|Cos\left[(\omega_1-\omega_2)t+\varphi_0\right]\right|}{r_1^2-2r_1 r_2 Cos\left[(\omega_1-\omega_2)t+\varphi_0\right]+r_2^2}} & if \quad a(t)<b(t)\end{cases} \qquad (Eccentricity)\ (81)$$

Note that $\vec{a}(t)$ (69) and $\vec{b}(t)$ (70) are sinusoidal functions of time t. Throughout a respective cycle of the points $\mathbf{P}_1$ and $\mathbf{P}_2$ moving around their own circles of revolution with ***fixed*** but ***different*** (60) angular velocities ω_1 (61) and ω_2 (62), respectively, the $(\vec{r}_1-\vec{r}_2)$ (75) vector has a vector value of $\left[\vec{a}(t)Cos(\omega_2 t)+\vec{b}(t)Sin(\omega_2 t)\right]$ (75) at each phase value of $\left[\phi(t)=\omega_2 t\right]$ (62), and moves in the plane formed by the $\vec{a}(t)$ (69) and $\vec{b}(t)$ (70) vectors, namely the $\vec{a}(t)-\vec{b}(t)$ plane, which is a variable moving plane in three dimensions for this case.

Based on (68) and depending on the values of ω_1 (61) and ω_2 (62), and the variation frequency of $\vec{\ell}\left[\phi(t)\right]$ (40) between the centers of these two circles, $d^2(\phi)$ (16) and therefore $d(\phi)$ (13) is expected to vary according to a sinusoid based on the higher of the angular frequencies ω_1 (61), ω_2 (62), $(\omega_1-\omega_2)$ (63), or angular frequencies in the variation of $\vec{\ell}\left[\phi(t)\right]$ (40), within a sinusoidal distance envelope varying according to the smaller of the angular frequencies ω_1 (61), ω_2 (62), $(\omega_1-\omega_2)$ (63), or angular frequencies in the variation of $\vec{\ell}\left[\phi(t)\right]$ (40). An example of such a d-curve as a function of time t is demonstrated in **Figure 3**.

Note here that based on the astronomical position data measured over time, the curve in **Figure 3** is identical to the variation of the Earth-Moon[3] distance over 700 days, which is an elliptical[5] motion curve with semi-major[6] and semi-minor[6] axis values varying in a harmonic style as the Moon comes closer to and goes farther away from the Earth, giving an idea of the cyclic relative Earth-Moon[3] motion. Therefore, we can conclude here that it is equivalently possible for the Earth and the Moon to be each revolving in a circular motion with ***fixed*** but ***different*** angular velocities with respect to the centers of their own individual circular orbits, where their orbital centers are displaced by an $\vec{\ell}(\phi)$ (7) vector from each other.

Over a 2π cycle of $\phi_0(t)=([\omega_1-\omega_2]t+\varphi_0)$ (63), squares of the semi-major[6] and semi-minor[6] axis magnitudes, namely $\left[a^2(t)\right]$ (71) and $\left[b^2(t)\right]$ (72), respectively, take their minimum $\left(a_{min}^2, b_{min}^2\right)$ (84) values when $\left\{Cos([\omega_1-\omega_2]t+\varphi_0)=+1\right\}$ (84), in other words when points $\mathbf{P}_1$ and $\mathbf{P}_2$ moving around their own circles with ***fixed*** but ***different*** (60) angular velocities ω_1 (61) and ω_2 (62) are *aligned* in phase, and they take their maximum $\left(a_{max}^2, b_{max}^2\right)$ (84) values when $\left\{Cos([\omega_1-\omega_2]t+\varphi_0)=-1\right\}$ (84), that is when points $\mathbf{P}_1$ and $\mathbf{P}_2$ moving around their own circles with ***fixed*** but ***different*** (60) angular velocities ω_1 (61) and ω_2 (62) are *opposite* in phase.

$$
\begin{cases}
Cos([\omega_1-\omega_2]t+\varphi_0)=+1 \;\Rightarrow\; \begin{cases} a_{min}^2 = r_1^2 - 2r_1 r_2 Cos\beta + r_2^2 \\ b_{min}^2 = r_1^2 - 2r_1 r_2 + r_2^2 = (r_1 - r_2)^2 \\ \phi_0(t) = (\omega_1-\omega_2)t+\varphi_0 = 2n\pi \qquad [n\ integer] \\ \mathbf{P}_1\ and\ \mathbf{P}_2\ are\ aligned\ in\ phase \end{cases} \\
Cos([\omega_1-\omega_2]t+\varphi_0)=-1 \;\Rightarrow\; \begin{cases} a_{max}^2 = r_1^2 + 2r_1 r_2 Cos\beta + r_2^2 \\ b_{max}^2 = r_1^2 + 2r_1 r_2 + r_2^2 = (r_1 + r_2)^2 \\ \phi_0(t) = (\omega_1-\omega_2)t+\varphi_0 = (2n+1)\pi \qquad [n\ integer] \\ \mathbf{P}_1\ and\ \mathbf{P}_2\ are\ opposite\ in\ phase \end{cases}
\end{cases} \qquad (84)
$$

Further to our analysis in (84), over a 2π cycle of $\phi_0(t)=([\omega_1-\omega_2]t+\varphi_0)$ (63), whenever $\{Cos([\omega_1-\omega_2]t+\varphi_0)=0\}$ (85), $[a^2(t)]$ (71) and $[b^2(t)]$ (72) have equal values, as for magnitudes of radius vectors of a circle.

$$Cos([\omega_1-\omega_2]t+\varphi_0)=0 \quad\Rightarrow\quad a^2(t)=b^2(t)=r_1^2+r_2^2 \qquad (85)$$

A more special case of (61) - (63) is expressed in (87) - (89), when the moving points $\mathbf{P}_1$ and $\mathbf{P}_2$ are moving with the ***same fixed*** angular velocity ω (86) - (88) with respect to the centers of their own circles of revolution, i.e. when $(\omega_1=\omega_2=\omega)$ (86), and are phased apart by a constant angle φ_0 (89) over all time t, the reference timestamp t_0 taken to be the point in time when $\mathbf{P}_2$ has a phase of $[\phi_2(t_0)=0]$ (88), in the configuration described in **Figure 2**.

$$\omega_1=\omega_2=\omega \qquad (\omega\ constant) \qquad (86)$$

$$\phi_1(t)=\omega t+\varphi_0=\phi_2(t)+\phi_0(t)=\phi(t)+\phi_0(t)=\phi+\phi_0 \;;\; \phi_1(t_0=0)=\varphi_0 \qquad (Phase\ of\ \mathbf{P}_1) \quad (87)$$

$$\phi_2(t)=\omega t=\phi(t)=\phi \;;\; \phi_2(t_0=0)=0 \qquad (Phase\ of\ \mathbf{P}_2) \quad (88)$$

$$\phi_0(t)=\phi_1(t)-\phi_2(t)=(\omega-\omega)t+\varphi_0=\varphi_0 \;;\; \varphi_0\ constant \;;\; (Phase\ difference\ of\ \mathbf{P}_1\ and\ \mathbf{P}_2) \quad (89)$$

In this more special case, at each $\phi(t)$ (88), vector radii $\vec{r}_1[\phi(t)]$ (38) and $\vec{r}_2[\phi(t)]$ (39) of the two circles, and the vector distance $\vec{\ell}[\phi(t)]$ (40) with magnitude $\ell[\phi(t)]$ (41) between the centers of these two circles, as well as the vector distance $\vec{d}[\phi(t)]$ (42) between the moving points $\mathbf{P}_1$ and $\mathbf{P}_2$ on two respective circles, semi-major[6] and semi-minor[6] axis vectors $\vec{a}(t)$ (43) and $\vec{b}(t)$ (44) of the vector ellipse[5], their Dot Product[4] $[\vec{a}(t)\cdot\vec{b}(t)]$ (47), their magnitudes squared $[a^2(t)]$ (45) and $[b^2(t)]$ (46), the elliptic representation of the vector distance $\vec{d}[\phi(t)]$ (53) - (54), elliptic coordinate vectors $\vec{\mathbf{X}}[\phi(t)]$ (55) - (56) and $\vec{\mathbf{Y}}[\phi(t)]$ (57) - (58), square of the instantaneous focal[5,7] distance $\left[c^2(t)\right]$ (51) of the vector ellipse[5], and instantaneous eccentricity[5] $e(t)$ (52) of the vector ellipse[5] become as in (90) - (107).

$$\vec{r}_1 = \vec{r}_1(\omega t) = \{\ \hat{x}\, r_1 Cos\,\beta\, Cos\,\varphi_0 + \hat{y}\, r_1 Sin\,\varphi_0 + \hat{z}\, r_1 Sin\,\beta\, Cos\,\varphi_0\}\, Cos(\omega t) + \{-\hat{x}\, r_1 Cos\,\beta\, Sin\,\varphi_0 + \hat{y}\, r_1 Cos\,\varphi_0 - \hat{z}\, r_1 Sin\,\beta\, Sin\,\varphi_0\}\, Sin(\omega t) \quad (90)$$

$$\vec{r}_2 = \vec{r}_2(\omega t) = \hat{x}\, r_2 Cos(\omega t) + \hat{y}\, r_2 Sin(\omega t) \quad (91)$$

$$\vec{\ell} = \vec{\ell}(\omega t) = \hat{x}\,\ell_x(\omega t) + \hat{y}\,\ell_y(\omega t) + \hat{z}\,\ell_z(\omega t) \quad (92)$$

$$\left|\vec{\ell}(\omega t)\right| = \ell(\omega t) = \sqrt{\ell^2(\omega t)} = \sqrt{\vec{\ell}(\omega t)\cdot\vec{\ell}(\omega t)} = \sqrt{\ell_x^{\,2}(\omega t) + \ell_y^{\,2}(\omega t) + \ell_z^{\,2}(\omega t)} \quad (93)$$

$$\vec{d}(\omega t) = \{\ \hat{x}(r_1 Cos\,\beta\, Cos\,\varphi_0 - r_2) + \hat{y}\, r_1 Sin\,\varphi_0 + \hat{z}\, r_1 Sin\,\beta\, Cos\,\varphi_0\}\, Cos(\omega t) + \{-\hat{x}\, r_1 Cos\,\beta\, Sin\,\varphi_0 + \hat{y}(r_1 Cos\,\varphi_0 - r_2) - \hat{z}\, r_1 Sin\,\beta\, Sin\,\varphi_0\}\, Sin(\omega t) + \{\hat{x}\,\ell_x(\omega t) + \hat{y}\,\ell_y(\omega t) + \hat{z}\,\ell_z(\omega t)\} \quad (94)$$

$$\vec{a} = \hat{x}\{r_1 Cos\,\beta\, Cos\varphi_0 - r_2\} + \hat{y}\, r_1 Sin\,\varphi_0 + \hat{z}\, r_1 Sin\,\beta\, Cos\varphi_0 \quad (95)$$

$$\vec{b} = -\hat{x}\, r_1 Cos\,\beta\, Sin\,\varphi_0 + \hat{y}\{r_1 Cos\varphi_0 - r_2\} - \hat{z}\, r_1 Sin\,\beta\, Sin\,\varphi_0 \quad (96)$$

$$a^2 = \vec{a}\cdot\vec{a} = r_1^{\,2} - 2 r_1 r_2\, Cos\beta\, Cos\,\varphi_0 + r_2^{\,2} \quad (97)$$

$$b^2 = \vec{b}\cdot\vec{b} = r_1^{\,2} - 2 r_1 r_2\, Cos\,\varphi_0 + r_2^{\,2} \quad (98)$$

$$\vec{a}\cdot\vec{b} = r_1 r_2 (Cos\beta - 1)\, Sin\,\varphi_0 \quad (99)$$

$$\vec{d}(\omega t) = \vec{\mathbf{X}}(\omega t) + \vec{\mathbf{Y}}(\omega t) + \vec{\ell}(\omega t) = \vec{r}_1(\omega t) - \vec{r}_2(\omega t) + \vec{\ell}(\omega t) \quad (100)$$

$$\vec{r}_1 - \vec{r}_2 = \vec{r}_1(\omega t) - \vec{r}_2(\omega t) = \vec{\mathbf{X}}(\omega t) + \vec{\mathbf{Y}}(\omega t) = \vec{a}(t) Cos(\omega t) + \vec{b}(t) Sin(\omega t) \quad (101)$$

$$\vec{\mathbf{X}}(\omega t) = \vec{a}\, Cos(\omega t) \quad ; \quad \left|\vec{\mathbf{X}}(\omega t)\right| = X(\omega t) \quad (102)$$

$$\vec{\mathbf{X}}(\omega t)\cdot\vec{\mathbf{X}}(\omega t) = X^2(\omega t) = \vec{a}\cdot\vec{a}\, Cos^2(\omega t) = a^2 Cos^2(\omega t) \quad (103)$$

$$\vec{\mathbf{Y}}(\omega t) = \vec{b}\, Sin(\omega t) \quad ; \quad \left|\vec{\mathbf{Y}}(\omega t)\right| = Y(\omega t) \quad (104)$$

$$\vec{\mathbf{Y}}(\omega t)\cdot\vec{\mathbf{Y}}(\omega t) = Y^2(\omega t) = \vec{b}\cdot\vec{b}\, Sin^2(\omega t) = b^2 Sin^2(\omega t) \quad (105)$$

$$c^2 = \left|a^2 - b^2\right| = 2\, r_1 r_2 (1 - Cos\beta)\left|Cos\,\varphi_0\right| \qquad (\mathit{Focal\ Distance\ Squared}) \quad (106)$$

$$e = \begin{cases} \dfrac{c}{a} = \sqrt{\dfrac{2\, r_1 r_2 \left(1 - Cos\beta\right)\left|Cos\ \varphi_0\right|}{r_1^2 - 2 r_1 r_2\, Cos\beta\, Cos\,\varphi_0 + r_2^2}} & if \quad a > b \\ \dfrac{c}{b} = \sqrt{\dfrac{2\, r_1 r_2 \left(1 - Cos\beta\right)\left|Cos\ \varphi_0\right|}{r_1^2 - 2 r_1 r_2\, Cos\,\varphi_0 + r_2^2}} & if \quad a < b \end{cases} \qquad (Eccentricity) \qquad (107)$$

In this special case $\vec{a}$ (95) and $\vec{b}$ (96) are constant vectors over time t. Throughout a respective cycle of the points $\mathbf{P}_1$ and $\mathbf{P}_2$ moving with the ***same fixed*** angular velocity ω (86) around the centers of their own circles of revolution, the $\left(\vec{r}_1 - \vec{r}_2\right)$ (101) vector has a vector value of $\left[\vec{a}(t)Cos(\omega t)+\vec{b}(t)Sin(\omega t)\right]$ (101) at each phase value of $\left[\phi(t)=\omega t\right]$ (88), and moves in the plane formed by the constant $\vec{a}$ (95) and $\vec{b}$ (96) vectors, namely the $\vec{a}-\vec{b}$ plane, which is a fixed plane in three dimensions for this case. Note that $\vec{a}$ (95) is the vector value of $\left[\vec{r}_1(\omega t)-\vec{r}_2(\omega t)\right]$ (101) when $\left(\omega t = 2n\pi\right)$, and $\vec{b}$ (96) is the vector value of $\left[\vec{r}_1(\omega t)-\vec{r}_2(\omega t)\right]$ (101) when $\left(\omega t = \dfrac{(4n+1)\pi}{2}\right)$, where n is an integer.

Based on (94), $d^2(\phi)$ (16) and therefore $d(\phi)$ (13) is expected to vary according to a sinusoid depending on the on the value of the angular frequency ω (86) - (88).

Based on Sun-Earth distance observation[8], this scenario seems to match that of the Sun and Earth couple, as the orbit of the Earth with respect to the Sun is a distinct fixed ellipse that completes its one cycle through the course of a year, with the Sun at a seemingly fixed point with respect to the Earth throughout this cycle. Therefore, it is highly possible that the Sun and the Earth are moving around their individual circular obits, with the ***same fixed*** angular velocity ω (86) - (88) with respect to the centers of their own circles of revolution, with their orbital centers displaced by an $\vec{\ell}(\phi)$ (7) vector from each other.

CONCLUSIONS

Based on the analysis demonstrated above, one can reach the conclusion that the cyclical distance-based observation of the Sun and the planets in the Solar System, which has lead to the empirical Kepler's Law[1,2] stating that the "orbit of a planet with respect to the Sun is elliptical, with the Sun at one of the two foci", may equivalently be due to "the Sun, the Planets, and their Moons each revolving in a circular motion with ***fixed*** but ***different*** angular velocities with respect to the centers of their own individual circular orbits, where those orbital centers are displaced by different $\vec{\ell}(\phi)$ (7) vectors from each other."

This mathematical revelation would have very far-reaching consequences in physics. It would lead to the formulation of a fundamental new model of motion for spatial bodies in the Universe, including the Sun, the Planets, and the Satellites in the Solar System and elsewhere, as well as at particle and sub-atomic level, where "all bodies move at some angular velocity around their own circular orbits of revolution with different radii and centers of revolution in space", but "each appear to be moving in elliptical motion with respect to each other, where they see the other body located at a fixed or variable point in their respective virtual plane of motion". This subsequently leads to more insight into forces in nature.

In the irregular case when centers of the two circles are displaced by an $\vec{\ell}\left[\phi(t)\right]$ (40) vector variable over time t, the relative elliptic behavior of the respective motions of points $\mathbf{P}_1$ and $\mathbf{P}_2$ on the two circles still holds, but the analysis of their relative motion just becomes more complicated, as they no longer see each other positioned at a fixed point in their respective plane of motion, but rather see the other moving according to the $\vec{\ell}\left[\phi(t)\right]$ (40) vector.

In this Article we also make the broader assertion that "Spatial bodies must be revolving around individual circular orbits of revolution in space, which determine their individual equatorial[17] planes, whose normal and the direction of motion of the body around its individual circular orbit determine the axis of self-revolution according to 'right hand rule'[19]. Direction of self-rotation axis[12,13] of each spatial body is also determined according to 'right hand rule'[19] based on the direction of self-rotation of the body. This *self-revolution axis and the self-rotation axis*[12,13] *of each spatial body must be aligned along the normal of its equatorial*[17] *plane*." Based on this assertion, we also infer that *the axial tilt*[13] *of a body in its motion relative to another body must be based on the angle between their planes of individual circular orbit of revolution in space.*

FIGURES

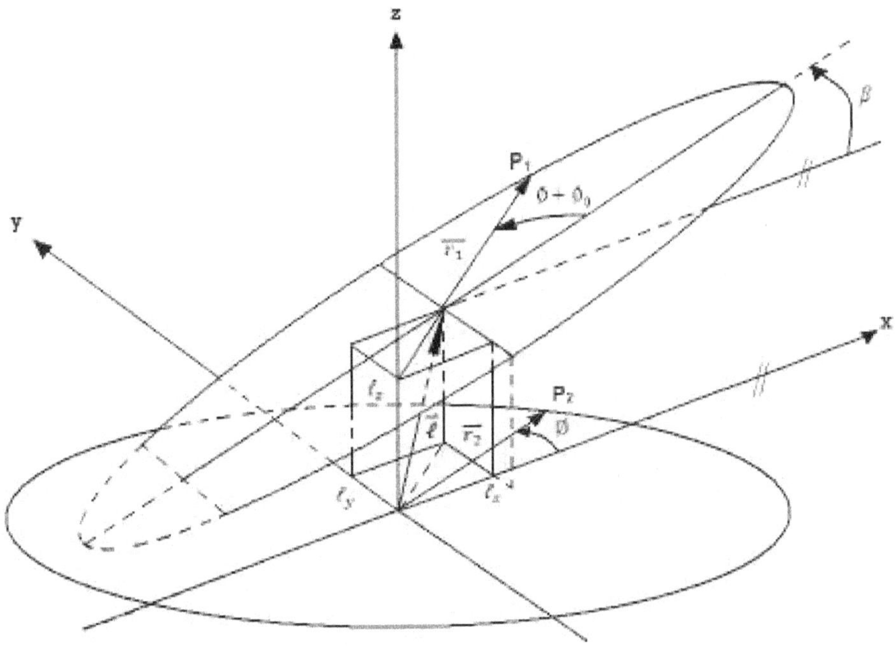

Figure 1 Points $\mathbf{P}_1$ and $\mathbf{P}_2$ on Two Circles in Space

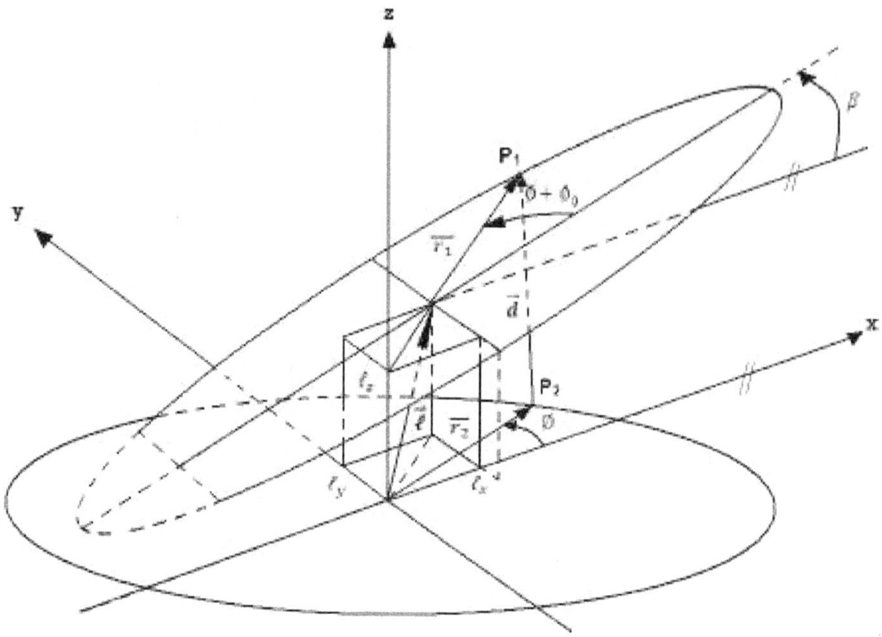

Figure 2 Distance Between Points $\mathbf{P}_1$ and $\mathbf{P}_2$ on Two Different Circles in Space

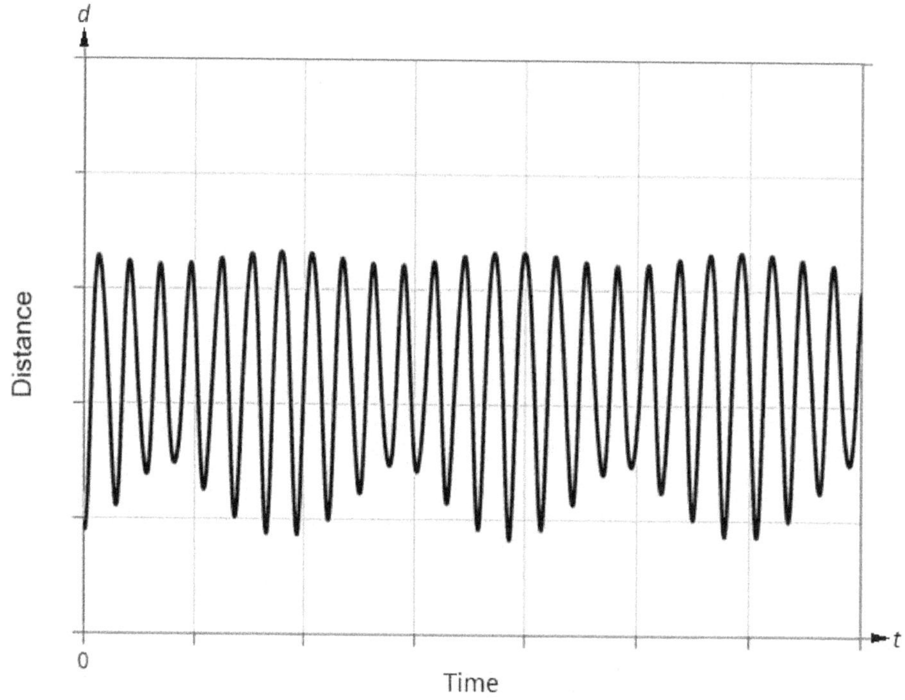

Figure 3 A Distance Curve d Between Moving Points $\mathbf{P_1}$ and $\mathbf{P_2}$ as a Function of Time t

REFERENCES

References in this Article can be any Physics, Electromagnetics, and Calculus textbook, as the physics equations and mathematical identities used as a basis for the proof are all currently accepted theory in existing textbooks.

1. Halliday, D., Resnick, R., *Fundamentals of Physics (3rd Edition)*, John Wiley & Sons, 1988, ISBN 0-471-63735-1
2. Kepler's laws of planetary motion, Wikipedia, https://en.wikipedia.org/wiki/Kepler%27s_laws_of_planetary_motion, (2020)
3. Lunar distance, Wikipedia, https://en.wikipedia.org/wiki/Lunar_distance_(astronomy), (2020)
4. Dot product, Wikipedia, https://en.wikipedia.org/wiki/Dot_product, (2020)
5. Ellipse, Wikipedia, http://en.wikipedia.org/wiki/Ellipse, (2020)
6. Semi-major and semi-minor axes, Wikipedia, http://en.wikipedia.org/wiki/Semi-major_and_semi-minor_axes, (2020)
7. Focus (geometry), Wikipedia, http://en.wikipedia.org/wiki/Focus_(geometry), (2020)
8. Earth's orbit, Wikipedia, https://en.wikipedia.org/wiki/Earth%27s_orbit, (2020)
9. Sun, Wikipedia, https://en.wikipedia.org/wiki/Sun, (2020)
10. Earth, Wikipedia, https://en.wikipedia.org/wiki/Earth, (2020)
11. Moon, Wikipedia, https://en.wikipedia.org/wiki/Moon, (2020)
12. Rotation around a fixed axis, Wikipedia, http://en.wikipedia.org/wiki/Rotation_around_a_fixed_axis, (2020)
13. Axial tilt, Wikipedia, https://en.wikipedia.org/wiki/Axial_tilt, (2020)
14. Geographical pole, Wikipedia, https://en.wikipedia.org/wiki/Geographical_pole, (2020)
15. North Pole, Wikipedia, https://en.wikipedia.org/wiki/North_Pole, (2020)
16. South Pole, Wikipedia, https://en.wikipedia.org/wiki/South_Pole, (2020)
17. Equator, Wikipedia, https://en.wikipedia.org/wiki/Equator, (2020)
18. Earth's orbit, Wikipedia, https://en.wikipedia.org/wiki/Earth%27s_orbit, (2020)
19. Right hand rule, Wikipedia, https://en.wikipedia.org/wiki/Right-hand_rule, (2020)
20. VLBA Finds Planet Orbiting Small Cool Star, National Radio Astronomy Observatory, https://public.nrao.edu/news/vlba-finds-planet/, August 4, 2020

ARTICLE 2

ANALYTICAL SOLUTION FOR TWO CIRCLES IN SPACE USING THEIR ELLIPTIC DISTANCES

Author: Aslı Pınar Tan[II]

SUMMARY

Based on measured astronomical position data of heavenly objects in the Solar System and other planetary systems, all bodies in space seem to move in some kind of elliptical motion with respect to each other. In a previous article, it is mathematically demonstrated and proven that the "distance between points on any two fixed circles in space" is equivalent to the "distance of points on a vector ellipse from another fixed or moving point, like in two-dimensional space". Equivalently, it amounts to showing that points moving on two different circular paths or orbits in space, vector-wise mimic an elliptical path with respect to each other. In this Article, an analytical method is developed to calculate orbital parameters of two heavenly bodies moving with the *same* angular velocity with respect to the centers of their own circular orbits of revolution, when the distance between their centers of revolution is also fixed.

ARTICLE

According to Kepler's 1st Law[1,2], "the orbit of a planet with respect to the Sun is an ellipse[4], with the Sun at one of the two foci", which is an empirical rule concluded through near estimation based on the observation of astronomical position data measured over time in the Solar System. In the **ARTICLE 1** "Points on two circles in space mimic an ellipse with respect to a point", it is mathematically demonstrated and proven that the "distance between points on any two different circles in three-dimensional space" is equivalent to the "distance of points on a vector ellipse from another fixed or moving point, similar to two-dimensional space". In other words, it is shown that moving points on two circles in space vector-wise mimic an elliptical path with respect to each other, whether they are moving with the *same* angular velocity, or *different* but *fixed* angular velocities, or even with *different* and *changing* angular velocities with respect to their own centers of revolution, virtually seeing each other as positioned at a instantaneously stationary point in space on their respective virtual ecliptic plane.

Based on the mathematical derivation and assertion in the **ARTICLE 1** "Points on two circles in space mimic an ellipse with respect to a point", in this Article, an analytical method is developed to calculate orbital parameters of two heavenly bodies moving with the ***same fixed constant*** angular velocity ω (1) - (2) with respect to the centers of their own circles of revolution, when the distance between their centers of revolution is also constant.

Consider a system of two circles in three-dimensional space, with the geometry of the system demonstrated as in **Figure 1** in Cartesian $(\hat{x},\hat{y},\hat{z})$ coordinates, where we choose $Circle_2$ to be in the horizontal $\hat{x}-\hat{y}$ plane, and $Circle_1$ to have an inclination angle of β (4) with respect to the horizontal $\hat{x}-\hat{y}$ plane. Moving points $\mathbf{P}_1$ and $\mathbf{P}_2$, are defined on these two circles $Circle_1$ and

[II] *ASLI PINAR TAN*
Linkedin Website: *https://www.linkedin.com./in/apinartan*

$Circle_2$, respectively, phased apart by a constant angle $\left[\phi_0(t)=\varphi_0\right]$ (3) over all time t, the reference timestamp t_0 taken to be the point in time when **P₂** has a phase of $\left[\phi_2(t_0)=0\right]$ (2).

$$\phi_1(t)=\phi+\phi_0=\phi(t)+\phi_0(t)=\phi_2(t)+\phi_0(t)=\omega t+\varphi_0 \quad ; \quad \phi_1(t_0=0)=\varphi_0 \qquad (\mathit{Phase\ of\ } \mathbf{P}_1) \quad (1)$$

$$\phi_2(t)=\phi=\phi(t)=\omega t \quad ; \quad \phi_2(t_0=0)=0 \qquad (\mathit{Phase\ of\ } \mathbf{P}_2) \quad (2)$$

$$\phi_0(t)=\phi_0=\phi_1(t)-\phi_2(t)=\varphi_0 \quad ; \quad \varphi_0\ \mathit{constant} \quad ; \quad (\mathit{Phase\ difference\ of\ } \mathbf{P}_1 \mathit{\ and\ } \mathbf{P}_2) \quad (3)$$

More explicitly, in the Cartesian $(\hat{x},\hat{y},\hat{z})$ coordinate configuration of **Figure 1**, at each ϕ (2), the location of $\mathbf{P}_1$ phased at $(\phi+\phi_0)$ (1) in its own circular revolution is defined by the vector $\left[\vec{r}_1(\phi)+\vec{\ell}\right]$ (26), where $\vec{r}_1(\phi)$ (4) - (5) is the vector radius of the circle around which $\mathbf{P}_1$ moves, and $\vec{\ell}$ (10) is the constant distance vector between the centers of the two circles with magnitude ℓ (11), directed from the center of the circle of $\mathbf{P}_2$ to the center of the circle of $\mathbf{P}_1$, and the location of $\mathbf{P}_2$ phased at ϕ (2) in its own circular revolution is defined by the vector $\vec{r}_2(\phi)$ (7), which is the vector radius of the circle around which $\mathbf{P}_2$ moves. Note that the inclination angle β (4) between the planes of these two circles is also taken to be constant.

Vector radii $\vec{r}_1(\phi)$ (4) - (5) and $\vec{r}_2(\phi)$ (7) of these two circles have constant magnitudes r_1 (9) and r_2 (9), respectively, where $\hat{r}_1(\phi)$ (6) is the unit vector in the direction of $\vec{r}_1(\phi)$ (4) - (5) and $\hat{r}_2(\phi)$ (8) is the unit vector in the direction of $\vec{r}_2$ (7). The scalar magnitude r_1 (9) of the radius vector $\vec{r}_1(\phi)$ (4) - (5) is calculated as the square root of r_1^2 (9), which in turn is calculated in terms of the Dot Product[3] $(\vec{r}_1\cdot\vec{r}_1)$ (9) of the vector $\vec{r}_1(\phi)$ (4) - (5) with itself. The scalar magnitude r_2 (9) of the radius vector $\vec{r}_2(\phi)$ (7) is calculated as the square root of r_2^2 (9), which in turn is calculated in terms of the Dot Product[3] $(\vec{r}_2\cdot\vec{r}_2)$ (9) of the vector $\vec{r}_2$ (7) with itself. The constant scalar distance between centers of the two circles, namely magnitude ℓ (11) of $\vec{\ell}$ (10), is calculated as the square root of ℓ^2 (11), which in turn is calculated in terms of the Dot Product[3] $(\vec{\ell}\cdot\vec{\ell})$ (11) of the vector $\vec{\ell}$ (10) with itself. Note that in all our operations, we take the "square of a vector" as the "Dot Product[3] of the vector with itself", which amounts to the scalar "square of the magnitude" for any vector.

$$\begin{aligned}\vec{r}_1=\vec{r}_1(\phi)=\{\ &\hat{x}\,r_1 Cos\,\beta\, Cos\,\varphi_0+\hat{y}\,r_1 Sin\,\varphi_0+\hat{z}\,r_1 Sin\,\beta\, Cos\,\varphi_0\}\,Cos(\omega t)+\\ &\{-\hat{x}\,r_1 Cos\,\beta\, Sin\,\varphi_0+\hat{y}\,r_1 Cos\,\varphi_0-\hat{z}\,r_1 Sin\,\beta\, Sin\,\varphi_0\}\,Sin(\omega t)\end{aligned} \quad (4)$$

$$\vec{r}_1=\vec{r}_1(\phi+\phi_0)=\hat{x}\,r_1 Cos(\phi+\phi_0)\,Cos\,\beta+\hat{y}\,r_1 Sin(\phi+\phi_0)+\hat{z}\,r_1 Cos(\phi+\phi_0)\,Sin\,\beta \quad (5)$$

$$\hat{r}_1=\hat{r}_1(\phi)=\hat{x}\,Cos(\phi+\phi_0)\,Cos\,\beta+\hat{y}\,{}_1 Sin(\phi+\phi_0)+\hat{z}\,Cos(\phi+\phi_0)\,Sin\,\beta \ \Rightarrow\ \vec{r}_1=\hat{r}_1(\phi)\,r_1 \quad (6)$$

$$\vec{r}_2=\vec{r}_2(\phi)=\hat{x}\,r_2 Cos(\omega t)+\hat{y}\,r_2 Sin(\omega t) \quad (7)$$

$$\hat{r}_2=\hat{r}_2(\phi)=\hat{x}\,Cos\,\phi+\hat{y}\,Sin\,\phi \ \Rightarrow\ \vec{r}_2=\hat{r}_2(\phi)\,r_2 \quad (8)$$

$$\vec{r}_1 \cdot \vec{r}_1 = r_1^2 \quad ; \quad |\vec{r}_1| = r_1 = \sqrt{\vec{r}_1 \cdot \vec{r}_1} \quad ; \quad \vec{r}_2 \cdot \vec{r}_2 = r_2^2 \quad ; \quad |\vec{r}_2| = r_2 = \sqrt{\vec{r}_2 \cdot \vec{r}_2} \tag{9}$$

$$\vec{\ell}(\phi) = \vec{\ell} = \hat{\boldsymbol{x}}\,\ell_x + \hat{\boldsymbol{y}}\,\ell_y + \hat{\boldsymbol{z}}\,\ell_z \tag{10}$$

$$|\vec{\ell}| = \ell = \sqrt{\ell^2} = \sqrt{\vec{\ell} \cdot \vec{\ell}} = \sqrt{\ell_x^{\,2} + \ell_y^{\,2} + \ell_z^{\,2}} \tag{11}$$

In this configuration of **Figure 1**, at each ϕ (2), the vector distance $\vec{\boldsymbol{d}}(\phi)$ (12) - (15) directed from moving point $\mathbf{P}_2$ to moving point $\mathbf{P}_1$ on two respective circles is defined, as well as its magnitude $d(\phi)$ (21) which is also the scalar distance between $\mathbf{P}_1$ and $\mathbf{P}_2$. Note that the scalar distance $d(\phi)$ (21) between $\mathbf{P}_1$ and $\mathbf{P}_2$ is calculated as the square root of $d^2(\phi)$ (22) - (24), which in turn is calculated in terms of the Dot Product[3] $\left[\vec{\boldsymbol{d}}(\phi) \cdot \vec{\boldsymbol{d}}(\phi)\right]$ (22) - (24) of the vector distance $\vec{\boldsymbol{d}}(\phi)$ (12) - (15) with itself. We have $\hat{\boldsymbol{d}}(\phi)$ (25) as the unit vector in the direction of $\vec{\boldsymbol{d}}(\phi)$ (12) - (15). In expressing $\vec{\boldsymbol{d}}(\phi)$ (15) using $\vec{\boldsymbol{d}}(\phi)$ (14), virtual vectors $\vec{\boldsymbol{a}}$ (16) and $\vec{\boldsymbol{b}}$ (17) are defined, whose magnitudes a (18) and b (19) are calculated as the square root of a^2 (18) and b^2 (19), respectively, which in turn are calculated in terms of the Dot Product[3] $(\vec{\boldsymbol{a}} \cdot \vec{\boldsymbol{a}})$ (18) of vector $\vec{\boldsymbol{a}}$ (16) with itself and the Dot Product[3] $(\vec{\boldsymbol{b}} \cdot \vec{\boldsymbol{b}})$ (19) of vector $\vec{\boldsymbol{b}}$ (17) with itself, respectively. Based on the definitions of $\vec{\boldsymbol{a}}$ (16) and $\vec{\boldsymbol{b}}$ (17), their Dot Product[3] $(\vec{\boldsymbol{a}} \cdot \vec{\boldsymbol{b}})$ (20) is also defined. Note that $\vec{\boldsymbol{a}}$ (16) is the value of vector difference $\left[\vec{r}_1(\phi) - \vec{r}_2(\phi)\right]$ (16) when $(\phi = 0)$ (2), and $\vec{\boldsymbol{b}}$ (17) is the value of vector difference $\left[\vec{r}_1(\phi) - \vec{r}_2(\phi)\right]$ (17) when $\left(\phi = \dfrac{\pi}{2}\right)$ (2). As the phase difference $(\phi_0 = \varphi_0)$ (3) is constant for all ϕ (2), $\vec{\boldsymbol{a}}$ (16) and $\vec{\boldsymbol{b}}$ (17) are constant vectors for all ϕ (2), whereas their magnitudes a (18) and b (19), respectively, are also constant. Throughout a respective cycle of the phased points $\mathbf{P}_1$ and $\mathbf{P}_2$ moving around their own circles, the $\left[\vec{r}_1(\phi) - \vec{r}_2(\phi)\right]$ (15) vector has a vector value of $\left[\vec{\boldsymbol{a}}\,Cos(\omega t) + \vec{\boldsymbol{b}}\,Sin(\omega t)\right]$ (15) at each phase value of $(\phi = \omega t)$ (2), and moves in the plane formed by the $\vec{\boldsymbol{a}}$ (16) and $\vec{\boldsymbol{b}}$ (17) vectors, namely the $\vec{\boldsymbol{a}} - \vec{\boldsymbol{b}}$ plane, which is a <u>fixed</u> plane in three dimensions as $(\phi_0 = \varphi_0)$ (3) is constant for all ϕ (2).

$$\begin{aligned} \vec{\boldsymbol{d}}(\phi) = \hat{\boldsymbol{x}}\,&[r_1 Cos(\phi+\phi_0)Cos\,\beta - r_2\,Cos\,\phi + \ell_x] + \hat{\boldsymbol{y}}\,[r_1 Sin(\phi+\phi_0) - r_2 Sin\,\phi + \ell_y] \\ &+ \hat{\boldsymbol{z}}\,[r_1 Cos(\phi+\phi_0)Sin\,\beta + \ell_z] \end{aligned} \tag{12}$$

$$\vec{\boldsymbol{d}}(\phi) = \hat{\boldsymbol{x}}\,d_x(\phi) + \hat{\boldsymbol{y}}\,d_y(\phi) + \hat{\boldsymbol{z}}\,d_z(\phi) \;\Rightarrow\; \begin{cases} d_x(\phi) = r_1 Cos(\phi+\phi_0)Cos\,\beta - r_2\,Cos\phi + \ell_x \\ d_y(\phi) = r_1 Sin(\phi+\phi_0) - r_2 Sin\,\phi + \ell_y \\ d_z(\phi) = r_1 Cos(\phi+\phi_0)Sin\,\beta + \ell_z \end{cases} \tag{13}$$

$$\vec{d}(\phi)=\left[\ \hat{x}(r_1 Cos\,\beta\, Cos\,\varphi_0 - r_2)+\hat{y}\,r_1 Sin\,\varphi_0+\hat{z}\,r_1 Sin\,\beta\, Cos\,\varphi_0\right] Cos(\omega t)$$
$$+\left[-\hat{x}\,r_1 Cos\,\beta\, Sin\,\varphi_0+\hat{y}(r_1 Cos\,\varphi_0 - r_2)-\hat{z}\,r_1 Sin\,\beta\, Sin\,\varphi_0\right] Sin(\omega t) \tag{14}$$
$$+\left[\hat{x}\,\ell_x+\hat{y}\,\ell_y+\hat{z}\,\ell_z\right]$$

$$\vec{d}(\phi)=\vec{r}_1(\phi)-\vec{r}_2(\phi)+\vec{\ell} \quad where \quad \vec{r}_1(\phi)-\vec{r}_2(\phi)=\vec{a}\, Cos(\omega t)+\vec{b}\, Sin(\omega t) \tag{15}$$

$$\vec{a}=\hat{x}\{r_1 Cos\,\beta\, Cos\varphi_0 - r_2\}+\hat{y}\,r_1 Sin\,\varphi_0+\hat{z}\,r_1 Sin\,\beta\, Cos\varphi_0 \quad\Rightarrow\quad \vec{a}=\left[\vec{r}_1(\phi)-\vec{r}_2(\phi)\right](\phi=0) \tag{16}$$

$$\vec{b}=-\hat{x}\,r_1 Cos\,\beta\, Sin\,\varphi_0+\hat{y}\{r_1 Cos\varphi_0 - r_2\}-\hat{z}\,r_1 Sin\,\beta\, Sin\,\varphi_0 \quad\Rightarrow\quad \vec{b}=\left[\vec{r}_1(\phi)-\vec{r}_2(\phi)\right]\left(\phi=\frac{\pi}{2}\right) \tag{17}$$

$$a^2=\vec{a}\cdot\vec{a}=r_1^2-2\,r_1\,r_2\,Cos\beta\, Cos\,\varphi_0+r_2^2 \quad ; \quad |\vec{a}|=a=\sqrt{a^2}=\sqrt{\vec{a}\cdot\vec{a}} \tag{18}$$

$$b^2=\vec{b}\cdot\vec{b}=r_1^2-2\,r_1\,r_2\,Cos\,\varphi_0+r_2^2 \quad ; \quad |\vec{b}|=b=\sqrt{b^2}=\sqrt{\vec{b}\cdot\vec{b}} \tag{19}$$

$$\vec{a}\cdot\vec{b}=r_1\,r_2\,(Cos\beta-1)\,Sin\,\varphi_0 \tag{20}$$

$$d(\phi)=|\vec{d}(\phi)|=\sqrt{d^2(\phi)}=\sqrt{\vec{d}(\phi)\cdot\vec{d}(\phi)}=\sqrt{\left[d_x(\phi)\right]^2+\left[d_y(\phi)\right]^2+\left[d_z(\phi)\right]^2} \tag{21}$$

$$d^2(\phi)=\vec{d}(\phi)\cdot\vec{d}(\phi)$$
$$=[r_1 Cos(\phi+\phi_0)Cos\,\beta-r_2\,Cos\,\phi+\ell_x(\phi)]^2+[r_1 Sin(\phi+\phi_0)-r_2\,Sin\phi+\ell_y(\phi)]^2+ \tag{22}$$
$$[r_1 Cos(\phi+\phi_0)\,Sin\,\beta+\ell_z(\phi)]^2$$

$$d^2(\phi)=\vec{d}(\phi)\cdot\vec{d}(\phi)=\left[d_x(\phi)\right]^2+\left[d_y(\phi)\right]^2+\left[d_z(\phi)\right]^2 \tag{23}$$

$$d^2(\phi)=\vec{d}(\phi)\cdot\vec{d}(\phi)=\left(\vec{r}_1-\vec{r}_2+\vec{\ell}\right)\cdot\left(\vec{r}_1-\vec{r}_2+\vec{\ell}\right)=\vec{r}_1\cdot\vec{r}_1-2\,\vec{r}_1\cdot\vec{r}_2+\vec{r}_2\cdot\vec{r}_2+2\left(\vec{r}_1-\vec{r}_2\right)\cdot\vec{\ell}+\vec{\ell}\cdot\vec{\ell} \tag{24}$$

$$\hat{d}(\phi)=\frac{\vec{d}(\phi)}{|\vec{d}(\phi)|}=\frac{\vec{d}(\phi)}{d(\phi)}=\hat{x}\frac{d_x(\phi)}{d(\phi)}+\hat{y}\frac{d_y(\phi)}{d(\phi)}+\hat{z}\frac{d_z(\phi)}{d(\phi)} \quad\Rightarrow\quad \vec{d}(\phi)=\hat{d}(\phi)|\vec{d}(\phi)|=\hat{d}(\phi)\,d(\phi) \tag{25}$$

Vector distance $\vec{d}(\phi)$ (14) - (15), which is between points $\mathbf{P}_1$ and $\mathbf{P}_2$ on two respective circles at any value of the phase ϕ (2), can be equivalently expressed as $\vec{d}(\phi)$ (26) in terms of the virtual vectors $\vec{\mathbf{X}}(\phi)$ (27) and $\vec{\mathbf{Y}}(\phi)$ (28) that we have defined, based on the $\vec{a}$ (16) and $\vec{b}$ (17). Magnitude $X(\phi)$ (27) of virtual vector $\vec{\mathbf{X}}(\phi)$ (27) is calculated as square root of $X^2(\phi)$ (27), which in turn is calculated in terms of the Dot Product[3] $\left[\vec{\mathbf{X}}(\phi)\cdot\vec{\mathbf{X}}(\phi)\right]$ (27) of vector $\vec{\mathbf{X}}(\phi)$ (27) with itself. Magnitude $Y(\phi)$ (28) of virtual vector $\vec{\mathbf{Y}}(\phi)$ (27) is calculated as square root of $Y^2(\phi)$ (27), which in turn is calculated in terms of the Dot Product[3] $\left[\vec{\mathbf{Y}}(\phi)\cdot\vec{\mathbf{Y}}(\phi)\right]$ (27) of vector $\vec{\mathbf{Y}}(\phi)$ (27) with itself.

$$\vec{d}(\phi)=\vec{r_1}(\phi)-\vec{r_2}(\phi)+\vec{\ell} \quad where \quad \vec{r_1}(\phi)-\vec{r_2}(\phi)=\bar{\mathbf{X}}(\phi)+\bar{\mathbf{Y}}(\phi)=\vec{a}\,Cos(\omega t)+\vec{b}\,Sin(\omega t) \quad (26)$$

$$\bar{\mathbf{X}}(\phi)=\vec{a}\,Cos(\omega t)\ ;\ \bar{\mathbf{X}}(\phi)\cdot\bar{\mathbf{X}}(\phi)=X^2(\phi)=\vec{a}\cdot\vec{a}\ Cos^2(\omega t)=a^2Cos^2(\omega t)\ ;\ \left|\bar{\mathbf{X}}(\phi)\right|=X(\phi) \quad (27)$$

$$\bar{\mathbf{Y}}(\phi)=\vec{b}\,Sin(\omega t)\ ;\ \bar{\mathbf{Y}}(\phi)\cdot\bar{\mathbf{Y}}(\phi)=Y^2(\phi)=\vec{b}\cdot\vec{b}\ Sin^2(\omega t)=b^2Sin^2(\omega t)\ ;\ \left|\bar{\mathbf{Y}}(\phi)\right|=Y(\phi) \quad (28)$$

According to the definitions of vectors $\bar{\mathbf{X}}(\phi)$ (27), $\bar{\mathbf{Y}}(\phi)$ (28), $\vec{a}$ (16), and $\vec{b}$ (17), the relation in (29) is valid and holds for all ϕ (2), leading to (30) - (31).

$$\frac{\bar{\mathbf{X}}(\phi)\cdot\bar{\mathbf{X}}(\phi)}{\vec{a}\cdot\vec{a}}+\frac{\bar{\mathbf{Y}}(\phi)\cdot\bar{\mathbf{Y}}(\phi)}{\vec{b}\cdot\vec{b}}=\frac{X^2(\phi)}{a^2}+\frac{Y^2(\phi)}{b^2}=Cos^2\phi+Sin^2\phi=1 \quad (29)$$

$$\Rightarrow \boxed{\frac{\bar{\mathbf{X}}(\phi)\cdot\bar{\mathbf{X}}(\phi)}{\vec{a}\cdot\vec{a}}+\frac{\bar{\mathbf{Y}}(\phi)\cdot\bar{\mathbf{Y}}(\phi)}{\vec{b}\cdot\vec{b}}=1} \quad (Definition\ of\ Vector\ Ellipse\ in\ 3-Dimensions) \quad (30)$$

$$\Rightarrow \boxed{\frac{X^2(\phi)}{a^2}+\frac{Y^2(\phi)}{b^2}=1} \quad (Definition\ of\ Scalar\ Ellipse\ in\ 2-Dimensions) \quad (31)$$

As (31) is the defining equation of an ellipse[4] in two dimensions, where a (18) is the semi-major[5] axis and b (19) is the semi-minor[5] axis of the ellipse[4] when $(a>b)$[4,5] (34) holds, and vice versa, with (31) reducing to the special case of a circle when $(a=b)$[4,5] (34) holds, we claim that (30) is the defining equation a "vector ellipse" in three-dimensional space, and it indicates that the vector pair $\left[\bar{\mathbf{X}}(\phi),\bar{\mathbf{Y}}(\phi)\right]$ (27) - (28) defines points on a vector ellipse (30) in three dimensions. *In other words, the vector distance* $\vec{d}(\phi)$ (14) *between two points* $\mathbf{P}_1$ *phased at* $(\phi+\phi_0)$ (1) *and* $\mathbf{P}_2$ *phased at* ϕ (2) *on two respective circles, whose centers are displaced by a constant vector* $\vec{\ell}$ (10), *can equivalently be mathematically expressed and interpreted as the distance* $\vec{d}(\phi)$ (26) *of points on a virtual vector ellipse* (30), *whose locations with respect to a virtual origin at each* ϕ (2) *are determined by the sum of vector pair* $\left[\bar{\mathbf{X}}(\phi),\bar{\mathbf{Y}}(\phi)\right]$ (27) - (28), *from another fixed point displaced from the same virtual origin of the ellipse by a constant vector* $\left(-\vec{\ell}\right)$ (10), *where* $\vec{a}$ (16) and $\vec{b}$ (17) *are fixed semi-major*[5] *and semi-minor*[5] *axis vectors of the vector ellipse* (30).

Based on trigonometry, $\left[(1-Cos\beta)\geq 0\right]$ (32) is always true. As a result, the sign of the difference (33) of a^2 (18) and b^2 (19) is based on the sign of $Cos\,\varphi_0$ (33), and thus the value of $(\phi_0=\varphi_0)$ (3). Thus, depending on the values of the parameters (r_1,r_2) (9) and β (4), and the value of $(\phi_0=\varphi_0)$ (3), the expression in (34) holds for the semi-major[5] and semi-minor[5] axes of the ellipse[4], or the instantaneous radii of the vector circle whenever $(a=b)$ (18) - (19).

$$-1\leq Cos\beta\leq 1 \quad\Rightarrow\quad (1-Cos\beta)\geq 0 \quad (32)$$

$$a^2 - b^2 = 2\,r_1 r_2 \left(1 - Cos\beta\right) Cos\,\varphi_0 \begin{cases} \geq 0 \;\; if \;\; Cos\,\varphi_0 \geq 0 \;\Rightarrow\; -\dfrac{\pi}{2} \leq \varphi_0 \leq \dfrac{\pi}{2} \\ \leq 0 \;\; if \;\; Cos\,\varphi_0 \leq 0 \;\Rightarrow\; \dfrac{\pi}{2} \leq \varphi_0 \leq \dfrac{3\pi}{2} \end{cases} \quad (33)$$

$$\begin{cases} a > b \;\; \Rightarrow \;\; \vec{a} \; is \; semi-major \; axis \; \& \; \vec{b} \; is \; semi-minor \; axis \; of \; vector \; ellipse \\ a < b \;\; \Rightarrow \;\; \vec{b} \; is \; semi-major \; axis \; \& \; \vec{a} \; is \; semi-minor \; axis \; of \; vector \; ellipse \\ a = b \;\; \Rightarrow \;\; \vec{a} \; and \; \vec{b} \; are \; radii \; of \; vector \; circle \end{cases} \quad (34)$$

As such, square of the constant focal[4,6] distance c^2 (35) and constant eccentricity[4] e (36) of this vector ellipse (30) are also defined.

$$c^2 = \left|a^2 - b^2\right| = 2\,r_1 r_2 \left(1 - Cos\beta\right)\left|Cos\,\varphi_0\right| \qquad \left(Focal \; Distance \; Squared\right) \quad (35)$$

$$e = \begin{cases} \dfrac{c}{a} = \sqrt{\dfrac{2\,r_1 r_2 \left(1 - Cos\beta\right)\left|Cos\,\varphi_0\right|}{r_1^2 - 2 r_1 r_2 \, Cos\beta \, Cos\,\varphi_0 + r_2^2}} & if \quad a > b \\ \dfrac{c}{b} = \sqrt{\dfrac{2\,r_1 r_2 \left(1 - Cos\beta\right)\left|Cos\,\varphi_0\right|}{r_1^2 - 2 r_1 r_2 \, Cos\,\varphi_0 + r_2^2}} & if \quad a < b \end{cases} \qquad \left(Eccentricity\right) \quad (36)$$

Based on $\vec{d}(\phi)$ (26), square of the distance $d^2(\phi)$ (22) - (24) between the points $\mathbf{P}_1$ and $\mathbf{P}_2$ can be expressed for all ϕ also as $d^2(\phi)$ (37) - (42), also utilizing (16) - (20) and (35).

$$d^2(\phi) = \vec{d}(\phi) \cdot \vec{d}(\phi) = \left[\vec{\mathbf{X}}(\phi) + \vec{\mathbf{Y}}(\phi) + \vec{\ell}\right] \cdot \left[\vec{\mathbf{X}}(\phi) + \vec{\mathbf{Y}}(\phi) + \vec{\ell}\right] \quad (37)$$

$$d^2(\phi) = \vec{\mathbf{X}}(\phi) \cdot \vec{\mathbf{X}}(\phi) + 2\vec{\mathbf{X}}(\phi) \cdot \vec{\mathbf{Y}}(\phi) + \vec{\mathbf{Y}}(\phi) \cdot \vec{\mathbf{Y}}(\phi) + 2\left[\vec{\mathbf{X}}(\phi) + \vec{\mathbf{Y}}(\phi)\right] \cdot \vec{\ell} + \vec{\ell} \cdot \vec{\ell} \quad (38)$$

$$d^2(\phi) = \vec{d}(\phi) \cdot \vec{d}(\phi) = \left[\vec{a}\,Cos(\omega t) + \vec{b}\,Sin(\omega t) + \vec{\ell}\right] \cdot \left[\vec{a}\,Cos(\omega t) + \vec{b}\,Sin(\omega t) + \vec{\ell}\right] \quad (39)$$

$$d^2(\phi) = \left[\vec{a}\,Cos\,\phi + \vec{b}\,Sin\phi\right]^2 + 2\left[\vec{a}\,Cos\,\phi + \vec{b}\,Sin\phi\right] \cdot \vec{\ell} + \ell^2 \quad (40)$$

$$d^2(\phi) = a^2 Cos^2\phi + 2\vec{a} \cdot \vec{b}\,Sin\phi\,Cos\,\phi + b^2 Sin^2\phi + 2\vec{a} \cdot \vec{\ell}\,Cos\,\phi + 2\vec{b} \cdot \vec{\ell}\,Sin\phi + \ell^2 \quad (41)$$

$$d^2(\phi) = b^2 + 2r_1 r_2 \left(1 - Cos\beta\right) Cos\phi \, Cos\left(\phi + \phi_0\right) + 2\left(\vec{a}Cos\phi + \vec{b}Sin\phi\right) \cdot \vec{\ell} + \ell^2 \quad (42)$$

This much was derived in the **ARTICLE 1** "Points on two circles in space mimic an ellipse with respect to a point" Based on (1) - (42), we will now derive an analytical method to calculate the orbital parameters $\left(r_1, r_2, \beta, \varphi_0, \ell_x, \ell_y, \ell_z, \ell\right)$ of the two circles and their respective positions in space, as well as the parameters (a, b, c, e) of the virtual ellipse[4] formed by the relative motion of the two heavenly bodies moving with the ***same fixed constant*** angular velocity ω (1) - (2) with respect to the centers of their own circles of revolution, when the distance $\vec{\ell}$ (10) between their centers of revolution is also constant.

For values of $d^2(\phi)$ (40) at quarter cycles, using trigonometric identities, we find that $d^2\left(-\frac{\pi}{2}+\phi\right)$ (43), $d^2(\pi+\phi)$ (44), and $d^2\left(\frac{\pi}{2}+\phi\right)$ (45) holds for any ϕ (2).

$$d^2\left(-\frac{\pi}{2}+\phi\right)=\left[\vec{a}\,Sin\,\phi-\vec{b}\,Cos\,\phi\right]^2+2\left[\vec{a}\,Sin\,\phi-\vec{b}\,Cos\,\phi\right]\cdot\vec{\ell}+\ell^2 \tag{43}$$

$$d^2(\pi+\phi)=\left[\vec{a}\,Cos\,\phi+\vec{b}\,Sin\,\phi\right]^2-2\left[\vec{a}\,Cos\,\phi+\vec{b}\,Sin\,\phi\right]\cdot\vec{\ell}+\ell^2 \tag{44}$$

$$d^2\left(\frac{\pi}{2}+\phi\right)=\left[\vec{a}\,Sin\,\phi-\vec{b}\,Cos\,\phi\right]^2-2\left[\vec{a}\,Sin\,\phi-\vec{b}\,Cos\,\phi\right]\cdot\vec{\ell}+\ell^2 \tag{45}$$

For values of $d^2(\phi)$ (42) at quarter cycles, using trigonometric identities, we find that $d^2\left(-\frac{\pi}{2}+\phi\right)$ (46), $d^2(\pi+\phi)$ (47), and $d^2\left(\frac{\pi}{2}+\phi\right)$ (48) also holds for any ϕ (2).

$$d^2\left(-\frac{\pi}{2}+\phi\right)=b^2+2r_1r_2\left(1-Cos\beta\right)Sin\,\phi\,Sin\left(\phi+\phi_0\right)+2\left(\vec{a}\,Sin\,\phi-\vec{b}\,Cos\,\phi\right)\cdot\vec{\ell}+\ell^2 \tag{46}$$

$$d^2(\pi+\phi)=b^2+2r_1r_2\left(1-Cos\beta\right)Cos\phi\,Cos\left(\phi+\phi_0\right)-2\left(\vec{a}Cos\phi+\vec{b}Sin\,\phi\right)\cdot\vec{\ell}+\ell^2 \tag{47}$$

$$d^2\left(\frac{\pi}{2}+\phi\right)=b^2+2r_1r_2\left(1-Cos\beta\right)Sin\,\phi\,Sin\left(\phi+\phi_0\right)-2\left(\vec{a}\,Sin\,\phi-\vec{b}\,Cos\,\phi\right)\cdot\vec{\ell}+\ell^2 \tag{48}$$

Utilizing the sum and difference of $d^2(\phi)$ (40) and $d^2(\pi+\phi)$ (44), as well as the sum and difference of $d^2(\phi)$ (42) and $d^2(\pi+\phi)$ (47), we define functions $\boldsymbol{k}_1(\phi)$ (49) and $\boldsymbol{k}_3(\phi)$ (51), respectively. Utilizing the sum and difference of $d^2\left(\frac{\pi}{2}+\phi\right)$ (45) and $d^2\left(-\frac{\pi}{2}+\phi\right)$ (43), as well as the sum and difference of $d^2\left(\frac{\pi}{2}+\phi\right)$ (48) and $d^2\left(-\frac{\pi}{2}+\phi\right)$ (46), we define the functions $\boldsymbol{k}_2(\phi)$ (50) and $\boldsymbol{k}_4(\phi)$ (52), respectively.

$$\begin{aligned}\boldsymbol{k}_1(\phi)=\frac{d^2(\phi)+d^2(\pi+\phi)}{2}&=\left[\vec{a}\,Cos\,\phi+\vec{b}\,Sin\,\phi\right]^2+\ell^2\\&=a^2Cos^2\phi+2\vec{a}\cdot\vec{b}\,Sin\,\phi\,Cos\,\phi+b^2Sin^2\phi\\&=b^2+2r_1r_2\left(1-Cos\beta\right)Cos\phi\,Cos\left(\phi+\phi_0\right)+\ell^2\end{aligned} \tag{49}$$

$$\begin{aligned}\boldsymbol{k}_2(\phi)=\frac{d^2\left(\frac{\pi}{2}+\phi\right)+d^2\left(-\frac{\pi}{2}+\phi\right)}{2}&=\left[\vec{a}\,Sin\,\phi-\vec{b}\,Cos\,\phi\right]^2+\ell^2\\&=a^2Sin^2\phi-2\vec{a}\cdot\vec{b}\,Sin\,\phi\,Cos\,\phi+b^2Cos^2\phi\\&=b^2+2r_1r_2\left(1-Cos\beta\right)Sin\,\phi\,Sin\left(\phi+\phi_0\right)+\ell^2\end{aligned} \tag{50}$$

$$\boldsymbol{k}_3(\phi)=\frac{d^2(\phi)-d^2(\pi+\phi)}{4}=\vec{a}\cdot\vec{\ell}\,Cos\,\phi+\vec{b}\cdot\vec{\ell}\,Sin\,\phi \tag{51}$$

$$k_4(\phi) = \frac{d^2\left(\frac{\pi}{2}+\phi\right) - d^2\left(-\frac{\pi}{2}+\phi\right)}{4} = -\vec{a}\cdot\vec{\ell}\,Sin\,\phi + \vec{b}\cdot\vec{\ell}\,Cos\,\phi \tag{52}$$

Inspecting closely, based on trigonometric identities, $k_2(\phi)$ (50) is a quarter cycle retarded (53) version of $k_1(\phi)$ (49), and $k_4(\phi)$ (52) is a quarter cycle retarded (54) version of $k_3(\phi)$ (51).

$$k_1\left(\frac{\pi}{2}+\phi\right) = k_2(\phi) \tag{53}$$

$$k_3\left(\frac{\pi}{2}+\phi\right) = k_4(\phi) \tag{54}$$

Sample plots of $k_1(\phi)$ (49), $k_2(\phi)$ (50), $k_3(\phi)$ (51), and $k_4(\phi)$ (52) over a relative $\mathbf{P}_1$ and $\mathbf{P}_2$ cycle can be seen in **Figure 3**, **Figure 4**, **Figure 5**, and **Figure 6**, respectively. The expected behavior in (53) - (54) can also be observed in these plots.

We see that $d^2(\phi)$ (40) can be expressed as $d^2(\phi)$ (55) in terms of $k_1(\phi)$ (49) and $k_3(\phi)$ (51).

$$d^2(\phi) = k_1(\phi) + 2k_3(\phi) \tag{55}$$

As $\vec{a}$ (16), $\vec{b}$ (17), and $\vec{\ell}$ (10) are constant vectors with fixed directions and magnitudes, their Dot Products[3] with each other, namely $\vec{a}\cdot\vec{\ell}$ (56) and $\vec{b}\cdot\vec{\ell}$ (57), are also constants.

$$\vec{a}\cdot\vec{\ell} = \{r_1 Cos\,\beta\,Cos\varphi_0 - r_2\}\,\ell_x + r_1 Sin\,\varphi_0\,\ell_y + r_1 Sin\,\beta\,Cos\varphi_0\ell_z \tag{56}$$

$$\vec{b}\cdot\vec{\ell} = -r_1 Cos\,\beta\,Sin\,\varphi_0\ell_x + \{r_1 Cos\varphi_0 - r_2\}\,\ell_y - r_1 Sin\,\beta\,Sin\,\varphi_0\ell_z \tag{57}$$

As such, from $k_3(\phi)$ (51) and $k_4(\phi)$ (52), we can define the $k_{3,0}$ (58) and $k_{4,0}$ (59) constants, respectively, for this two-circle system.

$$k_{3,0} = k_3(0) = \frac{d^2(0) - d^2(\pi)}{4} = \vec{a}\cdot\vec{\ell} \tag{58}$$

$$k_{4,0} = k_4(0) = \frac{d^2\left(\frac{\pi}{2}\right) - d^2\left(-\frac{\pi}{2}\right)}{4} = \vec{b}\cdot\vec{\ell} \tag{59}$$

Using $k_{3,0}$ (58) and $k_{4,0}$ (59) in $k_3(\phi)$ (51) and $k_4(\phi)$ (52) would yield the expressions of $k_3(\phi)$ (60) and $k_4(\phi)$ (61).

$$k_3(\phi) = k_{3,0}\,Cos\,\phi + k_{4,0}\,Sin\,\phi \tag{60}$$

$$k_4(\phi) = -k_{3,0}\,Sin\,\phi + k_{4,0}\,Cos\,\phi \tag{61}$$

An alternative way of expressing $k_3(\phi)$ (60) and $k_4(\phi)$ (61) is as $k_3(\phi)$ (62) and $k_4(\phi)$ (63).

$$k_3(\phi) = \sqrt{k_{3,0}^{\ 2} + k_{4,0}^{\ 2}}\left(\frac{k_{3,0}}{\sqrt{k_{3,0}^{\ 2} + k_{4,0}^{\ 2}}}Cos\,\phi + \frac{k_{4,0}}{\sqrt{k_{3,0}^{\ 2} + k_{4,0}^{\ 2}}}Sin\,\phi\right) = \sqrt{k_{3,0}^{\ 2} + k_{4,0}^{\ 2}}\,Cos(\phi - \xi) \tag{62}$$

$$\boldsymbol{k}_4(\phi)=\sqrt{\boldsymbol{k}_{3,0}{}^2+\boldsymbol{k}_{4,0}{}^2}\left(-\frac{\boldsymbol{k}_{3,0}}{\sqrt{\boldsymbol{k}_{3,0}{}^2+\boldsymbol{k}_{4,0}{}^2}}Sin\phi+\frac{\boldsymbol{k}_{4,0}}{\sqrt{\boldsymbol{k}_{3,0}{}^2+\boldsymbol{k}_{4,0}{}^2}}Cos\phi\right)=-\sqrt{\boldsymbol{k}_{3,0}{}^2+\boldsymbol{k}_{4,0}{}^2}\,Sin(\phi-\xi)\quad(63)$$

In this context, ξ (66) comes up as a constant angle based on $\boldsymbol{k}_{3,0}$ (58) and $\boldsymbol{k}_{4,0}$ (59), or equivalently based on $\vec{\boldsymbol{a}}\cdot\vec{\ell}$ (56) and $\vec{\boldsymbol{b}}\cdot\vec{\ell}$ (57), which is a significant identifier for this two-circle system, defined based on its sine $Sin(\xi)$ (64) and cosine $Cos(\xi)$ (65).

$$Cos(\xi)=\frac{\boldsymbol{k}_{3,0}}{\sqrt{\boldsymbol{k}_{3,0}{}^2+\boldsymbol{k}_{4,0}{}^2}}=\frac{\vec{\boldsymbol{a}}\cdot\vec{\ell}}{\sqrt{(\vec{\boldsymbol{a}}\cdot\vec{\ell})^2+(\vec{\boldsymbol{b}}\cdot\vec{\ell})^2}}\tag{64}$$

$$Sin(\xi)=\frac{\boldsymbol{k}_{4,0}}{\sqrt{\boldsymbol{k}_{3,0}{}^2+\boldsymbol{k}_{4,0}{}^2}}=\frac{\vec{\boldsymbol{b}}\cdot\vec{\ell}}{\sqrt{(\vec{\boldsymbol{a}}\cdot\vec{\ell})^2+(\vec{\boldsymbol{b}}\cdot\vec{\ell})^2}}\tag{65}$$

$$\xi=Cos^{-1}\left(\frac{\boldsymbol{k}_{3,0}}{\sqrt{\boldsymbol{k}_{3,0}{}^2+\boldsymbol{k}_{4,0}{}^2}}\right)=Sin^{-1}\left(\frac{\boldsymbol{k}_{4,0}}{\sqrt{\boldsymbol{k}_{3,0}{}^2+\boldsymbol{k}_{4,0}{}^2}}\right)\tag{66}$$

Based on $\boldsymbol{k}_3(\phi)$ (62) and $\boldsymbol{k}_4(\phi)$ (63), $Cos(\phi-\xi)$ (67) and $Sin(\phi-\xi)$ (68) are also defined for the angle $(\phi-\xi)$ (69), at every ϕ (2).

$$Cos(\phi-\xi)=\frac{\boldsymbol{k}_3(\phi)}{\sqrt{\boldsymbol{k}_{3,0}{}^2+\boldsymbol{k}_{4,0}{}^2}}\tag{67}$$

$$Sin(\phi-\xi)=-\frac{\boldsymbol{k}_4(\phi)}{\sqrt{\boldsymbol{k}_{3,0}{}^2+\boldsymbol{k}_{4,0}{}^2}}\tag{68}$$

$$(\phi-\xi)=Cos^{-1}\left(\frac{\boldsymbol{k}_3(\phi)}{\sqrt{\boldsymbol{k}_{3,0}{}^2+\boldsymbol{k}_{4,0}{}^2}}\right)=Sin^{-1}\left(-\frac{\boldsymbol{k}_4(\phi)}{\sqrt{\boldsymbol{k}_{3,0}{}^2+\boldsymbol{k}_{4,0}{}^2}}\right)\tag{69}$$

Based on $\boldsymbol{k}_3(\phi)$ (62) and $\boldsymbol{k}_4(\phi)$ (63), we obtain the results for $\boldsymbol{k}_3(\xi)$ (70), $\boldsymbol{k}_3\left(\frac{\pi}{2}+\xi\right)$ (71), $\boldsymbol{k}_3(\pi+\xi)$ (70), $\boldsymbol{k}_3\left(-\frac{\pi}{2}+\xi\right)$ (71), $\boldsymbol{k}_4(\xi)$ (71), $\boldsymbol{k}_4\left(\frac{\pi}{2}+\xi\right)$ (70), $\boldsymbol{k}_4(\pi+\xi)$ (71), and $\boldsymbol{k}_4\left(-\frac{\pi}{2}+\xi\right)$ (70).

$$\boldsymbol{k}_{3,\max}=\boldsymbol{k}_3(\xi)=-\boldsymbol{k}_3(\pi+\xi)=\boldsymbol{k}_{4,\max}=-\boldsymbol{k}_4\left(\frac{\pi}{2}+\xi\right)=\boldsymbol{k}_4\left(-\frac{\pi}{2}+\xi\right)=\sqrt{\boldsymbol{k}_{3,0}{}^2+\boldsymbol{k}_{4,0}{}^2}\tag{70}$$

$$\boldsymbol{k}_3\left(\frac{\pi}{2}+\xi\right)=\boldsymbol{k}_3\left(-\frac{\pi}{2}+\xi\right)=\boldsymbol{k}_4(\xi)=\boldsymbol{k}_4(\pi+\xi)=0\tag{71}$$

$$\boldsymbol{k}_{3,\max}=\boldsymbol{k}_{4,\max}=\sqrt{\boldsymbol{k}_{3,0}{}^2+\boldsymbol{k}_{4,0}{}^2}\quad;\quad\boldsymbol{k}_{3,\min}=\boldsymbol{k}_{4,\min}=-\sqrt{\boldsymbol{k}_{3,0}{}^2+\boldsymbol{k}_{4,0}{}^2}\tag{72}$$

Looking at our definition of $\boldsymbol{k}_3(\phi)$ (51), we see that it is a measure of the projection of $\left[\vec{r}_1(\phi)-\vec{r}_2(\phi)=\vec{a}Cos\phi+\vec{b}Sin\phi\right]$ (26) on $\vec{\ell}$ (10), which are components of $\vec{d}(\phi)$ (26). Inspecting closely, further to $\boldsymbol{k}_4(\phi)$ (52) being a quarter cycle retarded (54) version of $\boldsymbol{k}_3(\phi)$ (51), both $\boldsymbol{k}_3(\phi)$ (51) and $\boldsymbol{k}_4(\phi)$ (52) harmonically vary in the value range between $\left[-\sqrt{\boldsymbol{k}_{3,0}{}^2+\boldsymbol{k}_{4,0}{}^2}\right]$ (72) and $\sqrt{\boldsymbol{k}_{3,0}{}^2+\boldsymbol{k}_{4,0}{}^2}$ (72). When $(\phi=\xi)$ (66), $\boldsymbol{k}_3(\phi)$ (51) takes on its maximum value $\boldsymbol{k}_{3,\max}$ (70) over a cycle, which is $\sqrt{\boldsymbol{k}_{3,0}{}^2+\boldsymbol{k}_{4,0}{}^2}$ (70) based on $\boldsymbol{k}_3(\phi)$ (60). We therefore infer that ξ (66) is an angle-wise measure of how close the radii difference vector $\left[\vec{r}_1(\phi)-\vec{r}_2(\phi)\right]$ (26) is to the $\vec{\ell}$ (10) vector between the centers of the two circles. When $(\phi=\xi)$ (66), $\left[\vec{r}_1(\phi)-\vec{r}_2(\phi)\right]$ (26) vector takes on its closest position to the $\vec{\ell}$ (10) vector over a cycle, as their projection $\boldsymbol{k}_{3,\max}$ (70) onto each other is maximum.

Moving on, taking the sum and difference of $\boldsymbol{k}_1(\phi)$ (49) and $\boldsymbol{k}_2(\phi)$(50), we obtain the relations for $\left[\boldsymbol{k}_1(\phi)+\boldsymbol{k}_2(\phi)\right]$ (73) and $\left[\boldsymbol{k}_1(\phi)-\boldsymbol{k}_2(\phi)\right]$ (74), and taking sum and difference of $\boldsymbol{k}_3(\phi)$ (51) and $\boldsymbol{k}_4(\phi)$ (52), we obtain the relations for $\left[\boldsymbol{k}_3(\phi)+\boldsymbol{k}_4(\phi)\right]$ (75) and $\left[\boldsymbol{k}_3(\phi)-\boldsymbol{k}_4(\phi)\right]$ (76), at any ϕ (2). A sample plot of $\left[\boldsymbol{k}_1(\phi)-\boldsymbol{k}_2(\phi)\right]$ (74) over a relative $\mathbf{P}_1$ and $\mathbf{P}_2$ cycle is in **Figure 7**.

$$\boldsymbol{k}_1(\phi)+\boldsymbol{k}_2(\phi)=a^2+b^2+2\ell^2 \tag{73}$$

$$\boldsymbol{k}_1(\phi)-\boldsymbol{k}_2(\phi)=2r_1r_2(1-Cos\beta)Cos(2\phi+\phi_0) \tag{74}$$

$$\boldsymbol{k}_3(\phi)+\boldsymbol{k}_4(\phi)=(\boldsymbol{k}_{3,0}+\boldsymbol{k}_{4,0})Cos\phi-(\boldsymbol{k}_{3,0}-\boldsymbol{k}_{4,0})Sin\phi \tag{75}$$

$$\boldsymbol{k}_3(\phi)-\boldsymbol{k}_4(\phi)=(\boldsymbol{k}_{3,0}-\boldsymbol{k}_{4,0})Cos\phi+(\boldsymbol{k}_{3,0}+\boldsymbol{k}_{4,0})Sin\phi \tag{76}$$

Further, based on $\boldsymbol{k}_3(\phi)$ (51) and $\boldsymbol{k}_4(\phi)$ (52), as well as $\boldsymbol{k}_3(\phi)$ (60) and $\boldsymbol{k}_4(\phi)$ (61), we observe that $\left\{[\boldsymbol{k}_3(\phi)]^2+[\boldsymbol{k}_4(\phi)]^2\right\}$ (77) is constant over all ϕ (2).

$$[\boldsymbol{k}_3(\phi)]^2+[\boldsymbol{k}_4(\phi)]^2=(\vec{a}\cdot\vec{\ell})^2+(\vec{b}\cdot\vec{\ell})^2=\boldsymbol{k}_{3,0}{}^2+\boldsymbol{k}_{4,0}{}^2 \tag{77}$$

Solving (75) and (76) for ϕ (2), we obtain the identities for $Sin\phi$ (78) and $Cos\phi$ (79).

$$Sin\phi=\frac{\boldsymbol{k}_{4,0}\boldsymbol{k}_3(\phi)-\boldsymbol{k}_{3,0}\boldsymbol{k}_4(\phi)}{\boldsymbol{k}_{3,0}{}^2+\boldsymbol{k}_{4,0}{}^2} \tag{78}$$

$$Cos\phi=\frac{\boldsymbol{k}_{3,0}\boldsymbol{k}_3(\phi)+\boldsymbol{k}_{4,0}\boldsymbol{k}_4(\phi)}{\boldsymbol{k}_{3,0}{}^2+\boldsymbol{k}_{4,0}{}^2} \tag{79}$$

Based on $Sin\phi$ (78) and $Cos\phi$ (79), we also find $Sin\varphi_0$ (80) and $Cos\varphi_0$ (81) for $(\phi=\varphi_0)$ (3).

$$Sin\varphi_0=\frac{\boldsymbol{k}_{4,0}\boldsymbol{k}_3(\varphi_0)-\boldsymbol{k}_{3,0}\boldsymbol{k}_4(\varphi_0)}{\boldsymbol{k}_{3,0}{}^2+\boldsymbol{k}_{4,0}{}^2} \tag{80}$$

$$Cos\,\varphi_0 = \frac{\boldsymbol{k}_{3,0}\boldsymbol{k}_3(\varphi_0) + \boldsymbol{k}_{4,0}\boldsymbol{k}_4(\varphi_0)}{\boldsymbol{k}_{3,0}^{\;2} + \boldsymbol{k}_{4,0}^{\;2}} \tag{81}$$

Utilizing $Sin\,\phi$ (78), $Cos\,\phi$ (79), trigonometric identities, and (77), we obtain $Sin(\phi_p - \phi_q)$ (82) and $Cos(\phi_p - \phi_q)$ (83) as well, for any two phase angles $\phi = \phi_p$ (2) and $\phi = \phi_q$ (2) in a cycle.

$$Sin(\phi_p - \phi_q) = \frac{\boldsymbol{k}_3(\phi_p)\boldsymbol{k}_4(\phi_q) - \boldsymbol{k}_4(\phi_p)\boldsymbol{k}_3(\phi_q)}{\boldsymbol{k}_{3,0}^{\;2} + \boldsymbol{k}_{4,0}^{\;2}} \tag{82}$$

$$Cos(\phi_p - \phi_q) = \frac{\boldsymbol{k}_3(\phi_p)\boldsymbol{k}_3(\phi_q) + \boldsymbol{k}_4(\phi_p)\boldsymbol{k}_4(\phi_q)}{\boldsymbol{k}_{3,0}^{\;2} + \boldsymbol{k}_{4,0}^{\;2}} \tag{83}$$

We further solve for $\boldsymbol{k}_3(\phi_p)$ (84) and $\boldsymbol{k}_4(\phi_p)$ (85) in $Sin(\phi_p - \phi_q)$ (82) and $Cos(\phi_p - \phi_q)$ (83), also making use of (77) for simplification.

$$\boldsymbol{k}_3(\phi_p) = \boldsymbol{k}_4(\phi_q)Sin(\phi_p - \phi_q) + \boldsymbol{k}_3(\phi_q)Cos(\phi_p - \phi_q) \tag{84}$$

$$\boldsymbol{k}_4(\phi_p) = -\boldsymbol{k}_3(\phi_q)Sin(\phi_p - \phi_q) + \boldsymbol{k}_4(\phi_q)Cos(\phi_p - \phi_q) \tag{85}$$

Continuing our analysis, in the configuration of **Figure 1**, we define the unit vectors $\hat{\boldsymbol{u}}_{1\perp}$ (86) and $\hat{\boldsymbol{u}}_{2\perp}$ (87) normal to $Circle_1$ and $Circle_2$ in the direction of the self-rotation axes[7,8] of the moving heavenly bodies at these points $\mathbf{P}_1$ and $\mathbf{P}_2$ defined on these two circles $Circle_1$ and $Circle_2$, respectively, namely the unit vectors pointing in the directions of the North Poles[10,11] of these bodies along their rotational axes[8] respectively, and whose directions are determined according to 'right hand rule'[12] based on the direction of self-rotations of the bodies. Based on our assertion in the **ARTICLE 1** "Points on two circles in space mimic an ellipse with respect to a point", the self-rotation axis[7,8] and the axis of individual circular revolution of each body must be aligned along the normal of its equatorial[9] plane.

Further, all moving bodies in space must obey "Faraday's Law of Induction"[13] in their motions, which states that "Electromotive Force (EMF)[14] around a closed path is equal to the negative of time rate of change of the Magnetic Flux[15] enclosed by the path, implying that charged particles moving in an external magnetic field with changing flux[15] move in a direction to reverse the change in that magnetic flux[15]. Therefore, as our research results for the Sun-Earth-Moon system also support based on the results presented in the **ARTICLE 3** "Analytical calculation of earth and sun orbital parameters from distance data" and **ARTICLE 4** "An analysis of the circular orbital motion of the moon and the earth", we assert that all bodies in space possibly behave like macro-scale charged particles moving in external magnetic fields with changing flux[15] through the area of their individual circular orbit of revolution while moving in *clockwise*[16] direction from a vantage point above their North Pole[10,11], which we expect must be the main cause that triggers their self rotation in the reverse (*counterclockwise*[16]) direction, as seen from a vantage point above their North Pole[10,11], where we always accept the North Pole[10,11] of a body or revolution to be the pole rotating in *counterclockwise*[16] direction according to "right hand rule"[12].

As such, in the configuration of **Figure 1**, for the heavenly body at point $\mathbf{P}_2$ moving around $Circle_2$, as we choose the direction of increasing ϕ (2) to be the *clockwise*[16] direction from a vantage point above the $\hat{z}-$axis, which also determines the axis of individual circular revolution of the heavenly body at point $\mathbf{P}_2$ moving around $Circle_2$, we take the direction of $\hat{\boldsymbol{u}}_{2\perp}$ (87) aligned with the North Pole[10,11] of that body to be $(-\hat{z})$ (87), based on the "right hand rule"[12] and our assertion above that the individual circular orbit of revolution in *clockwise*[16] direction from a vantage point above the North Pole[10,11] of a body triggers self rotation in the reverse (*counterclockwise*[16]) direction. Similarly, in the configuration of **Figure 1**, for the heavenly body at point $\mathbf{P}_1$ moving around $Circle_1$, as we choose the direction of increasing ϕ (2) to be the *clockwise*[16] direction from a vantage point above the $(-\hat{x}Sin\beta+\hat{z}Cos\beta)-$axis, which also determines the axis of individual circular revolution of the heavenly body at point $\mathbf{P}_1$ moving around $Circle_1$, we take the direction of $\hat{\boldsymbol{u}}_{1\perp}$ (86) aligned with the North Pole[10,11] of that body to be $(\hat{x}Sin\beta-\hat{z}Cos\beta)$ (86), based on the "right hand rule"[12] and our assertion above that the individual circular orbit of revolution in *clockwise*[16] direction from a vantage point above the North Pole[10,11] of a body triggers self rotation in the reverse (*counterclockwise*[16]) direction.

$$\hat{\boldsymbol{u}}_{1\perp} = \hat{x}Sin\beta - \hat{z}Cos\beta \tag{86}$$

$$\hat{\boldsymbol{u}}_{2\perp} = -\hat{z} \tag{87}$$

With $\vec{\boldsymbol{d}}(\phi)$ (26) chosen to be directed from moving point $\mathbf{P}_2$ to moving point $\mathbf{P}_1$, Dot Product[3], or equivalently, the projection of $\left[-\vec{\boldsymbol{d}}(\phi)\right]$ (12) on $\hat{\boldsymbol{u}}_{1\perp}$ (86) and the projection of $\vec{\boldsymbol{d}}(\phi)$ (12) on $\hat{\boldsymbol{u}}_{2\perp}$ (87) would be as in (90) and (91), respectively, at each ϕ (2), where $\gamma_1(\phi)$ (88) is defined as the angle between $\hat{\boldsymbol{u}}_{1\perp}$ (86) and $\left[-\vec{\boldsymbol{d}}(\phi)\right]$ (26), and $\gamma_2(\phi)$ (89) is defined as the angle between $\hat{\boldsymbol{u}}_{2\perp}$ (86) and $\vec{\boldsymbol{d}}(\phi)$ (26), with $\gamma_{1,\Delta}(\phi)$ (88) defined as the tilt[8] angle between $\hat{\boldsymbol{u}}_{1\perp}$ (86) and the component of $\hat{\boldsymbol{u}}_{1\perp}$ (86) normal to $\left[-\vec{\boldsymbol{d}}(\phi)\right]$ (26), and $\gamma_{2,\Delta}(\phi)$ (89) defined as the tilt[8] angle between $\hat{\boldsymbol{u}}_{2\perp}$ (86) and the component of $\hat{\boldsymbol{u}}_{2\perp}$ (87) normal to $\vec{\boldsymbol{d}}(\phi)$ (26).

$$\gamma_1(\phi) = \frac{\pi}{2} - \gamma_{1,\Delta}(\phi) \tag{88}$$

$$\gamma_2(\phi) = \frac{\pi}{2} - \gamma_{2,\Delta}(\phi) \tag{89}$$

$$\left[-\vec{\boldsymbol{d}}(\phi)\right]\cdot\hat{\boldsymbol{u}}_{1\perp} = \left|\vec{\boldsymbol{d}}(\phi)\right| Cos\left[\gamma_1(\phi)\right] = r_2\, Cos\phi\, Sin\beta - \ell_x Sin\beta + \ell_z Cos\beta \tag{90}$$

$$\vec{\boldsymbol{d}}(\phi)\cdot\hat{\boldsymbol{u}}_{2\perp} = \left|\vec{\boldsymbol{d}}(\phi)\right| Cos\left[\gamma_2(\phi)\right] = -\left[r_1 Cos(\phi+\phi_0) Sin\beta + \ell_z\right] \tag{91}$$

Considering point $\mathbf{P}_2$'s $\vec{\boldsymbol{d}}(\phi)\cdot\hat{\boldsymbol{u}}_{2\perp}$ (91) case, minimum and maximum values that $\gamma_2(\phi)$ (89) would take over a relative cycle of $\mathbf{P}_1$ and $\mathbf{P}_2$, namely $\gamma_{2,\min}$ (92) and $\gamma_{2,\max}$ (93), respectively, can be geometrically visualized in the most simple form as in **Figure 2**, and they can also be expressed in terms of the minimum and maximum relative tilt[8] angles γ_{2,Δ_2} (92) and γ_{2,Δ_1} (93)between the North Pole[10,11] unit vector $\hat{\boldsymbol{u}}_{2\perp}$ (87) along the rotational axis[8] of the body at $\mathbf{P}_2$ and the component of $\hat{\boldsymbol{u}}_{2\perp}$ (87) normal to $\vec{\boldsymbol{d}}(\phi)$ (12) directed from $\mathbf{P}_2$ to $\mathbf{P}_1$. Based on the relation in (91), the minimum and maximum tilts[8] of the body at $\mathbf{P}_2$ occur at half a cycle, or $\left(\phi_{\gamma_{2,max}} - \phi_{\gamma_{2,min}} = \pi\right)$ (96), apart, at $\left(\phi = -\phi_0 = -\varphi_0\right)$ (94) and $\left(\phi = \pi - \phi_0 = \pi - \varphi_0\right)$ (95). We choose to take the point in the relative cycle of $\mathbf{P}_1$ and $\mathbf{P}_2$ when $\left(\phi = \phi_{\gamma_{2,min}}\right)$ (94) occurs to correspond to the angle $\left(\phi = -\varphi_0\right)$ (94), and the point when $\left(\phi = \phi_{\gamma_{2,max}}\right)$ (95) occurs to correspond to the angle $\left(\phi = \pi - \varphi_0\right)$ (95). Note in **Figure 2** that at the minimum tilt[8] of the body at $\mathbf{P}_2$ with respect to $\vec{\boldsymbol{d}}(\phi)$ (12) directed from $\mathbf{P}_2$ to $\mathbf{P}_1$, P_2 represents the position of $\mathbf{P}_2$ and P_1 represents the position of $\mathbf{P}_1$, whereas the maximum tilt[8] of the body at $\mathbf{P}_2$ with respect to $\vec{\boldsymbol{d}}(\phi)$ (12) directed from $\mathbf{P}_2$ to $\mathbf{P}_1$, $P_2^{'}$ represents the position of $\mathbf{P}_2$ and $P_1^{'}$ represents the position of $\mathbf{P}_1$. P_1 and $P_1^{'}$ are not co-located, but are sketched together in **Figure 2** just for simplicity of visualization.

$$\gamma_{2,\min} = \gamma_2\left(\phi_{\gamma_{2,\min}}\right) = \frac{\pi}{2} - \gamma_{2,\Delta}\left(\phi_{\gamma_{2,\min}}\right) = \frac{\pi}{2} - \gamma_{2,\Delta_2} \quad ; \quad \gamma_{2,\Delta}\left(\phi_{\gamma_{2,\min}}\right) = \gamma_{2,\Delta_2} \tag{92}$$

$$\gamma_{2,\max} = \gamma_2\left(\phi_{\gamma_{2,\max}}\right) = \frac{\pi}{2} - \gamma_{2,\Delta}\left(\phi_{\gamma_{2,\max}}\right) = \frac{\pi}{2} + \gamma_{2,\Delta_1} \quad ; \quad \gamma_{2,\Delta}\left(\phi_{\gamma_{2,\max}}\right) = -\gamma_{2,\Delta_1} \tag{93}$$

$$\phi_{\gamma_{2,\min}} = -\phi_0 = -\varphi_0 \tag{94}$$

$$\phi_{\gamma_{2,\max}} = \pi - \phi_0 = \pi - \varphi_0 \tag{95}$$

$$\phi_{\gamma_{2,\max}} = \pi + \phi_{\gamma_{2,\min}} \tag{96}$$

In similar lines, for $\mathbf{P}_1$'s $\left[-\vec{\boldsymbol{d}}(\phi)\right]\cdot\hat{\boldsymbol{u}}_{1\perp}$ (90) case, maximum and minimum values that $\gamma_1(\phi)$ (88) would take over a relative cycle of $\mathbf{P}_1$ and $\mathbf{P}_2$, namely $\gamma_{1,\max}$ (97) and $\gamma_{1,\min}$ (97), respectively, can be expressed in terms of the tilt[8] angles γ_{1,Δ_1} (97) and γ_{1,Δ_2} (97) between the North Pole[10] unit vector $\hat{\boldsymbol{u}}_{1\perp}$ (86) along the rotational axis[8] of the body at $\mathbf{P}_1$ and the normal component of $\hat{\boldsymbol{u}}_{1\perp}$ (86) to the negative of the distance vector $\left[-\vec{\boldsymbol{d}}(\phi)\right]$ (12) directed from $\mathbf{P}_1$ to $\mathbf{P}_2$.

$$\gamma_{1,\max} = \frac{\pi}{2} + \gamma_{1,\Delta_1} \quad ; \quad \gamma_{1,\min} = \frac{\pi}{2} - \gamma_{1,\Delta_2} \tag{97}$$

Based on (92) - (96), we obtain the relation in (98) from (91) at $\left(\phi = \phi_{\gamma_{2,\min}} = -\phi_0 = -\varphi_0\right)$ (94), and we obtain the relation in (99) from (91) at $\left(\phi = \phi_{\gamma_{2,\max}} = \pi - \phi_0 = \pi - \varphi_0\right)$ (95). As a result, we are able to solve for $\left(r_1 Sin\beta\right)$ (100) and ℓ_z (101) from (98) - (99).

$$\left|\vec{\boldsymbol{d}}\left(\phi_{\gamma_{2,\min}}\right)\right| Cos\left(\frac{\pi}{2} - \gamma_{2,\Delta_2}\right) = \left|\vec{\boldsymbol{d}}\left(\phi_{\gamma_{2,\min}}\right)\right| Sin\left(\gamma_{2,\Delta_2}\right) = -r_1\, Sin\beta - \ell_z \tag{98}$$

$$\left|\vec{\boldsymbol{d}}\left(\phi_{\gamma_{2,\max}}\right)\right| Cos\left(\frac{\pi}{2} + \gamma_{2,\Delta_1}\right) = -\left|\vec{\boldsymbol{d}}\left(\phi_{\gamma_{2,\max}}\right)\right| Sin\left(\gamma_{2,\Delta_1}\right) = r_1\, Sin\beta - \ell_z \tag{99}$$

$$r_1 Sin\,\beta = -\frac{\left\{\left|\vec{d}\left(\phi_{\gamma_{2,\max}}\right)\right| Sin\left(\gamma_{2,\Delta_1}\right) + \left|\vec{d}\left(\phi_{\gamma_{2,\min}}\right)\right| Sin\left(\gamma_{2,\Delta_2}\right)\right\}}{2} \tag{100}$$

$$\ell_z = \frac{\left\{\left|\vec{d}\left(\phi_{\gamma_{2,\max}}\right)\right| Sin\left(\gamma_{2,\Delta_1}\right) - \left|\vec{d}\left(\phi_{\gamma_{2,\min}}\right)\right| Sin\left(\gamma_{2,\Delta_2}\right)\right\}}{2} \tag{101}$$

Having gained as much insight into a system of two circles in space based on our analysis through (40) - (101), around which different heavenly bodies move with the ***same fixed constant*** angular velocity ω (1) - (2) with respect to the centers of their own circles of revolution, where the distance $\vec{\ell}$ (10) between their centers of revolution is also constant, we shall now move on to utilize this insight to derive an analytical method to calculate the orbital parameters $\left(r_1, r_2, \beta, \varphi_0, \ell_x, \ell_y, \ell_z, \ell\right)$ of the two circles and their respective positions in space, as well as the parameters (a, b, c, e) of the virtual ellipse[4] formed by the relative motion of these two heavenly bodies, provided we have the distance data over a full cycle between the two heavenly bodies on the two circles, namely the distance data between points $\mathbf{P}_1$ and $\mathbf{P}_2$ on the two respective circles at any value of the phase ϕ (2), as well as the minimum and maximum tilt[8] angles $\gamma_{2,\min}$ (92) and $\gamma_{2,\max}$ (93) between the North Pole[10,11] unit vector $\hat{\boldsymbol{u}}_{2\perp}$ (87) along the rotational axis[8] of the body at $\mathbf{P}_2$ and the component of $\hat{\boldsymbol{u}}_{2\perp}$ (87) normal to $\vec{\boldsymbol{d}}(\phi)$ (12) directed from $\mathbf{P}_2$ to $\mathbf{P}_1$.

Based on the scalar distance $d(\phi)$ (21) data we have between points $\mathbf{P}_1$ and $\mathbf{P}_2$ on the two respective circles over a full cycle, we start by calculating values of $\boldsymbol{k}_1(\phi)$ (49), $\boldsymbol{k}_2(\phi)$ (50), $\boldsymbol{k}_3(\phi)$ (51), and $\boldsymbol{k}_4(\phi)$ (52) for all data points that we have, as well as values of $\left[\boldsymbol{k}_1(\phi)+\boldsymbol{k}_2(\phi)\right]$ (73), $\left[\boldsymbol{k}_1(\phi)-\boldsymbol{k}_2(\phi)\right]$ (74), $\left[\boldsymbol{k}_3(\phi)+\boldsymbol{k}_4(\phi)\right]$ (75), and $\left[\boldsymbol{k}_3(\phi)-\boldsymbol{k}_4(\phi)\right]$ (76). In doing this, we first calculate $d^2(\phi)$ (40) for all data points that we have, and provided that the amount of data points we have are sufficiently many, also considering that points $\mathbf{P}_1$ and $\mathbf{P}_2$ on the two respective circles move with the ***same fixed constant*** angular velocity ω (1) - (2) with respect to the centers of their own circles of revolution, we make the assumption that data points half a cycle apart in a two-circle system are approximately $(\phi = \pi)$ apart, and that data points a quarter cycle apart in a two-circle system are approximately $\left(\phi = \frac{\pi}{2}\right)$ apart.

Inspecting calculated $\left[\boldsymbol{k}_1(\phi)-\boldsymbol{k}_2(\phi)\right]$ (74) values over a full cycle, we determine its maximum $\left[\boldsymbol{k}_1(\phi)-\boldsymbol{k}_2(\phi)\right]_{Max}$ (102), which must be when $\left[Cos\left(2\phi+\phi_0\right)=1\right]$ (102) and thus $\left[\phi = -\frac{\phi_0}{2} = -\frac{\varphi_0}{2}\right]$ (102), where the value of $\left(\phi_0 = \varphi_0\right)$ (3) is yet to be determined. The $\left\{V_1 = \left[\boldsymbol{k}_1(\phi)-\boldsymbol{k}_2(\phi)\right]_{Max}\right\}$ (102) value provides us the value of $\left[2r_1r_2\left(1-Cos\beta\right)\right]$ (102) all the same, at this point.

$$V_1 = \left[\boldsymbol{k}_1(\phi) - \boldsymbol{k}_2(\phi)\right]_{Max} = 2r_1 r_2 \left(1 - Cos\beta\right) \quad ; \quad Cos\left(2\phi + \phi_0\right) = 1 \;\Rightarrow\; \phi = -\frac{\phi_0}{2} = -\frac{\varphi_0}{2} \quad (102)$$

If $\left[2r_1 r_2\left(1 - Cos\beta\right) = V_1 = 0\right]$ (102) is found, we would understand that $\left(Cos\beta = 1\right)$ (137), meaning $\left(\beta = 0\right)$ (137). This is a special case, whose possible solution is investigated through (137) - (165).

We also take note of calculated values of $\left[\boldsymbol{k}_1(\phi) + \boldsymbol{k}_2(\phi)\right]$ (73) and $\left\{\left[\boldsymbol{k}_3(\phi)\right]^2 + \left[\boldsymbol{k}_4(\phi)\right]^2\right\}$ (77), which are constant over a full cycle, i.e. for all ϕ (2). Thus, $\left[V_2 = \boldsymbol{k}_1(\phi) + \boldsymbol{k}_2(\phi)\right]$ (103) value provides the value of $\left(a^2 + b^2 + 2\ell^2\right)$ (73), and $\left\{V_3 = \left[\boldsymbol{k}_3(\phi)\right]^2 + \left[\boldsymbol{k}_4(\phi)\right]^2\right\}$ (104) value provides us the value of $\left(\boldsymbol{k}_{3,0}{}^2 + \boldsymbol{k}_{4,0}{}^2\right)$ (77), and equivalently the value of $\left[\left(\vec{\boldsymbol{a}} \cdot \vec{\ell}\right)^2 + \left(\vec{\boldsymbol{b}} \cdot \vec{\ell}\right)^2\right]$ (77).

$$V_2 = \boldsymbol{k}_1(\phi) + \boldsymbol{k}_2(\phi) = a^2 + b^2 + 2\ell^2 \quad (103)$$

$$V_3 = \left[\boldsymbol{k}_3(\phi)\right]^2 + \left[\boldsymbol{k}_4(\phi)\right]^2 = \left(\vec{\boldsymbol{a}} \cdot \vec{\ell}\right)^2 + \left(\vec{\boldsymbol{b}} \cdot \vec{\ell}\right)^2 = \boldsymbol{k}_{3,0}{}^2 + \boldsymbol{k}_{4,0}{}^2 \quad (104)$$

Next, we determine the data point at which calculated $\boldsymbol{k}_3(\phi)$ (51) takes on its maximum value $\boldsymbol{k}_{3,\max}$ (105) over a full cycle, which is $\sqrt{\boldsymbol{k}_{3,0}{}^2 + \boldsymbol{k}_{4,0}{}^2}$ (70), and we expect it to be matching with the square root of the $\left(V_3 = \boldsymbol{k}_{3,0}{}^2 + \boldsymbol{k}_{4,0}{}^2\right)$ (77) value previously found. At this data point, we expect $\left(\phi = \xi\right)$ (105) to be true, based on (70).

$$\boldsymbol{k}_3(\phi) = \boldsymbol{k}_{3,\max} = \boldsymbol{k}_3(\xi) = \sqrt{\boldsymbol{k}_{3,0}{}^2 + \boldsymbol{k}_{4,0}{}^2} = \sqrt{V_3} \quad \Rightarrow \quad \phi = \xi \quad (105)$$

Therefore, at the same data point where we expect $\left(\phi = \xi\right)$ (105) to be the case, we check the value of $\left[\boldsymbol{k}_1(\phi) - \boldsymbol{k}_2(\phi)\right]$ (74) and take note of it as V_4 (106), where $\left(\phi_0 = \varphi_0\right)$ (3).

$$V_4 = \boldsymbol{k}_1(\xi) - \boldsymbol{k}_2(\xi) = 2r_1 r_2\left(1 - Cos\beta\right)Cos\left(2\xi + \varphi_0\right) \quad (106)$$

We had already found the value of $\left[2r_1 r_2\left(1 - Cos\beta\right)\right]$ (102) as V_1 (102). So, we can now determine that $Cos\left(2\xi + \varphi_0\right)$ (107) is V_4 (106) divided by V_1 (102), and thus $\left(2\xi + \varphi_0\right)$ (108) is either the positive or negative value of arccosine of "V_4 (106) divided by V_1 (102)".

$$Cos\left(2\xi + \phi_0\right) = \frac{V_4}{V_1} \quad (107)$$

$$\left(2\xi + \varphi_0\right) = \pm Cos^{-1}\left(\frac{V_4}{V_1}\right) \quad (108)$$

Further, knowing the minimum and maximum relative tilt[8] angles $\gamma_{2,\min}$ (92) and $\gamma_{2,\max}$ (93) between the North Pole[10,11] vector $\hat{\boldsymbol{u}}_{2\perp}$ (87) along the rotational axis[8] of the body at $\mathbf{P}_2$ and the component of $\hat{\boldsymbol{u}}_{2\perp}$ (87) normal to $\vec{\boldsymbol{d}}(\phi)$ (12) directed from $\mathbf{P}_2$ to $\mathbf{P}_1$, as well as the distance data

$\left|\vec{d}\left(\phi_{\gamma_{2,\min}}\right)\right|$ (98) and $\left|\vec{d}\left(\phi_{\gamma_{2,\max}}\right)\right|$ (99) beween $\mathbf{P}_1$ and $\mathbf{P}_2$ corresponding to those data points of minimum and maximum relative tilt[8] angles $\gamma_{2,\min}$ (92) and $\gamma_{2,\max}$ (93), respectively, we are able to calculate the values V_5 (109) and V_6 (110), which allows us to determine the values of $(r_1 Sin\,\beta)$ (100) and ℓ_z (101), respectively, at this point.

$$V_5 = -\frac{\left\{\left|\vec{d}\left(\phi_{\gamma_{2,\max}}\right)\right| Sin\left(\gamma_{2,\Delta_1}\right) + \left|\vec{d}\left(\phi_{\gamma_{2,\min}}\right)\right| Sin\left(\gamma_{2,\Delta_2}\right)\right\}}{2} = r_1 Sin\,\beta \qquad (109)$$

$$V_6 = \frac{\left\{\left|\vec{d}\left(\phi_{\gamma_{2,\max}}\right)\right| Sin\left(\gamma_{2,\Delta_1}\right) - \left|\vec{d}\left(\phi_{\gamma_{2,\min}}\right)\right| Sin\left(\gamma_{2,\Delta_2}\right)\right\}}{2} = \ell_z \qquad (110)$$

If $\left[2r_1 r_2\left(1-Cos\beta\right) = V_1 \neq 0\right]$ (102) but $\left(r_1 Sin\,\beta = V_5 = 0\right)$ (109) is found, we would then know that $\left(Cos\beta \neq 1\right)$ (166) but $\left(Sin\,\beta = 0\right)$ (166), meaning $\left(\beta = \pi\right)$ (166). This is another special case, whose possible solution is described in (166) - (186).

We had previously determined that the data points of minimum and maximum relative tilt[8] angles $\gamma_{2,\min}$ (92) and $\gamma_{2,\max}$ (93) correspond to a phase of $\mathbf{P}_2$ at $\left(\phi = -\phi_0 = -\varphi_0\right)$ (94) and $\left(\phi = \pi - \phi_0 = \pi - \varphi_0\right)$ (95), and we have chosen to take the point in the relative cycle of $\mathbf{P}_1$ and $\mathbf{P}_2$ when $\left(\phi = \phi_{\gamma_{2,\min}}\right)$ (94) occurs to correspond to phase angle $\left(\phi = -\varphi_0\right)$ (94), and we have chosen to take the point in the relative cycle of $\mathbf{P}_1$ and $\mathbf{P}_2$ when $\left(\phi = \phi_{\gamma_{2,\max}}\right)$ (95) occurs to correspond to phase angle $\left(\phi = \pi - \varphi_0\right)$ (95). Therefore, for the data point corresponding to minimum relative tilt[8] $\gamma_{2,\min}$ (92) corresponding phase angle $\left(\phi = \phi_{\gamma_{2,\min}} = -\varphi_0\right)$ (94), we are also able to calculate the $\left[\boldsymbol{k}_1(\phi) - \boldsymbol{k}_2(\phi)\right]$ (74) value of $[\boldsymbol{k}_1(-\varphi_0) - \boldsymbol{k}_2(-\varphi_0)]$ (111), where the $\left\{V_7 = [\boldsymbol{k}_1(-\varphi_0) - \boldsymbol{k}_2(-\varphi_0)]\right\}$ (111) value provides us the $\left[2r_1 r_2\left(1-Cos\beta\right)Cos\left(\varphi_0\right)\right]$ (111) value.

$$V_7 = \left[\boldsymbol{k}_1(-\varphi_0) - \boldsymbol{k}_2(-\varphi_0)\right] = 2r_1 r_2\left(1-Cos\beta\right)Cos\left(\varphi_0\right) \quad ; \quad \phi = \phi_{\gamma_{2,\min}} = -\varphi_0 \qquad (111)$$

Subsequently, the $\left[2r_1 r_2\left(1-Cos\beta\right)Cos\left(\varphi_0\right)\right]$ (111) value V_7 (111) provides us the value of the square of the constant focal[4,6] distance c^2 (35) of this vector ellipse (30) formed by this wo circle system, and thus the focal[4,6] distance c (112), taking its square root.

$$c = \sqrt{2\,r_1 r_2\left(1-Cos\beta\right)\left|Cos\,\varphi_0\right|} = \sqrt{\left|2\,r_1 r_2\left(1-Cos\beta\right)Cos\,\varphi_0\right|} = \sqrt{\left|V_7\right|} \qquad (Focal\ Distance)\ (112)$$

We had already found the value of $\left[2r_1 r_2\left(1-Cos\beta\right)\right]$ (102) as V_1 (102). So, we can now determine that $Cos\left(\varphi_0\right)$ (113) is V_7 (111) divided by V_1 (102), and thus φ_0 (114) is either the positive or negative value of arccosine of "V_7 (111) divided by V_1 (102)". We can further make the analysis in (115) - (116) regarding the possible value of φ_0 (114).

$$Cos(\varphi_0)=\frac{V_7}{V_1} \tag{113}$$

$$\varphi_0=\pm Cos^{-1}\left(\frac{V_7}{V_1}\right) \tag{114}$$

$$-1\leq Cos\beta\leq 1 \quad\Rightarrow\quad (1-Cos\beta)\geq 0 \quad\Rightarrow\quad 2r_1r_2(1-Cos\beta)=V_1\geq 0 \tag{115}$$

$$2r_1r_2(1-Cos\beta)Cos(\varphi_0)=V_7\begin{cases}\geq 0 \ \ if \ \ Cos(\varphi_0)\geq 0\Rightarrow -\frac{\pi}{2}\leq\varphi_0\leq\frac{\pi}{2}\\ \leq 0 \ \ if \ \ Cos(\varphi_0)\leq 0\Rightarrow \frac{\pi}{2}\leq\varphi_0\leq\frac{3\pi}{2}\end{cases} \tag{116}$$

Note that $\left[\boldsymbol{k}_1(\phi)-\boldsymbol{k}_2(\phi)\right]$ (74) at $(\phi=0)$ (2), namely $\left[\boldsymbol{k}_1(0)-\boldsymbol{k}_2(0)\right]$ (117), must have the same value as $\left[\boldsymbol{k}_1(\phi)-\boldsymbol{k}_2(\phi)\right]$ (74) at $(\phi=-\varphi_0)$ (94), namely $\{V_7=[\boldsymbol{k}_1(-\varphi_0)-\boldsymbol{k}_2(-\varphi_0)]\}$ (111), as well as the same value as $\left[\boldsymbol{k}_1(\phi)-\boldsymbol{k}_2(\phi)\right]$ (74) at $(\phi=\pi)$ (2) and at $(\phi=\pi-\varphi_0)$ (95). Looking closely at **Figure 7**, over a relative $\mathbf{P}_1$ and $\mathbf{P}_2$ cycle, these same values of $\left[\boldsymbol{k}_1(\phi)-\boldsymbol{k}_2(\phi)\right]$ (74) at $(\phi=0)$ (2), $(\phi=-\varphi_0)$ (94), $(\phi=\pi)$ (2), and at $(\phi=\pi-\varphi_0)$ (95) must occur along a horizontal line of same value V_7 (111) cutting across the figure at four points.

$$\left[\boldsymbol{k}_1(0)-\boldsymbol{k}_2(0)\right]=2r_1r_2(1-Cos\beta)Cos(\varphi_0)=V_7=[\boldsymbol{k}_1(-\varphi_0)-\boldsymbol{k}_2(-\varphi_0)] \tag{117}$$

As we know the data points which correspond to minimum and maximum relative tilt[8] angles $\gamma_{2,\min}$ (92) and $\gamma_{2,\max}$ (93), namely the data points corresponding to phase angles $(\phi=\phi_{\gamma_{2,\min}}=-\varphi_0)$ (94) and $(\phi=\phi_{\gamma_{2,\max}}=\pi-\varphi_0)$ (95) over a relative $\mathbf{P}_1$ and $\mathbf{P}_2$ cycle, we are able to determine which one of the two data points corresponding to $(\phi=\phi_{\gamma_{2,\min}}=-\varphi_0)$ (94) and $(\phi=\phi_{\gamma_{2,\max}}=\pi-\varphi_0)$ (95) occurs before the other in a relative $\mathbf{P}_1$ and $\mathbf{P}_2$ cycle, having the same $\left[\boldsymbol{k}_1(\phi)-\boldsymbol{k}_2(\phi)\right]$ (74) value of V_7 (111) on either the left side (uphill) or the right side (downhill) of the first peak in **Figure 7**. Subsequently, we determine the data points corresponding to $(\phi=0)$ (2) and $(\phi=\pi)$ (2) based on the following analysis, as well as making the assessment regarding the correct value of φ_0 (114).

- If the data point corresponding to $(\phi=\phi_{\gamma_{2,\min}}=-\varphi_0)$ (94) occurs before the data point corresponding to $(\phi=\phi_{\gamma_{2,\max}}=\pi-\varphi_0)$ (95), and has $\left[\boldsymbol{k}_1(\phi)-\boldsymbol{k}_2(\phi)\right]$ (74) value V_7 (111) on the left side (uphill) of the first peak in **Figure 7**, the data point corresponding to $(\phi=\phi_{\gamma_{2,\max}}=\pi-\varphi_0)$ (95) has $\left[\boldsymbol{k}_1(\phi)-\boldsymbol{k}_2(\phi)\right]$ (74) value V_7 (111) on the left side (uphill) of the second peak in **Figure 7**, and we can determine that the data point corresponding to $(\phi=0)$ (2) occurs after the data point corresponding to

$\left(\phi = \phi_{\gamma_{2,min}} = -\varphi_0\right)$ (94), and has $\left[\boldsymbol{k}_1(\phi) - \boldsymbol{k}_2(\phi)\right]$ (74) value V_7 (111) on the <u>right</u> side (downhill) of the <u>first</u> peak in **Figure 7**, and the data point corresponding to $(\phi = \pi)$ (2) occurs after the data point corresponding to $\left(\phi = \phi_{\gamma_{2,max}} = \pi - \varphi_0\right)$ (95), and has $\left[\boldsymbol{k}_1(\phi) - \boldsymbol{k}_2(\phi)\right]$ (74) value V_7 (111) on the <u>right</u> side (downhill) of the <u>second</u> peak in **Figure 7**. In this case, we can also reason that $(-\varphi_0 < 0 < \pi - \varphi_0 < \pi)$, and determine which of the two possible values of φ_0 (114), $\left[Cos^{-1}\left(\frac{V_7}{V_1}\right)\right]$ or $\left[-Cos^{-1}\left(\frac{V_7}{V_1}\right)\right]$, is the correct solution for φ_0 (114) in this two circle system.

- If the data point corresponding to $\left(\phi = \phi_{\gamma_{2,min}} = -\varphi_0\right)$ (94) occurs before the data point corresponding to $\left(\phi = \phi_{\gamma_{2,max}} = \pi - \varphi_0\right)$ (95), and has $\left[\boldsymbol{k}_1(\phi) - \boldsymbol{k}_2(\phi)\right]$ (74) value V_7 (111) on the <u>right</u> side (downhill) of the <u>first</u> peak in **Figure 7**, the data point corresponding to $\left(\phi = \phi_{\gamma_{2,max}} = \pi - \varphi_0\right)$ (95) has $\left[\boldsymbol{k}_1(\phi) - \boldsymbol{k}_2(\phi)\right]$ (74) value V_7 (111) on the <u>right</u> side (downhill) of the <u>second</u> peak in **Figure 7**, and we can determine that the data point corresponding to $(\phi = 0)$ (2) occurs before the data point corresponding to $\left(\phi = \phi_{\gamma_{2,min}} = -\varphi_0\right)$ (94), and has $\left[\boldsymbol{k}_1(\phi) - \boldsymbol{k}_2(\phi)\right]$ (74) value V_7 (111) on the <u>left</u> side (uphill) of the <u>first</u> peak in **Figure 7**, and the data point corresponding to $(\phi = \pi)$ (2) occurs before the data point corresponding to $\left(\phi = \phi_{\gamma_{2,max}} = \pi - \varphi_0\right)$ (95), and has $\left[\boldsymbol{k}_1(\phi) - \boldsymbol{k}_2(\phi)\right]$ (74) value V_7 (111) on the <u>left</u> side (uphill) of the <u>second</u> peak in **Figure 7**. In this case, we can also reason that $(0 < -\varphi_0 < \pi < \pi - \varphi_0)$, and determine which of the two possible values of φ_0 (114), $\left[Cos^{-1}\left(\frac{V_7}{V_1}\right)\right]$ or $\left[-Cos^{-1}\left(\frac{V_7}{V_1}\right)\right]$, is the correct solution for φ_0 (114) in this two circle system.
- If the data point corresponding to $\left(\phi = \phi_{\gamma_{2,min}} = -\varphi_0\right)$ (94) occurs after the data point corresponding to $\left(\phi = \phi_{\gamma_{2,max}} = \pi - \varphi_0\right)$ (95), and has $\left[\boldsymbol{k}_1(\phi) - \boldsymbol{k}_2(\phi)\right]$ (74) value V_7 (111) on the <u>left</u> side (uphill) of the <u>second</u> peak in **Figure 7**, the data point corresponding to $\left(\phi = \phi_{\gamma_{2,max}} = \pi - \varphi_0\right)$ (95) has $\left[\boldsymbol{k}_1(\phi) - \boldsymbol{k}_2(\phi)\right]$ (74) value V_7 (111) on the <u>left</u> side (uphill) of the <u>first</u> peak in **Figure 7**, and we can determine that the data point corresponding to $(\phi = 0)$ (2) occurs before the data point corresponding to $\left(\phi = \phi_{\gamma_{2,min}} = -\varphi_0\right)$ (94), and has $\left[\boldsymbol{k}_1(\phi) - \boldsymbol{k}_2(\phi)\right]$ (74) value V_7 (111) on the <u>right</u> side (downhill) of the <u>first</u> peak in **Figure 7**, and the data point corresponding to $(\phi = \pi)$ (2) occurs before the data point corresponding to $\left(\phi = \phi_{\gamma_{2,max}} = \pi - \varphi_0\right)$ (95), and has

$\left[\boldsymbol{k}_1(\phi)-\boldsymbol{k}_2(\phi)\right]$ (74) value V_7 (111) on the right side (downhill) of the second peak in **Figure 7**. In this case, we can also reason that $(\pi-\varphi_0<0<-\varphi_0<\pi)$, and determine which of the two possible values of φ_0 (114), $\left[Cos^{-1}\left(\frac{V_7}{V_1}\right)\right]$ or $\left[-Cos^{-1}\left(\frac{V_7}{V_1}\right)\right]$, is the correct solution for φ_0 (114) in this two circle system.

- If the data point corresponding to $\left(\phi=\phi_{\gamma_{2,\min}}=-\varphi_0\right)$ (94) occurs after the data point corresponding to $\left(\phi=\phi_{\gamma_{2,\max}}=\pi-\varphi_0\right)$ (95), and has $\left[\boldsymbol{k}_1(\phi)-\boldsymbol{k}_2(\phi)\right]$ (74) value V_7 (111) on the right side (downhill) of the second peak in **Figure 7**, the data point corresponding to $\left(\phi=\phi_{\gamma_{2,\max}}=\pi-\varphi_0\right)$ (95) has $\left[\boldsymbol{k}_1(\phi)-\boldsymbol{k}_2(\phi)\right]$ (74) value V_7 (111) on the right side (downhill) of the first peak in **Figure 7**, and we can determine that the data point corresponding to $(\phi=0)$ (2) occurs before the data point corresponding to $\left(\phi=\phi_{\gamma_{2,\max}}=\pi-\varphi_0\right)$ (95), and has $\left[\boldsymbol{k}_1(\phi)-\boldsymbol{k}_2(\phi)\right]$ (74) value V_7 (111) on the left side (downhill) of the first peak in **Figure 7**, and the data point corresponding to $(\phi=\pi)$ (2) occurs before the data point corresponding to $\left(\phi=\phi_{\gamma_{2,\min}}=-\varphi_0\right)$ (94), and has $\left[\boldsymbol{k}_1(\phi)-\boldsymbol{k}_2(\phi)\right]$ (74) value V_7 (111) on the left side (downhill) of the second peak in **Figure 7**. In this case, we can also reason that $(0<\pi-\varphi_0<\pi<-\varphi_0)$, and determine which of the two possible values of φ_0 (114), $\left[Cos^{-1}\left(\frac{V_7}{V_1}\right)\right]$ or $\left[-Cos^{-1}\left(\frac{V_7}{V_1}\right)\right]$, is the correct solution for φ_0 (114) in this two circle system.

Having determined the correct value of φ_0 (114) at this point, and already knowing the data point at which the calculated $\boldsymbol{k}_3(\phi)$ (51) takes on its maximum value $\boldsymbol{k}_{3,\max}$ (105) over a full cycle, which is $\sqrt{\boldsymbol{k}_{3,0}^{\;2}+\boldsymbol{k}_{4,0}^{\;2}}$ (70), we can now also determine which of the two possible values of $(2\xi+\varphi_0)$ (108), $\left[Cos^{-1}\left(\frac{V_4}{V_1}\right)\right]$ or $\left[-Cos^{-1}\left(\frac{V_4}{V_1}\right)\right]$, is the correct solution, and we thus determine the correct value of ξ (118).

$$\xi=\frac{\pm Cos^{-1}\left(\frac{V_4}{V_1}\right)-\varphi_0}{2} \tag{118}$$

As we have now also determined the data points corresponding to $(\phi=0)$ (2) and $(\phi=\pi)$ (2) based on the above analysis, we can also determine the values of $\boldsymbol{k}_3(\phi)$ (51) and $\boldsymbol{k}_4(\phi)$ (52) at the data point corresponding to $(\phi=0)$ (2), namely the values of $\boldsymbol{k}_{3,0}$ (58) and $\boldsymbol{k}_{4,0}$ (59), respectively, and equivalently the values of $\vec{\boldsymbol{a}}\cdot\vec{\ell}$ (56) and $\vec{\boldsymbol{b}}\cdot\vec{\ell}$ (57), respectively.

Now that we have the values of $\boldsymbol{k}_{3,0}$ (58) and $\boldsymbol{k}_{4,0}$ (59), as well as the values of $\boldsymbol{k}_3(\phi)$ (51) and $\boldsymbol{k}_4(\phi)$ (52) for all ϕ (2) corresponding to the data points we have for this two circle system, we can also calculate the values of $Sin\phi$ (78) and $Cos\phi$ (79) for all ϕ (2) corresponding to the data points we have. Thus we can also calculate the values of ϕ (119) for all the data points we have for this two circle system.

$$\phi = Sin^{-1}\left(\frac{\boldsymbol{k}_{4,0}\boldsymbol{k}_3(\phi)-\boldsymbol{k}_{3,0}\boldsymbol{k}_4(\phi)}{\boldsymbol{k}_{3,0}{}^2+\boldsymbol{k}_{4,0}{}^2}\right) = Cos^{-1}\left(\frac{\boldsymbol{k}_{3,0}\boldsymbol{k}_3(\phi)+\boldsymbol{k}_{4,0}\boldsymbol{k}_4(\phi)}{\boldsymbol{k}_{3,0}{}^2+\boldsymbol{k}_{4,0}{}^2}\right) \tag{119}$$

As we already have the values of $\boldsymbol{k}_1(\phi)$ (49) for all calculated values of ϕ (119) corresponding to the data points we have for this two circle system, and we have already determined the value of $(\phi_0=\varphi_0)$ (114), we are now able to determine the value of $(b^2+\ell^2)$ (120), namely the sum of b^2 (19) and ℓ^2 (11), from the V_8 (120) value that must be the same at any ϕ (119).

$$V_8 = \boldsymbol{k}_1(\phi) - 2r_1r_2(1-Cos\beta)Cos\phi\, Cos(\phi+\varphi_0) = b^2+\ell^2 \tag{120}$$

Moving on, using $d^2(\phi)$ (55) expressed in terms of $\boldsymbol{k}_1(\phi)$ (49) and $\boldsymbol{k}_3(\phi)$ (60), we have the $d^2(\phi_p)$ (121) expression using distance squared data value at any $(\phi=\phi_p)$ (119) data point, and the $d^2(\phi_q)$ (122) expression using distance squared data value at any other $(\phi=\phi_q)$ (119) data point, also utilizing the found values of $\boldsymbol{k}_3(\phi)$ (60) at the $(\phi=\phi_p)$ (119) and $(\phi=\phi_q)$ (119) data points, the found φ_0 (114) value, and the expressions for a^2 (18), b^2 (19), and $(\vec{\boldsymbol{a}}\cdot\vec{\boldsymbol{b}})$ (20).

$$\begin{aligned}
d^2(\phi_p) &= \boldsymbol{k}_1(\phi_p)+2\boldsymbol{k}_3(\phi_p)\\
&= a^2Cos^2\phi_p + 2\vec{\boldsymbol{a}}\cdot\vec{\boldsymbol{b}}\, Sin\phi_p Cos\phi_p + b^2Sin^2\phi_p + \ell^2 + 2\boldsymbol{k}_3(\phi_p)\\
&= \left[r_1^2 - 2r_1r_2Cos\beta Cos(\varphi_0)+r_2^2\right]Cos^2\phi_p - 2\left[r_1r_2(1-Cos\beta)Sin(\varphi_0)\right]Sin\phi_p Cos\phi_p\\
&\quad + \left[r_1^2-2r_1r_2Cos(\varphi_0)+r_2^2\right]Sin^2\phi_p + \ell^2 + 2\boldsymbol{k}_3(\phi_p)\\
&= (r_1^2+r_2^2+\ell^2) - 2r_1r_2Cos(\varphi_0)(Cos\beta Cos^2\phi_p + Sin^2\phi_p)\\
&\quad - 2r_1r_2(1-Cos\beta)Sin(\varphi_0)Sin\phi_p Cos\phi_p + 2\boldsymbol{k}_3(\phi_p)
\end{aligned} \tag{121}$$

$$\begin{aligned}
d^2(\phi_q) &= \boldsymbol{k}_1(\phi_q)+2\boldsymbol{k}_3(\phi_q)\\
&= a^2Cos^2\phi_q + 2\vec{\boldsymbol{a}}\cdot\vec{\boldsymbol{b}}\, Sin\phi_q Cos\phi_q + b^2Sin^2\phi_q + \ell^2 + 2\boldsymbol{k}_3(\phi_q)\\
&= \left[r_1^2 - 2r_1r_2Cos\beta Cos(\varphi_0)+r_2^2\right]Cos^2\phi_q - 2\left[r_1r_2(1-Cos\beta)Sin(\varphi_0)\right]Sin\phi_q Cos\phi_q\\
&\quad + \left[r_1^2-2r_1r_2Cos(\varphi_0)+r_2^2\right]Sin^2\phi_q + \ell^2 + 2\boldsymbol{k}_3(\phi_q)\\
&= (r_1^2+r_2^2+\ell^2) - 2r_1r_2Cos(\varphi_0)(Cos\beta Cos^2\phi_q + Sin^2\phi_q)\\
&\quad - 2r_1r_2(1-Cos\beta)Sin(\varphi_0)Sin\phi_q Cos\phi_q + 2\boldsymbol{k}_3(\phi_q)
\end{aligned} \tag{122}$$

Subtracting (122) from (121) and reorganizing, we obtain the relation in (123).

$$
\begin{aligned}
2r_1r_2Cos(\varphi_0)\left[Cos\beta\left(Cos^2\phi_p-Cos^2\phi_q\right)+\left(Sin^2\phi_p-Sin^2\phi_q\right)\right]&\\
=-\left[d^2(\phi_p)-d^2(\phi_q)\right]&\\
-2r_1r_2(1-Cos\beta)Sin(\varphi_0)\left(Sin\phi_pCos\phi_p-Sin\phi_qCos\phi_q\right)&\\
+2\left[\boldsymbol{k}_3(\phi_p)-\boldsymbol{k}_3(\phi_q)\right]&
\end{aligned}
\tag{123}
$$

Dividing all sides of (123) by $\left[2r_1r_2(1-Cos\beta)\right]$ (102), and reorganizing, we get an expression to obain the value of $Cos\beta$ (124) from the value of V_9 (124), a value we are now able to determine, as we already have the values of all parameters and variables involved in the expression (124) other than $Cos\beta$ (124), as well as the found value of V_1 (102).

$$
V_9=\frac{\left\{\begin{aligned}&-\left[d^2(\phi_p)-d^2(\phi_q)\right]-V_1Sin(\varphi_0)\left(Sin\phi_pCos\phi_p-Sin\phi_qCos\phi_q\right)\\&+2\left[\boldsymbol{k}_3(\phi_p)-\boldsymbol{k}_3(\phi_q)\right]-V_1Cos(\varphi_0)\left(Sin^2\phi_p-Sin^2\phi_q\right)\end{aligned}\right\}}{\left\{\begin{aligned}&-\left[d^2(\phi_p)-d^2(\phi_q)\right]-V_1Sin(\varphi_0)\left(Sin\phi_pCos\phi_p-Sin\phi_qCos\phi_q\right)\\&+2\left[\boldsymbol{k}_3(\phi_p)-\boldsymbol{k}_3(\phi_q)\right]+V_1Cos(\varphi_0)\left(Cos^2\phi_p-Cos^2\phi_q\right)\end{aligned}\right\}}=Cos\beta \tag{124}
$$

Based on $\left(Cos\beta=V_9\right)$ (124) and the sign of $\left(r_1Sin\beta=V_5\right)$ (109), we are also able to determine whether the positive or negative arccosine solution is the correct value for β (125).

$$
\beta=\begin{cases}Cos^{-1}(V_9) & if \quad V_5>0 \quad \Rightarrow \quad Sin\beta>0\\ -Cos^{-1}(V_9) & if \quad V_5<0 \quad \Rightarrow \quad Sin\beta<0\end{cases} \tag{125}
$$

Having obtained the value of β (125) for our two circle system as such, we can now also determine the value of r_1 (126) from $\left(r_1Sin\beta=V_5\right)$ (109), i.e. when $\left(Sin\beta\neq0\right)$ (126).

$$
r_1=\sqrt{\frac{V_5^2}{Sin^2\beta}} \quad ; \quad Sin\beta\neq0 \quad \Rightarrow \quad \beta\neq0 \ \ \& \ \ \beta\neq\pi \tag{126}
$$

We can now also determine the value of r_2 (127) from $\left[2r_1r_2(1-Cos\beta)=V_1\right]$ (102), when $\left(Cos\beta\neq\pm1\right)$ (127), making use of the found values of $Cos\beta$ (124) and r_1 (126).

$$
r_2=\frac{V_1}{2r_1(1-Cos\beta)} \quad ; \quad Cos\beta\neq\pm1 \quad \Rightarrow \quad \beta\neq0 \ \ \& \ \ \beta\neq\pi \tag{127}
$$

Subsequently, using the found values of r_1 (126), r_2 (127), $Cos\beta$ (124), and φ_0 (114), at this point we can determine the semi-major[5] and semi-minor[5] axis magnitudes a (18) and b (19), as well as semi-major[5] and semi-minor[5] axis vectors $\vec{a}$ (16) and $\vec{b}$ (17) of the ellipse[4] formed by the relative motion of points $\mathbf{P}_1$ and $\mathbf{P}_2$ on two respective circles in our system, where $\vec{a}$ (130) with magnitude a (128) is the semi-major[5] axis vector and $\vec{b}$ (131) with magnitude b (129) is the semi-minor[5] axis vector of the ellipse[4] when $(a>b)^{4,5}$ (34), and vice versa, with (31) reducing to the special case of a circle when $(a=b)^{4,5}$ (34).

$$a=\sqrt{a^2}=\sqrt{r_1^2-2r_1 r_2 Cos\beta Cos(\varphi_0)+r_2^2} \tag{128}$$

$$b=\sqrt{b^2}=\sqrt{r_1^2-2r_1 r_2 Cos(\varphi_0)+r_2^2} \tag{129}$$

$$\vec{a}=\hat{x}\left[r_1 Cos\beta Cos(\varphi_0)-r_2\right]+\hat{y}\,r_1 Sin(\varphi_0)+\hat{z}\,r_1 Sin\beta Cos(\varphi_0) \tag{130}$$

$$\vec{b}=-\hat{x}\,r_1 Cos\beta Sin(\varphi_0)+\hat{y}\left[r_1 Cos(\varphi_0)-r_2\right]-\hat{z}\,r_1 Sin\beta Sin(\varphi_0) \tag{131}$$

The eccentricity[4] e (132) of this vector ellipse (30) can also be found now based on (36), using the found values of the semi-major[5] and semi-minor[5] axes a (128) and b (129), and the found value of focal[4,6] distance c (112) of the ellipse[4] formed by the relative motion of points $\mathbf{P}_1$ and $\mathbf{P}_2$ on two respective circles in our system.

$$e=\begin{cases}\dfrac{c}{a} & if \quad a>b\\[2ex] \dfrac{c}{b} & if \quad a<b\end{cases} \qquad (Eccentricity) \tag{132}$$

Having determined the value of b (129), we can also determine the value of ℓ (133) from $(b^2+\ell^2=V_8)$ (120).

$$\ell=\sqrt{b^2-V_8} \tag{133}$$

We can solve for ℓ_x (134) and ℓ_y (135) from the expressions for $\vec{a}\cdot\vec{\ell}$ (56) and $\vec{b}\cdot\vec{\ell}$ (57), also replacing $\vec{a}\cdot\vec{\ell}$ (56) and $\vec{b}\cdot\vec{\ell}$ (57) with $\boldsymbol{k}_{3,0}$ (58) and $\boldsymbol{k}_{4,0}$ (59), respectively. As we already have found the values of r_1 (126), r_2 (127), $Cos\beta$ (124), φ_0 (114), $\boldsymbol{k}_{3,0}$ (58), $\boldsymbol{k}_{4,0}$ (59), and ℓ_z (110), we are now also able to determine the values of ℓ_x (134) and ℓ_y (135).

$$\ell_x=\frac{\boldsymbol{k}_{3,0}\{r_1 Cos\varphi_0-r_2\}-\boldsymbol{k}_{4,0}r_1 Sin\varphi_0}{\{r_1^2 Cos\beta-r_1 r_2 Cos\varphi_0(1+Cos\beta)+r_2^2\}}-\frac{r_1 Sin\beta(r_1-r_2 Cos\varphi_0)}{\{r_1^2 Cos\beta-r_1 r_2 Cos\varphi_0(1+Cos\beta)+r_2^2\}}\ell_z \tag{134}$$

$$\ell_y=\frac{\boldsymbol{k}_{3,0}r_1 Cos\beta Sin\varphi_0+\boldsymbol{k}_{4,0}\{r_1 Cos\beta Cos\varphi_0-r_2\}}{\{r_1^2 Cos\beta-r_1 r_2 Cos\varphi_0(1+Cos\beta)+r_2^2\}}-\frac{r_1 r_2 Sin\beta Sin\varphi_0}{\{r_1^2 Cos\beta-r_1 r_2 Cos\varphi_0(1+Cos\beta)+r_2^2\}}\ell_z \tag{135}$$

Note that we expect the ℓ (136) value obtained from ℓ (11) using the found values of ℓ_x (134), ℓ_y (135), and ℓ_z (110) to be the same as the ℓ (133) value we have found previously.

$$\ell=\sqrt{\ell^2}=\sqrt{\ell_x^{\,2}+\ell_y^{\,2}+\ell_z^{\,2}} \tag{136}$$

Therefore, at this point we have determined the values of r_1 (126), r_2 (127), β (125), φ_0 (114), ℓ_x (134), ℓ_y (135), ℓ_z (110), and ℓ (136) for this two circle system, as well as the values of the semi-major[5] and semi-minor[5] axes a (128) and b (129), the focal[4,6] distance c (112), and the eccentricity[4] e (132) of this vector ellipse[4] (30) formed by the relative motion of points $\mathbf{P}_1$ and $\mathbf{P}_2$ on two respective circles in our system, as we had asserted. These are the values we had initially set out to find in this Article, which we have achieved at this point, for which we have set forth a step by step analytical method as outlined above between (102) - (136).

If $\left[2r_1 r_2 \left(1-Cos\beta\right)=V_1=0\right]$ (102) is found, we would understand that $\left(Cos\beta=1\right)$ (137), meaning $\left(\beta=0\right)$ (137). This is a special case where the two circles around which points $\mathbf{P}_1$ and $\mathbf{P}_2$ move are parallel, the values of the semi-major[5] and semi-minor[5] axes a (128) and b (129) are the same $\left(a=b\right)$ (138), and the focal[4,6] distance c (112) and the eccentricity[4] e (132) of this vector ellipse[4] (30) formed by the relative motion of points $\mathbf{P}_1$ and $\mathbf{P}_2$ on two respective circles in our system are both zero (142) - (143), leading to a relative circular motion of $\mathbf{P}_1$ and $\mathbf{P}_2$, with vectors $\vec{\boldsymbol{a}}$ (16) and $\vec{\boldsymbol{b}}$ (17) reducing to $\vec{\boldsymbol{a}}$ (139) and $\vec{\boldsymbol{b}}$ (140), normal (141) to each other in the $\hat{\boldsymbol{x}}-\hat{\boldsymbol{y}}$ plane of **Figure 1**.

$$2r_1 r_2 \left(1-Cos\beta\right)=V_1=0 \quad \Rightarrow \quad Cos\beta=1 \quad \Rightarrow \quad \beta=0 \tag{137}$$

$$a=b=\sqrt{a^2}=\sqrt{b^2}=\sqrt{r_1^{\,2}-2r_1 r_2 Cos\left(\varphi_0\right)+r_2^{\,2}} \tag{138}$$

$$\vec{\boldsymbol{a}}=\hat{\boldsymbol{x}}\left[r_1 Cos\left(\varphi_0\right)-r_2\right]+\hat{\boldsymbol{y}}\, r_1 Sin\left(\varphi_0\right) \;\; ; \;\; \vec{\boldsymbol{a}}=\left[\vec{\boldsymbol{r}}_1\left(\phi\right)-\vec{\boldsymbol{r}}_2\left(\phi\right)\right]\left(\phi=0\right) \tag{139}$$

$$\vec{\boldsymbol{b}}=-\hat{\boldsymbol{x}}\, r_1 Sin\left(\varphi_0\right)+\hat{\boldsymbol{y}}\left[r_1 Cos\left(\varphi_0\right)-r_2\right] \;\; ; \;\; \vec{\boldsymbol{b}}=\left[\vec{\boldsymbol{r}}_1\left(\phi\right)-\vec{\boldsymbol{r}}_2\left(\phi\right)\right]\left(\phi=\frac{\pi}{2}\right) \tag{140}$$

$$\vec{\boldsymbol{a}}\cdot\vec{\boldsymbol{b}}=0 \tag{141}$$

$$c=0 \qquad \left(Focal\ Distance\right) \tag{142}$$

$$e=0 \qquad \left(Eccentricity\right) \tag{143}$$

Further, when $\left(\beta=0\right)$ (137), $\vec{\boldsymbol{r}}_1\left(\phi\right)$ (5) and $\vec{\boldsymbol{r}}_2\left(\phi\right)$ (7) simplify to $\vec{\boldsymbol{r}}_1\left(\phi\right)$ (144) and $\vec{\boldsymbol{r}}_2\left(\phi\right)$ (145).

$$\vec{\boldsymbol{r}}_1=\vec{\boldsymbol{r}}_1\left(\phi+\phi_0\right)=\hat{\boldsymbol{x}}\, r_1 Cos\left(\phi+\varphi_0\right)+\hat{\boldsymbol{y}}\, r_1 Sin\left(\phi+\varphi_0\right) \tag{144}$$

$$\vec{\boldsymbol{r}}_2=\vec{\boldsymbol{r}}_2\left(\phi\right)=\hat{\boldsymbol{x}}\, r_2 Cos\phi+\hat{\boldsymbol{y}}\, r_2 Sin\phi \tag{145}$$

In this special case, as $\left(\beta=0\right)$ (137), $\boldsymbol{k}_1\left(\phi\right)$ (49) and $\boldsymbol{k}_2\left(\phi\right)$ (50) must have the same value of V_{10} (146) for all ϕ (2), as expressed in (146) based on b (138) and ℓ (11), and the data point where the calculated $\boldsymbol{k}_3\left(\phi\right)$ (51) takes on its maximum value $\boldsymbol{k}_{3,\max}$ (105) over a full cycle, that is $\sqrt{\boldsymbol{k}_{3,0}^{\;2}+\boldsymbol{k}_{4,0}^{\;2}}$ (70), must be where $\left(\phi=\xi\right)$ (147).

$$V_{10}=\boldsymbol{k}_1\left(\phi\right)=\boldsymbol{k}_2\left(\phi\right)=b^2+\ell^2=r_1^{\,2}-2r_1 r_2 Cos\left(\varphi_0\right)+r_2^{\,2}+\ell_x^{\,2}+\ell_y^{\,2}+\ell_z^{\,2} \tag{146}$$

$$\boldsymbol{k}_3\left(\phi\right)=\boldsymbol{k}_{3,\max}=\boldsymbol{k}_3\left(\xi\right)=\sqrt{\boldsymbol{k}_{3,0}^{\;2}+\boldsymbol{k}_{4,0}^{\;2}} \quad \Rightarrow \quad \phi=\xi \tag{147}$$

Moving on, knowing the minimum and maximum relative tilt[8] angles $\gamma_{2,\min}$ (92) and $\gamma_{2,\max}$ (93) between the North Pole[10,11] vector $\hat{\boldsymbol{u}}_{2\perp}$ (87) along the rotational axis[8] of the body at $\mathbf{P}_2$ and the component of $\hat{\boldsymbol{u}}_{2\perp}$ (87) normal to $\vec{\boldsymbol{d}}(\phi)$ (12) directed from $\mathbf{P}_2$ to $\mathbf{P}_1$, as well as the distance data $\left|\vec{\boldsymbol{d}}\left(\phi_{\gamma_{2,\min}}\right)\right|$ (98) and $\left|\vec{\boldsymbol{d}}\left(\phi_{\gamma_{2,\max}}\right)\right|$ (99) beween $\mathbf{P}_1$ and $\mathbf{P}_2$ corresponding to those data points of minimum and maximum relative tilt[8] angles $\gamma_{2,\min}$ (92) and $\gamma_{2,\max}$ (93), respectively, we can calculate the values V_5 (148) and V_6 (149) in this special case as well, which should allow us to confirm that $\left(r_1 Sin\,\beta = 0\right)$ (100) as $\left(\beta = 0\right)$ (137), and determine the value of ℓ_z (101) in (149).

$$V_5 = -\frac{\left\{\left|\vec{\boldsymbol{d}}\left(\phi_{\gamma_{2,\max}}\right)\right| Sin\left(\gamma_{2,\Delta_1}\right) + \left|\vec{\boldsymbol{d}}\left(\phi_{\gamma_{2,\min}}\right)\right| Sin\left(\gamma_{2,\Delta_2}\right)\right\}}{2} = r_1 Sin\,\beta = 0 \tag{148}$$

$$V_6 = \frac{\left\{\left|\vec{\boldsymbol{d}}\left(\phi_{\gamma_{2,\max}}\right)\right| Sin\left(\gamma_{2,\Delta_1}\right) - \left|\vec{\boldsymbol{d}}\left(\phi_{\gamma_{2,\min}}\right)\right| Sin\left(\gamma_{2,\Delta_2}\right)\right\}}{2} = \ell_z \tag{149}$$

In this special case with $\left(\beta = 0\right)$ (137), ℓ_x (134) and ℓ_y (135) reduce to ℓ_x (150) and ℓ_y (151).

$$\ell_x = \frac{\boldsymbol{k}_{3,0}\left\{r_1 Cos\varphi_0 - r_2\right\} - \boldsymbol{k}_{4,0} r_1 Sin\,\varphi_0}{\left\{r_1^2 - 2r_1 r_2 Cos\varphi_0 + r_2^2\right\}} = \frac{\boldsymbol{k}_{3,0}\left\{r_1 Cos\varphi_0 - r_2\right\} - \boldsymbol{k}_{4,0} r_1 Sin\,\varphi_0}{b^2} \tag{150}$$

$$\ell_y = \frac{\boldsymbol{k}_{3,0} r_1 Sin\,\varphi_0 + \boldsymbol{k}_{4,0}\left\{r_1 Cos\varphi_0 - r_2\right\}}{\left\{r_1^2 - 2r_1 r_2 Cos\varphi_0 + r_2^2\right\}} = \frac{\boldsymbol{k}_{3,0} r_1 Sin\,\varphi_0 + \boldsymbol{k}_{4,0}\left\{r_1 Cos\varphi_0 - r_2\right\}}{b^2} \tag{151}$$

Placing ℓ_x (150) and ℓ_y (151), as well as the found value of ℓ_z (149) in (146) would lead to (152), and thus the quadratic equation (153) in b^2 (138), whose positive solution would give us the value of b^2 (154), as we already have the values of $\left(\boldsymbol{k}_{3,0}{}^2 + \boldsymbol{k}_{4,0}{}^2 = \boldsymbol{k}_{3,\max}{}^2\right)$ (147), V_{10} (146), and $\left(\ell_z = V_6\right)$ (149) for this parallel two circle system, from which ℓ^2 (155) can also be found.

$$\begin{aligned} V_{10} &= b^2 + \ell^2 = r_1^2 - 2r_1 r_2 Cos\left(\varphi_0\right) + r_2^2 \\ &+ \frac{\boldsymbol{k}_{3,0}{}^2\left\{r_1^2 - 2r_1 r_2 Cos\left(\varphi_0\right) + r_2^2\right\} + \boldsymbol{k}_{4,0}{}^2\left\{r_1^2 - 2r_1 r_2 Cos\left(\varphi_0\right) + r_2^2\right\}}{\left\{r_1^2 - 2r_1 r_2 Cos\left(\varphi_0\right) + r_2^2\right\}^2} + \ell_z^2 \\ &= r_1^2 - 2r_1 r_2 Cos\left(\varphi_0\right) + r_2^2 + \frac{\boldsymbol{k}_{3,0}{}^2 + \boldsymbol{k}_{4,0}{}^2}{\left\{r_1^2 - 2r_1 r_2 Cos\left(\varphi_0\right) + r_2^2\right\}} + \ell_z^2 = b^2 + \frac{\boldsymbol{k}_{3,0}{}^2 + \boldsymbol{k}_{4,0}{}^2}{b^2} + V_6^2 \end{aligned} \tag{152}$$

$$b^4 + \left(V_6^2 - V_{10}\right) b^2 + \left(\boldsymbol{k}_{3,0}{}^2 + \boldsymbol{k}_{4,0}{}^2\right) = 0 \tag{153}$$

$$b^2 = r_1^2 - 2r_1 r_2 Cos\left(\varphi_0\right) + r_2^2 = \frac{\left(V_{10} - V_6^2\right) \pm \sqrt{\left(V_6^2 - V_{10}\right)^2 - 4\left(\boldsymbol{k}_{3,0}{}^2 + \boldsymbol{k}_{4,0}{}^2\right)}}{2} > 0 \tag{154}$$

$$\ell^2 = b^2 - V_{10} > 0 \tag{155}$$

Based on (26), $\boldsymbol{k}_3(\phi)$ (51) can be restated as $\boldsymbol{k}_3(\phi)$ (156), and based on (54), and $\boldsymbol{k}_4(\phi)$ (52) can be restated as $\boldsymbol{k}_4(\phi)$ (157), also utilizing $\vec{r}_1(\phi)$ (144) and $\vec{r}_2(\phi)$ (145).

$$\begin{aligned}\boldsymbol{k}_3(\phi) &= \left[\vec{\boldsymbol{a}}Cos\,\phi + \vec{\boldsymbol{b}}Sin\,\phi\right]\cdot\vec{\ell} = \left[\vec{r}_1(\phi) - \vec{r}_2(\phi)\right]\cdot\vec{\ell} \\ &= \left[r_1 Cos(\phi+\varphi_0) - r_2 Cos\phi\right]\ell_x + \left[r_1 Sin(\phi+\varphi_0) - r_2 Sin\,\phi\right]\ell_y\end{aligned} \tag{156}$$

$$\boldsymbol{k}_4(\phi) = \left[-\vec{\boldsymbol{a}}Sin\,\phi + \vec{\boldsymbol{b}}Cos\,\phi\right]\cdot\vec{\ell} = \boldsymbol{k}_3\left(\frac{\pi}{2}+\phi\right) = \left[\vec{\boldsymbol{r}}_1\left(\frac{\pi}{2}+\phi\right) - \vec{\boldsymbol{r}}_2\left(\frac{\pi}{2}+\phi\right)\right]\cdot\vec{\ell} \tag{157}$$

Using $\vec{r}_1(\phi)$ (144), $\vec{r}_2(\phi)$ (145), and $\vec{\ell}$ (10), $\boldsymbol{k}_3(\phi)$ (156) and $\boldsymbol{k}_4(\phi)$ (157) can be found at $(\phi = 0)$ (2), which would enable us to express the values of $\boldsymbol{k}_{3,0}$ (58) and $\boldsymbol{k}_{4,0}$ (59) as $\boldsymbol{k}_{3,0}$ (158) and $\boldsymbol{k}_{4,0}$ (159), respectively.

$$\boldsymbol{k}_{3,0} = \boldsymbol{k}_3(0) = \left[\vec{r}_1(0) - \vec{r}_2(0)\right]\cdot\vec{\ell} = \left[r_1 Cos(\varphi_0) - r_2\right]\ell_x + r_1 Sin(\varphi_0)\ell_y \tag{158}$$

$$\boldsymbol{k}_{4,0} = \boldsymbol{k}_4(0) = \boldsymbol{k}_3\left(\frac{\pi}{2}\right) = \left[\vec{\boldsymbol{r}}_1\left(\frac{\pi}{2}\right) - \vec{\boldsymbol{r}}_2\left(\frac{\pi}{2}\right)\right]\cdot\vec{\ell} = -r_1 Sin(\varphi_0)\ell_x + \left[r_1 Cos(\varphi_0) - r_2\right]\ell_y \tag{159}$$

Differentiating the right side of the expression to find its maximum, mathematically we see that $\boldsymbol{k}_3(\phi)$ (156) takes on its maximum value when the relation in (160) holds.

$$\frac{\left[r_1 Cos(\phi+\varphi_0) - r_2 Cos\phi\right]}{\left[r_1 Sin(\phi+\varphi_0) - r_2 Sin\,\phi\right]} = \frac{\ell_x}{\ell_y} \tag{160}$$

As we already know that $\boldsymbol{k}_3(\phi)$ (51), and therefore $\boldsymbol{k}_3(\phi)$ (156), takes its maximum value $\sqrt{\boldsymbol{k}_{3,0}^{\;2} + \boldsymbol{k}_{4,0}^{\;2}}$ (70) over a full cycle at $(\phi = \xi)$ (147), the relation in (160) becomes as in (161), and we also know that this value of $\boldsymbol{k}_3(\xi)$ (163) is $\boldsymbol{k}_{3,\max}$ (147) for our two circle system, we can write the expression in (162) utilizing $\boldsymbol{k}_{3,0}$ (158) and $\boldsymbol{k}_{4,0}$ (159), as well as the expression in (164) utilizing (161).

$$\frac{\left[r_1 Cos(\xi+\varphi_0) - r_2 Cos(\xi)\right]}{\left[r_1 Sin(\xi+\varphi_0) - r_2 Sin(\xi)\right]} = \frac{\ell_x}{\ell_y} \tag{161}$$

$$\boldsymbol{k}_{3,\max}^{\;2} = \boldsymbol{k}_{3,0}^{\;2} + \boldsymbol{k}_{4,0}^{\;2} = \left[r_1^2 - 2r_1 r_2\,Cos(\varphi_0) + r_2^2\right]\left(\ell_x^{\;2} + \ell_y^{\;2}\right) = b^2\left(\ell_x^{\;2} + \ell_y^{\;2}\right) \tag{162}$$

$$\boldsymbol{k}_{3,\max} = \sqrt{\boldsymbol{k}_{3,0}^{\;2} + \boldsymbol{k}_{4,0}^{\;2}} = \boldsymbol{k}_3(\xi) = \left[r_1 Cos(\xi+\varphi_0) - r_2 Cos(\xi)\right]\ell_x + \left[r_1 Sin(\xi+\varphi_0) - r_2 Sin(\xi)\right]\ell_y \tag{163}$$

$$\begin{aligned}
\boldsymbol{k}_{3,\max} = \sqrt{\boldsymbol{k}_{3,0}^{\;2} + \boldsymbol{k}_{4,0}^{\;2}} &= \left[r_1 Cos(\xi+\varphi_0) - r_2 Cos(\xi) \right] \ell_x + \frac{\left[r_1 Sin(\xi+\varphi_0) - r_2 Sin(\xi) \right]^2}{\left[r_1 Cos(\xi+\varphi_0) - r_2 Cos(\xi) \right]} \ell_x \\
&= \frac{\left[r_1 Cos(\xi+\varphi_0) - r_2 Cos(\xi) \right]^2}{\left[r_1 Sin(\xi+\varphi_0) - r_2 Sin(\xi) \right]} \ell_y + \left[r_1 Sin(\xi+\varphi_0) - r_2 Sin(\xi) \right] \ell_y \\
&= \frac{\left\{ r_1^2 - 2 r_1 r_2 Cos(\varphi_0) + r_2^2 \right\}}{\left[r_1 Cos(\xi+\varphi_0) - r_2 Cos(\xi) \right]} \ell_x = \frac{b^2}{\left[r_1 Cos(\xi+\varphi_0) - r_2 Cos(\xi) \right]} \ell_x \\
&= \frac{\left\{ r_1^2 - 2 r_1 r_2 Cos(\varphi_0) + r_2^2 \right\}}{\left[r_1 Sin(\xi+\varphi_0) - r_2 Sin(\xi) \right]} \ell_y = \frac{b^2}{\left[r_1 Sin(\xi+\varphi_0) - r_2 Sin(\xi) \right]} \ell_y
\end{aligned} \qquad (164)$$

Subsequently, (164) yields the relation in (165).

$$\frac{\ell_x}{\left[r_1 Cos(\xi+\varphi_0) - r_2 Cos(\xi) \right]} = \frac{\ell_y}{\left[r_1 Sin(\xi+\varphi_0) - r_2 Sin(\xi) \right]} = \frac{\boldsymbol{k}_{3,\max}}{b^2} = \frac{\sqrt{\boldsymbol{k}_{3,0}^{\;2} + \boldsymbol{k}_{4,0}^{\;2}}}{b^2} \qquad (165)$$

The results in ℓ_x (150) and ℓ_y (151) are compliant with the relation in (165).

Due to symmetry in this two parallel circle special case where $(\beta = 0)$ (137), there are infinitely many value possibilities for r_1 (9), r_2 (9), and φ_0 (3) that obey the expression for b^2 (154), and thus infinitely many ℓ_x (150) and ℓ_y (151), or equivalently ℓ_x (165) and ℓ_y (165), value possibilities, and we cannot obtain any further information analytically. Additional physical observations related to this two circle system may yield the values for these parameters, such as distance data from points $\mathbf{P}_1$ and $\mathbf{P}_2$, moving on their respective circles, to point $\mathbf{P}_3$ moving around a third circle in space, as a reference.

If $\left[2 r_1 r_2 (1 - Cos\beta) = V_1 \neq 0 \right]$ (102) but $(r_1 Sin\,\beta = V_5 = 0)$ (109) is found, we would know that $(Cos\beta \neq 1)$ (166) but $(Sin\,\beta = 0)$ (166), meaning $(\beta = \pi)$ (166). This is another special case where the two circles are parallel, but points $\mathbf{P}_1$ and $\mathbf{P}_2$ revolve their circles in reverse directions.

$$2 r_1 r_2 (1 - Cos\beta) = V_1 \neq 0 \;\;\&\;\; r_1 Sin\,\beta = V_5 = 0 \;\;\Rightarrow\;\; Cos\beta \neq 1 \;\;\&\;\; Sin\,\beta = 0 \;\;\Rightarrow\;\; \beta = \pi \quad (166)$$

In this special case, the process through (102) - (136) is still valid, but we cannot utilize $(r_1 Sin\,\beta = V_5 = 0)$ (109) when $(Sin\,\beta = 0)$ (166) to find r_1 (126). Hence, we proceed as follows.

As we must have already determined the values of V_1 (102) and V_7 (111) from our data, in this special case where $(\beta = \pi)$ (166), utilizing $\left[2 r_1 r_2 (1 - Cos\beta) = V_1 \right]$ (102) we obtain the value of $(r_1 r_2)$ (167), and utilizing $\left[2 r_1 r_2 (1 - Cos\beta) Cos(\varphi_0) = V_7 \right]$ (111) we obtain the value of $\left[r_1 r_2 Cos(\varphi_0) \right]$ (168).

$$r_1 r_2 = \frac{V_1}{4} \qquad (167)$$

$$r_1 r_2 Cos(\varphi_0) = \frac{V_7}{4} \tag{168}$$

Further, when $(\beta = \pi)$ (166), $\bar{r}_1(\phi)$ (5) and $\bar{r}_2(\phi)$ (7) simplify to $\bar{r}_1(\phi)$ (169) and $\bar{r}_2(\phi)$ (170).

$$\bar{r}_1 = \bar{r}_1(\phi + \phi_0) = -\hat{x}\, r_1 Cos(\phi + \varphi_0) + \hat{y}\, r_1 Sin(\phi + \varphi_0) \tag{169}$$

$$\bar{r}_2 = \bar{r}_2(\phi) = \hat{x}\, r_2 Cos\phi + \hat{y}\, r_2 Sin\phi \tag{170}$$

Based on (26), $k_3(\phi)$ (51) can be restated as $k_3(\phi)$ (171), and based on (54), and $k_4(\phi)$ (52) can be restated as $k_4(\phi)$ (172), also utilizing $\bar{r}_1(\phi)$ (169) and $\bar{r}_2(\phi)$ (170).

$$\begin{aligned} k_3(\phi) &= \left[\bar{a} Cos\,\phi + \bar{b} Sin\,\phi\right] \cdot \bar{\ell} = \left[\bar{r}_1(\phi) - \bar{r}_2(\phi)\right] \cdot \bar{\ell} \\ &= -\left[r_1 Cos(\phi + \varphi_0) + r_2 Cos\phi\right]\ell_x + \left[r_1 Sin(\phi + \varphi_0) - r_2 Sin\phi\right]\ell_y \end{aligned} \tag{171}$$

$$\begin{aligned} k_4(\phi) &= \left[-\bar{a} Sin\,\phi + \bar{b} Cos\,\phi\right] \cdot \bar{\ell} = k_3\left(\frac{\pi}{2} + \phi\right) = \left[\bar{r}_1\left(\frac{\pi}{2} + \phi\right) - \bar{r}_2\left(\frac{\pi}{2} + \phi\right)\right] \cdot \bar{\ell} \\ &= \left[r_1 Sin(\phi + \varphi_0) + r_2 Sin\,\phi\right]\ell_x + \left[r_1 Cos(\phi + \varphi_0) - r_2 Cos\phi\right]\ell_y \end{aligned} \tag{172}$$

With $(\beta = \pi)$ (166), expressions for semi-major[5] and semi-minor[5] axes a (128) and b (129), the Dot Product[3] $(\bar{a} \cdot \bar{b})$ (20), and the focal[4,6] distance c (112) of this vector ellipse[4] (30) formed by the relative motion of points $\mathbf{P}_1$ and $\mathbf{P}_2$ on the two respective circles in our system, as well as ℓ_x (134) and ℓ_y (135), become a (173), b (174), $(\bar{a} \cdot \bar{b})$ (20), c (176), ℓ_x (177), and ℓ_y (178), respectively.

$$a = \sqrt{a^2} = \sqrt{\bar{a} \cdot \bar{a}} = \sqrt{r_1^2 + 2 r_1 r_2 Cos(\varphi_0) + r_2^2} \tag{173}$$

$$b = \sqrt{b^2} = \sqrt{\bar{b} \cdot \bar{b}} = \sqrt{r_1^2 - 2 r_1 r_2 Cos(\varphi_0) + r_2^2} \tag{174}$$

$$\bar{a} \cdot \bar{b} = -2 r_1 r_2 Sin\,\varphi_0 \tag{175}$$

$$c = \sqrt{4 r_1 r_2 \left|Cos\,\varphi_0\right|} = 2\sqrt{\left|r_1 r_2 Cos\,\varphi_0\right|} = \sqrt{\left|V_7\right|} \qquad (Focal\ Distance) \tag{176}$$

$$\ell_x = \frac{k_{3,0}\left[r_1 Cos(\varphi_0) - r_2\right] - k_{4,0} r_1 Sin(\varphi_0)}{\left(-r_1^2 + r_2^2\right)} \tag{177}$$

$$\ell_y = \frac{k_{3,0} r_1 Sin(\varphi_0) + k_{4,0}\left[r_1 Cos(\varphi_0) + r_2\right]}{\left(r_1^2 - r_2^2\right)} \tag{178}$$

As we already have the values of $k_1(\phi)$ (49) for all calculated values of ϕ (119) corresponding to the data points we have for this two circle system, and we have already determined the value of $(\phi_0 = \varphi_0)$ (114), we determine the value of $(b^2 + \ell^2)$ (120), namely the sum of b^2 (174) and

ℓ^2 (11), from the V_8 (120) value that must be the same at any ϕ (119). As we must have already found the value of $(\ell_z = V_6)$ (110), $(b^2 + \ell^2 = V_8)$ (120) expression can be reorganized as (179).

$$V_8 - V_6^2 = b^2 + (\ell^2 - \ell_z^2) = b^2 + (\ell_x^2 + \ell_y^2) = \left[r_1^2 - 2r_1r_2\,Cos(\varphi_0) + r_2^2\right] + (\ell_x^2 + \ell_y^2) \quad (179)$$

The $(\ell_x^2 + \ell_y^2)$ (180) expression is obtained using ℓ_x (177) and ℓ_y (178), also utilizing the expressions for a (173), b (174), $(\vec{a}\cdot\vec{b})$ (175), ξ (66), (26), and (172).

$$\begin{aligned}\ell_x^2 + \ell_y^2 &= \frac{\boldsymbol{k}_{3,0}^2\left[r_1^2 - 2r_1r_2Cos(\varphi_0) + r_2^2\right] + 4\boldsymbol{k}_{3,0}\boldsymbol{k}_{4,0}r_1r_2Sin(\varphi_0) + \boldsymbol{k}_{4,0}^2\left[r_1^2 + 2r_1r_2Cos(\varphi_0) + r_2^2\right]}{(r_1^2 - r_2^2)^2}\\ &= \frac{\boldsymbol{k}_{3,0}^2b^2 - 2\boldsymbol{k}_{3,0}\boldsymbol{k}_{4,0}\vec{a}\cdot\vec{b} + \boldsymbol{k}_{4,0}^2a^2}{(r_1^2 - r_2^2)^2} = \frac{(\boldsymbol{k}_{3,0}\vec{b} - \boldsymbol{k}_{4,0}\vec{a})^2}{(r_1^2 - r_2^2)^2}\\ &= \frac{\left\{\sqrt{\boldsymbol{k}_{3,0}^2 + \boldsymbol{k}_{4,0}^2}\left[-\vec{a}Sin(\xi) + \vec{b}Cos(\xi)\right]\right\}^2}{(r_1 - r_2)^2(r_1 + r_2)^2} = \frac{\left\{\sqrt{\boldsymbol{k}_{3,0}^2 + \boldsymbol{k}_{4,0}^2}\left[\vec{r_1}\left(\frac{\pi}{2} + \xi\right) - \vec{r_2}\left(\frac{\pi}{2} + \xi\right)\right]\right\}^2}{(r_1 - r_2)^2(r_1 + r_2)^2}\end{aligned} \quad (180)$$

Placing the obtained $(\ell_x^2 + \ell_y^2)$ (180) expression in (179) yields the expression in (181), also replacing $\left(2r_1r_2 = \frac{V_1}{2}\right)$ (167) and $\left[2r_1r_2Cos(\varphi_0) = \frac{V_7}{2}\right]$ (168) values.

$$\begin{aligned}V_8 - V_6^2 &= \left[r_1^2 - 2r_1r_2\,Cos(\varphi_0) + r_2^2\right] + \frac{\left\{\sqrt{\boldsymbol{k}_{3,0}^2 + \boldsymbol{k}_{4,0}^2}\left[\vec{r_1}\left(\frac{\pi}{2} + \xi\right) - \vec{r_2}\left(\frac{\pi}{2} + \xi\right)\right]\right\}^2}{(r_1 - r_2)^2(r_1 + r_2)^2}\\ &= \left[r_1^2 + r_2^2 - \frac{V_7}{2}\right] + \frac{(\boldsymbol{k}_{3,0}^2 + \boldsymbol{k}_{4,0}^2)\left\{r_1^2 + r_2^2 - \frac{V_1}{2}Cos(2\xi + \varphi_0)\right\}}{\left(r_1^2 + r_2^2 - \frac{V_1}{2}\right)\left(r_1^2 + r_2^2 + \frac{V_1}{2}\right)}\end{aligned} \quad (181)$$

Based on the process through (102) - (136), we must have already determined values of V_1 (102), V_6 (110), V_7 (111), and V_8 (120), as well as ξ (66), $\left[Cos(2\xi + \varphi_0) = \frac{V_4}{V_1}\right]$ (107), and $\sqrt{\boldsymbol{k}_{3,0}^2 + \boldsymbol{k}_{4,0}^2}$ (70), as we also know that $\boldsymbol{k}_3(\phi)$ (60) takes on its maximum value $\left(\boldsymbol{k}_{3,max} = \sqrt{\boldsymbol{k}_{3,0}^2 + \boldsymbol{k}_{4,0}^2}\right)$ (70) over a cycle at the data point corresponding to $(\phi = \xi)$ (119).

Reorganizing (181), we obtain a 3rd order equation (182) in $\left[r^2 = r_1^2 + r_2^2\right]$ (183) whose value V_{11} (183) can be found by solving (182).

$$\left(r^2 - \frac{V_7}{2}\right)\left(r^2 - \frac{V_1}{2}\right)\left(r^2 + \frac{V_1}{2}\right) + (V_8 - V_6^2)\left(r^2 - \frac{V_1}{2}\right)\left(r^2 + \frac{V_1}{2}\right) + (\boldsymbol{k}_{3,0}^2 + \boldsymbol{k}_{4,0}^2)\left(r^2 - \frac{V_4}{2}\right) = 0 \quad (182)$$

$$r^2 = r_1^2 + r_2^2 = V_{11} \tag{183}$$

Now we have two equations (167) and (183) in the two variables r_1 (9) and r_2 (9). Reorganizing (167), we obtain (184).

$$r_1 = \frac{V_1}{4r_2} \tag{184}$$

Placing the expression for r_1 (184) in (183) yields the quadratic equation (185) in r_2^2 (9), whose valid solution is the positive $\left(r_2^2 > 0\right)$ (186) one, using which we obtain the solution for r_1 (184).

$$16r_2^4 - 16V_{11}r_2^2 + V_1^2 = 0 \tag{185}$$

$$r_2^2 = \frac{2V_{11} \pm \sqrt{4V_{11}^2 - V_1^2}}{4} > 0 \quad \Rightarrow \quad r_2 = \sqrt{r_2^2} \tag{186}$$

Hence, we have found the r_1 (184) and r_2 (186) solutions via (166) - (186) for this special case with $(\beta = \pi)$ (166) where the two circles are parallel, with points $\mathbf{P}_1$ and $\mathbf{P}_2$ revolving their circles in reverse directions, applying the process through (102) - (136) except for r_1 (126), as we cannot utilize $(r_1 Sin\beta = V_5 = 0)$ (109) when $(Sin\beta = 0)$ (166).

As such, we have completed our research to set forth an analytical method, as outlined above between (102) - (186) including two special cases, to determine the values of orbital parameters for a two circle system in space, as well as the values of the parameters related to the vector ellipse[4] (30) formed by the relative motion of points $\mathbf{P}_1$ and $\mathbf{P}_2$ on two respective circles, moving with respect to centers of their own circles of revolution with the ***same fixed constant*** angular velocity ω (1) - (2), provided we have distance data over a full cycle between the moving points $\mathbf{P}_1$ and $\mathbf{P}_2$ on the two respective circles at sufficiently many data points, as well as minimum and maximum tilt[8] angle values $\gamma_{2,\min}$ (92) and $\gamma_{2,\max}$ (93) between the North Pole[10,11] unit vector $\hat{\boldsymbol{u}}_{2\perp}$ (87) of the body at $\mathbf{P}_2$ and the component of $\hat{\boldsymbol{u}}_{2\perp}$ (87) normal to $\vec{\boldsymbol{d}}(\phi)$ (12) from $\mathbf{P}_2$ to $\mathbf{P}_1$.

CONCLUSIONS

Based on the mathematical derivation and assertion in the **ARTICLE 1** "Points on two circles in space mimic an ellipse with respect to a point", in this Article, an analytical method is developed to calculate orbital parameters of two heavenly bodies moving with the ***same fixed constant*** angular velocity with respect to the centers of their own circles of revolution, when the distance between their centers of revolution is also constant.

Note that in all our operations, we take the "square of a vector" as the "Dot Product[3] of the vector with itself", which amounts to the scalar "square of the magnitude" for any vector.

FIGURES

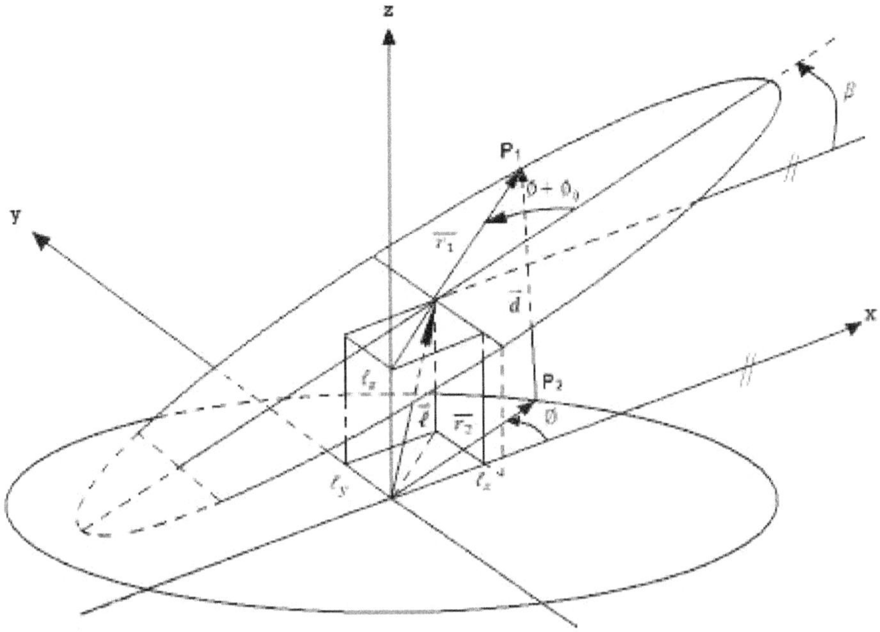

Figure 1 Distance Between Points $\mathbf{P}_1$ and $\mathbf{P}_2$ on Two Different Circles in Space

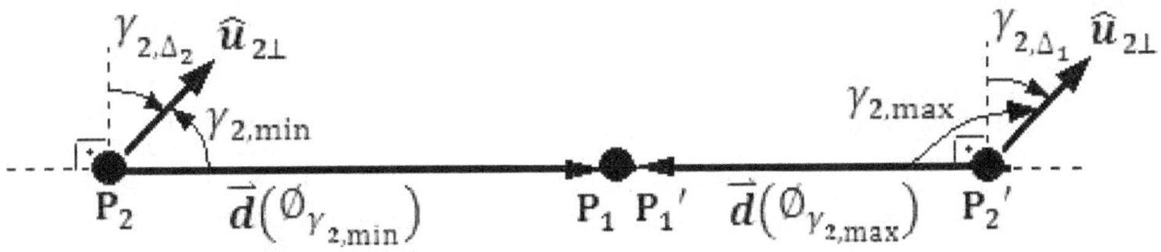

Figure 2 Simple geometric visualization of $\vec{d}(\phi)\cdot\hat{u}_{2\perp}$, $\gamma_{2,\max}$ and $\gamma_{2,\min}$

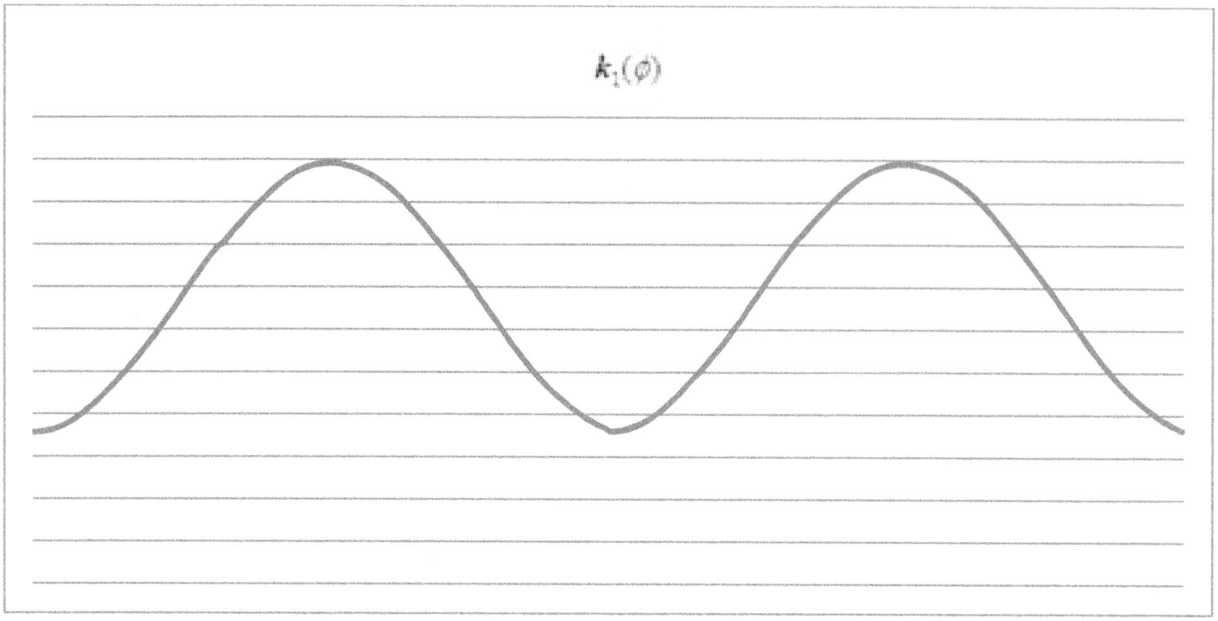

Figure 3 $k_1(\phi)$ over relative $\mathbf{P}_1$ and $\mathbf{P}_2$ cycle

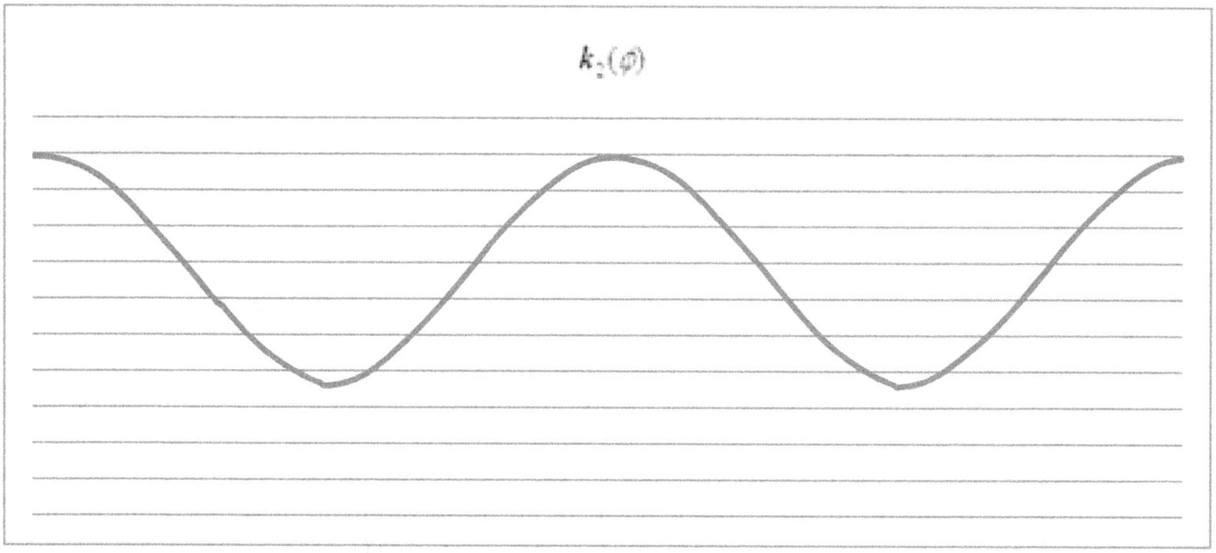

Figure 4 $k_2(\phi)$ over relative $\mathbf{P}_1$ and $\mathbf{P}_2$ cycle

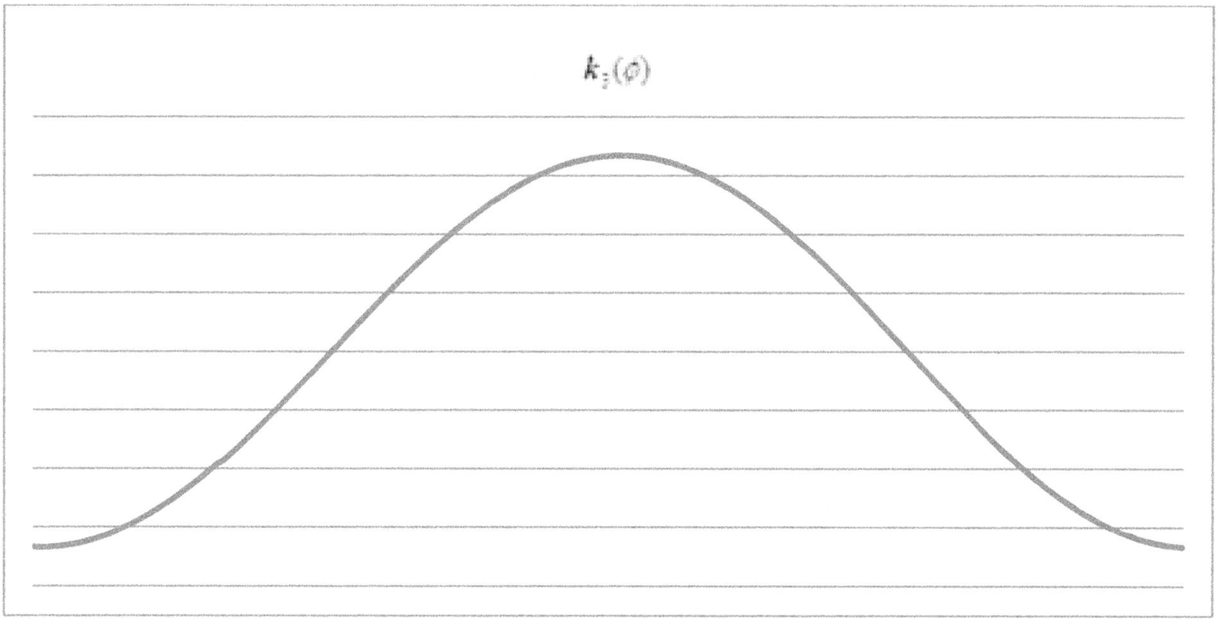

Figure 5 $k_3(\phi)$ over relative $\mathbf{P}_1$ and $\mathbf{P}_2$ cycle

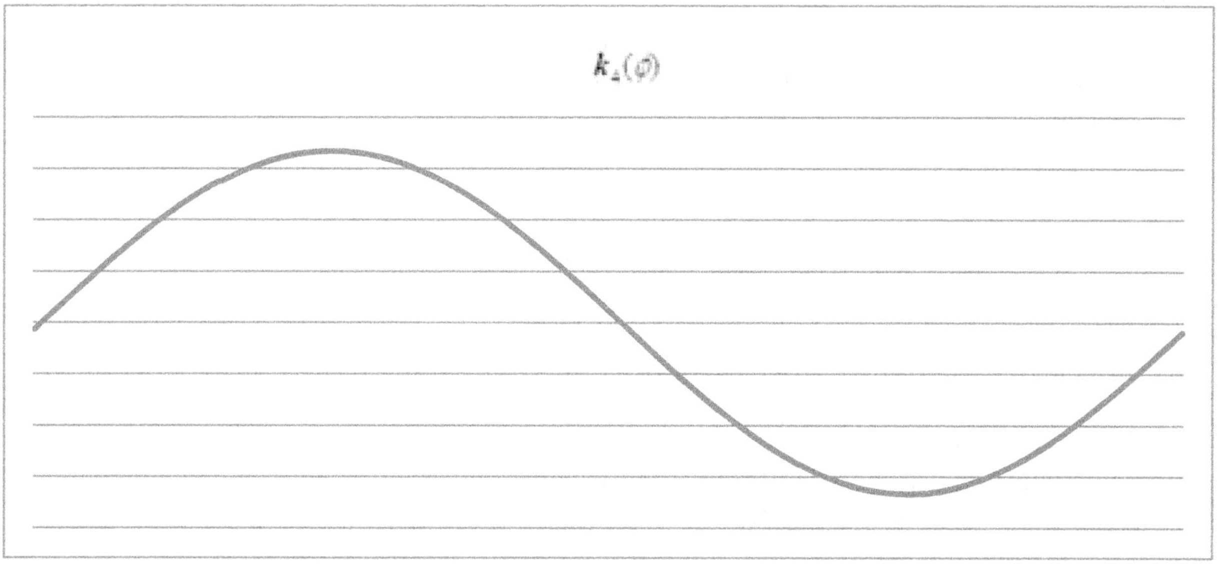

Figure 6 $k_4(\phi)$ over relative $\mathbf{P}_1$ and $\mathbf{P}_2$ cycle

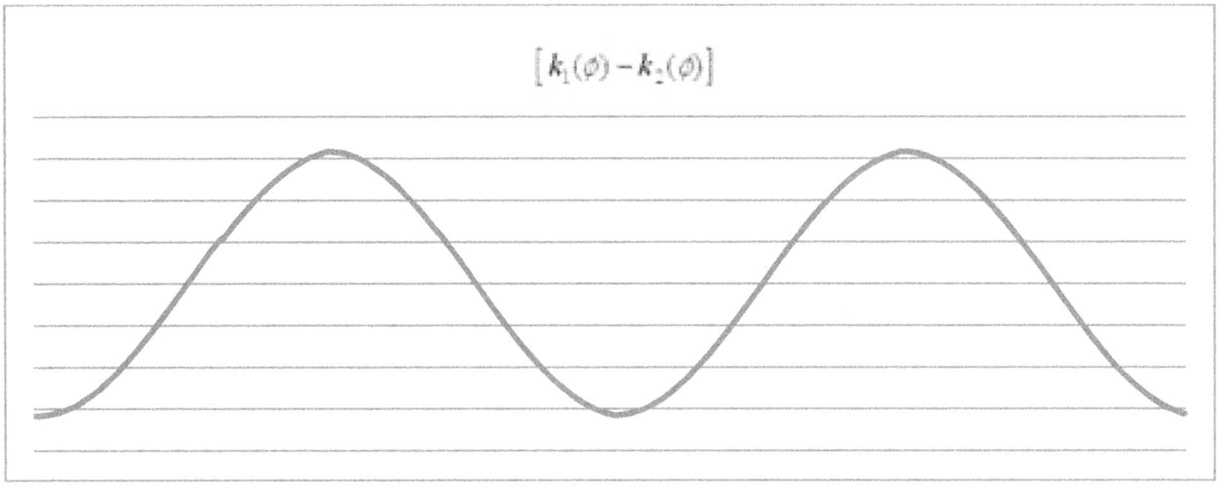

Figure 7 $[\boldsymbol{k}_1(\phi) - \boldsymbol{k}_2(\phi)]$ over relative $\mathbf{P}_1$ and $\mathbf{P}_2$ cycle

REFERENCES

References in this Article can be any Physics, Electromagnetics, and Calculus textbook, as the physics equations and mathematical identities used as a basis for the proof are all currently accepted theory in existing textbooks.

1. Halliday, D., Resnick, R., *Fundamentals of Physics (3rd Edition)*, John Wiley & Sons, 1988, ISBN 0-471-63735-1
2. Kepler's laws of planetary motion, Wikipedia, https://en.wikipedia.org/wiki/Kepler%27s_laws_of_planetary_motion, (2020)
3. Dot product, Wikipedia, https://en.wikipedia.org/wiki/Dot_product, (2020)
4. Ellipse, Wikipedia, http://en.wikipedia.org/wiki/Ellipse, (2020)
5. Semi-major and semi-minor axes, Wikipedia, http://en.wikipedia.org/wiki/Semi-major_and_semi-minor_axes, (2020)
6. Focus (geometry), Wikipedia, http://en.wikipedia.org/wiki/Focus_(geometry), (2020)
7. Rotation around a fixed axis, Wikipedia, http://en.wikipedia.org/wiki/Rotation_around_a_fixed_axis, (2020)
8. Axial tilt, Wikipedia, https://en.wikipedia.org/wiki/Axial_tilt, (2020)
9. Equator, Wikipedia, https://en.wikipedia.org/wiki/Equator, (2020)
10. Geographical pole, Wikipedia, https://en.wikipedia.org/wiki/Geographical_pole, (2020)
11. North Pole, Wikipedia, https://en.wikipedia.org/wiki/North_Pole, (2020)
12. Right hand rule, Wikipedia, https://en.wikipedia.org/wiki/Right-hand_rule, (2020)
13. Faraday's law of induction, Wikipedia, https://en.wikipedia.org/wiki/Faraday%27s_law_of_induction, (2020)
14. Electromotive force, Wikipedia, https://en.wikipedia.org/wiki/Electromotive_force, (2020)
15. Magnetic flux, Wikipedia, https://en.wikipedia.org/wiki/Magnetic_flux, (2020)
16. Clockwise, Wikipedia, https://en.wikipedia.org/wiki/Clockwise, (2020)
17. Solstice, Wikipedia, https://en.wikipedia.org/wiki/Solstice, (2020)

ARTICLE 3

ANALYTICAL CALCULATION OF EARTH & SUN ORBITAL PARAMETERS FROM DISTANCE DATA

Author: Aslı Pınar Tan[III]

SUMMARY

Based on observed astronomical position data of heavenly objects in the Solar System and other planetary systems, all bodies in space seem to move in some kind of elliptical motion with respect to each other. In previous Articles, it is mathematically demonstrated and proven that bodies moving around different circles in space mimic vector-wise elliptical paths with respect to each other, whether they are moving with the *same*, *fixed different*, or *changing* angular velocities with respect to their own centers of revolution, and an analytical method is developed to calculate orbital parameters of two heavenly bodies moving with the *same* angular velocity in their own circular orbital revolutions. In this Article, the individual orbital parameters of the Sun and Earth are calculated using that method analytically to an approximation, based on yearly Sun-Earth distance data observed by NASA Landsat, as they exhibit an almost fixed elliptic orbit around one another.

ARTICLE

According to Kepler's 1st Law[1], "the orbit of a planet with respect to the Sun is an ellipse[22] , with the Sun at one of the two foci[22,24]", which is an empirical rule concluded through near estimation based on the observation of astronomical position data measured over time in the Solar System.

It has been shown in the **ARTICLE 1** "Points on two circles in space mimic an ellipse with respect to a point" that the distances between moving points on two circles in space are equivalent to the distances of points on an ellipse[22] from a fixed or moving point in space, depending on the angular velocities of the points on two circles in their individual motion. It has also been shown that regarding the moving points around two different circles, the semi-major[23] and semi-minor[23] axis magnitudes of the relatively observed elliptical distances may also vary harmonically over a cycle, based on the individual angular velocities of the points on the two circles. Further, in the case when the moving points on two circles are equally phased, namely when the points on two different circles move with the *same* fixed angular velocity around their own centers of revolution, the semi-major and semi-minor axis magnitudes of the relatively observed elliptical orbit are found to be fixed over a cycle.

Considering that the Earth and other planets in the Solar System are observed to be moving in elliptical orbits with respect to the Sun, and the Sun is observed to be positioned at a relatively stationary point with respect to these orbits, it is possible that the Sun is moving in its own circular orbit around its own center of orbital revolution with its own angular velocity, and the other planets are also moving in their own circular orbits around their own centers of orbital revolution with their own angular velocities, with the centers of revolution of the Sun and all planets being at fixed (or harmonically varying) distances from each other, the individual angular

[III] *ASLI PINAR TAN*

Linkedin Website: *https://www.linkedin.com./in/apinartan*

velocities of the Sun and every planet being fixed (with the same or a different value) to ensure periodicity of their motions with respect to each other.

Specifically for the case of the Sun-Earth[2] pair, the distance data exhibits behavior close to a single ellipse over every yearly[26] cycle of duration $\left(T_{Year} = T_{Sun} = T_{Earth}\right)$ (1), with no variation of the semi-major and semi-minor axes. So, based on the mathematics demonstrated and developed in the **ARTICLE 1** "Points on two circles in space mimic an ellipse with respect to a point", it is just as well possible for the Earth and the Sun to be moving around different circular orbits, each with the *same fixed* angular velocity $\left(\omega = \omega_{Sun} = \omega_{Earth}\right)$ (2) with respect to their individual centers of revolution. With this assumption based on the given observations, it is possible to calculate the orbital parameters of the Sun and the Earth using the Sun-Earth distance data over a yearly[26] cycle and the analytical method formulated in the **ARTICLE 2** "Analytical solution for two circles in space using their elliptic distances". In this Article, this calculation is performed.

$$T_{Year} = T_{Sun} = T_{Earth} = 365.256363004 \; days \qquad \left(Sun\ and\ Earth's\ Orbital\ Period\right) \qquad (1)$$

$$\omega = \omega_{Sun} = \omega_{Earth} = \frac{2\pi}{T_{Year}} = \mathbf{0.017202124161519} \; radians/day \quad \left(Sun\ \&\ Earth\ Angular\ Velocity\right) \quad (2)$$

The following **Table 1** lists the Sun-Earth distances (d) in astronomical units (au) for each Day of the Year (DOY) over a yearly cycle, as listed by NASA on a 2011 website[3] document based on Landsat observation data. In this Article, the orbital parameters of the Earth and the Sun will be calculated to an approximation based on this data, step by step applying the analytical method and mathematical framework introduced in the **ARTICLE 2** "Analytical solution for two circles in space using their elliptic distances".

Note that DOY 1 and DOY 366 correspond to January 1st and DOY 365 corresponds to December 31st. The distance data is continuous over a year cycle, between a minimum value of 0.983297 au on January 3 (DOY 3) and a maximum value of 1.0167037 au on July 5 (DOY 186), and yearly periodic.

The astronomical unit[4] (au) is defined in terms of meters as in (3).

$$1 \; astronomical\ unit \; (au) \; = \; 149{,}597{,}870{,}700 \; meters \; (exactly) \qquad (3)$$

Table 1 Sun-Earth distance (d) in astronomical units (au) for Day of the Year (DOY)

DOY	d	DOY	d	DOY	d	DOY	d	DOY	d	DOY	d
1	0.9833098	**62**	0.9913291	**123**	1.0080649	**184**	1.0166955	**245**	1.0089833	**306**	0.9922752
2	0.9833010	**63**	0.9915781	**124**	1.0083131	**185**	1.0167021	**246**	1.0087419	**307**	0.9920197
3	0.9832970	**64**	0.9918293	**125**	1.0085586	**186**	1.0167037	**247**	1.0084978	**308**	0.9917663
4	0.9832983	**65**	0.9920831	**126**	1.0088021	**187**	1.0167006	**248**	1.0082511	**309**	0.9915152
5	0.9833045	**66**	0.9923392	**127**	1.0090426	**188**	1.0166928	**249**	1.0080019	**310**	0.9912669
6	0.9833159	**67**	0.9925975	**128**	1.0092806	**189**	1.0166801	**250**	1.0077506	**311**	0.9910212
7	0.9833324	**68**	0.9928579	**129**	1.0095160	**190**	1.0166628	**251**	1.0074971	**312**	0.9907780
8	0.9833541	**69**	0.9931204	**130**	1.0097481	**191**	1.0166408	**252**	1.0072409	**313**	0.9905376

DOY	d	DOY	d	DOY	d	DOY	d	DOY	d	DOY	d
9	0.9833806	**70**	0.9933850	**131**	1.0099779	**192**	1.0166137	**253**	1.0069829	**314**	0.9903000
10	0.9834126	**71**	0.9936513	**132**	1.0102047	**193**	1.0165821	**254**	1.0067228	**315**	0.9900654
11	0.9834494	**72**	0.9939198	**133**	1.0104283	**194**	1.0165458	**255**	1.0064605	**316**	0.9898338
12	0.9834914	**73**	0.9941899	**134**	1.0106487	**195**	1.0165049	**256**	1.0061965	**317**	0.9896051
13	0.9835384	**74**	0.9944617	**135**	1.0108663	**196**	1.0164591	**257**	1.0059305	**318**	0.9893795
14	0.9835906	**75**	0.9947352	**136**	1.0110806	**197**	1.0164087	**258**	1.0056626	**319**	0.9891571
15	0.9836476	**76**	0.9950101	**137**	1.0112919	**198**	1.0163537	**259**	1.0053931	**320**	0.9889379
16	0.9837098	**77**	0.9952865	**138**	1.0114996	**199**	1.0162939	**260**	1.0051219	**321**	0.9887220
17	0.9837771	**78**	0.9955642	**139**	1.0117042	**200**	1.0162295	**261**	1.0048492	**322**	0.9885094
18	0.9838491	**79**	0.9958431	**140**	1.0119051	**201**	1.0161604	**262**	1.0045749	**323**	0.9883003
19	0.9839263	**80**	0.9961233	**141**	1.0121030	**202**	1.0160865	**263**	1.0042992	**324**	0.9880945
20	0.9840083	**81**	0.9964046	**142**	1.0122969	**203**	1.0160080	**264**	1.0040220	**325**	0.9878922
21	0.9840953	**82**	0.9966869	**143**	1.0124871	**204**	1.0159249	**265**	1.0037439	**326**	0.9876938
22	0.9841869	**83**	0.9969699	**144**	1.0126740	**205**	1.0158373	**266**	1.0034641	**327**	0.9874987
23	0.9842835	**84**	0.9972539	**145**	1.0128571	**206**	1.0157452	**267**	1.0031835	**328**	0.9873076
24	0.9843850	**85**	0.9975385	**146**	1.0130363	**207**	1.0156484	**268**	1.0029017	**329**	0.9871202
25	0.9844911	**86**	0.9978237	**147**	1.0132118	**208**	1.0155472	**269**	1.0026189	**330**	0.9869368
26	0.9846018	**87**	0.9981096	**148**	1.0133833	**209**	1.0154413	**270**	1.0023354	**331**	0.9867573
27	0.9847172	**88**	0.9983957	**149**	1.0135511	**210**	1.0153309	**271**	1.0020510	**332**	0.9865816
28	0.9848373	**89**	0.9986824	**150**	1.0137149	**211**	1.0152164	**272**	1.0017661	**333**	0.9864101
29	0.9849620	**90**	0.9989694	**151**	1.0138747	**212**	1.0150971	**273**	1.0014802	**334**	0.9862425
30	0.9850912	**91**	0.9992566	**152**	1.0140301	**213**	1.0149738	**274**	1.0011941	**335**	0.9860792
31	0.9852250	**92**	0.9995437	**153**	1.0141820	**214**	1.0148461	**275**	1.0009078	**336**	0.9859204
32	0.9853632	**93**	0.9998310	**154**	1.0143296	**215**	1.0147139	**276**	1.0006208	**337**	0.9857655
33	0.9855058	**94**	1.0001183	**155**	1.0144727	**216**	1.0145779	**277**	1.0003334	**338**	0.9856152
34	0.9856527	**95**	1.0004053	**156**	1.0146120	**217**	1.0144373	**278**	1.0000464	**339**	0.9854692
35	0.9858041	**96**	1.0006922	**157**	1.0147470	**218**	1.0142927	**279**	0.9997590	**340**	0.9853275
36	0.9859597	**97**	1.0009789	**158**	1.0148779	**219**	1.0141439	**280**	0.9994717	**341**	0.9851904
37	0.9861195	**98**	1.0012652	**159**	1.0150043	**220**	1.0139912	**281**	0.9991842	**342**	0.9850579
38	0.9862837	**99**	1.0015508	**160**	1.0151263	**221**	1.0138344	**282**	0.9988971	**343**	0.9849299
39	0.9864517	**100**	1.0018362	**161**	1.0152442	**222**	1.0136734	**283**	0.9986105	**344**	0.9848065

DOY	d	DOY	d	DOY	d	DOY	d	DOY	d	DOY	d
40	0.9866241	**101**	1.0021209	**162**	1.0153576	**223**	1.0135089	**284**	0.9983239	**345**	0.9846876
41	0.9868004	**102**	1.0024048	**163**	1.0154668	**224**	1.0133402	**285**	0.9980379	**346**	0.9845736
42	0.9869809	**103**	1.0026881	**164**	1.0155715	**225**	1.0131679	**286**	0.9977524	**347**	0.9844641
43	0.9871652	**104**	1.0029706	**165**	1.0156716	**226**	1.0129914	**287**	0.9974675	**348**	0.9843594
44	0.9873533	**105**	1.0032519	**166**	1.0157675	**227**	1.0128114	**288**	0.9971833	**349**	0.9842595
45	0.9875454	**106**	1.0035324	**167**	1.0158585	**228**	1.0126277	**289**	0.9968998	**350**	0.9841643
46	0.9877412	**107**	1.0038117	**168**	1.0159453	**229**	1.0124402	**290**	0.9966171	**351**	0.9840739
47	0.9879406	**108**	1.0040900	**169**	1.0160273	**230**	1.0122491	**291**	0.9963353	**352**	0.9839886
48	0.9881439	**109**	1.0043670	**170**	1.0161048	**231**	1.0120544	**292**	0.9960544	**353**	0.9839079
49	0.9883506	**110**	1.0046426	**171**	1.0161779	**232**	1.0118562	**293**	0.9957749	**354**	0.9838321
50	0.9885609	**111**	1.0049165	**172**	1.0162460	**233**	1.0116545	**294**	0.9954963	**355**	0.9837613
51	0.9887745	**112**	1.0051893	**173**	1.0163095	**234**	1.0114495	**295**	0.9952192	**356**	0.9836953
52	0.9889913	**113**	1.0054603	**174**	1.0163685	**235**	1.0112408	**296**	0.9949431	**357**	0.9836344
53	0.9892117	**114**	1.0057296	**175**	1.0164225	**236**	1.0110291	**297**	0.9946685	**358**	0.9835784
54	0.9894352	**115**	1.0059972	**176**	1.0164719	**237**	1.0108141	**298**	0.9943955	**359**	0.9835274
55	0.9896619	**116**	1.0062629	**177**	1.0165166	**238**	1.0105957	**299**	0.9941240	**360**	0.9834815
56	0.9898914	**117**	1.0065266	**178**	1.0165565	**239**	1.0103742	**300**	0.9938541	**361**	0.9834406
57	0.9901240	**118**	1.0067884	**179**	1.0165917	**240**	1.0101498	**301**	0.9935861	**362**	0.9834046
58	0.9903595	**119**	1.0070481	**180**	1.0166219	**241**	1.0099220	**302**	0.9933199	**363**	0.9833739
59	0.9905978	**120**	1.0073056	**181**	1.0166476	**242**	1.0096917	**303**	0.9930556	**364**	0.9833483
60	0.9908390	**121**	1.0075610	**182**	1.0166683	**243**	1.0094584	**304**	0.9927934	**365**	0.9833277
61	0.9910827	**122**	1.0078142	**183**	1.0166844	**244**	1.0092224	**305**	0.9925331	**366**	0.9833098

To be used with the Sun-Earth distance data in **Table 1**, the solstice[5] dates[6] of the Earth are around June 21 (DOY 172) and December 21 (DOY 355), at which dates the Northern Hemisphere of the Earth is most and least inclined towards the Sun, respectively, and the axial tilt[7] angle $\gamma_{Earth,\Delta Max}$ (4) of the Earth with respect to the so called ecliptic[8] plane is approximately 23.44° (4). The equinox[9] dates[6] of the Earth with respect to Sun are around March 20 (DOY 79) and September 22 (DOY 265), on which dates the rotational axis[7] of the Earth is approximately perpendicular to the line joining the centers of the Earth and Sun.

$$\gamma_{Earth,\Delta Max} = 23.44° \quad (4)$$

Based on observation[10,11], there is a day out of the year when the Sun's North Pole is said to tip most towards the Earth during September and a day out of the year when the Sun's North Pole is said to tip most away from the Earth during March. There are also two days during the year, in

June and December, when the Sun's North and South Poles[28], as viewed from Earth, are not considered to tip toward or away from Earth. Under an animation[12] of the tilt of the Sun's axis as seen from the Earth during the span of one Earth year, the publishers give an explanation such as "the wide (side-to-side) variation in tilt is due to the variation of Earth's axis, whereas the small nodding back and forth tilt is the result of the Sun's $7.25°$ tilt of its rotational pole relative to the ecliptic[8]." The so called axial tilt[7,10,11] angle of the Sun with respect to the so called ecliptic[8] plane is said to be approximately $7.25°$ (5), which is said to be the observed[10] angle of Sun's North Pole tipping towards the Earth around the first week of September, and the observed[10] angle of Sun's North Pole tipping away from the Earth around the first week of March.

We shall discover later in this Article that this $\alpha_{Sun,wobble} = 7.25°$ (5) is *not* the maximum axial tilt $\gamma_{Sun,\Delta Max}$ (104) of the Sun's rotational pole with respect to the virtual ecliptic[8] plane, but rather the angle that the Sun's North Pole seemingly makes towards or away from the Earth during first weeks of September and March, respectively, namely its "wobble" angle (330), with respect to the normal to the distance vector between the Earth and the Sun, based on the line of sight of an observer aligned with the North Pole[29] vector of the Earth.

$$\alpha_{Sun,wobble} = 7.25° \qquad ; \qquad \alpha_{Sun,wobble} \neq \gamma_{Sun,\Delta Max} \tag{5}$$

Based on this preliminary information and data, we will now step by step try and calculate the orbital parameters of the Sun and the Earth using the analytical method formulated in the **ARTICLE 2** "Analytical solution for two circles in space using their elliptic distances".

In this calculation, we take the $Circle_1$ to be the circle of revolution of the Sun, which is tilted by an angle β (6) with respect to $Circle_2$, and $Circle_2$ in the selected horizontal plane to be the circle of revolution of the Earth. Please refer to **Figure 1** for the geometry of this system in Cartesian $(\hat{x},\hat{y},\hat{z})$ coordinates. As the relative Sun-Earth system cycle is pretty constant phase-wise, based on the fact that the Sun-Earth distance is about the same for the same date every year, we expect their phase difference ϕ_0 to be constant. Therefore, based on the **ARTICLE 1** "Points on two circles in space mimic an ellipse with respect to a point", $\vec{r}_2(\phi)$ (7) would be defined for the radius vector of the Earth's individual orbit at each angle ϕ, and $\vec{r}_1(\phi+\phi_0)$ (6) would be defined for the radius vector of the Sun's individual orbit at each angle ϕ, where the Sun is moving at a constant phase difference ϕ_0 in its own plane with respect to the Earth's motion in its individual plane, r_1 and r_2 being the radii of individual revolutions of the Earth and the Sun, respectively, and the Sun-Earth vector distance $\vec{d}(\phi)$ at each angle ϕ can be stated as in (8), where $\vec{\ell}(\phi)$ (9) is the vector distance between centers of individual revolutions of Sun and Earth at every ϕ, having $\ell(\phi)$ (9) as its magnitude. We do not know yet if $\vec{\ell}(\phi)$ (9) or $\ell(\phi)$ (9) is constant or variable for all ϕ, which is to be determined a little later based on an analysis of the data[13]. Normally, it would not matter which circle is selected for which body's individual revolution; however, especially in order to use the correct axial tilt[7,10,11] angles of the Earth and Sun's North Poles[28,29], namely $\gamma_{Earth,\Delta Max}$ (4) and $\gamma_{Sun,\Delta Max}$, with respect to the distance vector $\vec{d}(\phi)$ (8) between the two, it is important to decide in advance which circle determines the revolution plane for each. Note here that the Sun-Earth distance vector $\vec{d}(\phi)$ (8) is directed from the Earth towards the Sun, as also demonstrated in **Figure 2**.

$$\vec{r_1}(\phi+\phi_0)=\hat{x}r_1Cos(\phi+\phi_0)Cos\,\beta+\hat{y}r_1Sin(\phi+\phi_0)+\hat{z}r_1Cos(\phi+\phi_0)Sin\,\beta \;;\; \vec{r_1}\cdot\vec{r_1}=r_1^2 \;;\; |\vec{r_1}|=r_1 \quad (6)$$

$$\vec{r_2}(\phi)=\hat{x}r_2\,Cos\,\phi+\hat{y}r_2\,Sin\phi \;;\; \vec{r_2}\cdot\vec{r_2}=r_2^2 \;;\; |\vec{r_2}|=r_2 \quad (7)$$

$$\vec{d}(\phi)=\vec{r_1}(\phi+\phi_0)-\vec{r_2}(\phi)+\vec{\ell}(\phi) \quad (8)$$

$$\vec{\ell}(\phi)=\hat{x}\ell_x(\phi)+\hat{y}\ell_y(\phi)+\hat{z}\ell_z(\phi) \;;\; \ell^2(\phi)=\vec{\ell}(\phi)\cdot\vec{\ell}(\phi)=\ell_x^{\,2}(\phi)+\ell_y^{\,2}(\phi)+\ell_z^{\,2}(\phi) \quad (9)$$

Based on the analysis in the **ARTICLE 1** "Points on two circles in space mimic an ellipse with respect to a point", as the phase difference ϕ_0 is considered to be constant due to the reasoning above, note here that the semi-major[23] and semi-minor[23] axis vectors of the ellipse[22] of the Sun-Earth system formed by their relative motions, namely $\vec{a}$ (10) and $\vec{b}$ (11), are constant, where a (10) and b (11) are their constant magnitudes, respectively, with c (12) as the constant focal[22,24] distance of this ellipse, and the parameter $\vec{a}\cdot\vec{b}$ (12) their constant Dot Product[14], in which case the Sun-Earth distance vector $\vec{d}(\phi)$ (8) can be expressed in terms of them as in (13).

$$\vec{a}=\hat{x}(r_1Cos\,\beta\,Cos\,\phi_0-r_2)+\hat{y}r_1Sin\phi_0+\hat{z}r_1Sin\,\beta\,Cos\,\phi_0 \;;\; a^2=\vec{a}\cdot\vec{a}=r_1^2-2\,r_1r_2Cos\beta\,Cos\,\phi_0+r_2^2 \quad (10)$$

$$\vec{b}=-\hat{x}r_1Cos\,\beta\,Sin\phi_0+\hat{y}(r_1Cos\,\phi_0-r_2)-\hat{z}r_1Sin\,\beta\,Sin\phi_0 \;;\; b^2=\vec{b}\cdot\vec{b}=r_1^2-2\,r_1r_2Cos\,\phi_0+r_2^2 \quad (11)$$

$$c^2=|a^2-b^2|=2r_1r_2(1-Cos\beta)|Cos\,\phi_0| \;;\; \vec{a}\cdot\vec{b}=r_1r_2(Cos\beta-1)Sin\phi_0 \quad (12)$$

$$\vec{d}(\phi)=\vec{a}\,Cos\,\phi+\vec{b}\,Sin\,\phi+\vec{\ell}(\phi) \;;\; \vec{r_1}(\phi+\phi_0)-\vec{r_2}(\phi)=\vec{a}\,Cos\,\phi+\vec{b}\,Sin\,\phi \quad (13)$$

As demonstrated in the **ARTICLE 2** "Analytical solution for two circles in space using their elliptic distances", the Sun-Earth vector distance $\vec{d}(\phi)$ and the distance value $d(\phi)$ and its square $d^2(\phi)$ at each ϕ can be expanded as in (14) - (19), using the defined constants and variables in (6) - (13).

$$\begin{aligned}\vec{d}(\phi)=\;&\hat{x}\left[r_1Cos(\phi+\phi_0)Cos\,\beta-r_2\,Cos\phi+\ell_x(\phi)\right]+\hat{y}\left[r_1Sin(\phi+\phi_0)-r_2Sin\phi+\ell_y(\phi)\right]\\&+\hat{z}\left[r_1Cos(\phi+\phi_0)Sin\,\beta+\ell_z(\phi)\right]\end{aligned} \quad (14)$$

$$\vec{d}(\phi)=\hat{x}d_x(\phi)+\hat{y}d_y(\phi)+\hat{z}d_z(\phi) \Rightarrow \begin{cases}d_x(\phi)=r_1Cos(\phi+\phi_0)Cos\,\beta-r_2\,Cos\phi+\ell_x(\phi)\\ d_y(\phi)=r_1Sin(\phi+\phi_0)-r_2Sin\phi+\ell_y(\phi)\\ d_z(\phi)=r_1Cos(\phi+\phi_0)Sin\,\beta+\ell_z(\phi)\end{cases} \quad (15)$$

$$d(\phi)=|\vec{d}(\phi)|=\sqrt{\vec{d}(\phi)\cdot\vec{d}(\phi)}=\sqrt{d^2(\phi)}=\sqrt{\left[d_x(\phi)\right]^2+\left[d_y(\phi)\right]^2+\left[d_z(\phi)\right]^2} \quad (16)$$

$$d^2(\phi)=\vec{d}(\phi)\cdot\vec{d}(\phi)=(\vec{a}\,Cos\,\phi+\vec{b}\,Sin\phi)^2+2\,(\vec{a}\cdot\vec{\ell}(\phi)Cos\phi+\vec{b}\cdot\vec{\ell}(\phi)Sin\phi)+\ell^2(\phi) \quad (17)$$

$$d^2(\phi)=a^2Cos^2\phi+2\vec{a}\cdot\vec{b}\,Sin\phi Cos\phi+b^2Sin^2\phi+2\vec{a}\cdot\vec{\ell}(\phi)Cos\phi+2\vec{b}\cdot\vec{\ell}(\phi)Sin\phi+\ell^2(\phi) \quad (18)$$

$$d^2(\phi)=b^2+2r_1r_2(1-Cos\beta)Cos\phi\,Cos(\phi+\phi_0)+2(\vec{a}Cos\phi+\vec{b}Sin\phi)\cdot\vec{\ell}(\phi)+\ell^2(\phi) \quad (19)$$

The Excel sheet "Tan, A.P., Asli Pinar Tan Analysis Based on Earth-Sun distance (d) Landsat.xlsx"[13] in which detailed calculations are made based on the data listed by NASA on a 2011 website[3] document based on Landsat observation data, as described below, can be found in the REFERENCES section.

To be able to work on a more refined data set, we first quadruple the number of distance data points d from 365 data points to $4 \times 365 = 1460$ data points, through interpolation such that the distance value of each middle $Data\,Point_{i+1}$ between two consecutive $Data\,Point_i$ and $Data\,Point_{i+2}$ is the average of the distance values of the two, as described in (20), where d_i and d_{i+4} are Sun-Earth distance values of any two consecutive data points among the original 365 data points, i being an integer from 1 to 1460. Data points must be considered continuous in circular fashion, such that for instance when $i = 1460$, $d_{i+1} = d_1$.

$$d_{i+2} = \frac{d_i + d_{i+4}}{2} \quad ; \quad d_{i+1} = \frac{d_i + d_{i+2}}{2} \quad ; \quad d_{i+3} = \frac{d_{i+2} + d_{i+4}}{2} \tag{20}$$

Using these 1460 distance data points, we have calculated distance squared $d^2(\phi)$, and then $\boldsymbol{k}_1(\phi)$, $\boldsymbol{k}_2(\phi)$, $\boldsymbol{k}_3(\phi)$, and $\boldsymbol{k}_4(\phi)$ defined in the **ARTICLE 2** "Analytical solution for two circles in space using their elliptic distances", based on the simple algorithm in (21) - (25), with an assumption that data points half a cycle apart in a Sun-Earth system cycle are approximately $(\phi = \pi)$ apart, and that data points a quarter cycle apart in a Sun-Earth system cycle are approximately $\left(\phi = \frac{\pi}{2}\right)$ apart.

$$d(\phi) = d_i \quad ; \quad d(\pi + \phi) = d_{i+730} \quad ; \quad d\left(\frac{\pi}{2} + \phi\right) = d_{i+365} \quad ; \quad d\left(-\frac{\pi}{2} + \phi\right) = d_{i-365} \tag{21}$$

$$\boldsymbol{k}_1(\phi) = \frac{d^2(\phi) + d^2(\pi + \phi)}{2} \quad \Rightarrow \quad \boldsymbol{k}_{1,i} = \frac{d_i^2 + d_{i+730}^2}{2} \tag{22}$$

$$\boldsymbol{k}_2(\phi) = \frac{d^2\left(\frac{\pi}{2} + \phi\right) + d^2\left(-\frac{\pi}{2} + \phi\right)}{2} \quad \Rightarrow \quad \boldsymbol{k}_{2,i} = \frac{d_{i+365}^2 + d_{i-365}^2}{2} \tag{23}$$

$$\boldsymbol{k}_3(\phi) = \frac{d^2(\phi) - d^2(\pi + \phi)}{4} \quad \Rightarrow \quad \boldsymbol{k}_{3,i} = \frac{d_i^2 - d_{i+730}^2}{4} \tag{24}$$

$$\boldsymbol{k}_4(\phi) = \frac{d^2\left(\frac{\pi}{2} + \phi\right) - d^2\left(-\frac{\pi}{2} + \phi\right)}{4} \quad \Rightarrow \quad \boldsymbol{k}_{4,i} = \frac{d_{i+365}^2 - d_{i-365}^2}{4} \tag{25}$$

Using the data calculated in the Excel sheet "Tan, A.P., Asli Pinar Tan Analysis Based on Earth-Sun distance (d) Landsat.xlsx"[13], Sun-Earth distance value $d(\phi)$ (21) and distance value squared $d^2(\phi)$ are plotted in **Figure 3** and **Figure 4**, respectively, as well as $\boldsymbol{k}_1(\phi)$ (22), $\boldsymbol{k}_2(\phi)$(23), $\boldsymbol{k}_3(\phi)$ (24), and $\boldsymbol{k}_4(\phi)$ (25) plotted in **Figure 5**, **Figure 6**, **Figure 7**, and **Figure 8**, respectively, over a yearly Sun-Earth relative motion cycle. Similarly, using these values, $\boldsymbol{k}_1(\phi) + \boldsymbol{k}_2(\phi)$,

$k_1(\phi)-k_2(\phi)$, $[k_3(\phi)]^2+[k_4(\phi)]^2$, and its square root $\sqrt{[k_3(\phi)]^2+[k_4(\phi)]^2}$ are calculated and plotted in **Figure 9**, **Figure 10**, **Figure 11**, and **Figure 12**, respectively. All these plots give us an idea about the behavior of change of these parameters over a year.

Note though that the calculation method presented in the **ARTICLE 2** "Analytical solution for two circles in space using their elliptic distances" is based on a case in which $\vec{\ell}(\phi)=\vec{\ell}$ (26) is a constant vector for all ϕ, with a constant magnitude ℓ.

$$\vec{\ell}(\phi)=\vec{\ell}=\hat{x}\ell_x+\hat{y}\ell_y+\hat{z}\ell_z \quad ; \quad \vec{\ell}\cdot\vec{\ell}=\ell^2=\ell_x^{\,2}+\ell_y^{\,2}+\ell_z^{\,2} \quad ; \quad |\vec{\ell}|=\ell \tag{26}$$

In such a specific case, $k_1(\phi)$, $k_2(\phi)$, $k_3(\phi)$, and $k_4(\phi)$ analytically reduce to (27) - (30), in which case $k_1(\phi)+k_2(\phi)$ (31) and $[k_3(\phi)]^2+[k_4(\phi)]^2$ (33) turn out to be constant for all ϕ, when $\vec{a}$ (10) and $\vec{b}$ (11) are also constant as in our Sun-Earth system, and $k_1(\phi)-k_2(\phi)$ (32), cleared from the effect of $\vec{\ell}$ (26), becomes only a phased $Cos(2\phi+\phi_0)$ curve.

$$\begin{aligned} k_1(\phi)&=\frac{d^2(\phi)+d^2(\pi+\phi)}{2}=\left(\vec{a}\,Cos\,\phi+\vec{b}\,Sin\phi\right)^2+\ell^2 \\ &=b^2+2r_1r_2\left(1-Cos\beta\right)Cos\phi\,Cos\left(\phi+\phi_0\right)+\ell^2 \end{aligned} \tag{27}$$

$$\begin{aligned} k_2(\phi)&=\frac{d^2\left(\frac{\pi}{2}+\phi\right)+d^2\left(-\frac{\pi}{2}+\phi\right)}{2}=\left(-\vec{a}\,Sin\,\phi+\vec{b}\,Cos\,\phi\right)^2+\ell^2 \\ &=b^2+2r_1r_2\left(1-Cos\beta\right)Sin\,\phi\,Sin\left(\phi+\phi_0\right)+\ell^2 \end{aligned} \tag{28}$$

$$k_3(\phi)=\frac{d^2(\phi)-d^2(\pi+\phi)}{4}=\vec{a}\cdot\vec{\ell}\,Cos\,\phi+\vec{b}\cdot\vec{\ell}\,Sin\,\phi \tag{29}$$

$$k_4(\phi)=\frac{d^2\left(\frac{\pi}{2}+\phi\right)-d^2\left(-\frac{\pi}{2}+\phi\right)}{4}=-\;\vec{a}\cdot\vec{\ell}\,Sin\phi+\vec{b}\cdot\vec{\ell}\,Cos\,\phi \tag{30}$$

$$k_1(\phi)+k_2(\phi)=a^2+b^2+2\ell^2 \tag{31}$$

$$k_1(\phi)-k_2(\phi)=2r_1r_2\left(1-Cos\beta\right)Cos\left(2\phi+\phi_0\right) \tag{32}$$

$$[k_3(\phi)]^2+[k_4(\phi)]^2=\left(\vec{a}\cdot\vec{\ell}\right)^2+\left(\vec{b}\cdot\vec{\ell}\right)^2 \tag{33}$$

However, note in **Figure 9** and **Figure 11** that $k_1(\phi)+k_2(\phi)$ and $[k_3(\phi)]^2+[k_4(\phi)]^2$ calculated from the data in the Excel sheet "Tan, A.P., Asli Pinar Tan Analysis Based on Earth-Sun distance (d) Landsat.xlsx"[13] are *not constant* but rather *harmonically varying* over a year. Therefore, based on **Figure 9** and **Figure 11**, one can conclude that $\ell(\phi)$ (9) and $\vec{\ell}(\phi)$ (9), namely the vector distance and its magnitude between centers of individual revolutions of Sun and Earth is *not constant* but rather *harmonically varying* over a year. ***This is one of the main findings in this Article about the Sun-Earth system.***

Based on this finding, for a more realistic analysis of the Sun-Earth system, let us re-formulate $\boldsymbol{k}_1(\phi)$, $\boldsymbol{k}_2(\phi)$, $\boldsymbol{k}_3(\phi)$, and $\boldsymbol{k}_4(\phi)$ analytically for a more general case when $\vec{\ell}(\phi)$ (9) is variable, based on the generic definition of distance value squared $d^2(\phi)$ in (17) - (19). This is performed in (34) - (41), first by expanding $d^2(\phi)$ (34), $d^2(\pi+\phi)$ (35), $d^2\left(\frac{\pi}{2}+\phi\right)$ (36), and $d^2\left(-\frac{\pi}{2}+\phi\right)$ (37), and then using these, re-formulating the generic $\boldsymbol{k}_1(\phi)$ (38), $\boldsymbol{k}_2(\phi)$ (39), $\boldsymbol{k}_3(\phi)$ (40), and $\boldsymbol{k}_4(\phi)$ (41).

$$d^2(\phi)=b^2+2r_1r_2(1-Cos\beta)Cos\phi\,Cos(\phi+\phi_0)+2\left(\vec{\boldsymbol{a}}Cos\phi+\vec{\boldsymbol{b}}Sin\phi\right)\cdot\vec{\ell}(\phi)+\ell^2(\phi) \quad (34)$$

$$d^2(\pi+\phi)=b^2+2r_1r_2(1-Cos\beta)Cos\phi\,Cos(\phi+\phi_0)-2\left(\vec{\boldsymbol{a}}Cos\phi+\vec{\boldsymbol{b}}Sin\phi\right)\cdot\vec{\ell}(\pi+\phi)+\ell^2(\pi+\phi) \quad (35)$$

$$d^2\left(\frac{\pi}{2}+\phi\right)=b^2+2r_1r_2(1-Cos\beta)Sin\phi\,Sin(\phi+\phi_0)-2\left(\vec{\boldsymbol{a}}Sin\phi-\vec{\boldsymbol{b}}Cos\phi\right)\cdot\vec{\ell}\left(\frac{\pi}{2}+\phi\right)+\ell^2\left(\frac{\pi}{2}+\phi\right) \quad (36)$$

$$\begin{aligned}d^2\left(-\frac{\pi}{2}+\phi\right)&=b^2+2r_1r_2(1-Cos\beta)Sin\phi\,Sin(\phi+\phi_0)\\&+2\left(\vec{\boldsymbol{a}}Sin\phi-\vec{\boldsymbol{b}}Cos\phi\right)\cdot\vec{\ell}\left(-\frac{\pi}{2}+\phi\right)+\ell^2\left(-\frac{\pi}{2}+\phi\right)\end{aligned} \quad (37)$$

$$\begin{aligned}\boldsymbol{k}_1(\phi)&=\frac{d^2(\phi)+d^2(\pi+\phi)}{2}=b^2+2r_1r_2(1-Cos\beta)Cos\phi\,Cos(\phi+\phi_0)\\&+\left(\vec{\boldsymbol{a}}Cos\phi+\vec{\boldsymbol{b}}Sin\phi\right)\cdot\left[\vec{\ell}(\phi)-\vec{\ell}(\pi+\phi)\right]+\frac{\ell^2(\phi)+\ell^2(\pi+\phi)}{2}\end{aligned} \quad (38)$$

$$\begin{aligned}\boldsymbol{k}_2(\phi)&=\frac{d^2\left(\frac{\pi}{2}+\phi\right)+d^2\left(-\frac{\pi}{2}+\phi\right)}{2}=b^2+2r_1r_2(1-Cos\beta)Sin\phi\,Sin(\phi+\phi_0)\\&+\left(-\vec{\boldsymbol{a}}Sin\phi+\vec{\boldsymbol{b}}Cos\phi\right)\cdot\left[\vec{\ell}\left(\frac{\pi}{2}+\phi\right)-\vec{\ell}\left(-\frac{\pi}{2}+\phi\right)\right]+\frac{\ell^2\left(\frac{\pi}{2}+\phi\right)+\ell^2\left(-\frac{\pi}{2}+\phi\right)}{2}\\&=b^2+2r_1r_2(1-Cos\beta)Cos\left(\frac{\pi}{2}+\phi\right)Cos\left(\frac{\pi}{2}+\phi+\phi_0\right)\\&+\left[\vec{\boldsymbol{a}}Cos\left(\frac{\pi}{2}+\phi\right)+\vec{\boldsymbol{b}}Sin\left(\frac{\pi}{2}+\phi\right)\right]\cdot\left[\vec{\ell}\left(\frac{\pi}{2}+\phi\right)-\vec{\ell}\left(-\frac{\pi}{2}+\phi\right)\right]+\frac{\ell^2\left(\frac{\pi}{2}+\phi\right)+\ell^2\left(-\frac{\pi}{2}+\phi\right)}{2}\end{aligned} \quad (39)$$

$$\boldsymbol{k}_3(\phi)=\frac{d^2(\phi)-d^2(\pi+\phi)}{4}=\left(\vec{\boldsymbol{a}}Cos\phi+\vec{\boldsymbol{b}}Sin\phi\right)\cdot\frac{\left[\vec{\ell}(\phi)+\vec{\ell}(\pi+\phi)\right]}{2}+\frac{\ell^2(\phi)-\ell^2(\pi+\phi)}{4} \quad (40)$$

$$k_4(\phi)=\frac{d^2\left(\frac{\pi}{2}+\phi\right)-d^2\left(-\frac{\pi}{2}+\phi\right)}{4}$$

$$=-\left(\vec{a}Sin\phi-\vec{b}Cos\phi\right)\cdot\frac{\left[\vec{\ell}\left(\frac{\pi}{2}+\phi\right)+\vec{\ell}\left(-\frac{\pi}{2}+\phi\right)\right]}{2}+\frac{\ell^2\left(\frac{\pi}{2}+\phi\right)-\ell^2\left(-\frac{\pi}{2}+\phi\right)}{4} \tag{41}$$

$$=\left[\vec{a}Cos\left(\frac{\pi}{2}+\phi\right)+\vec{b}Sin\left(\frac{\pi}{2}+\phi\right)\right]\cdot\frac{\left[\vec{\ell}\left(\frac{\pi}{2}+\phi\right)+\vec{\ell}\left(-\frac{\pi}{2}+\phi\right)\right]}{2}+\frac{\ell^2\left(\frac{\pi}{2}+\phi\right)-\ell^2\left(-\frac{\pi}{2}+\phi\right)}{4}$$

Using the generic $k_1(\phi)$ (38), $k_2(\phi)$ (39), $k_3(\phi)$ (40), and $k_4(\phi)$ (41), one can find the generic formulas for $k_1(\phi)+k_2(\phi)$ (42), $k_1(\phi)-k_2(\phi)$ (43), and $[k_3(\phi)]^2+[k_4(\phi)]^2$ (44).

$$\begin{aligned}k_1(\phi)+k_2(\phi)=a^2+b^2+&\frac{\ell^2(\phi)+\ell^2(\pi+\phi)+\ell^2\left(\frac{\pi}{2}+\phi\right)+\ell^2\left(-\frac{\pi}{2}+\phi\right)}{2}\\&+\left(\vec{a}Cos\phi+\vec{b}Sin\phi\right)\cdot\left[\vec{\ell}(\phi)-\vec{\ell}(\pi+\phi)\right]\\&+\left[\vec{a}Cos\left(\frac{\pi}{2}+\phi\right)+\vec{b}Sin\left(\frac{\pi}{2}+\phi\right)\right]\cdot\left[\vec{\ell}\left(\frac{\pi}{2}+\phi\right)-\vec{\ell}\left(-\frac{\pi}{2}+\phi\right)\right]\end{aligned} \tag{42}$$

$$\begin{aligned}k_1(\phi)-k_2(\phi)=2r_1r_2(1-Cos\beta)Cos(2\phi+\phi_0)+&\frac{\ell^2(\phi)+\ell^2(\pi+\phi)-\ell^2\left(\frac{\pi}{2}+\phi\right)-\ell^2\left(-\frac{\pi}{2}+\phi\right)}{2}\\+\left(\vec{a}Cos\phi+\vec{b}Sin\phi\right)\cdot\left[\vec{\ell}(\phi)-\vec{\ell}(\pi+\phi)\right]&\\-\left[\vec{a}Cos\left(\frac{\pi}{2}+\phi\right)+\vec{b}Sin\left(\frac{\pi}{2}+\phi\right)\right]\cdot\left[\vec{\ell}\left(\frac{\pi}{2}+\phi\right)-\vec{\ell}\left(-\frac{\pi}{2}+\phi\right)\right]&\end{aligned} \tag{43}$$

$$\begin{aligned}[k_3(\phi)]^2+[k_4(\phi)]^2=&\left\{\left(\vec{a}Cos\phi+\vec{b}Sin\phi\right)\cdot\frac{\left[\vec{\ell}(\phi)+\vec{\ell}(\pi+\phi)\right]}{2}+\frac{\ell^2(\phi)-\ell^2(\pi+\phi)}{4}\right\}^2\\&+\left\{\left[\vec{a}Cos\left(\frac{\pi}{2}+\phi\right)+\vec{b}Sin\left(\frac{\pi}{2}+\phi\right)\right]\cdot\frac{\left[\vec{\ell}\left(\frac{\pi}{2}+\phi\right)+\vec{\ell}\left(-\frac{\pi}{2}+\phi\right)\right]}{2}+\frac{\ell^2\left(\frac{\pi}{2}+\phi\right)-\ell^2\left(-\frac{\pi}{2}+\phi\right)}{4}\right\}^2\end{aligned} \tag{44}$$

The behavior exhibited in **Figure 10** for $k_1(\phi)-k_2(\phi)$ of the Sun-Earth system is a phased $Cos(2\phi+\phi_0)$ curve with two cycles over 2π, as if in the case of (32), seemingly cleared from the effect of $\vec{\ell}(\phi)$ (9) for all ϕ. Considering the additional factors in (43) compared to (32), the simplest case for this to be possible is when (45) is valid for all ϕ. If this is true, $\vec{\ell}(\phi)$ (9) must be a harmonic vector function that repeats itself at least every half-cycle of the Sun-Earth system, and it must also at least have the same magnitude every quarter of each Sun-Earth cycle, with the same or reversed direction.

$$\vec{\ell}(\phi)=\vec{\ell}(\pi+\phi) \quad ; \quad \vec{\ell}\left(\frac{\pi}{2}+\phi\right)=\vec{\ell}\left(-\frac{\pi}{2}+\phi\right) \quad ; \quad \vec{\ell}(\phi)=\pm\vec{\ell}\left(\frac{\pi}{2}+\phi\right) \quad \textit{for all } \phi \tag{45}$$

In such a case, $\boldsymbol{k}_1(\phi)$ (38), $\boldsymbol{k}_2(\phi)$ (39), $\boldsymbol{k}_3(\phi)$ (40), $\boldsymbol{k}_4(\phi)$ (41), $\boldsymbol{k}_1(\phi)+\boldsymbol{k}_2(\phi)$ (42), $\boldsymbol{k}_1(\phi)-\boldsymbol{k}_2(\phi)$ (43), and $[\boldsymbol{k}_3(\phi)]^2+[\boldsymbol{k}_4(\phi)]^2$ (44) reduce to $\boldsymbol{k}_1(\phi)$ (46), $\boldsymbol{k}_2(\phi)$ (47), $\boldsymbol{k}_3(\phi)$ (48), $\boldsymbol{k}_4(\phi)$ (49), $\boldsymbol{k}_1(\phi)+\boldsymbol{k}_2(\phi)$ (50), $\boldsymbol{k}_1(\phi)-\boldsymbol{k}_2(\phi)$ (51), and $[\boldsymbol{k}_3(\phi)]^2+[\boldsymbol{k}_4(\phi)]^2$ (52), also utilizing (10) - (13).

$$\begin{aligned}\boldsymbol{k}_1(\phi)=\frac{d^2(\phi)+d^2(\pi+\phi)}{2}&=b^2+2r_1r_2(1-Cos\beta)Cos\phi Cos(\phi+\phi_0)+\ell(\phi)^2\\&=(\vec{\boldsymbol{a}}\,Cos\phi+\vec{\boldsymbol{b}}\,Sin\phi)^2+\ell(\phi)^2\end{aligned} \tag{46}$$

$$\begin{aligned}\boldsymbol{k}_2(\phi)=\frac{d^2\left(\frac{\pi}{2}+\phi\right)+d^2\left(-\frac{\pi}{2}+\phi\right)}{2}&=b^2+2r_1r_2(1-Cos\beta)Sin\phi\, Sin(\phi+\phi_0)+\ell(\phi)^2\\&=b^2+2r_1r_2(1-Cos\beta)Cos\left(\frac{\pi}{2}+\phi\right)Cos\left(\frac{\pi}{2}+\phi+\phi_0\right)+\ell(\phi)^2\\&=(-\vec{\boldsymbol{a}}\,Sin\phi+\vec{\boldsymbol{b}}\,Cos\phi)^2+\ell(\phi)^2\end{aligned} \tag{47}$$

$$\boldsymbol{k}_3(\phi)=\frac{d^2(\phi)-d^2(\pi+\phi)}{4}=\vec{\boldsymbol{a}}\cdot\vec{\ell}(\phi)Cos\phi+\vec{\boldsymbol{b}}\cdot\vec{\ell}(\phi)Sin\phi \tag{48}$$

$$\begin{aligned}\boldsymbol{k}_4(\phi)=\frac{d^2\left(\frac{\pi}{2}+\phi\right)-d^2\left(-\frac{\pi}{2}+\phi\right)}{4}&=(-\vec{\boldsymbol{a}}Sin\phi+\vec{\boldsymbol{b}}Cos\phi)\cdot\vec{\ell}\left(\frac{\pi}{2}+\phi\right) \quad ; \quad \vec{\ell}\left(\frac{\pi}{2}+\phi\right)=\pm\vec{\ell}(\phi)\\&=\left[\vec{\boldsymbol{a}}Cos\left(\frac{\pi}{2}+\phi\right)+\vec{\boldsymbol{b}}Sin\left(\frac{\pi}{2}+\phi\right)\right]\cdot\vec{\ell}\left(\frac{\pi}{2}+\phi\right)\end{aligned} \tag{49}$$

$$\boldsymbol{k}_1(\phi)+\boldsymbol{k}_2(\phi)=a^2+b^2+2\ell^2(\phi) \tag{50}$$

$$\boldsymbol{k}_1(\phi)-\boldsymbol{k}_2(\phi)=2r_1r_2(1-Cos\beta)Cos(2\phi+\phi_0) \tag{51}$$

$$[\boldsymbol{k}_3(\phi)]^2+[\boldsymbol{k}_4(\phi)]^2=[\vec{\boldsymbol{a}}\cdot\vec{\ell}(\phi)]^2+[\vec{\boldsymbol{b}}\cdot\vec{\ell}(\phi)]^2 \tag{52}$$

When inspected more closely, it can be seen in more detail in the Excel sheet "Tan, A.P., Asli Pinar Tan Analysis Based on Earth-Sun distance (d) Landsat.xlsx"[13] that **Figure 9** for $\boldsymbol{k}_1(\phi)+\boldsymbol{k}_2(\phi)$ of the Sun-Earth system is a group sinusoid curve of multiple frequencies with an offset, varying based on mainly 3 harmonic oscillations over a year, one with a frequency of 4 times a year, a second one that oscillates 12 times a year, and yet another small oscillation occuring approximately every Earth day (every 4 data points over 1460 data points over a yearly cycle). Therefore, the **Figure 9** curve, as well as the $\boldsymbol{k}_1(\phi)+\boldsymbol{k}_2(\phi)$ data in the Excel sheet "Tan, A.P., Asli Pinar Tan Analysis Based on Earth-Sun distance (d) Landsat.xlsx"[13], exhibit behavior making it is possible for $\boldsymbol{k}_1(\phi)+\boldsymbol{k}_2(\phi)$ of the Sun-Earth system to be in the form of (50), having a constant offset of $\left(a^2+b^2\right)$, where $\vec{\ell}(\phi)$ (9) is a harmonic vector function whose magnitude squared $\ell^2(\phi)$ (9) varies sinusoidally with a multitude of frequencies based on different factors that impact the distances between centers of individual orbits of revolution of the Earth and the Sun over a year, which takes on its same magnitude value every $\pi/2$ phase change, i.e. every quarter of each Sun-Earth cycle, supporting (45) extracted from **Figure 10** for $\boldsymbol{k}_1(\phi)-\boldsymbol{k}_2(\phi)$ to be valid for the Sun-Earth system. We still have to determine whether the direction of $\vec{\ell}\left(\frac{\pi}{2}+\phi\right)$ (45) is the same as or reversed with respect to $\vec{\ell}(\phi)$ (9) for the Sun-Earth system, which we shall be able to find out later in the analysis presented this Article.

With this analysis based on **Figure 9** and the Excel sheet "Tan, A.P., Asli Pinar Tan Analysis Based on Earth-Sun distance (d) Landsat.xlsx"[13] data for $\boldsymbol{k}_1(\phi)+\boldsymbol{k}_2(\phi)$ of the Sun-Earth system, we can conclude that one oscillation of the magnitude $\ell(\phi)$ (9) of the distance between the centers of individual orbits of revolutions of the Earth and the Sun is about 12 times a year, which is apparently based on the impact of the relative Earth-Moon motion, and a second oscillation of $\ell(\phi)$ (9) is seemingly based on the daily self-rotation of the Earth around its own axis, with a third oscillation of $\ell(\phi)$ (9) that is observed to occur about four times a year, the possible cause of which we shall now determine based on the following analysis.

As this third oscillation of $\ell(\phi)$ (9) over a Sun-Earth year is more prominent compared to the other two, having a bigger impact on the magnitude change of $\ell(\phi)$ (9) over a Sun-Earth relative year, it may possibly be caused by the relative orbital motion of another of the planets in the Solar System with the Sun, causing the $\ell(\phi)$ (9) wobble four times a year with respect to the Earth. The data[15] published by NASA Goddard Space Flight Center on the Orbital Periods of the main planets in the Solar System can be found in the following **Table 2**. It seems that the Sun and Mercury form a twin or binary[25] system of bodies in space, in which Mercury behaves as the Sun's satellite, similar to Earth-Moon system, and implies the possibility that the individual circular orbits of revolution of the Sun and Mercury may also be revolving around each other.

Taking these orbital period approximations based on observation, and considering that the relative orbit of Mercury[16] is the closest one to the Sun among the planets in the Solar system, we may conclude that the possible cause of the third oscillation of $\ell(\phi)$ (9), which is about four times over a Sun-Earth year, is due to the impact of relative Sun-Mercury motion which has about four cycles over a Sun-Earth year. It seems that the Sun and Mercury form a twin or binary[25] system of bodies in space, in which Mercury behaves as the Sun's satellite, similar to Earth-Moon system, and implies the possibility that the individual circular orbits of revolution of the Sun and Mercury may also be revolving around each other. Further, this analysis also implies that it is very well possible that the relative motions of the other planets and their satellites, as well as other moving bodies, in the Solar system impact the relative motion of the Sun and Earth in terms of additional minor oscillations of $\ell(\phi)$ (9) over longer periods than a Sun-Earth year, but their impact on Sun-Earth distance can be observed only over longer periods than a yearly cycle. These conclusions are also among the fundamental findings in this Article.

Table 2 Orbital Periods[15] of Planets in the Solar System

PLANET	**ORBITAL PERIOD** (Days)
Mercury	88.0
Venus	224.7
Earth	365.2
Mars	687.0
Jupiter	4,331.0
Saturn	10,747.0
Uranus	30,589.0
Neptune	59,800.0
Pluto	90,560.0

Comparing the curves in **Figure 9** and **Figure 10**, it can clearly be observed that all oscillations of $\ell^2(\phi)$ (9), and therefore $\ell(\phi)$ (9), are aligned with and phased in the same way with the $Cos(2\phi+\phi_0)$ (51) curve of **Figure 10**, i.e. shifted according to the same phase ϕ_0, where the minimums of the **Figure 9** curve match with the min-max values of the **Figure 10** curve on January 1, April 1, July 1, and September 30 / October 1, also verified based on the data value changes in the Excel sheet "Tan, A.P., Asli Pinar Tan Analysis Based on Earth-Sun distance (d) Landsat.xlsx"[13].

At this point, it is worth noticing that ancient people did *not* select the first days of months in the Solar Calendar cycle based on Min-Max Sun-Earth distance dates (Jan 3 and July 5), or the dates of Solstices (June 21 and December 21)[5,6], or Equinoxes (March 20 and September 23)[5,6], which slightly vary every year. Apparently, the selection of dates for the first days of months in the Solar Calendar by ancient people was not coincidental, but rather based on the phase ϕ_0 between individual revolutions of the Earth and the Sun in their own orbits, and also the aligned cycles of the harmonic $\ell(\phi)$ (9), the distance between centers of individual revolutions of the Sun and Earth, which subsequently has resulted in different duration months of 28, 30, and 31 days. Our ancestors possibly based these date selections on accurate Sun-Earth distance observations and the intermediate harmonic variations in it based on phase ϕ_0 and $\ell(\phi)$ (9), even if they were not aware of the existence of phase ϕ_0 or $\ell(\phi)$ (9).

Moving on, according to (46) - (47) and (45), the relations in (53) must hold for all ϕ. Note that $\boldsymbol{k}_2(\phi)$ sketched in **Figure 6** is a $\pm\pi/2$ - shifted version of $\boldsymbol{k}_1(\phi)$ sketched in **Figure 5**, and both exhibit the form of a $Cos(2\phi+\phi_0)$ curve with an offset, shifted according to the same phase ϕ_0 as in **Figure 9** and **Figure 10**, an observation which also supports the validity of (45), and thus the validity of $k_1(\phi)$ (46) and $k_2(\phi)$ (47) for the Sun-Earth system. These results in (53) are in agreement with (54) according to the data in the Excel sheet "Tan, A.P., Asli Pinar Tan Analysis Based on Earth-Sun distance (d) Landsat.xlsx"[13] calculated according to the algorithm in (21) - (25), confirming our analysis for the Sun-Earth system so far.

$$k_1(\pi+\phi)=k_1(\phi)=k_2\left(\frac{\pi}{2}+\phi\right) \quad ; \quad k_2(\pi+\phi)=k_2(\phi)=k_1\left(\frac{\pi}{2}+\phi\right) \qquad \textit{for all } \phi \quad (53)$$

$$k_{1,i+730}=k_{1,i}=k_{2,i+365} \quad ; \quad k_{2,i+730}=k_{2,i}=k_{1,i+365} \qquad i \textit{ circulating integer from } 1 \textit{ to } 1460 \quad (54)$$

It can similarly be seen that $k_4(\phi)$ sketched in **Figure 8** is a $\pm\pi/2$ - shifted version of $k_3(\phi)$ sketched in **Figure 7**, as also stated in (55) - (56), supporting the validity of $k_3(\phi)$ (48), which is a Dot Product[14] of $(\bar{a}\,Cos\,\phi+\bar{b}Sin\,\phi)$ vector with $\bar{\ell}(\phi)$ (9), and the validity of $k_4(\phi)$ (49), which is a Dot Product[14] of $(-\bar{a}Sin\phi+\bar{b}Cos\,\phi)$ vector with $\bar{\ell}\left(\frac{\pi}{2}+\phi\right)$ (45), where $\bar{a}$ (10) and $\bar{b}$ (11) are constant. These results in (55) - (56) are in agreement with (57) - (58) according to the data in the Excel sheet "Tan, A.P., Asli Pinar Tan Analysis Based on Earth-Sun distance (d) Landsat.xlsx"[13] calculated according to the algorithm in (21) - (25), confirming our analysis for the Sun-Earth system so far.

$$k_3(\pi+\phi)=-k_3(\phi) \quad ; \quad k_4(\pi+\phi)=-k_4(\phi) \qquad \textit{for all } \phi \qquad (55)$$

$$k_3\left(\frac{\pi}{2}+\phi\right)=k_4(\phi) \quad ; \quad k_4\left(\frac{\pi}{2}+\phi\right)=-k_3(\phi) \qquad \textit{for all } \phi \qquad (56)$$

$$k_{3,i+730}=-k_{3,i} \quad ; \quad k_{4,i+730}=-k_{4,i} \qquad i \textit{ circulating integer from } 1 \textit{ to } 1460 \qquad (57)$$

$$k_{3,i+365}=k_{4,i} \quad ; \quad k_{4,i+365}=-k_{3,i} \qquad i \textit{ circulating integer from } 1 \textit{ to } 1460 \qquad (58)$$

On the other hand, **Figure 11** for $[k_3(\phi)]^2+[k_4(\phi)]^2$ of the Sun-Earth system, and therefore **Figure 12** for $\sqrt{[k_3(\phi)]^2+[k_4(\phi)]^2}$ of the Sun-Earth system, exhibit behavior in line with the four-times-a-year oscillation observed in **Figure 9** for $\ell(\phi)$ (9). Taking into consideration that $[k_3(\phi)]^2+[k_4(\phi)]^2$ (52) is a square equation varying based on the sum of squares of the projections of the constant $\bar{a}$ (10) and $\bar{b}$ (11) on varying $\bar{\ell}(\phi)$ (9), namely $\left[\bar{a}\cdot\bar{\ell}(\phi)\right]^2+\left[\bar{b}\cdot\bar{\ell}(\phi)\right]^2$, a value which is variably proportional to $\ell^2(\phi)$ (9), again we can conclude that *an oscillation* of $\ell(\phi)$ (9) is <u>four times a year</u>. The second oscillation of $\ell(\phi)$ (9) observed in **Figure 9**, which is 12 times a year, is not observed in **Figure 11** or **Figure 12**. This may be because the Earth-Moon motion effect on $\left[\bar{a}\cdot\bar{\ell}(\phi)\right]^2+\left[\bar{b}\cdot\bar{\ell}(\phi)\right]^2$ is negligible due to the angles of projections. Again, comparing the **Figure 11** curve with the **Figure 10** curve, it can clearly be observed that the oscillation of $\left[\bar{a}\cdot\bar{\ell}(\phi)\right]^2+\left[\bar{b}\cdot\bar{\ell}(\phi)\right]^2$, and therefore $\bar{\ell}(\phi)$ (9), is phased in the same way with the $Cos(2\phi+\phi_0)$ curve, i.e. according to the same phase ϕ_0, and with polarity changes on January 1, April 1, July 1, and September 30 / October 1, also verified based on the data value changes in the Excel sheet "Tan, A.P., Asli Pinar Tan Analysis Based on Earth-Sun distance (d) Landsat.xlsx"[13].

As a result of our analysis so far, the data in the Excel sheet "Tan, A.P., Asli Pinar Tan Analysis Based on Earth-Sun distance (d) Landsat.xlsx"[13] support the validity of (45) - (52), and in our further analysis and calculation of the orbital parameters of the Sun and the Earth, we will be utilizing (45) - (52) with (3) - (25).

Continuing our analysis, based on (17), (46), and (48), $d^2(\phi)$ can also be expressed as in (59), in compact form, for the Sun-Earth system.

$$d^2(\phi)=\boldsymbol{k}_1(\phi)+2\boldsymbol{k}_3(\phi) \tag{59}$$

The maximum Sun-Earth distance value of 1.0167037 au encountered on July 5 (DOY 186) and the minimum Sun-Earth distance value of 0.983297 au encountered on January 3 (DOY 3) would therefore be logically expected to occur around the maximum and minimum values of $\boldsymbol{k}_1(\phi)$ (46) and/or $\boldsymbol{k}_3(\phi)$ (48), whichever has a more dominant value variation over a cycle.

The maximum and minimum values of $\boldsymbol{k}_1(\phi)$ calculated in the Excel sheet "Tan, A.P., Asli Pinar Tan Analysis Based on Earth-Sun distance (d) Landsat.xlsx"[13] are listed in (60) - (61).

$$\left[\boldsymbol{k}_1(\phi)\right]_{Max}=\mathbf{1.00059564301763}\,au^2 \quad on \quad 3\,April\ and\ 3\,October \tag{60}$$

$$\left[\boldsymbol{k}_1(\phi)\right]_{Min}=\mathbf{1.00027830860352}\,au^2 \quad on \quad 1\,January\ and\ 2\,July \tag{61}$$

Based on (61), both the maximum and minimum Sun-Earth distance values occur around a minimum value of $\boldsymbol{k}_1(\phi)$. So, looking at (59), it seems that $\boldsymbol{k}_1(\phi)$ (46) does not have too much impact on the maximum Sun-Earth distance, but it may or may not have an impact on the minimum Sun-Earth distance.

The maximum and minimum values of $\boldsymbol{k}_3(\phi)$ calculated in the Excel sheet "Tan, A.P., Asli Pinar Tan Analysis Based on Earth-Sun distance (d) Landsat.xlsx"[13] are listed in (62) - (63).

$$\left[\boldsymbol{k}_3(\phi)\right]_{Max}=\mathbf{0.016703036274542}\,au^2 \quad on \quad 5\,July \tag{62}$$

$$\left[\boldsymbol{k}_3(\phi)\right]_{Min}=\mathbf{-0.016703036274542}\,au^2 \quad on \quad 3\,January \tag{63}$$

Based on (62), the maximum Sun-Earth distance value occurs exactly on the same date as the maximum value of $\boldsymbol{k}_3(\phi)$, and based on (63), the minimum Sun-Earth distance value occurs exactly on the same date as the minimum value of $\boldsymbol{k}_3(\phi)$. This result can also be inferred by observing that the difference between the maximum and minimum values of $\boldsymbol{k}_1(\phi)$ in (60) - (61) is small throughout a whole cycle compared to the difference between the maximum and minimum values of $\boldsymbol{k}_3(\phi)$ in (62) - (63), whose impact on $d^2(\phi)$ is also doubled according to (59), and thus the value variation of $\boldsymbol{k}_3(\phi)$ (48) over a Sun-Earth cycle is what determines the maximum and minimum values of $d^2(\phi)$ (59), and $\boldsymbol{k}_1(\phi)$ (46) does not seem to have too much determining impact on the maximum or minimum Sun-Earth distance occurrence.

The maximum and minimum values of $\boldsymbol{k}_2(\phi)$ calculated in the Excel sheet "Tan, A.P., Asli Pinar Tan Analysis Based on Earth-Sun distance (d) Landsat.xlsx"[13] are listed in (64) - (65), which is a $\pm\pi/2$ - shifted occurrence of the maximum and minimum values of $\boldsymbol{k}_1(\phi)$ in (60) - (61), in alignment with what is expected according to (53).

$$\left[\boldsymbol{k}_2(\phi)\right]_{Max} = \mathbf{1.00059564301763}\ au^2 \quad on \quad 2\ January\ and\ 3\ July \tag{64}$$

$$\left[\boldsymbol{k}_2(\phi)\right]_{Min} = \mathbf{1.00027830860352}\ au^2 \quad on \quad 2\ April\ and\ 1\ October \tag{65}$$

The maximum and minimum values of $\boldsymbol{k}_4(\phi)$ and $\sqrt{[\boldsymbol{k}_3(\phi)]^2 + [\boldsymbol{k}_4(\phi)]^2}$ calculated in the Excel sheet "Tan, A.P., Asli Pinar Tan Analysis Based on Earth-Sun distance (d) Landsat.xlsx"[13] are listed in (66) - (69), those of $\boldsymbol{k}_4(\phi)$ being is a $\pm\pi/2$ - shifted occurrence of the maximum and minimum values of $\boldsymbol{k}_3(\phi)$ in (62) - (63), in alignment with what is expected according to (56).

$$\left[\boldsymbol{k}_4(\phi)\right]_{Max} = \mathbf{0.016703036274542}\ au^2 \quad on \quad 4\ April \tag{66}$$

$$\left[\boldsymbol{k}_4(\phi)\right]_{Min} = \mathbf{-0.016703036274542}\ au^2 \quad on \quad 4\ October \tag{67}$$

$$\left\{\sqrt{[\boldsymbol{k}_3(\phi)]^2 + [\boldsymbol{k}_4(\phi)]^2}\right\}_{Max} = \mathbf{0.0167040064731237}\ au^2 \quad on \quad 1\ January, 2\ April, 2\ July, 1\ October \tag{68}$$

$$\left\{\sqrt{[\boldsymbol{k}_3(\phi)]^2 + [\boldsymbol{k}_4(\phi)]^2}\right\}_{Min} = \mathbf{0.016692622192379}\ au^2 \quad on \quad 17\ February, 20\ May, 19\ August, 18\ November \tag{69}$$

Based on (51), as $(1 - Cos\beta) \geq 0$ mathematically, Min-Max values of $[\boldsymbol{k}_1(\phi) - \boldsymbol{k}_2(\phi)]$, namely $[\boldsymbol{k}_1(\phi) - \boldsymbol{k}_2(\phi)]_{Min-Max}$ (70), occur at $\left(\phi = \frac{n\pi}{2} - \frac{\phi_0}{2}\right)$ (71), where n is an integer.

$$[\boldsymbol{k}_1(\phi) - \boldsymbol{k}_2(\phi)]_{Max} = 2r_1 r_2 (1 - Cos\beta) \quad ; \quad [\boldsymbol{k}_1(\phi) - \boldsymbol{k}_2(\phi)]_{Min} = -2r_1 r_2 (1 - Cos\beta) \tag{70}$$

$$[\boldsymbol{k}_1(\phi) - \boldsymbol{k}_2(\phi)]_{Min-Max} \quad occur \quad at \quad \phi = \frac{n\pi}{2} - \frac{\phi_0}{2} \quad ; \quad n\ integer \tag{71}$$

Based on the data in the Excel sheet "Tan, A.P., Asli Pinar Tan Analysis Based on Earth-Sun distance (d) Landsat.xlsx"[13], the Min-Max values (70) of $[\boldsymbol{k}_1(\phi) - \boldsymbol{k}_2(\phi)]$ determine $[2r_1 r_2 (1 - Cos\beta)]$ (72).

$$[\boldsymbol{k}_1(\phi) - \boldsymbol{k}_2(\phi)]_{Max} = 2r_1 r_2 (1 - Cos\beta) = \mathbf{0.000317144894687038}\ au^2 \quad on \quad 2\ April\ \&\ 1\ October \tag{72}$$

$$[\boldsymbol{k}_1(\phi) - \boldsymbol{k}_2(\phi)]_{Min} = -2r_1 r_2 (1 - Cos\beta) = \mathbf{-0.000317144894687038}\ au^2 \quad on \quad 1\ January\ \&\ 2\ July \tag{73}$$

Hence, $\left\{\sqrt{[\boldsymbol{k}_3(\phi)]^2+[\boldsymbol{k}_4(\phi)]^2}\right\}_{Max}$ (68) occurence coincides with the same dates as that of either $[\boldsymbol{k}_1(\phi)-\boldsymbol{k}_2(\phi)]_{Max}$ (72) or $[\boldsymbol{k}_1(\phi)-\boldsymbol{k}_2(\phi)]_{Min}$ (73), as an additional observation, affirming the ϕ_0 phase alignment.

Based on (51) and the data in the Excel sheet "Tan, A.P., Asli Pinar Tan Analysis Based on Earth-Sun distance (d) Landsat.xlsx"[13], the zero crossings of $[\boldsymbol{k}_1(\phi)-\boldsymbol{k}_2(\phi)]$ (74) occur at $\left(\phi=\frac{n\pi}{4}-\frac{\phi_0}{2}\right)$ (75), where n is an odd integer, on February 18 (DOY 49), May 20 (DOY 140), August 19 (DOY 231), and November 18-19 (DOY 322-323), which are the dates $[\boldsymbol{k}_1(\phi)-\boldsymbol{k}_2(\phi)]$ takes on its closest data value to zero, and changes polarity. These are also the dates on and around which $\left\{\sqrt{[\boldsymbol{k}_3(\phi)]^2+[\boldsymbol{k}_4(\phi)]^2}\right\}_{Min}$ (69) occur.

$$[\boldsymbol{k}_1(\phi)-\boldsymbol{k}_2(\phi)]_0=2r_1r_2\left(1-Cos\beta\right)Cos\left(2\phi+\phi_0\right)=\pm 0.0000012183270556853\,6\approx 0 \qquad (74)$$

$$[\boldsymbol{k}_1(\phi)-\boldsymbol{k}_2(\phi)]_0 \quad occur\ at \quad 2\phi+\phi_0=\frac{n\pi}{2} \quad \Rightarrow \quad \phi=\frac{n\pi}{4}-\frac{\phi_0}{2} \quad ; \quad n\ odd\ integer \qquad (75)$$

At this point, based on $[\boldsymbol{k}_1(\phi)-\boldsymbol{k}_2(\phi)]$ (51) and the found value of $\left[2r_1r_2\left(1-Cos\beta\right)\right]$ (72), we have also calculated the value of $Cos\left(2\phi+\phi_0\right)$ at each data point (for all ϕ) for the Sun-Earth system, in the Excel sheet "Tan, A.P., Asli Pinar Tan Analysis Based on Earth-Sun distance (d) Landsat.xlsx"[13].

Moving on, $\boldsymbol{k}_1(\phi)+\boldsymbol{k}_2(\phi)$ (50) is composed of a fixed component, which is the sum of some constant components, and a variable component based on $\vec{\ell}(\phi)$ (9), for the Sun-Earth system. A generic approach can be made by expressing $\ell^2(\phi)$ (76) in terms of the sum of its constant and variable components such that, $L_{\langle \ell^2(\phi)\rangle}$ (76) is its constant <u>minimum</u> component throughout all ϕ and $f(\phi)_{\langle \ell^2(\phi)\rangle}$ (76) is its additional variable component which is a function of ϕ.

$$\ell^2(\phi)=\vec{\ell}(\phi)\cdot\vec{\ell}(\phi)=\ell_x^{\,2}(\phi)+\ell_y^{\,2}(\phi)+\ell_z^{\,2}(\phi)=L_{\langle \ell^2(\phi)\rangle}+f(\phi)_{\langle \ell^2(\phi)\rangle} \qquad (76)$$

Using the definition in (76), $\boldsymbol{k}_1(\phi)+\boldsymbol{k}_2(\phi)$ (50) can be expressed as in (77).

$$\boldsymbol{k}_1(\phi)+\boldsymbol{k}_2(\phi)=a^2+b^2+2\ell^2(\phi)=a^2+b^2+2L_{\langle \ell^2(\phi)\rangle}+2f(\phi)_{\langle \ell^2(\phi)\rangle} \qquad (77)$$

Taking into consideration the conclusion we have reached in (45) based on our previous analysis, $\vec{\ell}(\phi)$ (9) must be a vector whose magnitude is harmonically varying between $\ell_{\max}\left(\phi_{\ell_{\max}}\right)$ and $\ell_{\min}\left(\phi_{\ell_{\min}}\right)$, where $\ell_{\max}\left(\phi_{\ell_{\max}}\right)$ is the maximum magnitude value of $\vec{\ell}(\phi)$ (9) over a Sun-Earth cycle at the phase angles $\left(\phi=\phi_{\ell_{\max}}\right)$ and $\left(\phi=\pi+\phi_{\ell_{\max}}\right)$, and $\ell_{\min}\left(\phi_{\ell_{\min}}\right)$ is the minimum magnitude value of $\vec{\ell}(\phi)$ (9) over a Sun-Earth cycle at the phase angles $\left(\phi=\phi_{\ell_{\min}}\right)$ and

$\left(\phi = \pi + \phi_{\ell_{\min}}\right)$. Based on this observation and (76) - (77), we can expect (78) to hold, where $f(\phi)_{\langle \ell^2(\phi)\rangle, Max}$ is the maximum value that the $f(\phi)_{\langle \ell^2(\phi)\rangle}$ (76) takes over a Sun-Earth cycle.

$$\ell_{\min}{}^2\left(\phi_{\ell_{\min}}\right) = L_{\langle \ell^2(\phi)\rangle} \quad ; \quad \ell_{\max}{}^2\left(\phi_{\max}\right) = L_{\langle \ell^2(\phi)\rangle} + f(\phi)_{\langle \ell^2(\phi)\rangle, Max} \quad ; \quad f(\phi)_{\langle \ell^2(\phi)\rangle, Min} = 0 \quad (78)$$

Working with the data in the Excel sheet "Tan, A.P., Asli Pinar Tan Analysis Based on Earth-Sun distance (d) Landsat.xlsx"[13], we observe that $\boldsymbol{k}_1(\phi) + \boldsymbol{k}_2(\phi)$ varies in a harmonic fashion between its maximum and minimum values expressed in (79) and (80). $\left[\boldsymbol{k}_1(\phi) + \boldsymbol{k}_2(\phi)\right]_{Max}$ (79) occurs on February 23 (DOY 54), May 26 (DOY 146), August 25 (DOY 237), and November 24 (DOY 328), whereas $\left[\boldsymbol{k}_1(\phi) + \boldsymbol{k}_2(\phi)\right]_{Min}$ (80) occurs on January 1 (DOY 1), April 2 (DOY 92), July 2 (DOY 183), and October 1 (DOY 274). It is also worth noting here as an observation that the occurence dates of $\left\{\sqrt{\left[\boldsymbol{k}_3(\phi)\right]^2 + \left[\boldsymbol{k}_4(\phi)\right]^2}\right\}_{Max}$ (68) and $\left[\boldsymbol{k}_1(\phi) + \boldsymbol{k}_2(\phi)\right]_{Min}$ (80) coincide with each other. Based on the definitions of $\left[\boldsymbol{k}_3(\phi)\right]^2 + \left[\boldsymbol{k}_4(\phi)\right]^2$ (52) and $\left[\boldsymbol{k}_1(\phi) + \boldsymbol{k}_2(\phi)\right]$(50), this observation allows us to conclude that the sum of squares of the projections of the constant $\vec{\boldsymbol{a}}$ (10) and $\vec{\boldsymbol{b}}$ (11) on varying $\vec{\ell}(\phi)$ (9), namely $\left[\vec{\boldsymbol{a}} \cdot \vec{\ell}(\phi)\right]^2 + \left[\vec{\boldsymbol{b}} \cdot \vec{\ell}(\phi)\right]^2$, takes on its maximum value when the magnitude of $\ell(\phi)$ (9) is at its minimum value $\ell_{\min}\left(\phi_{\ell_{\min}}\right)$ (78).

$$\left[\boldsymbol{k}_1(\phi) + \boldsymbol{k}_2(\phi)\right]_{Max} = \mathbf{2.00089354882923}\ au^2 \ \textit{on}\ 23\ \textit{February}, 26\ \textit{May}, 25\ \textit{August}, 24\ \textit{November} \quad (79)$$

$$\left[\boldsymbol{k}_1(\phi) + \boldsymbol{k}_2(\phi)\right]_{Min} = \mathbf{2.00087376210173}\ au^2 \quad \textit{on} \quad 1\ \textit{January}, 2\ \textit{April}, 2\ \textit{July}, 1\ \textit{October} \quad (80)$$

The above definitions and analysis in (76) - (80) would enable us to conclude that the constant $\left[2b^2 + 2r_1 r_2\left(1 - Cos\beta\right)Cos\phi_0 + 2L_{\langle \ell^2(\phi)\rangle}\right]$ (81) component of $\left[\boldsymbol{k}_1(\phi) + \boldsymbol{k}_2(\phi)\right]$ (50) can be determined by the value of $\left[\boldsymbol{k}_1(\phi) + \boldsymbol{k}_2(\phi)\right]_{Min}$ (80), and that its variable $\left[2f(\phi)_{\langle \ell^2(\phi)\rangle}\right]$ (82) component varies in a range from 0 (78) to $\left[2f(\phi)_{\langle \ell^2(\phi)\rangle, Max}\right]$ (83), which is the difference of $\left[\boldsymbol{k}_1(\phi) + \boldsymbol{k}_2(\phi)\right]_{Max}$ (79) and $\left[\boldsymbol{k}_1(\phi) + \boldsymbol{k}_2(\phi)\right]_{Min}$ (80), varying in a cyclic fashion over all ϕ for the Sun-Earth system, leading to the results in (81) - (84).

$$b^2 + r_1 r_2\left(1 - Cos\beta\right)Cos\phi_0 + L_{\langle \ell^2(\phi)\rangle} = \frac{\left[\boldsymbol{k}_1(\phi) + \boldsymbol{k}_2(\phi)\right]_{Min}}{2} = \mathbf{1.00043688105086}\ au^2 \quad \textit{for all}\ \phi \quad (81)$$

$$\begin{aligned} f(\phi)_{\langle \ell^2(\phi)\rangle} &= \frac{\left\{\left[\boldsymbol{k}_1(\phi) + \boldsymbol{k}_2(\phi)\right] - \left[\boldsymbol{k}_1(\phi) + \boldsymbol{k}_2(\phi)\right]_{Min}\right\}}{2} && \textit{over all}\ \phi \\ &= \frac{\left\{\left[\boldsymbol{k}_1(\phi) + \boldsymbol{k}_2(\phi)\right] - \mathbf{2.00087376210173}\ au^2\right\}}{2} && \textit{over all}\ \phi \end{aligned} \quad (82)$$

$$f(\phi)_{\langle \ell^2(\phi)\rangle, Max} = \frac{\left\{\left[\boldsymbol{k}_1(\phi) + \boldsymbol{k}_2(\phi)\right]_{Max} - \left[\boldsymbol{k}_1(\phi) + \boldsymbol{k}_2(\phi)\right]_{Min}\right\}}{2} = \mathbf{0.00000989336375}\ au^2 \quad (83)$$

$$Range\ of\ f(\phi)_{\langle \ell^2(\phi)\rangle} \quad : \quad \{\mathbf{0}\ \ to\ \ \mathbf{0.00000989336375}\, au^2\} \tag{84}$$

The variable component $f(\phi)_{\langle \ell^2(\phi)\rangle}$ (82) of $\ell^2(\phi)$ (76) for the Sun-Earth system is calculated for all ϕ based on (82), and can be found in the Excel sheet "Tan, A.P., Asli Pinar Tan Analysis Based on Earth-Sun distance (d) Landsat.xlsx"[13]. Subtracting the variable component $f(\phi)_{\langle \ell^2(\phi)\rangle}$ (82) of $\ell^2(\phi)$ (76) from $\boldsymbol{k}_1(\phi)$ (46) and $\boldsymbol{k}_2(\phi)$ (47), the obtained function values in (85) and (86), respectively, are also calculated over all ϕ for the Sun-Earth system in the Excel sheet "Tan, A.P., Asli Pinar Tan Analysis Based on Earth-Sun distance (d) Landsat.xlsx"[13].

$$\begin{aligned}\boldsymbol{k}_1(\phi) - f(\phi)_{\langle \ell^2(\phi)\rangle} &= b^2 + 2r_1 r_2 (1-Cos\beta) Cos\phi\, Cos(\phi+\phi_0) + \ell(\phi)^2 - f(\phi)_{\langle \ell^2(\phi)\rangle} \\ &= b^2 + L_{\langle \ell^2(\phi)\rangle} + 2r_1 r_2 (1-Cos\beta) Cos\phi\, Cos(\phi+\phi_0)\end{aligned} \tag{85}$$

$$\begin{aligned}\boldsymbol{k}_2(\phi) - f(\phi)_{\langle \ell^2(\phi)\rangle} &= b^2 + 2r_1 r_2 (1-Cos\beta) Sin\phi\, Sin(\phi+\phi_0) + \ell(\phi)^2 - f(\phi)_{\langle \ell^2(\phi)\rangle} \\ &= b^2 + L_{\langle \ell^2(\phi)\rangle} + 2r_1 r_2 (1-Cos\beta) Sin\phi\, Sin(\phi+\phi_0)\end{aligned} \tag{86}$$

Based on (85) and (86), $\left[\boldsymbol{k}_1(\phi) - f(\phi)_{\langle \ell^2(\phi)\rangle}\right]_{Min-Max}$ and $\left[\boldsymbol{k}_2(\phi) - f(\phi)_{\langle \ell^2(\phi)\rangle}\right]_{Min-Max}$ occur as in (87) - (88) at $\left(\phi = \frac{n\pi}{2} - \frac{\phi_0}{2}\right)$, where n is any integer, and they both take the same maximum and minimum values in (89) and (90), respectively, $\left(\phi = \frac{\pi}{2}\right)$ apart from each other.

$$\left[\boldsymbol{k}_1(\phi) - f(\phi)_{\langle \ell^2(\phi)\rangle}\right]_{Max} \ \&\ \left[\boldsymbol{k}_2(\phi) - f(\phi)_{\langle \ell^2(\phi)\rangle}\right]_{Min} \quad occur\ \ at\ \ \phi = \frac{n\pi}{2} - \frac{\phi_0}{2} \ ;\ n\ even\ integer \tag{87}$$

$$\left[\boldsymbol{k}_1(\phi) - f(\phi)_{\langle \ell^2(\phi)\rangle}\right]_{Min} \ \&\ \left[\boldsymbol{k}_2(\phi) - f(\phi)_{\langle \ell^2(\phi)\rangle}\right]_{Max} \quad occur\ \ at\ \ \phi = \frac{n\pi}{2} - \frac{\phi_0}{2} \ ;\ n\ odd\ integer \tag{88}$$

$$\left[\boldsymbol{k}_1(\phi) - f(\phi)_{\langle \ell^2(\phi)\rangle}\right]_{Max} = \left[\boldsymbol{k}_2(\phi) - f(\phi)_{\langle \ell^2(\phi)\rangle}\right]_{Max} = b^2 + L_{\langle \ell^2(\phi)\rangle} + 2r_1 r_2 (1-Cos\beta) Cos^2\left(\frac{\phi_0}{2}\right) \tag{89}$$

$$\left[\boldsymbol{k}_1(\phi) - f(\phi)_{\langle \ell^2(\phi)\rangle}\right]_{Min} = \left[\boldsymbol{k}_2(\phi) - f(\phi)_{\langle \ell^2(\phi)\rangle}\right]_{Min} = b^2 + L_{\langle \ell^2(\phi)\rangle} - 2r_1 r_2 (1-Cos\beta) Sin^2\left(\frac{\phi_0}{2}\right) \tag{90}$$

Checking in the Excel sheet "Tan, A.P., Asli Pinar Tan Analysis Based on Earth-Sun distance (d) Landsat.xlsx"[13], $\left[\boldsymbol{k}_1(\phi) - f(\phi)_{\langle \ell^2(\phi)\rangle}\right]$ (85) varies between its maximum value $\left[\boldsymbol{k}_1(\phi) - f(\phi)_{\langle \ell^2(\phi)\rangle}\right]_{Max}$ (91) of $1.00059545349821\, au^2$, which occurs on April 2 (DOY 92) and October 1 (DOY 274), and its minimum value $\left[\boldsymbol{k}_1(\phi) - f(\phi)_{\langle \ell^2(\phi)\rangle}\right]_{Min}$ (92) of $1.00027830860352\, au^2$, which occurs on January 1 (DOY 1) and July 2 (DOY 183). Shifted by

$\phi = \frac{\pi}{2}$ apart, $\left[k_2(\phi) - f(\phi)_{\langle \ell^2(\phi)\rangle}\right]$ (86) varies between its maximum value $\left[k_2(\phi) - f(\phi)_{\langle \ell^2(\phi)\rangle}\right]_{Max}$ (93) of $1.00059545349821\, au^2$, which occurs on January 1 (DOY 1) and July 2 (DOY 183), and its minimum value $\left[k_2(\phi) - f(\phi)_{\langle \ell^2(\phi)\rangle}\right]_{Min}$ (94) of $1.00027830860352\, au^2$, which occurs on April 2 (DOY 92) and October 1 (DOY 274). Hence, along the yearly Sun-Earth cycle, for some ϕ, $\left[k_1(\phi) - f(\phi)_{\langle \ell^2(\phi)\rangle}\right]$ (85) and $\left[k_2(\phi) - f(\phi)_{\langle \ell^2(\phi)\rangle}\right]$ (86) take on the value of $\left[b^2 + r_1 r_2 (1 - Cos\beta) Cos\phi_0 + L_{\langle \ell^2(\phi)\rangle}\right]$ (81), which is $1.00043688105086\, au^2$.

$$\left[k_1(\phi) - f(\phi)_{\langle \ell^2(\phi)\rangle}\right]_{Max} = \mathbf{1.00059545349821}\, au^2 \quad on \quad 2\ April\ and\ 1\ October \tag{91}$$

$$\left[k_1(\phi) - f(\phi)_{\langle \ell^2(\phi)\rangle}\right]_{Min} = \mathbf{1.00027830860352}\, au^2 \quad on \quad 1\ January\ and\ 2\ July \tag{92}$$

$$\left[k_2(\phi) - f(\phi)_{\langle \ell^2(\phi)\rangle}\right]_{Max} = \mathbf{1.00059545349821}\, au^2 \quad on \quad 1\ January\ and\ 2\ July \tag{93}$$

$$\left[k_2(\phi) - f(\phi)_{\langle \ell^2(\phi)\rangle}\right]_{Min} = \mathbf{1.00027830860352}\, au^2 \quad on \quad 2\ April\ and\ 1\ October \tag{94}$$

Subtracting (90) from (89) and utilizing the values in (91) - (94), we obtain (95) - (96), which yield the same value obtained in (72) - (73) for $\left[2r_1 r_2 (1 - Cos\beta)\right]$.

$$\begin{aligned}&\left[k_1(\phi) - f(\phi)_{\langle \ell^2(\phi)\rangle}\right]_{Max} - \left[k_1(\phi) - f(\phi)_{\langle \ell^2(\phi)\rangle}\right]_{Min} \\ &\qquad = 2r_1 r_2 (1 - Cos\beta) = \mathbf{0.000317144894687038}\, au^2\end{aligned} \tag{95}$$

$$\begin{aligned}&\left[k_2(\phi) - f(\phi)_{\langle \ell^2(\phi)\rangle}\right]_{Max} - \left[k_2(\phi) - f(\phi)_{\langle \ell^2(\phi)\rangle}\right]_{Min} \\ &\qquad = 2r_1 r_2 (1 - Cos\beta) = \mathbf{0.000317144894687038}\, au^2\end{aligned} \tag{96}$$

Continuing our analysis, along the lines presented in the **ARTICLE 2** "Analytical solution for two circles in space using their elliptic distances", we define the unit vectors normal to the $Circle_1$ of the Sun and the $Circle_2$ of the Earth in the direction of their self-rotation axes[30,7] as $\hat{u}_{1\perp}$ (97) and $\hat{u}_{2\perp}$ (97), respectively, where the self-revolution axis and the self-rotation axis[30,7] of each body must be aligned along the normal of its equatorial[31] plane based on our assertion in the **ARTICLE 1** "Points on two circles in space mimic an ellipse with respect to a point", which are also the unit vectors pointing in the directions of the North Poles[28,29] of the Sun and the Earth along their rotational axes[7], respectively, and whose directions are determined according to 'right hand rule'[18] based on the direction of self-rotation of the body. The correct directions of $\hat{u}_{1\perp}$ (97) and $\hat{u}_{2\perp}$ (97) in the given topology are yet to be determined based on the Sun-Earth yearly data.

With $\vec{d}(\phi)$ (14) chosen to be directed from the Earth to the Sun, the Dot Product[14], or equivalently, the projection of $\left[-\vec{d}(\phi)\right]$ (14) on $\hat{u}_{1\perp}$ (97) and the projection of $\vec{d}(\phi)$ (14) on $\hat{u}_{2\perp}$ (97) would be as in (99) and (100), respectively, where $\gamma_1(\phi)=\gamma_{Sun}(\phi)$ (98) is the angle between $\hat{u}_{1\perp}$ (97) and $\left[-\vec{d}(\phi)\right]$ (14) and $\gamma_2(\phi)=\gamma_{Earth}(\phi)$ (98) is the angle between $\hat{u}_{2\perp}$ (97) and $\vec{d}(\phi)$ (14), at each ϕ, with $\gamma_{1,\Delta}(\phi)$ (98) defined as the tilt[7] angle between $\hat{u}_{1\perp}$ (97) and the component of $\hat{u}_{1\perp}$ (97) normal to $\left[-\vec{d}(\phi)\right]$ (14), and $\gamma_{2,\Delta}(\phi)$ (98) defined as the tilt[7] angle between $\hat{u}_{2\perp}$ (97) and the component of $\hat{u}_{2\perp}$ (97) normal to $\vec{d}(\phi)$ (14).

$$\hat{u}_{1\perp}=\pm\left(-\hat{x}Sin\,\beta+\hat{z}Cos\,\beta\right) \quad ; \quad \hat{u}_{2\perp}=\pm\hat{z} \tag{97}$$

$$\gamma_{Sun}(\phi)=\gamma_1(\phi)=\frac{\pi}{2}-\gamma_{1,\Delta}(\phi) \quad ; \quad \gamma_{Earth}(\phi)=\gamma_2(\phi)=\frac{\pi}{2}-\gamma_{2,\Delta}(\phi) \tag{98}$$

$$\begin{aligned}\left[-\vec{d}(\phi)\right]\cdot\hat{u}_{1\perp}&=\left|\vec{d}(\phi)\right|Cos\left[\gamma_1(\phi)\right]\\&=\begin{cases}-\left[r_2\,Cos\phi Sin\,\beta-\ell_x(\phi)Sin\,\beta+\ell_z(\phi)Cos\,\beta\right] & if \quad \hat{u}_{1\perp}=-\hat{x}Sin\,\beta+\hat{z}Cos\,\beta\\ r_2\,Cos\phi Sin\,\beta-\ell_x(\phi)Sin\,\beta+\ell_z(\phi)Cos\,\beta & if \quad \hat{u}_{1\perp}=-\left(-\hat{x}Sin\,\beta+\hat{z}Cos\,\beta\right)\end{cases}\end{aligned} \tag{99}$$

$$\vec{d}(\phi)\cdot\hat{u}_{2\perp}=\left|\vec{d}(\phi)\right|Cos\left[\gamma_2(\phi)\right]=\begin{cases} r_1Cos(\phi+\phi_0)Sin\,\beta+\ell_z(\phi) & if \quad \hat{u}_{2\perp}=\hat{z}\\ -[r_1Cos(\phi+\phi_0)Sin\,\beta+\ell_z(\phi)] & if \quad \hat{u}_{2\perp}=-\hat{z}\end{cases} \tag{100}$$

Considering the Earth's $\vec{d}(\phi)\cdot\hat{u}_{2\perp}$ (100) case, for instance, the maximum and minimum values that $\gamma_2(\phi)$ (98) would take over a relative Sun-Earth cycle, namely $\gamma_{2,\max}$ (101) and $\gamma_{2,\min}$ (101), respectively, can be geometrically visualized in the most simple form as in **Figure 13**, and they can also be expressed in terms of the tilt[7] angles γ_{2,Δ_1} (102) and γ_{2,Δ_2} (102) between the North Pole[28,29] vector $\hat{u}_{2\perp}$ (97) along the rotational axis[7] of the Earth and the component of $\hat{u}_{2\perp}$ (97) normal to $\vec{d}(\phi)$ (14) from the Earth to the Sun, at maximum and minimum relative tilt[7] of the Earth's North Pole[28,29] with respect to the Sun over a yearly cycle, respectively, which are observed equal for the Sun-Earth cycle and have the same $\gamma_{Earth,\Delta Max}$ (4) value. Based on observation, the maximum and minimum tilt[7] of the Earth occur at half a cycle, or around $\phi=\pi$ (103), apart, where $\phi=\phi_{Winter\ Solstice}=\phi_{\gamma_{2,\max}}$ (103) occurs at Winter Solstice[5,6] on December 21 and $\phi=\phi_{Summer\ Solstice}=\phi_{\gamma_{2,\min}}$ (103) occurs at Summer Solstice[5,6] on June 21, with $\vec{d}(\phi)$ (14) chosen to be directed from the Earth to the Sun. Note in **Figure 13** that at Earth's minimum tilt[7], P_2 represents the position of the Earth and P_1 represents the position of the Sun, whereas at Earth's maximum tilt[7], P_2' represents the position of the Earth and P_1' represents the position of the Sun. P_1 and P_1' are not co-located, but are sketched together in **Figure 13** just for simplicity of visualization.

$$\gamma_{2,\max} = \gamma_2\left(\phi_{\gamma_{2,\max}}\right) = \frac{\pi}{2} - \gamma_{2,\Delta}\left(\phi_{\gamma_{2,\max}}\right) = \frac{\pi}{2} + \gamma_{2,\Delta_1}$$
$$\gamma_{2,\min} = \gamma_2\left(\phi_{\gamma_{2,\min}}\right) = \frac{\pi}{2} - \gamma_{2,\Delta}\left(\phi_{\gamma_{2,\min}}\right) = \frac{\pi}{2} - \gamma_{2,\Delta_2} \quad (101)$$

$$-\gamma_{2,\Delta}\left(\phi_{\gamma_{2,\max}}\right) = \gamma_{2,\Delta_1} = \gamma_{2,\Delta_2} = \gamma_{2,\Delta}\left(\phi_{\gamma_{2,\min}}\right) = \gamma_{Earth,\Delta Max} = 23.44° \quad \textit{for the Sun-Earth cycle} \quad (102)$$

$$\phi_{Winter\ Solstice} = \phi_{\gamma_{2,\max}} = \pi + \phi_{\gamma_{2,\min}} = \pi + \phi_{Summer\ Solstice} \quad (103)$$

In similar lines, for the Sun's $\left[-\vec{d}(\phi)\right]\cdot\hat{u}_{1\perp}$ (99) case, the maximum and minimum values that $\gamma_1(\phi)$ (98) would take over a relative Sun-Earth cycle, namely $\gamma_{1,\max}$ (104) and $\gamma_{1,\min}$ (104), respectively, can be expressed in terms of the tilt[7] angles γ_{1,Δ_1} (104) and γ_{1,Δ_2} (104) between the North Pole[28] vector $\hat{u}_{1\perp}$ (97) of the Sun along the rotational axis[7] of the Sun and the normal component of $\hat{u}_{1\perp}$ (97) to the negative of the distance vector $\left[-\vec{d}(\phi)\right]$ (14) from the Sun to the Earth, at maximum and minimum relative tilt[7] of the Sun's North Pole[28] with respect to the Earth over a yearly cycle, respectively.

$$\gamma_{1,\max} = \frac{\pi}{2} + \gamma_{1,\Delta_1} \quad ; \quad \gamma_{1,\min} = \frac{\pi}{2} - \gamma_{1,\Delta_2} \quad (104)$$

At this point, for the purposes of our analytical method, based on our definition in (98) and along the lines of observation leading to (101) - (103), we can choose to make the approximation in (105) regarding the tilt[7] of the Earth, for all ϕ.

$$\gamma_{2,\Delta}(\pi+\phi) \simeq -\gamma_{2,\Delta}(\phi) \quad \Rightarrow \quad \begin{cases} \gamma_2(\phi) = \frac{\pi}{2} - \gamma_{2,\Delta}(\phi) \\ \gamma_2(\pi+\phi) = \frac{\pi}{2} - \gamma_{2,\Delta}(\pi+\phi) \simeq \frac{\pi}{2} + \gamma_{2,\Delta}(\phi) \end{cases} \quad \textit{for all } \phi \quad (105)$$

Based on (100) and (105), and also utilizing (45), we can solve for $\left[r_1 Cos(\phi+\phi_0)Sin\beta\right]$ (106) and $\ell_z(\phi)$ (107) analytically to an approximation, for the Sun-Earth system, for all ϕ.

$$r_1 Cos(\phi+\phi_0)Sin\beta \simeq \begin{cases} \frac{\left\{\left|\vec{d}(\phi)\right| + \left|\vec{d}(\pi+\phi)\right|\right\}}{2} Sin\left[\gamma_{2,\Delta}(\phi)\right] & if \quad \hat{u}_{2\perp} = \hat{z} \\ -\frac{\left\{\left|\vec{d}(\phi)\right| + \left|\vec{d}(\pi+\phi)\right|\right\}}{2} Sin\left[\gamma_{2,\Delta}(\phi)\right] & if \quad \hat{u}_{2\perp} = -\hat{z} \end{cases} \quad (106)$$

$$\ell_z(\phi) = \ell_z(\phi+\pi) \simeq \begin{cases} \frac{\left\{\left|\vec{d}(\phi)\right| - \left|\vec{d}(\pi+\phi)\right|\right\}}{2} Sin\left[\gamma_{2,\Delta}(\phi)\right] & if \quad \hat{u}_{2\perp} = \hat{z} \\ -\frac{\left\{\left|\vec{d}(\phi)\right| - \left|\vec{d}(\pi+\phi)\right|\right\}}{2} Sin\left[\gamma_{2,\Delta}(\phi)\right] & if \quad \hat{u}_{2\perp} = -\hat{z} \end{cases} \quad (107)$$

When we find the values of $\left[r_1 Sin\beta\right]$ and $Cos(\phi+\phi_0)$ for all ϕ of the Sun-Earth system later in this Article, we can also calculate the values of $Sin\left[\gamma_{2,\Delta}(\phi)\right]$ (108) to an approximation, obtained using (106).

$$Sin\left[\gamma_{2,\Delta}(\phi)\right]=\begin{cases}\dfrac{2r_1 Sin\beta Cos(\phi+\phi_0)}{\left\{\left|\vec{d}(\phi)\right|+\left|\vec{d}(\pi+\phi)\right|\right\}} & if \quad \hat{u}_{2\perp}=\hat{z}\\ -\dfrac{2r_1 Sin\beta Cos(\phi+\phi_0)}{\left\{\left|\vec{d}(\phi)\right|+\left|\vec{d}(\pi+\phi)\right|\right\}} & if \quad \hat{u}_{2\perp}=-\hat{z}\end{cases} \tag{108}$$

At the Spring Equinox[9] (at $\phi=\phi_{Equinox[E1]}$) and Autumn Equinox[9] (at $\phi=\phi_{Equinox[E2]}$) of the Earth with respect to the Sun, the rotational axis[7] of the Earth is approximately perpendicular to the line joining the centers of the Earth and Sun, in which case, based on (100), the projection of the distance vector $\vec{d}\left(\phi_{Equinox[E1]}\right)$ on $\hat{u}_{2\perp}$ would be zero at $\phi=\phi_{Equinox[E1]}$ and the projection of the distance vector $\vec{d}\left(\phi_{Equinox[E2]}\right)$ on $\hat{u}_{2\perp}$ would be zero at $\phi=\phi_{Equinox[E2]}$, which is also expressed in (109) - (110), leading to (111).

$$\begin{aligned}\vec{d}\left(\phi_{Equinox[E1]}\right)\cdot\hat{u}_{2\perp}&=\pm\left\{r_1 Cos\left(\phi_{Equinox[E1]}+\phi_0\right)Sin\beta+\ell_z\left(\phi_{Equinox[E1]}\right)\right\}\\&=\left|\vec{d}\left(\phi_{Equinox[E1]}\right)\right|Cos\left[\gamma_2\left(\phi_{Equinox[E1]}\right)\right]=0\end{aligned} \tag{109}$$

$$\begin{aligned}\vec{d}\left(\phi_{Equinox[E2]}\right)\cdot\hat{u}_{2\perp}&=\pm\left\{r_1 Cos\left(\phi_{Equinox[E2]}+\phi_0\right)Sin\beta+\ell_z\left(\phi_{Equinox[E2]}\right)\right\}\\&=\left|\vec{d}\left(\phi_{Equinox[E2]}\right)\right|Cos\left[\gamma_2\left(\phi_{Equinox[E2]}\right)\right]=0\end{aligned} \tag{110}$$

$$\Rightarrow \quad \gamma_2\left(\phi_{Equinox[E1]}\right)\simeq\pm\frac{\pi}{2} \quad and \quad \gamma_2\left(\phi_{Equinox[E2]}\right)\simeq\pm\frac{\pi}{2} \tag{111}$$

Based on (97) - (98) and (100) - (102), when the minimum tilt of the Earth occurs at Summer Solstice[5,6] on June 21 with $\phi=\phi_{\gamma_{2,\min}}$ (103) and the maximum tilt of the Earth occurs at Winter Solstice[5,6] on December 21 with $\phi=\phi_{\gamma_{2,\max}}=\pi+\phi_{\gamma_{2,\min}}$ (103), also utilizing (45), one would obtain (112) - (114).

$$\vec{\ell}(\phi)=\vec{\ell}(\pi+\phi) \quad \Rightarrow \quad \ell_z\left(\phi_{\gamma_{2,\max}}\right)=\ell_z\left(\phi_{\gamma_{2,\min}}\right) \tag{112}$$

$$\vec{d}\left(\phi_{\gamma_{2,\max}}\right)\cdot\hat{u}_{2\perp}=-\left|\vec{d}\left(\phi_{\gamma_{2,\max}}\right)\right|Sin\left(\gamma_{Earth,\Delta Max}\right)=\begin{cases}-r_1 Cos\left(\phi_{\gamma_{2,\min}}+\phi_0\right)Sin\beta+\ell_z\left(\phi_{\gamma_{2,\min}}\right) if \ \hat{u}_{2\perp}=\hat{z}\\ -[-r_1 Cos\left(\phi_{\gamma_{2,\min}}+\phi_0\right)Sin\beta+\ell_z\left(\phi_{\gamma_{2,\min}}\right)] if \ \hat{u}_{2\perp}=-\hat{z}\end{cases} \tag{113}$$

$$\vec{d}\left(\phi_{\gamma_{2,\min}}\right)\cdot\hat{u}_{2\perp}=\left|\vec{d}\left(\phi_{\gamma_{2,\min}}\right)\right|Sin\left(\gamma_{Earth,\Delta Max}\right)=\begin{cases}r_1 Cos\left(\phi_{\gamma_{2,\min}}+\phi_0\right)Sin\beta+\ell_z\left(\phi_{\gamma_{2,\min}}\right) & if \quad \hat{u}_{2\perp}=\hat{z}\\ -[r_1 Cos\left(\phi_{\gamma_{2,\min}}+\phi_0\right)Sin\beta+\ell_z\left(\phi_{\gamma_{2,\min}}\right)] & if \quad \hat{u}_{2\perp}=-\hat{z}\end{cases} \tag{114}$$

The values for $\left|\vec{d}\left(\phi_{\gamma_{2,\max}}\right)\right|$ (115) and $\left|\vec{d}\left(\phi_{\gamma_{2,\min}}\right)\right|$ (116) are taken from the data in the Excel sheet "Tan, A.P., Asli Pinar Tan Analysis Based on Earth-Sun distance (d) Landsat.xlsx"[13], which are 730 data points apart, or $\phi=\pi$ (103) apart, for December 21 (DOY 355) and June 21 (DOY 172), respectively.

$$\left|\vec{d}\left(\phi_{\gamma_{2,\max}}\right)\right| = 0.9837613\,au \qquad on\ December\ 21\ \left(DOY\ 355\right) \tag{115}$$

$$\left|\vec{d}\left(\phi_{\gamma_{2,\min}}\right)\right| = 1.01627775\,au \qquad on\ June\ 21\left(DOY\ 172\right) \tag{116}$$

Utilizing the values of $\gamma_{Earth,\Delta Max}$ (4), $\left|\vec{d}\left(\phi_{\gamma_{2,\max}}\right)\right|$ (115), and $\left|\vec{d}\left(\phi_{\gamma_{2,\min}}\right)\right|$ (116) with (105) - (107), or equivalenty subtracting (113) from (114), we would obtain the result in (117) - (118), whereas (119) - (120) would be obtained from the sum of (113) and (114), also taking into account (112).

$$r_1 Cos\left(\phi_{\gamma_{2,\min}}+\phi_0\right) Sin\,\beta = \begin{cases} \dfrac{\left\{\left|\vec{d}\left(\phi_{\gamma_{2,\max}}\right)\right|+\left|\vec{d}\left(\phi_{\gamma_{2,\min}}\right)\right|\right\}}{2} Sin\left(\gamma_{Earth,\Delta Max}\right) & if \quad \hat{u}_{2\perp}=\hat{z} \\ -\dfrac{\left\{\left|\vec{d}\left(\phi_{\gamma_{2,\max}}\right)\right|+\left|\vec{d}\left(\phi_{\gamma_{2,\min}}\right)\right|\right\}}{2} Sin\left(\gamma_{Earth,\Delta Max}\right) & if \quad \hat{u}_{2\perp}=-\hat{z} \end{cases} \tag{117}$$

$$r_1 Cos\left(\phi_{\gamma_{2,\min}}+\phi_0\right) Sin\,\beta = \begin{cases} \mathbf{0.397796274218557}\,au & if \quad \hat{u}_{2\perp}=\hat{z} \\ \mathbf{-0.397796274218557}\,au & if \quad \hat{u}_{2\perp}=-\hat{z} \end{cases} \tag{118}$$

$$\ell_z\left(\phi_{\gamma_{2,\max}}\right)=\ell_z\left(\phi_{\gamma_{2,\min}}\right)= \begin{cases} \dfrac{\left\{\left|\vec{d}\left(\phi_{\gamma_{2,\max}}\right)\right|-\left|\vec{d}\left(\phi_{\gamma_{2,\min}}\right)\right|\right\}}{2} Sin\left(\gamma_{Earth,\Delta Max}\right) & if \quad \hat{u}_{2\perp}=\hat{z} \\ -\dfrac{\left\{\left|\vec{d}\left(\phi_{\gamma_{2,\max}}\right)\right|-\left|\vec{d}\left(\phi_{\gamma_{2,\min}}\right)\right|\right\}}{2} Sin\left(\gamma_{Earth,\Delta Max}\right) & if \quad \hat{u}_{2\perp}=-\hat{z} \end{cases} \tag{119}$$

$$\ell_z\left(\phi_{\gamma_{2,\max}}\right)=\ell_z\left(\phi_{\gamma_{2,\min}}\right)= \begin{cases} \mathbf{0.00646733505568997}\,au & if \quad \hat{u}_{2\perp}=\hat{z} \\ \mathbf{-0.00646733505568997}\,au & if \quad \hat{u}_{2\perp}=-\hat{z} \end{cases} \tag{120}$$

Based on (113) - (114), the maximum and minimum tilt of the Earth over a Sun-Earth cycle would be expected to occur at the maximum $(\pm)$ swing of $\left[r_1\,Cos(\phi+\phi_0)\,Sin\,\beta\right]$ from $\ell_z(\phi)$, which occurs when $\phi=\phi_{\gamma_{2,\min}}$ (103) and $\phi=\phi_{\gamma_{2,\max}}=\pi+\phi_{\gamma_{2,\min}}$ (103). Making the observation that the variation range (84) of the variable component $f(\phi)_{\langle\ell^2(\phi)\rangle}$ (82) of $\ell^2(\phi)$ (76) is very small compared to the found absolute value of $\left|r_1 Cos\left(\phi_{\gamma_{2,\min}}+\phi_0\right) Sin\,\beta\right|$ (118), we can infer that the ϕ values for which the occurrence of (113) - (114), and therefore (118), happen at the maximum and minimum tilt of the Earth, are determined by the ϕ values when the maximum and minimum values of $\left[r_1 Cos(\phi+\phi_0)\,Sin\,\beta\right]$ (121) are attained at $\phi=-\phi_0$ and $\phi=\pi-\phi_0$ (122), and not determined by the variation of $\left|\ell_z(\phi)\right|$, which allows us to determine the value of $\pm r_1 Sin\,\beta$ (121) as well, at this point. Based on this important inference, we can further make the analyses in

(123) - (124), and combining these results with those in (118) and (122), we would further obtain the analysis result in (125). Subsequently, we can reach the conclusion that the value of (114) for the minimum tilt of the Earth at Summer Solstice[5,6] on June 21 occurs either at $\phi_{\gamma_{2,\min}} = -\phi_0$ (126) if $Sin\beta \geq 0$ (126) or at $\phi_{\gamma_{2,\min}} = \pi - \phi_0$ (126) if $Sin\beta \leq 0$ (126) when $\hat{\boldsymbol{u}}_{2\perp} = \hat{\boldsymbol{z}}$, and when $\hat{\boldsymbol{u}}_{2\perp} = -\hat{\boldsymbol{z}}$ it occurs at $\phi_{\gamma_{2,\min}} = \pi - \phi_0$ (126) if $Sin\beta \geq 0$ (126) or at $\phi_{\gamma_{2,\min}} = -\phi_0$ (126) if $Sin\beta \leq 0$ (126). Similarly, the value of of (113) for its maximum tilt at Winter Solstice[5,6] on December 21 would occur either at $\phi_{\gamma_{2,\max}} = \pi - \phi_0$ (126) if $Sin\beta \geq 0$ (126) or at $\phi_{\gamma_{2,\max}} = -\phi_0$ (126) if $Sin\beta \leq 0$ (126) when $\hat{\boldsymbol{u}}_{2\perp} = \hat{\boldsymbol{z}}$, and when $\hat{\boldsymbol{u}}_{2\perp} = -\hat{\boldsymbol{z}}$ it occurs at $\phi_{\gamma_{2,\max}} = -\phi_0$ (126) if $Sin\beta \geq 0$ (126) or at $\phi_{\gamma_{2,\max}} = \pi - \phi_0$ (126) if $Sin\beta \leq 0$ (126).

$$\left[r_1 Cos(\phi + \phi_0) Sin\beta\right]_{Max-Min} = \pm r_1 Sin\beta = \pm\mathbf{0.397796274218557}\ au \tag{121}$$

$$\left[r_1 Cos(\phi + \phi_0) Sin\beta\right]_{Max-Min} = \pm r_1 Sin\beta \quad at \quad Cos(\phi + \phi_0) = \pm 1 \;\Rightarrow\; \phi = -\phi_0 \quad or \quad \phi = \pi - \phi_0 \tag{122}$$

$$\left[r_1 Cos(\phi + \phi_0) Sin\beta\right]_{Max} = \begin{cases} r_1 Sin\beta & at \quad Cos(\phi + \phi_0) = 1 \Rightarrow Sin\beta \geq 0 \;\&\; \phi = -\phi_0 \\ -r_1 Sin\beta & at \quad Cos(\phi + \phi_0) = -1 \Rightarrow Sin\beta \leq 0 \;\&\; \phi = \pi - \phi_0 \end{cases} \tag{123}$$
$$= \mathbf{0.397796274218557}\ au$$

$$\left[r_1 Cos(\phi + \phi_0) Sin\beta\right]_{Min} = \begin{cases} -r_1 Sin\beta & at \quad Cos(\phi + \phi_0) = -1 \Rightarrow Sin\beta \geq 0 \;\&\; \phi = \pi - \phi_0 \\ r_1 Sin\beta & at \quad Cos(\phi + \phi_0) = 1 \Rightarrow Sin\beta \leq 0 \;\&\; \phi = -\phi_0 \end{cases} \tag{124}$$
$$= -\mathbf{0.397796274218557}\ au$$

$$r_1 Sin\beta = \begin{cases} \mathbf{0.397796274218557}\ au \Rightarrow Sin\beta \geq 0 \begin{cases} \hat{\boldsymbol{u}}_{2\perp} = \hat{\boldsymbol{z}} \Rightarrow Cos\left(\phi_{\gamma_{2,\min}} + \phi_0\right) = 1 \Rightarrow \phi_{\gamma_{2,\min}} = -\phi_0 \\ \hat{\boldsymbol{u}}_{2\perp} = -\hat{\boldsymbol{z}} \Rightarrow Cos\left(\phi_{\gamma_{2,\max}} + \phi_0\right) = 1 \Rightarrow \phi_{\gamma_{2,\max}} = -\phi_0 \end{cases} \\ -\mathbf{0.397796274218557}\ au \Rightarrow Sin\beta \leq 0 \begin{cases} \hat{\boldsymbol{u}}_{2\perp} = \hat{\boldsymbol{z}} \Rightarrow Cos\left(\phi_{\gamma_{2,\max}} + \phi_0\right) = 1 \Rightarrow \phi_{\gamma_{2,\max}} = -\phi_0 \\ \hat{\boldsymbol{u}}_{2\perp} = -\hat{\boldsymbol{z}} \Rightarrow Cos\left(\phi_{\gamma_{2,\min}} + \phi_0\right) = 1 \Rightarrow \phi_{\gamma_{2,\min}} = -\phi_0 \end{cases} \end{cases} \tag{125}$$

$$r_1 Sin\beta = \begin{cases} \mathbf{0.397796274218557}\ au \Rightarrow Sin\beta \geq 0 \begin{cases} \hat{\boldsymbol{u}}_{2\perp} = \hat{\boldsymbol{z}} \Rightarrow \phi_{\gamma_{2,\min}} = -\phi_0 \;\&\; \phi_{\gamma_{2,\max}} = \pi - \phi_0 \\ \hat{\boldsymbol{u}}_{2\perp} = -\hat{\boldsymbol{z}} \Rightarrow \phi_{\gamma_{2,\min}} = \pi - \phi_0 \;\&\; \phi_{\gamma_{2,\max}} = -\phi_0 \end{cases} \\ -\mathbf{0.397796274218557}\ au \Rightarrow Sin\beta \leq 0 \begin{cases} \hat{\boldsymbol{u}}_{2\perp} = \hat{\boldsymbol{z}} \Rightarrow \phi_{\gamma_{2,\min}} = \pi - \phi_0 \;\&\; \phi_{\gamma_{2,\max}} = -\phi_0 \\ \hat{\boldsymbol{u}}_{2\perp} = -\hat{\boldsymbol{z}} \Rightarrow \phi_{\gamma_{2,\min}} = -\phi_0 \;\&\; \phi_{\gamma_{2,\max}} = \pi - \phi_0 \end{cases} \end{cases} \tag{126}$$

As we now know that the phase angle is either $\phi = \phi_{\gamma_{2,\min}} = -\phi_0$ (126) or $\phi = \phi_{\gamma_{2,\min}} = \pi - \phi_0$ (126) at Summer Solstice[5,6] on June 21, or it is either $\phi = \phi_{\gamma_{2,\max}} = -\phi_0$ (126) or $\phi = \phi_{\gamma_{2,\max}} = \pi - \phi_0$ (126) at Winter Solstice[5,6] on December 21, we find out that $\boldsymbol{k}_1(\phi) - \boldsymbol{k}_2(\phi)$ (51) becomes $2r_1 r_2\left(1 - Cos\beta\right)Cos\phi_0$ as in (127) on those dates, in either case. We can thus obtain the previously calculated value of $\boldsymbol{k}_1(\phi_{\gamma_{2,\min}}) - \boldsymbol{k}_2(\phi_{\gamma_{2,\min}})$ (128) for the data point on June 21 or

$\boldsymbol{k}_1(\phi_{\gamma_{2,max}})-\boldsymbol{k}_2(\phi_{\gamma_{2,max}})$ (128) for the data point on December 21 from the Excel sheet "Tan, A.P., Asli Pinar Tan Analysis Based on Earth-Sun distance (d) Landsat.xlsx"[13], which would correspond to either $[\boldsymbol{k}_1(-\phi_0)-\boldsymbol{k}_2(-\phi_0)]$ (128) or $[\boldsymbol{k}_1(\pi-\phi_0)-\boldsymbol{k}_2(\pi-\phi_0)]$ (128), to be determined later. This would provide us the value of $\left[2r_1 r_2\left(1-Cos\beta\right)Cos\,\phi_0\right]$ (128).

$$\begin{aligned}2r_1 r_2\left(1-Cos\beta\right)Cos\left(2\phi_{\gamma_{2,min}}+\phi_0\right)&=2r_1 r_2\left(1-Cos\beta\right)Cos\left(2\phi_{\gamma_{2,max}}+\phi_0\right)\\&=\boldsymbol{k}_1(\phi=-\phi_0)-\boldsymbol{k}_2(\phi=-\phi_0)=\boldsymbol{k}_1(\phi=\pi-\phi_0)-\boldsymbol{k}_2(\phi=\pi-\phi_0)\quad(127)\\&=2r_1 r_2\left(1-Cos\beta\right)Cos\left(2\phi+\phi_0\right)=2r_1 r_2\left(1-Cos\beta\right)Cos\,\phi_0\end{aligned}$$

$$\begin{aligned}\boldsymbol{k}_1(-\phi_0)-\boldsymbol{k}_2(-\phi_0)&=\boldsymbol{k}_1(\pi-\phi_0)-\boldsymbol{k}_2(\pi-\phi_0)=2r_1 r_2\left(1-Cos\beta\right)Cos\,\phi_0\\&=\boldsymbol{k}_1(\phi_{\gamma_{2,min}})-\boldsymbol{k}_2(\phi_{\gamma_{2,min}})=\boldsymbol{k}_1(\phi_{\gamma_{2,max}})-\boldsymbol{k}_2(\phi_{\gamma_{2,max}})=\mathbf{-0.000277676302109064}\,au^2\end{aligned}\quad(128)$$

Dividing $\left[2r_1 r_2\left(1-Cos\beta\right)Cos\,\phi_0\right]$ (128) by $\left[2r_1 r_2\left(1-Cos\beta\right)\right]$ (72), we obtain the value of $Cos\,\phi_0$ (129) for the Sun-Earth system, as is a significant finding of this Article, which also reveals the value of ϕ_0 (130), one of the main parameters of the relative elliptic motion of the Sun-Earth system, that can either be the positive or negative of this value, which we will also determine a little later.

$$\boldsymbol{Cos\,\phi_0}=\mathbf{-0.875550282412959}\quad(129)$$

$$\phi_0=\mathbf{\pm 2.63736997198951}\,radians=\mathbf{\pm 151^\circ.11}\quad(130)$$

In addition to this, based on the **ARTICLE 1** "Points on two circles in space mimic an ellipse with respect to a point" and (10) - (11), as $\left(Cos\,\phi_0<0\right)$ (129), $\left(a^2<b^2\right)$ (131), which indicates that $\vec{\boldsymbol{a}}$ (10) is the semi-minor[23] axis vector and $\vec{\boldsymbol{b}}$ (11) is the semi-major[23] axis vector of the ellipse[22] of the Sun-Earth system formed by their relative motions, yet another finding of this Article, with the value of $\left(a^2-b^2\right)$ found as in (132) obtained using (128).

$$Cos\,\phi_0<0\quad\Rightarrow\quad a^2-b^2=2r_1 r_2\left(1-Cos\beta\right)Cos\,\phi_0<0\quad\Rightarrow\quad a^2<b^2\quad(131)$$

$$a^2-b^2=2r_1 r_2\left(1-Cos\beta\right)Cos\,\phi_0=\mathbf{-0.000277676302109064}\,au^2\quad(132)$$

We are thus able to determine the value of the focal[22,24] distance squared c^2 (133) now from (12) and (132), which allows us to find the value of the focal[22,24] distance c (134) for the relative ellipse[22] of the Sun-Earth system, which is another one of the main parameters of this Sun-Earth relative ellipse[22] , and is found to be $\mathbf{2.492.842,4}\,km$ using (3). Hence, the average Sun-Earth distance of $1\,au$ (3) turns out to be about 60 times (135) the focal[22,24] distance c (134) of this ellipse[22].

$$c^2=\left|a^2-b^2\right|=2r_1 r_2\left(1-Cos\beta\right)\left|Cos\,\phi_0\right|=\mathbf{0.000277676302109064}\,au^2\quad(133)$$

$$c=\mathbf{0.0166636221185271}\,au=\mathbf{2{,}492{,}842.4}\,km\quad(134)$$

$$1/c=\mathbf{60{,}01}\,au^{-1}\quad(135)$$

At this point, with the value of $\left(\phi_0 = \pm\mathbf{2.63736997198951}\,rad\right)$ (130) found for the Sun-Earth system, utilizing (72) and the values of $\boldsymbol{k}_1(\phi)-\boldsymbol{k}_2(\phi)$ (51) for all ϕ, the value of ϕ (136) corresponding to each data point is calculated in the Excel sheet "Tan, A.P., Asli Pinar Tan Analysis Based on Earth-Sun distance (d) Landsat.xlsx"[13], in *increasing order* according to the configuration suggested in the **ARTICLE 1** "Points on two circles in space mimic an ellipse with respect to a point", for both the positive and negative value possibilities of ϕ_0 (130).

$$\phi = \frac{Cos^{-1}\left[\dfrac{\boldsymbol{k}_1(\phi)-\boldsymbol{k}_2(\phi)}{2r_1 r_2\left(1-Cos\beta\right)}\right]-\phi_0}{2} \qquad \textit{for all } \phi\textit{, increasing order} \qquad (136)$$

Checking these ϕ (136) series over a Sun-Earth cycle, calculated for both the positive and negative value possibilities of ϕ_0 (130) in the Excel sheet "Tan, A.P., Asli Pinar Tan Analysis Based on Earth-Sun distance (d) Landsat.xlsx"[13], we observe the clear result in (137). It reveals that if $\left(\phi_0 = \mathbf{2.63736997198951}\,rad = \mathbf{151^\circ.11}\right)$, $\phi = 0$ at Summer Solstice[5,6] on June 21 and $\phi = \pi$ at Winter Solstice[5,6] on December 21, indicating that $\left(\phi_0 = \mathbf{2.63736997198951}\,rad = \mathbf{151^\circ.11}\right)$ <u>cannot</u> be a valid solution according to our previous conclusion that either $\phi = \phi_{\gamma_{2,\min}} = -\phi_0$ (126) or $\phi = \phi_{\gamma_{2,\min}} = \pi - \phi_0$ (126) must occur at Summer Solstice[5,6] on June 21, whereas $\phi_{\gamma_{2,\max}} = \pi + \phi_{\gamma_{2,\min}}$ (103) at Winter Solstice[5,6] on December 21. On the other hand, when $\left(\phi_0 = \mathbf{-2.63736997198951}\,rad = \mathbf{-151^\circ.11}\right)$ (138), the calculations reveal that $\phi = -\phi_0$ (139) at Summer Solstice[5,6] on June 21 and $\phi = \pi - \phi_0$ (140) at Winter Solstice[5,6] on December 21, indicating that $\left(\phi_0 = \mathbf{-2.63736997198951}\,rad = \mathbf{-151^\circ.11}\right)$ (138) must be the valid solution for the Sun-Earth system.

$$\phi_0 = \begin{cases} \mathbf{2.63736997198951}\,radians = \mathbf{151^\circ.11} \Rightarrow \phi(June\ 21) = 0\,;\ \phi(December\ 21) = \pi \\ \mathbf{-2.63736997198951}\,radians = \mathbf{-151^\circ.11} \Rightarrow \phi(June\ 21) = -\phi_0\,;\ \phi(December\ 21) = \pi - \phi_0 \end{cases} \qquad (137)$$

We have thus determined another one of the main parameters of the Sun-Earth system ellipse, namely the value of ϕ_0 (138), which indicates that in its own cycle of individual revolution, the Sun is lagging with a phase difference of $\left(\phi_0 = \mathbf{-2.63736997198951}\,rad = \mathbf{-151^\circ.11}\right)$ (138) behind Earth in its own cycle of individual revolution, or equivalently, the Earth in its own cycle of individual revolution is moving ahead of the Sun in its own cycle of individual revolution, by a phase difference of $\left(-\phi_0 = \mathbf{2.63736997198951}\,rad = \mathbf{151^\circ.11}\right)$. This conclusion is another one of the fundamental findings of this Article.

$$\phi_0 = \mathbf{-2.63736997198951}\,rad = \mathbf{-151^\circ.11} \qquad \textit{for the Sun} - \textit{Earth system} \qquad (138)$$

$$\phi_{Summer\ Solstice} = \phi_{\gamma_{2,\min}} = -\phi_0 = \mathbf{2.63736997198951}\,rad = \mathbf{151^\circ.11} \qquad \textit{on June } 21 \qquad (139)$$

$$\phi_{Winter\ Solstice} = \phi_{\gamma_{2,\max}} = \pi - \phi_0 = \mathbf{5.7789626255793}\,rad = \mathbf{331^\circ.11} = \mathbf{-28^\circ.89} \qquad \textit{on December } 21 \qquad (140)$$

Furthermore, analyzing (126) again in light of the result we have obtained in (138) - (140), we are now able to narrow the analysis in (126) down to the conclusion in (141).

$$\phi_{\gamma_{2,\min}} = -\phi_0 \,\&\, \phi_{\gamma_{2,\max}} = \pi - \phi_0 \begin{cases} Sin\beta \geq 0 \Rightarrow \hat{\boldsymbol{u}}_{2\perp} = \hat{z} \,\&\, r_1\, Sin\beta = \mathbf{0.397796274218557}\ au \\ Sin\beta \leq 0 \Rightarrow \hat{\boldsymbol{u}}_{2\perp} = -\hat{z} \,\&\, r_1\, Sin\beta = \mathbf{-0.397796274218557}\ au \end{cases} \quad (141)$$

Having determined the value of ϕ_0 (138), we can also calculate $Sin\phi_0$ (142), and using this, the value of the parameter $\vec{\boldsymbol{a}}\cdot\vec{\boldsymbol{b}}$ (143) for the Sun-Earth system, another a finding of this Article, which is the Dot Product[14] of the semi-minor[23] axis vector $\vec{\boldsymbol{a}}$ (10) with the semi-major[23] axis vector $\vec{\boldsymbol{b}}$ (11) of the ellipse for Sun-Earth system, as $\left(Cos\phi_0 < 0\right)$ (129) and $\left(a^2 < b^2\right)$ (131).

$$Sin\phi_0 = \mathbf{-0.483127005006538} \quad (142)$$

$$\vec{\boldsymbol{a}}\cdot\vec{\boldsymbol{b}} = -r_1 r_2\left(1 - Cos\beta\right)Sin\phi_0 = \mathbf{0.0000766106315616312}\ au^2 \quad (143)$$

Checking the data in the Excel sheet "Tan, A.P., Asli Pinar Tan Analysis Based on Earth-Sun distance (d) Landsat.xlsx"[13] for the Earth's individual circular orbital revolution, the data point that is closest to $\left(\phi = 0\right)$ is the data point $\phi_{i=65}$ (145) with $i = 65$ on January 17, the data point that is closest to $\left(\phi = \dfrac{\pi}{2}\right)$ is the data point $\phi_{i=430}$ (147) with $i = 430$ on April 18, the data point that is closest to $\left(\phi = \pi\right)$ is the data point $\phi_{i=795}$ (149) with $i = 795$ on July 18, and the data point that is closest to $\left(\phi = -\dfrac{\pi}{2}\right)$ is the data point $\phi_{i=1160}$ (151) with $i = 1160$ on October 17. Phases of Earth at equinox[9] dates[6] of Earth with respect to Sun are $\left(\phi_{Spring\ Equinox} = 58^\circ.829\right)$ (146) around March 20 (DOY 79) and $\left(\phi_{Autumn\ Equinox} = \mathbf{-117^\circ.736}\right)$ (150) around September 22 (DOY 265). The phase of Earth at minimum Sun-Earth distance date on January 3 is $\left(\phi_{Min\ Sun-Earth\ Distance} = \mathbf{-11^\circ.072}\right)$ (144) corresponding to the data point $i = 9$, and the phase of Earth at maximum Sun-Earth distance date on July 5 is $\left(\phi_{Max\ Sun-Earth\ Distance} = \mathbf{168^\circ.928}\right)$ (148) corresponding to the data point $i = 741$. The yearly dates corresponding to the phases of the Earth's self-revolution is another very fundamental finding of this Article, and $\phi\left[Degrees\right]$ over a yearly Sun-Earth cycle is plotted in **Figure 14**.

$$\phi_{Min\ Sun-Earth\ Distance} = \phi_{i=9} = \mathbf{-0.193236295254927}\ radians = \mathbf{-11^\circ.072} \qquad on\ January\ 3 \quad (144)$$

$$\phi_{i=65} = \mathbf{0.00058673254949726}\ radians = \mathbf{0^\circ.034} \simeq 0 \qquad on\ January\ 17 \quad (145)$$

$$\phi_{Spring\ Equinox} = \mathbf{1.02676771671574}\ radians = \mathbf{58^\circ.829} \qquad on\ March\ 20 \quad (146)$$

$$\phi_{i=(65+365)} = \phi_{i=430} = \mathbf{1.57138305934439}\ radians = \mathbf{90^\circ.034} \simeq \frac{\pi}{2} \qquad on\ April\ 18 \quad (147)$$

$$\phi_{Max\ Sun-Earth\ Distance} = \phi_{i=741} = \mathbf{2.94835635833487}\ radians = \mathbf{168^\circ.928} \qquad on\ July\ 5 \quad (148)$$

$$\phi_{i=(65+730)} = \phi_{i=795} = \mathbf{3.14217938613929}\ radians = \mathbf{180^\circ.034} \simeq \pi \qquad on\ July\ 18 \quad (149)$$

$$\phi_{Autumn\ Equinox} = \mathbf{-2.05488176708229}\ radians = \mathbf{242^\circ.264} = \mathbf{-117^\circ.736} \qquad on\ September\ 22 \quad (150)$$

$$\phi_{i=(65+1095)} = \phi_{i=1160} = \mathbf{-1.5702095942454}\ radians = \mathbf{-89^\circ.966} \simeq -\frac{\pi}{2} \qquad on\ October\ 17 \quad (151)$$

Having determined the value of ϕ_0 (138) and all ϕ (136) series of the Sun-Earth system, we now also calculate $(\phi+\phi_0)$ for all ϕ in the Excel sheet "Tan, A.P., Asli Pinar Tan Analysis Based on Earth-Sun distance (d) Landsat.xlsx"[13].

Moving on with our analysis, based on (34), (36), and (9), at the same time utilizing (45), (48), (49), and (51), we obtain the result in (152) for all ϕ.

$$\begin{aligned}\frac{d^2(\phi)-d^2\left(\frac{\pi}{2}+\phi\right)}{2} &= r_1 r_2\left(1-Cos\beta\right)Cos\left(2\phi+\phi_0\right)\\ &+\left[\vec{a}\cdot\vec{\ell}(\phi)Cos\phi+\vec{b}\cdot\vec{\ell}(\phi)Sin\phi\right]+\left[\vec{a}\cdot\vec{\ell}\left(\frac{\pi}{2}+\phi\right)Sin\phi-\vec{b}\cdot\vec{\ell}\left(\frac{\pi}{2}+\phi\right)Cos\phi\right]\\ &=\frac{\left[\boldsymbol{k}_1(\phi)-\boldsymbol{k}_2(\phi)\right]}{2}+\boldsymbol{k}_3(\phi)-\boldsymbol{k}_4\left(\frac{\pi}{2}+\phi\right)\\ &=\begin{cases}\frac{\left[\boldsymbol{k}_1(\phi)-\boldsymbol{k}_2(\phi)\right]}{2}+\left[\boldsymbol{k}_3(\phi)-\boldsymbol{k}_4(\phi)\right] & ; \quad if \quad \vec{\ell}\left(\frac{\pi}{2}+\phi\right)=\vec{\ell}(\phi)\\ \frac{\left[\boldsymbol{k}_1(\phi)-\boldsymbol{k}_2(\phi)\right]}{2}+\left[\boldsymbol{k}_3(\phi)+\boldsymbol{k}_4(\phi)\right] & ; \quad if \quad \vec{\ell}\left(\frac{\pi}{2}+\phi\right)=-\vec{\ell}(\phi)\end{cases}\end{aligned} \tag{152}$$

In line with the result obtained in (152), and the data in the Excel sheet "Tan, A.P., Asli Pinar Tan Analysis Based on Earth-Sun distance (d) Landsat.xlsx"[13] previously calculated according to the algorithm in (21) - (25), we would expect a comparison of calculations based on (153) to help us reach a reliable conclusion on whether $\left[\vec{\ell}\left(\frac{\pi}{2}+\phi\right)=\vec{\ell}(\phi)\right]$ or $\left[\vec{\ell}\left(\frac{\pi}{2}+\phi\right)=-\vec{\ell}(\phi)\right]$.

$$\frac{d_i^2-d_{(i+365)}^{\;\;2}}{2}=\begin{cases}\frac{\left[\boldsymbol{k}_{1,i}-\boldsymbol{k}_{2,i}\right]}{2}+\left[\boldsymbol{k}_{3,i}-\boldsymbol{k}_{4,i}\right] & ; \quad if \quad \vec{\ell}_{(i+365)}=\vec{\ell}_i\\ \frac{\left[\boldsymbol{k}_{1,i}-\boldsymbol{k}_{2,i}\right]}{2}+\left[\boldsymbol{k}_{3,i}+\boldsymbol{k}_{4,i}\right] & ; \quad if \quad \vec{\ell}_{(i+365)}=-\vec{\ell}_i\end{cases} \tag{153}$$

After calculating $\left[\frac{d_i^2-d_{(i+365)}^{\;\;2}}{2}\right]$, $\left\{\frac{\left[\boldsymbol{k}_{1,i}-\boldsymbol{k}_{2,i}\right]}{2}+\left[\boldsymbol{k}_{3,i}-\boldsymbol{k}_{4,i}\right]\right\}$, and $\left\{\frac{\left[\boldsymbol{k}_{1,i}-\boldsymbol{k}_{2,i}\right]}{2}+\left[\boldsymbol{k}_{3,i}+\boldsymbol{k}_{4,i}\right]\right\}$ in the Excel sheet "Tan, A.P., Asli Pinar Tan Analysis Based on Earth-Sun distance (d) Landsat.xlsx"[13] for all data points i over a year, and comparing the results, we observe the definite result in (154), which leads us to make the conclusion in (155) for the Sun-Earth system based on (152), another very fundamental result presented in this Article regarding a feature of the distance vector $\vec{\ell}(\phi)$ (9) between the individual centers of circular revolution of the Sun and the Earth, making us understand that $\vec{\ell}(\phi)$ (9) repeats itself harmonically every quarter cycle vector-wise as well as magnitude-wise (156), eliminating the uncertainty in (45) about the quarterly direction changes of $\vec{\ell}(\phi)$ (9).

$$\frac{d_i^2 - d_{(i+365)}^2}{2} = \frac{[\boldsymbol{k}_{1,i} - \boldsymbol{k}_{2,i}]}{2} + [\boldsymbol{k}_{3,i} - \boldsymbol{k}_{4,i}] \qquad for\ all\ i \Rightarrow \bar{\ell}_{(i+365)} = \bar{\ell}_i \tag{154}$$

$$\frac{d^2(\phi) - d^2\left(\frac{\pi}{2}+\phi\right)}{2} = \frac{[\boldsymbol{k}_1(\phi) - \boldsymbol{k}_2(\phi)]}{2} + [\boldsymbol{k}_3(\phi) - \boldsymbol{k}_4(\phi)] \qquad for\ all\ \phi \Rightarrow \bar{\ell}\left(\frac{\pi}{2}+\phi\right) = \bar{\ell}(\phi) \tag{155}$$

$$\bar{\ell}(\phi) = \bar{\ell}\left(\frac{\pi}{2}+\phi\right) = \bar{\ell}(\pi+\phi) = \bar{\ell}\left(-\frac{\pi}{2}+\phi\right) \qquad for\ all\ \phi \tag{156}$$

Based on the result we have obtained in (156), we can now finalize $\boldsymbol{k}_4(\phi)$ (49) for the Sun-Earth system to be as $\boldsymbol{k}_4(\phi)$ (157).

$$\begin{aligned}\boldsymbol{k}_4(\phi) = \frac{d^2\left(\frac{\pi}{2}+\phi\right) - d^2\left(-\frac{\pi}{2}+\phi\right)}{4} &= \left[\vec{\boldsymbol{a}}Cos\left(\frac{\pi}{2}+\phi\right) + \vec{\boldsymbol{b}}Sin\left(\frac{\pi}{2}+\phi\right)\right]\cdot\bar{\ell}\left(\frac{\pi}{2}+\phi\right) \\ &= \left(-\vec{\boldsymbol{a}}Sin\phi + \vec{\boldsymbol{b}}Cos\phi\right)\cdot\bar{\ell}(\phi)\end{aligned} \tag{157}$$

We can now solve for $\vec{\boldsymbol{a}}\cdot\bar{\ell}(\phi)$ (158) and $\vec{\boldsymbol{b}}\cdot\bar{\ell}(\phi)$ (159) in terms of $\boldsymbol{k}_3(\phi)$ (48) and $\boldsymbol{k}_4(\phi)$(157).

$$\vec{\boldsymbol{a}}\cdot\bar{\ell}(\phi) = \boldsymbol{k}_3(\phi)Cos\phi - \boldsymbol{k}_4(\phi)Sin\phi \tag{158}$$

$$\vec{\boldsymbol{b}}\cdot\bar{\ell}(\phi) = \boldsymbol{k}_3(\phi)Sin\phi + \boldsymbol{k}_4(\phi)Cos\phi \tag{159}$$

Having already calculated the values of $\boldsymbol{k}_3(\phi)$ (48), $\boldsymbol{k}_4(\phi)$(157), and ϕ (136) for all data points in the Excel sheet "Tan, A.P., Asli Pinar Tan Analysis Based on Earth-Sun distance (d) Landsat.xlsx"[13] according to the algorithm in (21) - (25), we now calculate the values of $\vec{\boldsymbol{a}}\cdot\bar{\ell}(\phi)$ (158) and $\vec{\boldsymbol{b}}\cdot\bar{\ell}(\phi)$ (159) for all data points in the Excel sheet "Tan, A.P., Asli Pinar Tan Analysis Based on Earth-Sun distance (d) Landsat.xlsx"[13] as well.

Further, our finding in (156) would also yield the results in (160) - (161), which would lead to the results in (55) - (56) based on (48) and (157), in agreement with (57) - (58), also confirmed by checking the data in the Excel sheet "Tan, A.P., Asli Pinar Tan Analysis Based on Earth-Sun distance (d) Landsat.xlsx"[13] calculated according to the algorithm in (21) - (25), verifying our analysis for the Sun-Earth system so far.

$$\vec{\boldsymbol{a}}\cdot\bar{\ell}(\phi) = \vec{\boldsymbol{a}}\cdot\bar{\ell}(\pi+\phi) = \vec{\boldsymbol{a}}\cdot\bar{\ell}\left(\frac{\pi}{2}+\phi\right) = \vec{\boldsymbol{a}}\cdot\bar{\ell}\left(-\frac{\pi}{2}+\phi\right) \qquad for\ all\ \phi \tag{160}$$

$$\vec{\boldsymbol{b}}\cdot\bar{\ell}(\phi) = \vec{\boldsymbol{b}}\cdot\bar{\ell}(\pi+\phi) = \vec{\boldsymbol{b}}\cdot\bar{\ell}\left(\frac{\pi}{2}+\phi\right) = \vec{\boldsymbol{b}}\cdot\bar{\ell}\left(-\frac{\pi}{2}+\phi\right) \qquad for\ all\ \phi \tag{161}$$

The calculated $\vec{\boldsymbol{a}}\cdot\bar{\ell}(\phi)$ (158) and $\vec{\boldsymbol{b}}\cdot\bar{\ell}(\phi)$ (159) have been plotted for all ϕ (136) over a yearly cycle in **Figure 15** and **Figure 16**, respectively. Both graphs demonstrate interesting behaviour overall. The results reveal a rippled graph in both cases, with $\vec{\boldsymbol{a}}\cdot\bar{\ell}(\phi)$ (158) *negative* and $\vec{\boldsymbol{b}}\cdot\bar{\ell}(\phi)$ (159) *positive* for all ϕ (136) over a yearly cycle, and both having sharp jumps quarterly around January 1 (DOY 1), April 2 (DOY 92), July 2 (DOY 183), and October 1 (DOY 274), coinciding with the occurence dates of $\left\{\sqrt{[\boldsymbol{k}_3(\phi)]^2+[\boldsymbol{k}_4(\phi)]^2}\right\}_{Max}$ (68) and $[\boldsymbol{k}_1(\phi)+\boldsymbol{k}_2(\phi)]_{Min}$ (80).

We can also make the observation that throughout the year, $\vec{\ell}(\phi)$ (9) moves spatially at a closer angle to the negative of the semi-minor axis vector $-\vec{\boldsymbol{a}}$ (10) compared to the semi-major axis vector $\vec{\boldsymbol{b}}$ (11), as $-\vec{\boldsymbol{a}}\cdot\vec{\ell}(\phi)$ (158) projection is always greater than $\vec{\boldsymbol{b}}\cdot\vec{\ell}(\phi)$ (159) magnitude-wise.

Inspecting $\boldsymbol{k}_3(\phi)$ (48) more closely and utilizing (52) to obtain (162), we can reorder and express $\boldsymbol{k}_3(\phi)$ (48) as in (164) and $\boldsymbol{k}_4(\phi)$ (157) as in (165), by introducing and defining a virtual angle $\xi(\phi)$ (163) which is *not* the same for every ϕ, in contrast to a constant ξ for all ϕ that provides an indication of the phase in orbit where $\vec{\boldsymbol{d}}(\phi)$ (13) is closest to a constant $\vec{\ell}(\phi)=\vec{\ell}$ (26), as was the case in the analogous analysis in the **ARTICLE 2** "Analytical solution for two circles in space using their elliptic distances".

$$\sqrt{\left[\vec{\boldsymbol{a}}\cdot\vec{\ell}(\phi)\right]^2+\left[\vec{\boldsymbol{b}}\cdot\vec{\ell}(\phi)\right]^2}=\sqrt{[\boldsymbol{k}_3(\phi)]^2+[\boldsymbol{k}_4(\phi)]^2} \tag{162}$$

$$Cos\left[\xi(\phi)\right]=\frac{\left[\vec{\boldsymbol{a}}\cdot\vec{\ell}(\phi)\right]}{\sqrt{\left[\vec{\boldsymbol{a}}\cdot\vec{\ell}(\phi)\right]^2+\left[\vec{\boldsymbol{b}}\cdot\vec{\ell}(\phi)\right]^2}} \quad ; \quad Sin\left[\xi(\phi)\right]=\frac{\left[\vec{\boldsymbol{b}}\cdot\vec{\ell}(\phi)\right]}{\sqrt{\left[\vec{\boldsymbol{a}}\cdot\vec{\ell}(\phi)\right]^2+\left[\vec{\boldsymbol{b}}\cdot\vec{\ell}(\phi)\right]^2}} \tag{163}$$

$$\begin{aligned}\boldsymbol{k}_3(\phi)&=\vec{\boldsymbol{a}}\cdot\vec{\ell}(\phi)Cos\,\phi+\vec{\boldsymbol{b}}\cdot\vec{\ell}(\phi)Sin\,\phi\\&=\sqrt{\left[\vec{\boldsymbol{a}}\cdot\vec{\ell}(\phi)\right]^2+\left[\vec{\boldsymbol{b}}\cdot\vec{\ell}(\phi)\right]^2}\left\{Cos\left[\xi(\phi)\right]Cos\,\phi+Sin\left[\xi(\phi)\right]Sin\,\phi\right\}\\&=\sqrt{[\boldsymbol{k}_3(\phi)]^2+[\boldsymbol{k}_4(\phi)]^2}\;Cos\left[\xi(\phi)-\phi\right]\end{aligned} \tag{164}$$

$$\begin{aligned}\boldsymbol{k}_4(\phi)&=-\vec{\boldsymbol{a}}\cdot\vec{\ell}(\phi)Sin\phi+\vec{\boldsymbol{b}}\cdot\vec{\ell}(\phi)Cos\,\phi\\&=\sqrt{\left[\vec{\boldsymbol{a}}\cdot\vec{\ell}(\phi)\right]^2+\left[\vec{\boldsymbol{b}}\cdot\vec{\ell}(\phi)\right]^2}\left\{-Cos\left[\xi(\phi)\right]Sin\phi+Sin\left[\xi(\phi)\right]Cos\,\phi\right\}\\&=\sqrt{[\boldsymbol{k}_3(\phi)]^2+[\boldsymbol{k}_4(\phi)]^2}\;Sin\left[\xi(\phi)-\phi\right]\end{aligned} \tag{165}$$

Apparently, the values of $Cos\left[\xi(\phi)\right]$ (163) and $Sin\left[\xi(\phi)\right]$ (163) provide a measure of how much more or less is the projection of the semi-minor axis vector $\vec{\boldsymbol{a}}$ (10) on $\vec{\ell}(\phi)$ (9) with respect to the projection of the semi-major axis vector $\vec{\boldsymbol{b}}$ (11) on $\vec{\ell}(\phi)$ (9), for each phase ϕ (136) of the ellipse formed by the relative motions of the Sun-Earth system, and $\xi(\phi)$ (163) provides a sense of how close or far $\vec{\boldsymbol{d}}(\phi)$ (13) direction is to $\vec{\ell}(\phi)$ (9), at each ϕ (136), in a Sun-Earth cycle.

Utilizing (164) and (165), we also obtain (166), providing more insight into the virtual ellipse formed by the Sun-Earth system, due to individual revolutions around their own circular orbits.

$$Cos\left[\xi(\phi)-\phi\right]=\frac{\boldsymbol{k}_3(\phi)}{\sqrt{[\boldsymbol{k}_3(\phi)]^2+[\boldsymbol{k}_4(\phi)]^2}} \quad ; \quad Sin\left[\xi(\phi)-\phi\right]=\frac{\boldsymbol{k}_4(\phi)}{\sqrt{[\boldsymbol{k}_3(\phi)]^2+[\boldsymbol{k}_4(\phi)]^2}} \quad (166)$$

At this point, values of $Cos\left[\xi(\phi)\right]$ (163), $Sin\left[\xi(\phi)\right]$ (163), $\xi(\phi)$ (163), $Cos\left[\xi(\phi)-\phi\right]$ (166), $Sin\left[\xi(\phi)-\phi\right]$ (166), and $\left[\xi(\phi)-\phi\right]$ (166) are calculated for all ϕ (136), that is for every data point in the Excel sheet "Tan, A.P., Asli Pinar Tan Analysis Based on Earth-Sun distance (d) Landsat.xlsx"[13], based on already calculated data according to the algorithm in (21) - (25), and the values calculated for $\xi(\phi)$ (163) and $\left[\xi(\phi)-\phi\right]$ (166) are plotted over a yearly Sun-Earth cycle in **Figure 17** and **Figure 18**, respectively.

Based on the obtained data and **Figure 17**, $\xi(\phi)$ (163) varies harmonically in a small angle range between $\mathbf{169^{\circ}.860}$ (167) and $\mathbf{163^{\circ}.621}$ (168), with sharp jumps or value changes around December 31 (DOY 365), April 1 (DOY 91), July 2 (DOY 183), and October 1 (DOY 274), coinciding with the occurence dates of $\left\{\sqrt{[\boldsymbol{k}_3(\phi)]^2+[\boldsymbol{k}_4(\phi)]^2}\right\}_{Max}$ (68) and $\left[\boldsymbol{k}_1(\phi)+\boldsymbol{k}_2(\phi)\right]_{Min}$ (80), just as $\vec{\boldsymbol{a}}\cdot\vec{\ell}(\phi)$ (158) and $\vec{\boldsymbol{b}}\cdot\vec{\ell}(\phi)$ (159). This important observation that $\xi(\phi)$ (163) varies harmonically in a small angle range between $\mathbf{169^{\circ}.860}$ (167) and $\mathbf{163^{\circ}.621}$ (168), as well as the definition of $\left[\xi(\phi)-\phi\right]$ (166) with (164), allows us to reach the understanding that throughout a year, $\vec{\ell}(\phi)$ (9) must be varying in a narrow angle range, which is closer to the sector of Earth's self-revolution when the Earth moves between $\phi=\mathbf{163^{\circ}.621}$ and $\phi=\mathbf{169^{\circ}.860}$.

$$\xi_{Max}(\phi)=\mathbf{2.96462117433723}\,radians=\mathbf{169^{\circ}.860} \quad January\ 1,\ April\ 2,\ July\ 3,\ October\ 2 \quad (167)$$

$$\xi_{Min}(\phi)=\mathbf{2.8557213606605}\,radians=\mathbf{163^{\circ}.621} \quad March\ 31,\ July\ 1,\ September\ 30,\ December\ 30 \quad (168)$$

Again based on the obtained data and **Figure 18**, $\xi(\phi)$ (163) is equal to ϕ (136) on July 5 (DOY 186), and $\xi(\phi)$ (163) is equal to $(\pi+\phi)$ (136) on January 3 (DOY 3), which are the dates of maximum and minimum Sun-Earth distance over a year, respectively. This result is compliant with the expectation based on the definition of $\left[\xi(\phi)-\phi\right]$ (166), taking into consideration (164) and $d^2(\phi)$ (59).

Continuing our analysis, utilizing (13), (16), and (156), as well as (21), the Dot Product[14] of distance vectors $\vec{\boldsymbol{d}}(\phi)$ and $\vec{\boldsymbol{d}}(\pi+\phi)$ would become as in (169), where $\alpha_{[\vec{d}(\phi)\measuredangle\vec{d}(\pi+\phi)]}$ (169) is the angle between $\vec{\boldsymbol{d}}(\phi)$ and $\vec{\boldsymbol{d}}(\pi+\phi)$ vectors, which are half a cycle, or $(\phi=\pi)$, apart.

$$\begin{aligned}\vec{\boldsymbol{d}}(\phi)\cdot\vec{\boldsymbol{d}}(\pi+\phi)&=d(\phi)\,d(\pi+\phi)\,Cos\,\alpha_{[\vec{d}(\phi)\measuredangle\vec{d}(\pi+\phi)]}=d_i\,d_{i+730}\,Cos\,\alpha_{[\vec{d}(\phi)\measuredangle\vec{d}(\pi+\phi)]}\\&=\left[\vec{\boldsymbol{a}}\,Cos\,\phi+\vec{\boldsymbol{b}}\,Sin\,\phi+\vec{\ell}(\phi)\right]\cdot\left[-\left(\vec{\boldsymbol{a}}\,Cos\,\phi+\vec{\boldsymbol{b}}\,Sin\,\phi\right)+\vec{\ell}(\phi)\right]\\&=\left[\ell^2(\phi)-\left(\vec{\boldsymbol{a}}\,Cos\,\phi+\vec{\boldsymbol{b}}\,Sin\,\phi\right)^2\right]\end{aligned} \quad (169)$$

As previously stated, on the solstice[5] dates[6] of the Earth, which are June 21 (DOY 172) and December 21 (DOY 355), that are $(\phi=\pi)$ (103) apart and at which dates the Northern Hemisphere of the Earth is most and least inclined towards the Sun, respectively, the maximum and minimum values that the angle $\gamma_2(\phi)$ (98) between $\hat{\boldsymbol{u}}_{2\perp}$ (97) and $\vec{\boldsymbol{d}}(\phi)$ (14) would take over a relative Sun-Earth cycle, namely $\gamma_{2,\max}$ (101) and $\gamma_{2,\min}$ (101), respectively, would sum up to $\pi\ radians = 180^\circ$ (170), as the tilt angles γ_{2,Δ_1} (102) and γ_{2,Δ_2} (102) are both observed[7] approximately equal to the axial tilt[7] angle $\gamma_{Earth,\Delta Max}$ (4) of the Earth. The angle $\gamma_2\left(\phi_{\gamma_{2,\min}}\right)=\gamma_{2,\min}$ (101) between $\hat{\boldsymbol{u}}_{2\perp}$ (97) and $\vec{\boldsymbol{d}}\left(\phi_{\gamma_{2,\min}}\right)$ at Summer Solstice[5,6] on June 21, and the angle $\gamma_2\left(\phi_{\gamma_{2,\max}}\right)=\gamma_{2,\max}$ (101) between $\hat{\boldsymbol{u}}_{2\perp}$ (97) and $\vec{\boldsymbol{d}}\left(\phi_{\gamma_{2,\max}}\right)$ at Winter Solstice[5,6] on December 21, sum up to $\pi\ radians = 180^\circ$ (170), and considering the relative orientation of the Earth at the two solstices[5], this observation allows us to reason that the angle between $\vec{\boldsymbol{d}}\left(\phi_{\gamma_{2,\min}}\right)$ and $\vec{\boldsymbol{d}}\left(\phi_{\gamma_{2,\max}}\right)$, which is $\alpha_{\left[\vec{d}\left(\phi_{\gamma_{2,\min}}\right)\measuredangle\vec{d}\left(\phi_{\gamma_{2,\max}}\right)\right],1^{st}\ iteration}$ (171), must also be around $\pi\ radians = 180^\circ$, even if not exactly, and we take it as its 1st iteration initial value in our pursuit to find accurate results.

$$\gamma_{2,\max}+\gamma_{2,\min}=\left(\frac{\pi}{2}+\gamma_{2,\Delta_1}\right)+\left(\frac{\pi}{2}-\gamma_{2,\Delta_2}\right)=\left(\frac{\pi}{2}+\gamma_{Earth,\Delta Max}\right)+\left(\frac{\pi}{2}-\gamma_{Earth,\Delta Max}\right)=\pi\ radians = 180^\circ \quad (170)$$

$$\alpha_{\left[\vec{d}\left(\phi_{\gamma_{2,\min}}\right)\measuredangle\vec{d}\left(\phi_{\gamma_{2,\max}}\right)\right],1^{st}\ iteration} \simeq \pi\ radians = 180^\circ \quad (171)$$

Based on this $\alpha_{\left[\vec{d}\left(\phi_{\gamma_{2,\min}}\right)\measuredangle\vec{d}\left(\phi_{\gamma_{2,\max}}\right)\right],1^{st}\ iteration}=180^\circ$ (171) approximation as a starting point, we will make an iterative calculation of $\ell\left(\phi_{\gamma_{2,\min}}\right)$ in (179), (222), (242), (262), and (282), and an iterative calculation of b in (181), (227), (247), and (267), to reach more accurate resultant values for $\ell\left(\phi_{\gamma_{2,\min}}\right)$ (284), b (285), a , β (277), and (r_1,r_2) (278).

Utilizing (139) - (140), (169), and (171), where $\left[\ell^2\left(\phi_{\gamma_{2,\min}}\right)\right]_{1^{st}\ iteration}$ is the 1st iteration value we set to find for $\ell^2\left(\phi_{\gamma_{2,\min}}\right)$ using $\alpha_{\left[\vec{d}\left(\phi_{\gamma_{2,\min}}\right)\measuredangle\vec{d}\left(\phi_{\gamma_{2,\max}}\right)\right],1^{st}\ iteration}$ (171), we would obtain the result in (172), $d_{(i=689)}$ (173) being the distance data on June 21, and $d_{(i+730=1419)}$ (174) being the distance data on December 21, taken from the Excel sheet "Tan, A.P., Asli Pinar Tan Analysis Based on Earth-Sun distance (d) Landsat.xlsx"[13].

$$\vec{d}\left(\phi_{\gamma_{2,\min}}\right)\cdot\vec{d}\left(\pi+\phi_{\gamma_{2,\min}}\right)=\vec{d}\left(\phi_{\gamma_{2,\min}}\right)\cdot\vec{d}\left(\phi_{\gamma_{2,\max}}\right)$$
$$=d\left(\phi_{\gamma_{2,\min}}\right)d\left(\phi_{\gamma_{2,\max}}\right)Cos\,\alpha_{\left[\vec{d}\left(\phi_{\gamma_{2,\min}}\right)\measuredangle\vec{d}\left(\phi_{\gamma_{2,\max}}\right)\right],1^{st}\ iteration}$$
$$=-d\left(\phi_{\gamma_{2,\min}}\right)d\left(\phi_{\gamma_{2,\max}}\right)=-d_{(i=689)}\,d_{(i+730=1419)} \tag{172}$$
$$=\left[\ell^2\left(\phi_{\gamma_{2,\min}}\right)\right]_{1^{st}\ iteration}-\left(\vec{a}\,Cos\,\phi_{\gamma_{2,\min}}+\vec{b}\,Sin\,\phi_{\gamma_{2,\min}}\right)^2$$

$$d\left(\phi_{\gamma_{2,\min}}\right)=d_{(i=689)}=\mathbf{1.01627775}\ au=\mathbf{152{,}032{,}987.4}\ km \qquad on\ June\ 21 \tag{173}$$

$$d\left(\phi_{\gamma_{2,\max}}\right)=d_{(i+730=1419)}=\mathbf{0.9837613}\ au=\mathbf{147{,}168{,}595.8}\ km \qquad on\ December\ 21 \tag{174}$$

Combining the result in (172) together with $\boldsymbol{k}_1(\phi)$ (46) for $\left(\phi=\phi_{\gamma_{2,\min}}=-\phi_0\right)$ (139), whose value is $\boldsymbol{k}_{1,(i=689)}$ (175) based on the Excel sheet "Tan, A.P., Asli Pinar Tan Analysis Based on Earth-Sun distance (d) Landsat.xlsx"[13] calculated according to the algorithm in (22), would yield (176).

$$\boldsymbol{k}_1\left(\phi_{\gamma_{2,\min}}\right)=\boldsymbol{k}_1\left(-\phi_0\right)=\boldsymbol{k}_{1,(i=689)}=\mathbf{1.00030338026138}\ au^2 \qquad on\ June\ 21 \tag{175}$$

$$\boldsymbol{k}_1\left(\phi_{\gamma_{2,\min}}\right)+\vec{d}\left(\phi_{\gamma_{2,\min}}\right)\cdot\vec{d}\left(\pi+\phi_{\gamma_{2,\min}}\right)=\boldsymbol{k}_{1,(i=689)}-d_{(i=689)}\,d_{(i+730=1419)}$$
$$=2\left[\ell^2\left(\phi_{\gamma_{2,\min}}\right)\right]_{1^{st}\ iteration}=2\left[\ell^2\left(-\phi_0\right)\right]_{1^{st}\ iteration} \tag{176}$$

Along the lines of the algorithm devised in (21) - (25), $\ell_i^{\,2}$ (177) corresponding to each data point in the Excel sheet "Tan, A.P., Asli Pinar Tan Analysis Based on Earth-Sun distance (d) Landsat.xlsx"[13] for the Sun-Earth system can be expressed based on the definition of $\ell^2(\phi)$ (9).

$$\ell_x(\phi)=\ell_{x,i}\ \ ;\ \ \ell_y(\phi)=\ell_{y,i}\ \ ;\ \ \ell_z(\phi)=\ell_{z,i}\ \ ;\ \ \ell^2(\phi)=\ell_i^{\,2}\ \ ;\ \ \ell_i^{\,2}=\ell_{x,i}^{\,2}+\ell_{y,i}^{\,2}+\ell_{z,i}^{\,2} \tag{177}$$

We can now determine the value of $\left[\ell^2_{(i=689)}\right]_{1^{st}\ iteration}$ (178) based on (176), and therefore also the value of $\left[\ell_{(i=689)}\right]_{1^{st}\ iteration}$ (179).

$$\left[\ell^2\left(\phi_{\gamma_{2,\min}}\right)\right]_{1^{st}\ iteration}=\left[\ell^2\left(-\phi_0\right)\right]_{1^{st}\ iteration}=\left[\ell^2_{(i=689)}\right]_{1^{st}\ iteration}$$
$$=\frac{\boldsymbol{k}_{1,(i=689)}-d_{(i=689)}\,d_{(i+730=1419)}}{2}=\mathbf{0.000264329880150682}\ au^2 \tag{178}$$

$$\left[\ell\left(\phi_{\gamma_{2,\min}}\right)\right]_{1^{st}\ iteration}=\left[\ell\left(-\phi_0\right)\right]_{1^{st}\ iteration}=\left[\ell_{(i=689)}\right]_{1^{st}\ iteration}$$
$$=\mathbf{0.0162582250000018}\ au=\mathbf{2{,}432{,}195.841362}\ km \tag{179}$$

Based on the definition of $\boldsymbol{k}_1(\phi)$ (46), we can also find the value of $\left[b^2\right]_{1^{st}\ iteration}$ (180) at this point, making use of the already found values of $\boldsymbol{k}_1(-\phi_0)$ (175), $\left[2r_1r_2\left(1-Cos\beta\right)Cos\phi_0\right]$ (132), and $\left[\ell^2(-\phi_0)\right]_{1^{st}\ iteration}$ (178). This allows us to determine the magnitude $\left[b\right]_{1^{st}\ iteration}$ (181) of the semi-major axis vector $\vec{b}$ (11) of the ellipse for the Sun-Earth system formed by their relative motions, which is the the semi-major axis vector based on the **ARTICLE 1** "Points on two circles in space mimic an ellipse with respect to a point" and (10) - (11), as $(Cos\,\phi_0 < 0)$ (129) and $(a^2 < b^2)$ (131). Based on (11) and (180), we have also obtained the identity (182) at this point.

$$\left[b^2\right]_{1^{st}\ iteration} = \boldsymbol{k}_1(-\phi_0) - 2r_1 r_2 (1 - Cos\beta) Cos\phi_0 - \left[\ell^2(-\phi_0)\right]_{1^{st}\ iteration} = \mathbf{1.00031672668333}\ au^2 \quad (180)$$

$$[b]_{1^{st}\ iteration} = \mathbf{1.00015835080418}\ au = \mathbf{149,621,559.643129}\ km \quad (181)$$

$$\left[b^2\right]_{1^{st}\ iteration} = \left[r_1^2\right]_{1^{st}\ iteration} - 2\left[r_1\right]_{1^{st}\ iteration}\left[r_2\right]_{1^{st}\ iteration} Cos\,\phi_0 + \left[r_2^2\right]_{1^{st}\ iteration} = \mathbf{1.00031672668333}\ au^2 \quad (182)$$

We have reached a point in our computations where we have three equations (72), (141), and (11), which are restated in (183), (184), and (185), respectively, in the three variables r_1 (6), r_2 (7), and β (6), that we are set to solve for, where $\left[m_3\right]_{1^{st}\ iteration}$ (186) is given by (182).

$$2r_1 r_2 (1 - Cos\beta) = \mathbf{0.000317144894687038}\ au^2 = m_1 \quad (183)$$

$$r_1\, Sin\,\beta = \pm\mathbf{0.397796274218557}\ au = m_2 \quad (184)$$

$$b^2 = r_1^2 - 2\, r_1 r_2 Cos\,\phi_0 + r_2^2 = m_3 \quad (185)$$

$$\left[m_3\right]_{1^{st}\ iteration} = \left[b^2\right]_{1^{st}\ iteration} = \mathbf{1.00031672668333}\ au^2 \quad (186)$$

Using (183) we would obtain (187).

$$2r_1 r_2 = \frac{m_1}{(1 - Cos\beta)} \quad (187)$$

Taking the squares of (183) and (184), we would have (188) and (189), respectively.

$$r_1^{\,2} r_2^{\,2}(1 - Cos\beta)(1 - Cos\beta) = \frac{m_1^{\,2}}{4} \quad (188)$$

$$r_1^{\,2} Sin^2\beta = r_1^{\,2}\left(1 - Cos^2\beta\right) = r_1^{\,2}(1 - Cos\beta)(1 + Cos\beta) = m_2^{\,2} \quad (189)$$

Dividing (188) by (189) would yield (190), whereas reordering (189) would give (191).

$$r_2^{\,2}\frac{(1 - Cos\beta)}{(1 + Cos\beta)} = \frac{m_1^2}{4m_2^{\,2}} \quad \Rightarrow \quad r_2^{\,2} = \frac{m_1^{\,2}}{4m_2^{\,2}}\frac{(1 + Cos\beta)}{(1 - Cos\beta)} \quad (190)$$

$$r_1^{\,2} = \frac{m_2^{\,2}}{(1 - Cos\beta)(1 + Cos\beta)} \quad (191)$$

Placing $r_1^{\,2}$ (191), $r_2^{\,2}$ (190), and $(2r_1 r_2)$ (187) into (185) and reordering, we would obtain a second order equation in β (192), which can equivalently be stated as in (193) in terms of its quadratic equation coefficients $\mathbf{A}_\beta$ (194), $\mathbf{B}_\beta$ (195), and $\mathbf{C}_\beta$ (196), from which $Cos\beta$ (197) can be solved for.

$$\left(\frac{m_1^2}{4m_2^2}+m_3\right)Cos^2\beta+\left(\frac{m_1^2}{2m_2^2}-m_1Cos\phi_0\right)Cos\beta+\left(\frac{m_1^2}{4m_2^2}+m_2^2-m_1Cos\phi_0-m_3\right)=0 \quad (192)$$

$$\mathbf{A}_\beta Cos^2\beta+\mathbf{B}_\beta Cos\beta+\mathbf{C}_\beta=0 \quad (193)$$

$$\mathbf{A}_\beta=\frac{m_1^2}{4m_2^2}+m_3 \quad (194)$$

$$\mathbf{B}_\beta=\frac{m_1^2}{2m_2^2}-m_1Cos\phi_0 \quad (195)$$

$$\mathbf{C}_\beta=\frac{m_1^2}{4m_2^2}+m_2^2-m_1Cos\phi_0-m_3 \quad (196)$$

$$Cos\beta=\frac{-\mathbf{B}_\beta\pm\sqrt{\mathbf{B}_\beta^2-4\mathbf{A}_\beta\mathbf{C}_\beta}}{2\mathbf{A}_\beta} \quad (197)$$

Based on the values of m_1 (183), m_2 (184), and $\left[m_3\right]_{1^{st}\ iteration}$ (186), we would obtain the values for $\left[\mathbf{A}_\beta\right]_{1^{st}\ iteration}$ (198), $\mathbf{B}_\beta$ (199), and $\left[\mathbf{C}_\beta\right]_{1^{st}\ iteration}$ (200), using (194) - (196).

$$\left[\mathbf{A}_\beta\right]_{1^{st}\ iteration}=\frac{m_1^2}{4m_2^2}+\left[m_3\right]_{1^{st}\ iteration}=\mathbf{1.00031688558704}\ au^2 \quad (198)$$

$$\mathbf{B}_\beta=\frac{m_1^2}{2m_2^2}-m_1Cos\phi_0=\mathbf{0.00027799410952802}\ au^2 \quad (199)$$

$$\left[\mathbf{C}_\beta\right]_{1^{st}\ iteration}=\frac{m_1^2}{4m_2^2}+m_2^2-m_1Cos\phi_0-\left[m_3\right]_{1^{st}\ iteration}=\mathbf{-0.841797015695351}\ au^2 \quad (200)$$

Using (197), we can thus solve for $\left[Cos\beta\right]_{1^{st}\ iteration}$ (201) now to obtain its possible values from this quadratic equation in (192), or in (193), based on the calculated values of $\left[\mathbf{A}_\beta\right]_{1^{st}\ iteration}$ (198), $\mathbf{B}_\beta$ (199), and $\left[\mathbf{C}_\beta\right]_{1^{st}\ iteration}$ (200).

$$\left[Cos\beta\right]_{1^{st}\ iteration}=\frac{-\mathbf{B}_\beta\pm\sqrt{\mathbf{B}_\beta^2-4\left[\mathbf{A}_\beta\right]_{1^{st}\ iteration}\left[\mathbf{C}_\beta\right]_{1^{st}\ iteration}}}{2\left[\mathbf{A}_\beta\right]_{1^{st}\ iteration}}=\begin{cases}\mathbf{0.917210689243313}\\\mathbf{-0.917488595288421}\end{cases} \quad (201)$$

Based on the two possible solutions $\left[Cos\beta_1\right]_{1^{st}\ iteration}$ (202) and $\left[Cos\beta_2\right]_{1^{st}\ iteration}$ (202) obtained for the value of $\left[Cos\beta\right]_{1^{st}\ iteration}$ (201), there are four possible values $\left[\beta_{1,1}\right]_{1^{st}\ iteration}$ (202), $\left[\beta_{1,2}\right]_{1^{st}\ iteration}$ (202), $\left[\beta_{2,1}\right]_{1^{st}\ iteration}$ (202), and $\left[\beta_{2,2}\right]_{1^{st}\ iteration}$ (202) that $\left[\beta\right]_{1^{st}\ iteration}$ (201) can take.

$$[Cos\beta]_{1^{st}\ iteration} = [Cos\beta_1]_{1^{st}\ iteration} = \mathbf{0.917210689243313}$$

$$\Rightarrow \begin{cases} [\beta]_{1^{st}\ iteration} = [\beta_{1,1}]_{1^{st}\ iteration} = \mathbf{0.409774491611697}\ radians = \mathbf{23^\circ.48} \\ [\beta]_{1^{st}\ iteration} = [\beta_{1,2}]_{1^{st}\ iteration} = \mathbf{-0.409774491611697}\ radians = \mathbf{-23^\circ.48} \end{cases}$$

(202)

$$[Cos\beta]_{1^{st}\ iteration} = [Cos\beta_2]_{1^{st}\ iteration} = \mathbf{-0.917488595288421}$$

$$\Rightarrow \begin{cases} [\beta]_{1^{st}\ iteration} = [\beta_{2,1}]_{1^{st}\ iteration} = \mathbf{2.73251627399843}\ radians = \mathbf{156^\circ.56} \\ [\beta]_{1^{st}\ iteration} = [\beta_{2,2}]_{1^{st}\ iteration} = \mathbf{-2.73251627399843}\ radians = \mathbf{-156^\circ.56} \end{cases}$$

For these four possible values (202) that $[\beta]_{1^{st}\ iteration}$ (201) can take, the possible values $[r_1]_{1^{st}\ iteration}$ (6) and $\hat{\boldsymbol{u}}_{2\perp}$ (97) can take based on (141) are listed in (203), and subsequently the possible values that $[r_2]_{1^{st}\ iteration}$ (7) can take based on (183) are also listed in (203).

$$\left.\begin{array}{l} [\beta_{1,1}]_{1^{st}\ iteration} = \mathbf{23^\circ.48} \Rightarrow Sin[\beta_{1,1}]_{1^{st}\ iteration} = \mathbf{0.398402499412098} \Rightarrow \hat{\boldsymbol{u}}_{2\perp} = \hat{z} \\ [\beta_{1,2}]_{1^{st}\ iteration} = \mathbf{-23^\circ.48} \Rightarrow Sin[\beta_{1,2}]_{1^{st}\ iteration} = \mathbf{-0.398402499412098} \Rightarrow \hat{\boldsymbol{u}}_{2\perp} = -\hat{z} \end{array}\right\}$$

$$\Rightarrow [r_1]_{1^{st}\ iteration} = [r_{1,1}]_{1^{st}\ iteration} = \mathbf{0.998478359964017}\ au$$

$$\Rightarrow [r_2]_{1^{st}\ iteration} = [r_{2,1}]_{1^{st}\ iteration} = \mathbf{0.00191829239539248}\ au$$

(203)

$$\left.\begin{array}{l} [\beta_{2,1}]_{1^{st}\ iteration} = \mathbf{156^\circ.56} \Rightarrow Sin[\beta_{2,1}]_{1^{st}\ iteration} = \mathbf{0.397762086573973} \Rightarrow \hat{\boldsymbol{u}}_{2\perp} = \hat{z} \\ [\beta_{2,2}]_{1^{st}\ iteration} = \mathbf{-156^\circ.56} \Rightarrow Sin[\beta_{2,2}]_{1^{st}\ iteration} = \mathbf{-0.397762086573973} \Rightarrow \hat{\boldsymbol{u}}_{2\perp} = -\hat{z} \end{array}\right\}$$

$$\Rightarrow [r_1]_{1^{st}\ iteration} = [r_{1,2}]_{1^{st}\ iteration} = \mathbf{1.00008594998301}\ au$$

$$\Rightarrow [r_2]_{1^{st}\ iteration} = [r_{2,2}]_{1^{st}\ iteration} = \mathbf{0.0000826908799381157}\ au$$

Calculating the value of $\left(r_1^2 - 2\, r_1 r_2 Cos\,\phi_0 + r_2^2\right)$ (204) with both possible (r_1, r_2) (203) solution alternatives fo a validity check, we see that all match the expected $\left[b^2\right]_{1^{st}\ iteration}$ (182) value we have found before.

$$\left.\begin{array}{l}[r_1]_{1^{st}\ iteration}=\left[r_{1,1}\right]_{1^{st}\ iteration}=\mathbf{0.998478359964017}\ au\\ [r_2]_{1^{st}\ iteration}=\left[r_{2,1}\right]_{1^{st}\ iteration}=\mathbf{0.00191829239539248}\ au\end{array}\right\}\Rightarrow$$

$$\left[r_{1,1}^{\ 2}\right]_{1^{st}\ iteration}-2\left[r_{1,1}\right]_{1^{st}\ iteration}\left[r_{2,1}\right]_{1^{st}\ iteration}Cos\,\phi_0+\left[r_{2,1}^{\ 2}\right]_{1^{st}\ iteration}$$
$$=\mathbf{1.00031672668333}\ au^2=\left[b^2\right]_{1^{st}\ iteration}$$

(204)

$$\left.\begin{array}{l}[r_1]_{1^{st}\ iteration}=\left[r_{1,2}\right]_{1^{st}\ iteration}=\mathbf{1.00008594998301}\ au\\ [r_2]_{1^{st}\ iteration}=\left[r_{2,2}\right]_{1^{st}\ iteration}=\mathbf{0.0000826908799381157}\ au\end{array}\right\}\Rightarrow$$

$$\left[r_{1,2}^{\ 2}\right]_{1^{st}\ iteration}-2\left[r_{1,2}\right]_{1^{st}\ iteration}\left[r_{2,2}\right]_{1^{st}\ iteration}Cos\,\phi_0+\left[r_{2,2}^{\ 2}\right]_{1^{st}\ iteration}$$
$$=\mathbf{1.00031672668333}\ au^2=\left[b^2\right]_{1^{st}\ iteration}$$

Based on the already calculated value of $\ell_z\left(\phi_{\gamma_{2,\min}}\right)=\ell_{(z,i=689)}$ (120), and the possible solution alternatives found for $[\beta]_{1^{st}\ iteration}$ (202) and (r_1,r_2) (203), let us continue to find the values of $\left(\left[\ell_{(x,i=689)}\right]_{1^{st}\ iteration},\left[\ell_{(y,i=689)}\right]_{1^{st}\ iteration},\left[\ell_{(i=689)}\right]_{1^{st}\ iteration}\right)$ (177), to justify whether the result is consistent with the value of $\left[\ell_{(i=689)}\right]_{1^{st}\ iteration}$ (179).

Using (9) - (11) and taking the Dot Products[14] of $\vec{a}$ (10) and $\vec{b}$ (11) with $\vec{\ell}(\phi)$ (9) would yield $\vec{a}\cdot\vec{\ell}(\phi)$ (205) and $\vec{b}\cdot\vec{\ell}(\phi)$ (206), respectively. Subsequently, we can solve for $\ell_x(\phi)$ (207) and $\ell_y(\phi)$ (208) from (205) - (206), in terms of $\vec{a}\cdot\vec{\ell}(\phi)$, $\vec{b}\cdot\vec{\ell}(\phi)$, and $\ell_z(\phi)$.

$$\vec{a}\cdot\vec{\ell}(\phi)=\left(r_1Cos\,\beta\,Cos\,\phi_0-r_2\right)\ell_x(\phi)+r_1Sin\phi_0\ell_y(\phi)+r_1Sin\,\beta\,Cos\,\phi_0\ell_z(\phi)\qquad(205)$$

$$\vec{b}\cdot\vec{\ell}(\phi)=-r_1Cos\,\beta\,Sin\phi_0\ell_x(\phi)+\left(r_1Cos\,\phi_0-r_2\right)\ell_y(\phi)-r_1Sin\,\beta\,Sin\phi_0\ell_z(\phi)\qquad(206)$$

$$\ell_x(\phi)=\frac{\left(r_1Cos\,\phi_0-r_2\right)\left[\vec{a}\cdot\vec{\ell}(\phi)\right]-r_1Sin\phi_0\left[\vec{b}\cdot\vec{\ell}(\phi)\right]-r_1Sin\,\beta\left(r_1-r_2Cos\,\phi_0\right)\ell_z(\phi)}{r_1^{\ 2}Cos\,\beta-r_1r_2\,Cos\,\phi_0\left(1+Cos\,\beta\right)+r_2^{\ 2}}\qquad(207)$$

$$\ell_y(\phi)=\frac{r_1Cos\,\beta\,Sin\phi_0\left[\vec{a}\cdot\vec{\ell}(\phi)\right]+\left(r_1Cos\,\beta\,Cos\,\phi_0-r_2\right)\left[\vec{b}\cdot\vec{\ell}(\phi)\right]-r_1r_2Sin\,\beta Sin\phi_0\ell_z(\phi)}{r_1^{\ 2}Cos\,\beta-r_1r_2\,Cos\,\phi_0\left(1+Cos\,\beta\right)+r_2^{\ 2}}\qquad(208)$$

We can simplify (207) as in (213) using coefficients $c_1(\phi)$ (209) and c_2 (210), and we can also simplify (208) as in (213) using coefficients $c_3(\phi)$ (211) and c_4 (212). We would therefore obtain the simplified quadratic equation (214) in $\ell_z(\phi)$ based on $\left[\ell^2(\phi)=\ell_x^{\ 2}(\phi)+\ell_y^{\ 2}(\phi)+\ell_z^{\ 2}(\phi)\right]$ (9), having the coefficients c_5 (215), $c_6(\phi)$ (215), and $c_7(\phi)$ (215), whose solution for $\ell_z(\phi)$ would be as in (216).

$$c_1(\phi)=\frac{(r_1 Cos\,\phi_0-r_2)\left[\vec{\boldsymbol{a}}\cdot\vec{\ell}(\phi)\right]-r_1 Sin\phi_0\left[\vec{\boldsymbol{b}}\cdot\vec{\ell}(\phi)\right]}{r_1^2 Cos\,\beta-r_1 r_2\,Cos\,\phi_0\,(1+Cos\,\beta)+r_2^2} \tag{209}$$

$$c_2=\frac{-r_1 Sin\,\beta\,(r_1-r_2 Cos\,\phi_0)}{r_1^2 Cos\,\beta-r_1 r_2\,Cos\,\phi_0\,(1+Cos\,\beta)+r_2^2} \tag{210}$$

$$c_3(\phi)=\frac{r_1 Cos\,\beta\,Sin\phi_0\left[\vec{\boldsymbol{a}}\cdot\vec{\ell}(\phi)\right]+(r_1 Cos\,\beta\,Cos\,\phi_0-r_2)\left[\vec{\boldsymbol{b}}\cdot\vec{\ell}(\phi)\right]}{r_1^2 Cos\,\beta-r_1 r_2\,Cos\,\phi_0\,(1+Cos\,\beta)+r_2^2} \tag{211}$$

$$c_4=\frac{-r_1 r_2 Sin\,\beta Sin\phi_0}{r_1^2 Cos\,\beta-r_1 r_2\,Cos\,\phi_0\,(1+Cos\,\beta)+r_2^2} \tag{212}$$

$$\ell_x(\phi)=c_1(\phi)+c_2\ell_z(\phi) \quad ; \quad \ell_y(\phi)=c_3(\phi)+c_4\ell_z(\phi) \tag{213}$$

$$c_5\ell_z^{\,2}(\phi)+2c_6(\phi)\ell_z(\phi)+c_7(\phi)=0 \tag{214}$$

$$c_5=c_2^{\,2}+c_4^{\,2}+1 \quad ; \quad c_6(\phi)=c_1(\phi)c_2+c_3(\phi)c_4 \quad ; \quad c_7(\phi)=c_1^2(\phi)+c_3^2(\phi)-\ell^2(\phi) \tag{215}$$

$$\ell_z(\phi)=\frac{-c_6(\phi)\pm\sqrt{c_6^{\,2}(\phi)-c_5c_7(\phi)}}{c_5} \tag{216}$$

As we now have the values of all $Cos\,\phi$ and $Sin\,\phi$ calculated for the ϕ (136) series based on the solution $\left(\phi_0=\mathbf{-2.63736997198951}\,rad=\mathbf{-151^\circ.11}\right)$ (138), in addition to having the values of all corresponding $\boldsymbol{k}_3(\phi)$ (48) and $\boldsymbol{k}_4(\phi)$ (157), we can now also calculate the values of $\vec{\boldsymbol{a}}\cdot\vec{\ell}(\phi)$ and $\vec{\boldsymbol{b}}\cdot\vec{\ell}(\phi)$ for all ϕ in the Excel sheet "Tan, A.P., Asli Pinar Tan Analysis Based on Earth-Sun distance (d) Landsat.xlsx"[13] using (158) and (159).

Based on these calculated values of $\vec{\boldsymbol{a}}\cdot\vec{\ell}(\phi)$ and $\vec{\boldsymbol{b}}\cdot\vec{\ell}(\phi)$ in the Excel sheet "Tan, A.P., Asli Pinar Tan Analysis Based on Earth-Sun distance (d) Landsat.xlsx"[13], the values of $\vec{\boldsymbol{a}}\cdot\vec{\ell}_{(i=689)}$ (217) and $\vec{\boldsymbol{b}}\cdot\vec{\ell}_{(i=689)}$ (218) corresponding to $\left(\ell_{(x,i=689)},\ell_{(y,i=689)},\ell_{(z,i=689)}\right)$ (177) at the Summer Solstice[5,6] on June 21 can be used together with the already calculated value of $\ell_z\left(\phi_{\gamma_{2,\min}}\right)$ (120), which is restated below as $\ell_{(z,i=689)}$ (219), to calculate the values of $\ell_{(x,i=689)}$ and $\ell_{(y,i=689)}$ based on the solution alternatives we have found for $[\beta]_{1^{st}\ iteration}$ (202) and (r_1,r_2) (203).

$$\vec{\boldsymbol{a}}\cdot\vec{\ell}\left(\phi_{\gamma_{2,\min}}\right)=\vec{\boldsymbol{a}}\cdot\vec{\ell}_{(i=689)}=\mathbf{-0.0123948725020029}\,au^2 \tag{217}$$

$$\vec{\boldsymbol{b}}\cdot\vec{\ell}\left(\phi_{\gamma_{2,\min}}\right)=\vec{\boldsymbol{b}}\cdot\vec{\ell}_{(i=689)}=\mathbf{0.011190035469387}\,au^2 \tag{218}$$

$$\ell_z\left(\phi_{\gamma_{2,\min}}\right)=\ell_{(z,i=689)}=\begin{cases}\mathbf{0.00646733505568997}\,au & if\quad \hat{\boldsymbol{u}}_{2\perp}=\hat{z}\\ \mathbf{-0.00646733505568997}\,au & if\quad \hat{\boldsymbol{u}}_{2\perp}=-\hat{z}\end{cases} \tag{219}$$

The possible values of $\left[\ell_{(x,i=689)}\right]_{1^{st}\ iteration}$ and $\left[\ell_{(y,i=689)}\right]_{1^{st}\ iteration}$ are calculated in (220) - (221)based on (207) - (208), or equivalently (213), using the values of $\ell_{(z,i=689)}$ (219) and the solution alternatives we have found for $[\beta]_{1^{st}\ iteration}$ (202) and (r_1, r_2) (203).

$$\left.\left.\begin{array}{l} \left.\begin{array}{l} \left[\beta_{1,1}\right]_{1^{st}\ iteration} = \mathbf{23^{\circ}.48} \\ \Rightarrow \hat{\boldsymbol{u}}_{2\perp} = \hat{z} \\ \Rightarrow \ell_{(z,i=689)} = \mathbf{0.00646733505568997}\ au \end{array}\right\} \\ \left.\begin{array}{l} \left[\beta_{1,2}\right]_{1^{st}\ iteration} = \mathbf{-23^{\circ}.48} \\ \Rightarrow \hat{\boldsymbol{u}}_{2\perp} = -\hat{z} \\ \Rightarrow \ell_{(z,i=689)} = \mathbf{-0.00646733505568997}\ au \end{array}\right\} \end{array}\right\}\left\{\begin{array}{l} \left.\begin{array}{l} \left[r_{1,1}\right]_{1^{st}\ iteration} = \mathbf{0.998478359964017}\ au \\ \left[r_{2,1}\right]_{1^{st}\ iteration} = \mathbf{0.00191829239539248}\ au \end{array}\right\} \\ \left.\begin{array}{l} \left[\ell_{(x,i=689),1}\right]_{1^{st}\ iteration} = \mathbf{0.0149203905061881}\ au \\ \left[\ell_{(y,i=689),1}\right]_{1^{st}\ iteration} = \mathbf{0.00379471391702057}\ au \end{array}\right\} \end{array}\right. \tag{220}$$

$$\left.\left.\begin{array}{l} \left.\begin{array}{l} \left[\beta_{2,1}\right]_{1^{st}\ iteration} = \mathbf{156^{\circ}.56} \\ \Rightarrow \hat{\boldsymbol{u}}_{2\perp} = \hat{z} \\ \Rightarrow \ell_{(z,i=689)} = \mathbf{0.00646733505568997}\ au \end{array}\right\} \\ \left.\begin{array}{l} \left[\beta_{2,2}\right]_{1^{st}\ iteration} = \mathbf{-156^{\circ}.56} \\ \Rightarrow \hat{\boldsymbol{u}}_{2\perp} = -\hat{z} \\ \Rightarrow \ell_{(z,i=689)} = \mathbf{-0.00646733505568997}\ au \end{array}\right\} \end{array}\right\}\left\{\begin{array}{l} \left.\begin{array}{l} \left[r_{1,2}\right]_{1^{st}\ iteration} = \mathbf{1.00008594998301}\ au \\ \left[r_{2,2}\right]_{1^{st}\ iteration} = \mathbf{0.0000826908799381157}\ au \end{array}\right\} \\ \left.\begin{array}{l} \left[\ell_{(x,i=689),2}\right]_{1^{st}\ iteration} = \mathbf{-0.0149167156570298}\ au \\ \left[\ell_{(y,i=689),2}\right]_{1^{st}\ iteration} = \mathbf{0.00380913383546712}\ au \end{array}\right\} \end{array}\right. \tag{221}$$

The resultant value of $\left[\ell_{(i=689)}\right]_{2^{nd}\ iteration}$ (222), which is calculated based on (177), is the same and is $\mathbf{2,498,079.874210}\ km$ for all the possible solution alternatives in (220) and (221).

$$\begin{aligned} \left[\ell_{(i=689)}\right]_{2^{nd}\ iteration} &= \left[\ell\left(\phi_{\gamma 2,\min}\right)\right]_{2^{nd}\ iteration} = \left[\ell\left(-\phi_0\right)\right]_{2^{nd}\ iteration} \\ &= \mathbf{0.01669863255754}\ au = \mathbf{2,498,079.874210}\ km \end{aligned} \tag{222}$$

As clearly expressed in (223), the value we have found for $\left[\ell_{(i=689)}\right]_{2^{nd}\ iteration}$ (222), which is $\mathbf{2,498,079.874210}\ km$, is close but deviated by around $\mathbf{65,884.0}\ km$ from the value we have found earlier for $\left[\ell_{(i=689)}\right]_{1^{st}\ iteration}$ (179) value, which is $\mathbf{2,432,195.841362}\ km$, with the assumption that $\alpha_{\left[\vec{d}\left(\phi_{\gamma 2,\min}\right)\measuredangle\vec{d}\left(\phi_{\gamma 2,\max}\right)\right],1^{st}\ iteration}$ (171) between $\vec{\boldsymbol{d}}\left(\phi_{\gamma 2,\min}\right)$ and $\vec{\boldsymbol{d}}\left(\phi_{\gamma 2,\max}\right)$ must be approximately $\pi\ radians = 180^{\circ}$.

$$\begin{aligned} \left[\ell_{(i=689)}\right]_{2^{nd}\ iteration} &= \mathbf{0.01669863255754}\ au = \mathbf{2,498,079.874210}\ km \\ &\neq \mathbf{0.0162582250000018}\ au = \mathbf{2,432,195.841362}\ km = \left[\ell_{(i=689)}\right]_{1^{st}\ iteration} \end{aligned} \tag{223}$$

As we are looking for an analytical result to an approximation, we can go on and apply an iterative approach, and as far as our analytical calculations are concerned, we can safely take the new found value of $\left[\ell_{(i=689)}\right]_{2^{nd}\ iteration} = \mathbf{0.01669863255754}\ au$ (222) as a new starting point for our 2nd iteration of calculations along the lines of (171) - (221).

It is just as possible that the exact angle $\alpha_{\left[\vec{d}\left(\phi_{\gamma 2,\min}\right)\measuredangle \vec{d}\left(\phi_{\gamma 2,\max}\right)\right]}$ (169) between $\vec{\boldsymbol{d}}\left(\phi_{\gamma 2,\min}\right)$ and $\vec{\boldsymbol{d}}\left(\phi_{\gamma 2,\max}\right)$ is close to $\pi\ radians = 180^\circ$, but <u>not exactly</u> $\pi\ radians = 180^\circ$. Based on (169) and having the new found value of $\left[\ell\left(\phi_{\gamma 2,\min}\right)\right]_{2^{nd}\ iteration} = \mathbf{0.01669863255754}\ au$ (222), we can rewrite (172) as in (224) to find a more accurate value for $\alpha_{\left[\vec{d}\left(\phi_{\gamma 2,\min}\right)\measuredangle \vec{d}\left(\phi_{\gamma 2,\max}\right)\right],2^{nd}\ iteration}$ (224), that may be closer to the actual value of $\alpha_{\left[\vec{d}\left(\phi_{\gamma 2,\min}\right)\measuredangle \vec{d}\left(\phi_{\gamma 2,\max}\right)\right]}$ (169).

$$\begin{aligned}
\vec{\boldsymbol{d}}\left(\phi_{\gamma 2,\min}\right)\cdot\vec{\boldsymbol{d}}\left(\pi+\phi_{\gamma 2,\min}\right) &= \vec{\boldsymbol{d}}\left(\phi_{\gamma 2,\min}\right)\cdot\vec{\boldsymbol{d}}\left(\phi_{\gamma 2,\max}\right) \\
&= d\left(\phi_{\gamma 2,\min}\right)d\left(\phi_{\gamma 2,\max}\right)Cos\,\alpha_{\left[\vec{d}\left(\phi_{\gamma 2,\min}\right)\measuredangle \vec{d}\left(\phi_{\gamma 2,\max}\right)\right],2^{nd}\ iteration} \\
&= d_{(i=689)}\,d_{(i+730=1419)}Cos\,\alpha_{\left[\vec{d}\left(\phi_{\gamma 2,\min}\right)\measuredangle \vec{d}\left(\phi_{\gamma 2,\max}\right)\right],2^{nd}\ iteration} \\
&= \left[\ell^2\left(\phi_{\gamma 2,\min}\right)\right]_{2^{nd}\ iteration} - \left(\vec{\boldsymbol{a}}\,Cos\,\phi_{\gamma 2,\min} + \vec{\boldsymbol{b}}\,Sin\,\phi_{\gamma 2,\min}\right)^2
\end{aligned} \tag{224}$$

In this case, utilizing (224) and $\boldsymbol{k}_1(\phi)$ (46) for $\left(\phi = \phi_{\gamma 2,\min} = -\phi_0\right)$ (139), which takes on the value $\boldsymbol{k}_{1,(i=689)}$ (175) based on the Excel sheet "Tan, A.P., Asli Pinar Tan Analysis Based on Earth-Sun distance (d) Landsat.xlsx"[13], and using $\left[\ell_{(i=689)}\right]_{2^{nd}\ iteration}$ (222), the angle $\alpha_{\left[\vec{d}\left(\phi_{\gamma 2,\min}\right)\measuredangle \vec{d}\left(\phi_{\gamma 2,\max}\right)\right],2^{nd}\ iteration}$ (225) between $\vec{\boldsymbol{d}}\left(\phi_{\gamma 2,\min}\right)$ and $\vec{\boldsymbol{d}}\left(\phi_{\gamma 2,\max}\right)$ can be calculated, which turns out to be a value close but not equal to $\pi\ radians = 180^\circ$.

$$\begin{aligned}
\alpha_{\left[\vec{d}\left(\phi_{\gamma 2,\min}\right)\measuredangle \vec{d}\left(\phi_{\gamma 2,\max}\right)\right],2^{nd}\ iteration} &= Cos^{-1}\left(\frac{2\left[\ell_{(i=689)}\right]_{2^{nd}\ iteration} - \boldsymbol{k}_{1,(i=689)}}{d_{(i=689)}\,d_{(i+730=1419)}}\right) \\
&= \mathbf{3.13397221004863}\ radians = \mathbf{179^\circ.56}
\end{aligned} \tag{225}$$

Based on the definition of $\boldsymbol{k}_1(\phi)$ (46), we can find the value of $\left[b^2\right]_{2^{nd}\ iteration}$ (226), making use of the already found values of $\boldsymbol{k}_1(-\phi_0)$ (175), $\left[2r_1 r_2\left(1-Cos\beta\right)Cos\phi_0\right]$ (132), and $\left[\ell^2(-\phi_0)\right]_{2^{nd}\ iteration}$ (222). This allows us to determine the magnitude $\left[b\right]_{2^{nd}\ iteration}$ (227) of the semi-major axis vector $\vec{\boldsymbol{b}}$ (11) of the ellipse for the Sun-Earth system formed by their relative motions, which is the the semi-major axis vector based on the **ARTICLE 1** "Points on two circles in space mimic an ellipse with respect to a point" and (10) - (11), as $\left(Cos\,\phi_0 < 0\right)$ (129)

and $\left(a^2 < b^2\right)$ (131). Based on (11) and (226), we have also obtained the identity (228) at this point.

$$\left[b^2\right]_{2^{nd}\ iteration} = \boldsymbol{k}_1\left(-\phi_0\right) - 2r_1 r_2\left(1 - Cos\beta\right)Cos\phi_0 - \left[\ell^2\left(-\phi_0\right)\right]_{2^{nd}\ iteration} = \mathbf{1.00030221223419}\ au^2 \quad (226)$$

$$\left[b\right]_{2^{nd}\ iteration} = \mathbf{1.00015109470229}\ au = \mathbf{149{,}620{,}474.145737}\ km \quad (227)$$

$$\begin{aligned}\left[b^2\right]_{2^{nd}\ iteration} &= \left[r_1^{\ 2}\right]_{2^{nd}\ iteration} - 2\left[r_1\right]_{2^{nd}\ iteration}\left[r_2\right]_{2^{nd}\ iteration} Cos\,\phi_0 + \left[r_2^{\ 2}\right]_{2^{nd}\ iteration} \\ &= \mathbf{1.00030221223419}\ au^2\end{aligned} \quad (228)$$

In this 2nd iteration of calculations, our three equations (72), (141), and (11) are as restated in (229), (230), and (231), respectively, in the three variables r_1 (6), r_2 (7), and β (6), that we are set to solve for, where $\left[m_3\right]_{2^{nd}\ iteration}$ (232) is given by (228).

$$2r_1 r_2\left(1 - Cos\beta\right) = \mathbf{0.000317144894687038}\ au^2 = m_1 \quad (229)$$

$$r_1\, Sin\beta = \pm\mathbf{0.397796274218557}\ au = m_2 \quad (230)$$

$$b^2 = r_1^{\ 2} - 2\ r_1 r_2 Cos\,\phi_0 + r_2^{\ 2} = m_3 \quad (231)$$

$$\left[m_3\right]_{2^{nd}\ iteration} = \left[b^2\right]_{2^{nd}\ iteration} = \mathbf{1.00030221223419}\ au^2 \quad (232)$$

Using the analysis in (187) - (197), and based on the values of m_1 (229), m_2 (230), and $\left[m_3\right]_{2^{nd}\ iteration}$ (232), we would obtain the values for $\left[\mathbf{A}_\beta\right]_{2^{nd}\ iteration}$ (233), $\mathbf{B}_\beta$ (234), and $\left[\mathbf{C}_\beta\right]_{2^{nd}\ iteration}$ (235).

$$\left[\mathbf{A}_\beta\right]_{2^{nd}\ iteration} = \frac{m_1^{\ 2}}{4m_2^{\ 2}} + \left[m_3\right]_{2^{nd}\ iteration} = \mathbf{1.0003023711379}\ au^2 \quad (233)$$

$$\mathbf{B}_\beta = \frac{m_1^{\ 2}}{2m_2^{\ 2}} - m_1 Cos\,\phi_0 = \mathbf{0.00027799410952802}\ au^2 \quad (234)$$

$$\left[\mathbf{C}_\beta\right]_{2^{nd}\ iteration} = \frac{m_1^{\ 2}}{4m_2^{\ 2}} + m_2^{\ 2} - m_1 Cos\,\phi_0 - \left[m_3\right]_{2^{nd}\ iteration} = \mathbf{-0.84178250124621}\ au^2 \quad (235)$$

Using (197), we can now solve for $\left[Cos\beta\right]_{2^{nd}\ iteration}$ (236) to obtain its possible values from the quadratic equation in (192), or in (193), based on the calculated values of $\left[\mathbf{A}_\beta\right]_{2^{nd}\ iteration}$ (233), $\mathbf{B}_\beta$ (234), and $\left[\mathbf{C}_\beta\right]_{2^{nd}\ iteration}$ (235).

$$\left[Cos\beta\right]_{2^{nd}\ iteration} = \frac{-\mathbf{B}_\beta \pm \sqrt{\mathbf{B}_\beta^{\ 2} - 4\left[\mathbf{A}_\beta\right]_{2^{nd}\ iteration}\left[\mathbf{C}_\beta\right]_{2^{nd}\ iteration}}}{2\left[\mathbf{A}_\beta\right]_{2^{nd}\ iteration}} = \begin{cases}\mathbf{0.917209433939717} \\ \mathbf{-0.917487344017259}\end{cases} \quad (236)$$

Based on the two possible solutions $\left[Cos\beta_1\right]_{2^{nd}\ iteration}$ (237) and $\left[Cos\beta_2\right]_{2^{nd}\ iteration}$ (237) obtained for the value of $\left[Cos\beta\right]_{2^{nd}\ iteration}$ (236), there are four possible values $\left[\beta_{1,1}\right]_{2^{nd}\ iteration}$ (237),

$\left[\beta_{1,2}\right]_{2^{nd}\ iteration}$ (237), $\left[\beta_{2,1}\right]_{2^{nd}\ iteration}$ (237), and $\left[\beta_{2,2}\right]_{2^{nd}\ iteration}$ (237) that $\left[\beta\right]_{2^{nd}\ iteration}$ (236) can take.

$$\left[Cos\beta\right]_{2^{nd}\ iteration} = \left[Cos\beta_1\right]_{2^{nd}\ iteration} = \mathbf{0.917209433939717}$$

$$\Rightarrow \begin{cases} \left[\beta\right]_{2^{nd}\ iteration} = \left[\beta_{1,1}\right]_{2^{nd}\ iteration} = \mathbf{0.409777642442942}\ radians = \mathbf{23^\circ.48} \\ \left[\beta\right]_{2^{nd}\ iteration} = \left[\beta_{1,2}\right]_{2^{nd}\ iteration} = \mathbf{-0.409777642442942}\ radians = \mathbf{-23^\circ.48} \end{cases}$$

$$\left[Cos\beta\right]_{2^{nd}\ iteration} = \left[Cos\beta_2\right]_{2^{nd}\ iteration} = \mathbf{-0.917487344017259}$$

$$\Rightarrow \begin{cases} \left[\beta\right]_{2^{nd}\ iteration} = \left[\beta_{2,1}\right]_{2^{nd}\ iteration} = \mathbf{2.73251312823199}\ radians = \mathbf{156^\circ.56} \\ \left[\beta\right]_{2^{nd}\ iteration} = \left[\beta_{2,2}\right]_{2^{nd}\ iteration} = \mathbf{-2.73251312823199}\ radians = \mathbf{-156^\circ.56} \end{cases} \quad (237)$$

For these four possible values (237) that $\left[\beta\right]_{2^{nd}\ iteration}$ (236) can take, the possible values $\left[r_1\right]_{2^{nd}\ iteration}$ (6) and $\hat{\boldsymbol{u}}_{2\perp}$ (97) can take based on (141) are listed in (238), and subsequently the possible values that $\left[r_2\right]_{2^{nd}\ iteration}$ (7) can take based on (183) are also listed in (238).

$$\left.\begin{matrix} \left[\beta_{1,1}\right]_{2^{nd}\ iteration} = \mathbf{23^\circ.48} \Rightarrow Sin\left[\beta_{1,1}\right]_{2^{nd}\ iteration} = \mathbf{0.398405389386218} \Rightarrow \hat{\boldsymbol{u}}_{2\perp} = \hat{z} \\ \left[\beta_{1,2}\right]_{2^{nd}\ iteration} = \mathbf{-23^\circ.48} \Rightarrow Sin\left[\beta_{1,2}\right]_{2^{nd}\ iteration} = \mathbf{-0.398405389386218} \Rightarrow \hat{\boldsymbol{u}}_{2\perp} = -\hat{z} \end{matrix}\right\}$$

$$\Rightarrow \left[r_1\right]_{2^{nd}\ iteration} = \left[r_{1,1}\right]_{2^{nd}\ iteration} = \mathbf{0.998471117148792}\ au$$

$$\Rightarrow \left[r_2\right]_{2^{nd}\ iteration} = \left[r_{2,1}\right]_{2^{nd}\ iteration} = \mathbf{0.00191827722437874}\ au \quad (238)$$

$$\left.\begin{matrix} \left[\beta_{2,1}\right]_{2^{nd}\ iteration} = \mathbf{156^\circ.56} \Rightarrow Sin\left[\beta_{2,1}\right]_{2^{nd}\ iteration} = \mathbf{0.397764972776835} \Rightarrow \hat{\boldsymbol{u}}_{2\perp} = \hat{z} \\ \left[\beta_{2,2}\right]_{2^{nd}\ iteration} = \mathbf{-156^\circ.56} \Rightarrow Sin\left[\beta_{2,2}\right]_{2^{nd}\ iteration} = \mathbf{-0.397764972776835} \Rightarrow \hat{\boldsymbol{u}}_{2\perp} = -\hat{z} \end{matrix}\right\}$$

$$\Rightarrow \left[r_1\right]_{2^{nd}\ iteration} = \left[r_{1,2}\right]_{2^{nd}\ iteration} = \mathbf{1.00007869330852}\ au$$

$$\Rightarrow \left[r_2\right]_{2^{nd}\ iteration} = \left[r_{2,2}\right]_{2^{nd}\ iteration} = \mathbf{0.0000826915339126612}\ au$$

Calculating the value of $\left(r_1^2 - 2\, r_1 r_2 Cos\,\phi_0 + r_2^2\right)$ (239) with both possible $\left(r_1, r_2\right)$ (238) solution alternatives fo a validity check, we see that all match the expected $\left[b^2\right]_{2^{nd}\ iteration}$ (228) value we have found before.

$$\left.\begin{array}{l}[r_1]_{2^{nd}\ iteration} = [r_{1,1}]_{2^{nd}\ iteration} = \mathbf{0.998471117148792}\ au \\ [r_2]_{2^{nd}\ iteration} = [r_{2,1}]_{2^{nd}\ iteration} = \mathbf{0.00191827722437874}\ au\end{array}\right\} \Rightarrow$$

$$[r_{1,1}^2]_{2^{nd}\ iteration} - 2[r_{1,1}]_{2^{nd}\ iteration}[r_{2,1}]_{2^{nd}\ iteration} Cos\,\phi_0 + [r_{2,1}^2]_{2^{nd}\ iteration}$$
$$= \mathbf{1.00030221223419}\ au^2 = [b^2]_{2^{nd}\ iteration} \tag{239}$$

$$\left.\begin{array}{l}[r_1]_{2^{nd}\ iteration} = [r_{1,2}]_{2^{nd}\ iteration} = \mathbf{1.00007869330852}\ au \\ [r_2]_{2^{nd}\ iteration} = [r_{2,2}]_{2^{nd}\ iteration} = \mathbf{0.0000826915339126612}\ au\end{array}\right\} \Rightarrow$$

$$[r_{1,2}^2]_{2^{nd}\ iteration} - 2[r_{1,2}]_{2^{nd}\ iteration}[r_{2,2}]_{2^{nd}\ iteration} Cos\,\phi_0 + [r_{2,2}^2]_{2^{nd}\ iteration}$$
$$= \mathbf{1.00030221223419}\ au^2 = [b^2]_{2^{nd}\ iteration}$$

The possible values of $[\ell_{(x,i=689)}]_{2^{nd}\ iteration}$ and $[\ell_{(y,i=689)}]_{2^{nd}\ iteration}$ are calculated in (240) - (241)based on (207) and (208), using the already calculated value of $\ell_z(\phi_{\gamma_{2,\min}})$ (120) restated as $\ell_{(z,i=689)}$ (219), and the solution alternatives we have found for $[\beta]_{2^{nd}\ iteration}$ (237) and (r_1, r_2) (238), and the calculated values of $\vec{a}\cdot\vec{\ell}_{(i=689)}$ (217) and $\vec{b}\cdot\vec{\ell}_{(i=689)}$ (218) in the Excel sheet "Tan, A.P., Asli Pinar Tan Analysis Based on Earth-Sun distance (d) Landsat.xlsx"[13] corresponding to $(\ell_{(x,i=689)}, \ell_{(y,i=689)}, \ell_{(z,i=689)})$ (177) at the Summer Solstice[5,6] on June 21.

$$\left.\begin{array}{l}\left.\begin{array}{l}[\beta_{1,1}]_{2^{nd}\ iteration} = \mathbf{23^\circ.48} \\ \Rightarrow \hat{\boldsymbol{u}}_{2\perp} = \hat{\boldsymbol{z}} \\ \Rightarrow \ell_{(z,i=689)} = \mathbf{0.00646733505568997}\ au\end{array}\right\} \\ \left.\begin{array}{l}[\beta_{1,2}]_{2^{nd}\ iteration} = \mathbf{-23^\circ.48} \\ \Rightarrow \hat{\boldsymbol{u}}_{2\perp} = -\hat{\boldsymbol{z}} \\ \Rightarrow \ell_{(z,i=689)} = \mathbf{-0.00646733505568997}\ au\end{array}\right\}\end{array}\right\} \left\{\begin{array}{l}[r_{1,1}]_{2^{nd}\ iteration} = \mathbf{0.998471117148792}\ au \\ [r_{2,1}]_{2^{nd}\ iteration} = \mathbf{0.00191827722437874}\ au \\ [\ell_{(x,i=689),1}]_{2^{nd}\ iteration} = \mathbf{0.0149205191356486}\ au \\ [\ell_{(y,i=689),1}]_{2^{nd}\ iteration} = \mathbf{0.00379474143784028}\ au\end{array}\right. \tag{240}$$

$$\left.\begin{array}{l}\left.\begin{array}{l}[\beta_{2,1}]_{2^{nd}\ iteration} = \mathbf{156^\circ.56} \\ \Rightarrow \hat{\boldsymbol{u}}_{2\perp} = \hat{\boldsymbol{z}} \\ \Rightarrow \ell_{(z,i=689)} = \mathbf{0.00646733505568997}\ au\end{array}\right\} \\ \left.\begin{array}{l}[\beta_{2,2}]_{2^{nd}\ iteration} = \mathbf{-156^\circ.56} \\ \Rightarrow \hat{\boldsymbol{u}}_{2\perp} = -\hat{\boldsymbol{z}} \\ \Rightarrow \ell_{(z,i=689)} = \mathbf{-0.00646733505568997}\ au\end{array}\right\}\end{array}\right\} \left\{\begin{array}{l}[r_{1,2}]_{2^{nd}\ iteration} = \mathbf{1.00007869330852}\ au \\ [r_{2,2}]_{2^{nd}\ iteration} = \mathbf{0.0000826915339126612}\ au \\ [\ell_{(x,i=689),2}]_{2^{nd}\ iteration} = \mathbf{-0.0149168442598292}\ au \\ [\ell_{(y,i=689),2}]_{2^{nd}\ iteration} = \mathbf{0.00380916148060411}\ au\end{array}\right. \tag{241}$$

The resultant value of $\left[\ell_{(i=689)}\right]_{3^{rd}\ iteration}$ (242), which is calculated based on (177), is the same and is $\mathbf{2,498,098.003344}\ km$ for all the possible solution values of $\left[\ell_{(x,i=689)}\right]_{2^{nd}\ iteration}$ and $\left[\ell_{(y,i=689)}\right]_{2^{nd}\ iteration}$ in (240) - (241), and the value of $\ell_z\left(\phi_{\Upsilon 2,\min}\right)$ (120) restated as $\ell_{(z,i=689)}$ (219).

$$\begin{aligned}\left[\ell_{(i=689)}\right]_{3^{rd}\ iteration} &= \left[\ell\left(\phi_{\Upsilon 2,\min}\right)\right]_{3^{rd}\ iteration} = \left[\ell\left(-\phi_0\right)\right]_{3^{rd}\ iteration} \\ &= \mathbf{0.0166987537433149}\ au = \mathbf{2,498,098.003344}\ km\end{aligned} \tag{242}$$

As clearly expressed in (243), the value we have found for $\left[\ell_{(i=689)}\right]_{3^{rd}\ iteration}$ (242), which is $\mathbf{2,498,098.003344}\ km$, is close but deviated by around $\mathbf{18.1}\ km$ from the value we have found earlier for $\left[\ell_{(i=689)}\right]_{2^{nd}\ iteration}$ (222), which is $\mathbf{2,498,079.874210}\ km$ and was deviated by around $\mathbf{65,884.0}\ km$ from the value we have found earlier for $\left[\ell_{(i=689)}\right]_{1^{st}\ iteration}$ (179) value, which is $\mathbf{2,432,195.841362}\ km$, based on an initial assumption that $\alpha_{\left[\vec{d}\left(\phi_{\Upsilon 2,\min}\right)\measuredangle\vec{d}\left(\phi_{\Upsilon 2,\max}\right)\right],1^{st}\ iteration} = 180^\circ$ (171).

$$\begin{aligned}\left[\ell_{(i=689)}\right]_{3^{rd}\ iteration} &= \mathbf{0.0166987537433149}\ au = \mathbf{2,498,098.003344}\ km \\ &\neq \mathbf{0.01669863255754}\ au = \mathbf{2,498,079.874210}\ km = \left[\ell_{(i=689)}\right]_{2^{nd}\ iteration} \\ &\neq \mathbf{0.0162582250000018}\ au = \mathbf{2,432,195.841362}\ km = \left[\ell_{(i=689)}\right]_{1^{st}\ iteration}\end{aligned} \tag{243}$$

Although we have achieved a better result with $\left[\ell_{(i=689)}\right]_{3^{rd}\ iteration}$ (242) compared to that of $\left[\ell_{(i=689)}\right]_{2^{nd}\ iteration}$ (222), we can go on applying this iterative approach until we find sufficiently more accurate results, and as far as our analytical approximation is concerned, we can now take the new found value of $\left[\ell_{(i=689)}\right]_{3^{rd}\ iteration} = \mathbf{0.0166987537433149}\ au$ (242) as a new starting point for our 3rd iteration of calculations along the lines of (224) - (241).

Based on (169) and with the new value of $\left[\ell\left(\phi_{\Upsilon 2,\min}\right)\right]_{3^{rd}\ iteration} = \mathbf{0.0166987537433149}\ au$ (242), we can rewrite (224) as in (244) to find a more accurate value for $\alpha_{\left[\vec{d}\left(\phi_{\Upsilon 2,\min}\right)\measuredangle\vec{d}\left(\phi_{\Upsilon 2,\max}\right)\right],3^{rd}\ iteration}$ (245), that may be closer to the actual value of $\alpha_{\left[\vec{d}\left(\phi_{\Upsilon 2,\min}\right)\measuredangle\vec{d}\left(\phi_{\Upsilon 2,\max}\right)\right]}$ (169).

$$\vec{d}\left(\phi_{\gamma_{2,\min}}\right)\cdot\vec{d}\left(\pi+\phi_{\gamma_{2,\min}}\right)=\vec{d}\left(\phi_{\gamma_{2,\min}}\right)\cdot\vec{d}\left(\phi_{\gamma_{2,\max}}\right)$$
$$=d\left(\phi_{\gamma_{2,\min}}\right)d\left(\phi_{\gamma_{2,\max}}\right)Cos\,\alpha_{\left[\vec{d}\left(\phi_{\gamma_{2,\min}}\right)\measuredangle\vec{d}\left(\phi_{\gamma_{2,\max}}\right)\right],3^{rd}\ iteration}$$
$$=d_{(i=689)}\,d_{(i+730=1419)}Cos\,\alpha_{\left[\vec{d}\left(\phi_{\gamma_{2,\min}}\right)\measuredangle\vec{d}\left(\phi_{\gamma_{2,\max}}\right)\right],3^{rd}\ iteration} \tag{244}$$
$$=\left[\ell^2\left(\phi_{\gamma_{2,\min}}\right)\right]_{3^{rd}\ iteration}-\left(\vec{a}\,Cos\,\phi_{\gamma_{2,\min}}+\vec{b}\,Sin\,\phi_{\gamma_{2,\min}}\right)^2$$

In this case, utilizing (244) and $\boldsymbol{k}_1(\phi)$ (46) for $\left(\phi=\phi_{\gamma_{2,\min}}=-\phi_0\right)$ (139), which takes on the value $\boldsymbol{k}_{1,(i=689)}$ (175) based on the Excel sheet "Tan, A.P., Asli Pinar Tan Analysis Based on Earth-Sun distance (d) Landsat.xlsx"[13], and using $\left[\ell_{(i=689)}\right]_{3^{rd}\ iteration}$ (242), the angle $\alpha_{\left[\vec{d}\left(\phi_{\gamma_{2,\min}}\right)\measuredangle\vec{d}\left(\phi_{\gamma_{2,\max}}\right)\right],3^{rd}\ iteration}$ (245) between $\vec{d}\left(\phi_{\gamma_{2,\min}}\right)$ and $\vec{d}\left(\phi_{\gamma_{2,\max}}\right)$ is calculated, which turns out to be a value almost the same as that of $\alpha_{\left[\vec{d}\left(\phi_{\gamma_{2,\min}}\right)\measuredangle\vec{d}\left(\phi_{\gamma_{2,\max}}\right)\right],2^{nd}\ iteration}$ (225), close but not equal to $\pi\ radians=180^\circ$.

$$\alpha_{\left[\vec{d}\left(\phi_{\gamma_{2,\min}}\right)\measuredangle\vec{d}\left(\phi_{\gamma_{2,\max}}\right)\right],3^{rd}\ iteration}=Cos^{-1}\left(\frac{2\left[\ell_{(i=689)}\right]_{3^{rd}\ iteration}-\boldsymbol{k}_{1,(i=689)}}{d_{(i=689)}\,d_{(i+730=1419)}}\right) \tag{245}$$
$$=\mathbf{3.13397114765456}\,radians=\mathbf{179^\circ.563}$$

Based on the definition of $\boldsymbol{k}_1(\phi)$ (46), we can find the value of $\left[b^2\right]_{3^{rd}\ iteration}$ (246), making use of the already found values of $\boldsymbol{k}_1(-\phi_0)$ (175), $\left[2r_1r_2\left(1-Cos\beta\right)Cos\phi_0\right]$ (132), and $\left[\ell^2\left(-\phi_0\right)\right]_{3^{rd}\ iteration}$ (242). This allows us to determine the magnitude $\left[b\right]_{3^{rd}\ iteration}$ (247) of the semi-major axis vector $\vec{b}$ (11) of the ellipse for the Sun-Earth system formed by their relative motions, which is the the semi-major axis vector based on the **ARTICLE 1** "Points on two circles in space mimic an ellipse with respect to a point" and (10) - (11), as $\left(Cos\,\phi_0<0\right)$ (129) and $\left(a^2<b^2\right)$ (131). Based on (11) and (246), we also obtain the identity (248) at this point.

$$\left[b^2\right]_{3^{rd}\ iteration}=\boldsymbol{k}_1\left(-\phi_0\right)-2r_1r_2\left(1-Cos\beta\right)Cos\phi_0-\left[\ell^2\left(-\phi_0\right)\right]_{3^{rd}\ iteration}=\mathbf{1.00030220818691}\,au^2 \tag{246}$$

$$\left[b\right]_{3^{rd}\ iteration}=\mathbf{1.00015109267895}\,au=\mathbf{149,620,473.843050}\,km \tag{247}$$

$$\left[b^2\right]_{3^{rd}\ iteration}=\left[r_1^2\right]_{3^{rd}\ iteration}-2\left[r_1\right]_{3^{rd}\ iteration}\left[r_2\right]_{3^{rd}\ iteration}Cos\,\phi_0+\left[r_2^2\right]_{3^{rd}\ iteration}$$
$$=\mathbf{1.00030220818691}\,au^2 \tag{248}$$

In this 3rd iteration of calculations, our three equations (72), (141), and (11) are as restated in (249), (250), and (251), respectively, in the three variables r_1 (6), r_2 (7), and β (6), that we are set to solve for, where $\left[m_3\right]_{3^{rd}\ iteration}$ (252) is given by (248).

$$2r_1r_2\left(1-Cos\beta\right)=\mathbf{0.000317144894687038}\,au^2=m_1 \tag{249}$$

$$r_1 \, Sin\,\beta = \pm\mathbf{0.397796274218557}\; au = m_2 \tag{250}$$

$$b^2 = r_1^{\,2} - 2\, r_1 r_2 Cos\,\phi_0 + r_2^{\,2} = m_3 \tag{251}$$

$$[m_3]_{3^{rd}\ iteration} = \left[b^2\right]_{3^{rd}\ iteration} = \mathbf{1.00030220818691}\; au^2 \tag{252}$$

Using the analysis in (187) - (197), and based on the values of m_1 (249), m_2 (250), and $[m_3]_{3^{rd}\ iteration}$ (252), we would obtain the values for $\left[\mathbf{A}_\beta\right]_{3^{rd}\ iteration}$ (253), $\mathbf{B}_\beta$ (254), and $\left[\mathbf{C}_\beta\right]_{3^{rd}\ iteration}$ (255).

$$\left[\mathbf{A}_\beta\right]_{3^{rd}\ iteration} = \frac{m_1^{\,2}}{4m_2^{\,2}} + [m_3]_{3^{rd}\ iteration} = \mathbf{1.00030236709062}\; au^2 \tag{253}$$

$$\mathbf{B}_\beta = \frac{m_1^{\,2}}{2m_2^{\,2}} - m_1 Cos\,\phi_0 = \mathbf{0.00027799410952802}\; au^2 \tag{254}$$

$$\left[\mathbf{C}_\beta\right]_{3^{rd}\ iteration} = \frac{m_1^{\,2}}{4m_2^{\,2}} + m_2^{\,2} - m_1 Cos\,\phi_0 - [m_3]_{3^{rd}\ iteration} = \mathbf{-0.841782497198922}\; au^2 \tag{255}$$

Using (197), we can now solve for $[Cos\beta]_{3^{rd}\ iteration}$ (256) to obtain its possible values from the quadratic equation in (192), or in (193), based on the calculated values of $\left[\mathbf{A}_\beta\right]_{3^{rd}\ iteration}$ (253), $\mathbf{B}_\beta$ (254), and $\left[\mathbf{C}_\beta\right]_{3^{rd}\ iteration}$ (255).

$$[Cos\beta]_{3^{rd}\ iteration} = \frac{-\mathbf{B}_\beta \pm \sqrt{\mathbf{B}_\beta^{\,2} - 4\left[\mathbf{A}_\beta\right]_{3^{rd}\ iteration}\left[\mathbf{C}_\beta\right]_{3^{rd}\ iteration}}}{2\left[\mathbf{A}_\beta\right]_{3^{rd}\ iteration}} = \begin{cases} \mathbf{0.917209433589676} \\ \mathbf{-0.917487343668342} \end{cases} \tag{256}$$

Based on the two possible solutions $[Cos\beta_1]_{3^{rd}\ iteration}$ (257) and $[Cos\beta_2]_{3^{rd}\ iteration}$ (257) for the value of $[Cos\beta]_{3^{rd}\ iteration}$ (256), there are four possible values $\left[\beta_{1,1}\right]_{3^{rd}\ iteration}$ (257), $\left[\beta_{1,2}\right]_{3^{rd}\ iteration}$ (257), $\left[\beta_{2,1}\right]_{3^{rd}\ iteration}$ (257), and $\left[\beta_{2,2}\right]_{3^{rd}\ iteration}$ (257) that $[\beta]_{3^{rd}\ iteration}$ (256) can take.

$$[Cos\beta]_{3^{rd}\ iteration} = [Cos\beta_1]_{3^{rd}\ iteration} = \mathbf{0.917209433589676}$$

$$\Rightarrow \begin{cases} [\beta]_{3^{rd}\ iteration} = \left[\beta_{1,1}\right]_{3^{rd}\ iteration} = \mathbf{0.409777643321547}\ radians = \mathbf{23^\circ.48} \\ [\beta]_{3^{rd}\ iteration} = \left[\beta_{1,2}\right]_{3^{rd}\ iteration} = \mathbf{-0.409777643321547}\ radians = \mathbf{-23^\circ.48} \end{cases}$$

(257)

$$[Cos\beta]_{3^{rd}\ iteration} = [Cos\beta_2]_{3^{rd}\ iteration} = \mathbf{-0.917487343668342}$$

$$\Rightarrow \begin{cases} [\beta]_{3^{rd}\ iteration} = \left[\beta_{2,1}\right]_{3^{rd}\ iteration} = \mathbf{2.7325131273548}\ radians = \mathbf{156^\circ.56} \\ [\beta]_{3^{rd}\ iteration} = \left[\beta_{2,2}\right]_{3^{rd}\ iteration} = \mathbf{-2.7325131273548}\ radians = \mathbf{-156^\circ.56} \end{cases}$$

For these four possible values (257) that $[\beta]_{3^{rd}\ iteration}$ (256) can take, the possible values $[r_1]_{3^{rd}\ iteration}$ (6) and $\hat{\boldsymbol{u}}_{2\perp}$ (97) can take based on (141) are listed in (258), and subsequently the possible values that $[r_2]_{3^{rd}\ iteration}$ (7) can take based on (183) are also listed in (258).

$$\left.\begin{array}{l} \left[\beta_{1,1}\right]_{3^{rd}\ iteration} = \mathbf{23^\circ.48} \Rightarrow Sin\left[\beta_{1,1}\right]_{3^{rd}\ iteration} = \mathbf{0.398405390192083} \Rightarrow \hat{\boldsymbol{u}}_{2\perp} = \hat{z} \\ \left[\beta_{1,2}\right]_{3^{rd}\ iteration} = \mathbf{-23^\circ.48} \Rightarrow Sin\left[\beta_{1,2}\right]_{3^{rd}\ iteration} = \mathbf{-0.398405390192083} \Rightarrow \hat{\boldsymbol{u}}_{2\perp} = -\hat{z} \end{array}\right\}$$

$$\Rightarrow \ [r_1]_{3^{rd}\ iteration} = \left[r_{1,1}\right]_{3^{rd}\ iteration} = \mathbf{0.998471115129159}\ au$$

$$\Rightarrow \ [r_2]_{3^{rd}\ iteration} = \left[r_{2,1}\right]_{3^{rd}\ iteration} = \mathbf{0.00191827722014836}\ au$$

(258)

$$\left.\begin{array}{l} \left[\beta_{2,1}\right]_{3^{rd}\ iteration} = \mathbf{156^\circ.56} \Rightarrow Sin\left[\beta_{2,1}\right]_{3^{rd}\ iteration} = \mathbf{0.397764973581649} \Rightarrow \hat{\boldsymbol{u}}_{2\perp} = \hat{z} \\ \left[\beta_{2,2}\right]_{3^{rd}\ iteration} = \mathbf{-156^\circ.56} \Rightarrow Sin\left[\beta_{2,2}\right]_{3^{rd}\ iteration} = \mathbf{-0.397764973581649} \Rightarrow \hat{\boldsymbol{u}}_{2\perp} = -\hat{z} \end{array}\right\}$$

$$\Rightarrow \ [r_1]_{3^{rd}\ iteration} = \left[r_{1,2}\right]_{3^{rd}\ iteration} = \mathbf{1.00007869128502}\ au$$

$$\Rightarrow \ [r_2]_{3^{rd}\ iteration} = \left[r_{2,2}\right]_{3^{rd}\ iteration} = \mathbf{0.0000826915340950211}\ au$$

Calculating the value of $\left(r_1^2 - 2\, r_1 r_2 Cos\,\phi_0 + r_2^2\right)$ (259) with both possible (r_1, r_2) (258) solution alternatives fo a validity check, we see that all match the expected $\left[b^2\right]_{3^{rd}\ iteration}$ (248) value we have found before.

$$\left.\begin{array}{l} [r_1]_{3^{rd}\ iteration} = \left[r_{1,1}\right]_{3^{rd}\ iteration} = \mathbf{0.998471115129159}\ au \\ [r_2]_{3^{rd}\ iteration} = \left[r_{2,1}\right]_{3^{rd}\ iteration} = \mathbf{0.00191827722014836}\ au \end{array}\right\} \Rightarrow$$

$$\left[r_{1,1}^{\ 2}\right]_{3^{rd}\ iteration} - 2\left[r_{1,1}\right]_{3^{rd}\ iteration}\left[r_{2,1}\right]_{3^{rd}\ iteration} Cos\,\phi_0 + \left[r_{2,1}^{\ 2}\right]_{3^{rd}\ iteration}$$

$$= \mathbf{1.00030220818691}\ au^2 = \left[b^2\right]_{3^{rd}\ iteration}$$

(259)

$$\left.\begin{array}{l} [r_1]_{3^{rd}\ iteration} = \left[r_{1,2}\right]_{3^{rd}\ iteration} = \mathbf{1.00007869128502}\ au \\ [r_2]_{3^{rd}\ iteration} = \left[r_{2,2}\right]_{3^{rd}\ iteration} = \mathbf{0.0000826915340950211}\ au \end{array}\right\} \Rightarrow$$

$$\left[r_{1,2}^{\ 2}\right]_{3^{rd}\ iteration} - 2\left[r_{1,2}\right]_{3^{rd}\ iteration}\left[r_{2,2}\right]_{3^{rd}\ iteration} Cos\,\phi_0 + \left[r_{2,2}^{\ 2}\right]_{3^{rd}\ iteration}$$

$$= \mathbf{1.00030220818691}\ au^2 = \left[b^2\right]_{3^{rd}\ iteration}$$

The possible values of $\left[\ell_{(x,i=689)}\right]_{3^{rd}\ iteration}$ and $\left[\ell_{(y,i=689)}\right]_{3^{rd}\ iteration}$ are calculated in (260) - (261) based on (207) and (208), using the already calculated value of $\ell_z\left(\phi_{\gamma_{2,\min}}\right)$ (120) restated as $\ell_{(z,i=689)}$ (219), the solution alternatives we have found for $[\beta]_{3^{rd}\ iteration}$ (257) and (r_1, r_2) (258), and the calculated values of $\vec{\boldsymbol{a}}\cdot\vec{\ell}_{(i=689)}$ (217) and $\vec{\boldsymbol{b}}\cdot\vec{\ell}_{(i=689)}$ (218) in the Excel sheet "Tan, A.P., Asli Pinar Tan Analysis Based on Earth-Sun distance (d) Landsat.xlsx"[13] corresponding to $\left(\ell_{(x,i=689)},\ell_{(y,i=689)},\ell_{(z,i=689)}\right)$ (177) at the Summer Solstice[5,6] on June 21.

$$\left.\begin{array}{l}\left.\begin{array}{l}\left[\beta_{1,1}\right]_{3^{rd}\ iteration}=\mathbf{23^\circ.48}\\ \Rightarrow\hat{\boldsymbol{u}}_{2\perp}=\hat{\boldsymbol{z}}\\ \Rightarrow\ell_{(z,i=689)}=\mathbf{0.00646733505568997}\ au\end{array}\right\}\begin{array}{l}\left[r_{1,1}\right]_{3^{rd}\ iteration}=\mathbf{0.998471115129159}\ au\\ \left[r_{2,1}\right]_{3^{rd}\ iteration}=\mathbf{0.00191827722014836}\ au\end{array}\\ \left.\begin{array}{l}\left[\beta_{1,2}\right]_{3^{rd}\ iteration}=\mathbf{-23^\circ.48}\\ \Rightarrow\hat{\boldsymbol{u}}_{2\perp}=-\hat{\boldsymbol{z}}\\ \Rightarrow\ell_{(z,i=689)}=\mathbf{-0.00646733505568997}\ au\end{array}\right\}\begin{array}{l}\left[\ell_{(x,i=689),1}\right]_{3^{rd}\ iteration}=\mathbf{0.0149205191715168}\ au\\ \left[\ell_{(y,i=689),1}\right]_{3^{rd}\ iteration}=\mathbf{0.00379474144551441}\ au\end{array}\end{array}\right.\quad(260)$$

$$\left.\begin{array}{l}\left.\begin{array}{l}\left[\beta_{2,1}\right]_{3^{rd}\ iteration}=\mathbf{156^\circ.56}\\ \Rightarrow\hat{\boldsymbol{u}}_{2\perp}=\hat{\boldsymbol{z}}\\ \Rightarrow\ell_{(z,i=689)}=\mathbf{0.00646733505568997}\ au\end{array}\right\}\begin{array}{l}\left[r_{1,2}\right]_{3^{rd}\ iteration}=\mathbf{1.00007869128502}\ au\\ \left[r_{2,2}\right]_{3^{rd}\ iteration}=\mathbf{0.0000826915340950211}\ au\end{array}\\ \left.\begin{array}{l}\left[\beta_{2,2}\right]_{3^{rd}\ iteration}=\mathbf{-156^\circ.56}\\ \Rightarrow\hat{\boldsymbol{u}}_{2\perp}=-\hat{\boldsymbol{z}}\\ \Rightarrow\ell_{(z,i=689)}=\mathbf{-0.00646733505568997}\ au\end{array}\right\}\begin{array}{l}\left[\ell_{(x,i=689),2}\right]_{3^{rd}\ iteration}=\mathbf{-0.01491684429569}\ au\\ \left[\ell_{(y,i=689),2}\right]_{3^{rd}\ iteration}=\mathbf{0.00380916148060411}\ au\end{array}\end{array}\right.\quad(261)$$

The resultant value of $\left[\ell_{(i=689)}\right]_{4^{th}\ iteration}$ (262), which is calculated based on (177), is the same and is $\mathbf{2,498,098.008399}\ km$ for all the possible solution values of $\left[\ell_{(x,i=689)}\right]_{3^{rd}\ iteration}$ and $\left[\ell_{(y,i=689)}\right]_{3^{rd}\ iteration}$ in (260) - (261), and the value of $\ell_z\left(\phi_{\gamma_{2,\min}}\right)$ (120) restated as $\ell_{(z,i=689)}$ (219).

$$\begin{aligned}\left[\ell_{(i=689)}\right]_{4^{th}\ iteration}&=\left[\ell\left(\phi_{\gamma_{2,\min}}\right)\right]_{4^{th}\ iteration}=\left[\ell\left(-\phi_0\right)\right]_{4^{th}\ iteration}\\ &=\mathbf{0.0166987537771075}\ au=\mathbf{2,498,098.008399}\ km\end{aligned}\quad(262)$$

As expressed in (263), the value we have found for $\left[\ell_{(i=689)}\right]_{4^{th}\ iteration}$ (262), which is $\mathbf{2,498,098.008399}\ km$, is close but deviated by around $\mathbf{5}\ m$ from the value we have found earlier for $\left[\ell_{(i=689)}\right]_{3^{rd}\ iteration}$ (242), which is $\mathbf{2,498,098.003344}\ km$ and was deviated by around $\mathbf{18.1}\ km$ from the value we have found earlier for $\left[\ell_{(i=689)}\right]_{2^{nd}\ iteration}$ (222), which is $\mathbf{2,498,079.874210}\ km$ and was deviated by around $\mathbf{65,884.0}\ km$ from the value we have found earlier for $\left[\ell_{(i=689)}\right]_{1^{st}\ iteration}$ (179) value, which is $\mathbf{2,432,195.841362}\ km$, calculated based on an initial assumption that $\alpha_{\left[\vec{d}\left(\phi_{\gamma 2,\min}\right)\measuredangle\vec{d}\left(\phi_{\gamma 2,\max}\right)\right],1^{st}\ iteration} = 180^\circ$ (171).

$$\begin{aligned}\left[\ell_{(i=689)}\right]_{4^{th}\ iteration} &= \mathbf{0.0166987537771075}\ au = \mathbf{2,498,098.008399}\ km \\ &\neq \mathbf{0.0166987537433149}\ au = \mathbf{2,498,098.003344}\ km = \left[\ell_{(i=689)}\right]_{3^{rd}\ iteration} \\ &\neq \mathbf{0.01669863255754}\ au = \mathbf{2,498,079.874210}\ km = \left[\ell_{(i=689)}\right]_{2^{nd}\ iteration} \\ &\neq \mathbf{0.0162582250000018}\ au = \mathbf{2,432,195.841362}\ km = \left[\ell_{(i=689)}\right]_{1^{st}\ iteration}\end{aligned} \quad (263)$$

Although we have achieved a much better result with $\left[\ell_{(i=689)}\right]_{4^{th}\ iteration}$ (262) compared to that of $\left[\ell_{(i=689)}\right]_{3^{rd}\ iteration}$ (242), we can go on applying this iterative approach one more iteration for a more accurate result, and as far as our analytical approximation is concerned, we can now take the new found value of $\left[\ell_{(i=689)}\right]_{4^{th}\ iteration} = \mathbf{0.0166987537771075}\ au$ (262) as a new starting point for our 4th iteration of calculations along the lines of (244) - (261).

Based on (169) and with the new value of $\left[\ell\left(\phi_{\gamma 2,\min}\right)\right]_{4^{th}\ iteration} = \mathbf{0.0166987537771075}\ au$ (262), we can rewrite (244) as in (264) to find a more accurate value for $\alpha_{\left[\vec{d}\left(\phi_{\gamma 2,\min}\right)\measuredangle\vec{d}\left(\phi_{\gamma 2,\max}\right)\right],4^{th}\ iteration}$ (265), that may be closer to the actual value of $\alpha_{\left[\vec{d}\left(\phi_{\gamma 2,\min}\right)\measuredangle\vec{d}\left(\phi_{\gamma 2,\max}\right)\right]}$ (169).

$$\begin{aligned}\vec{d}\left(\phi_{\gamma 2,\min}\right)\cdot\vec{d}\left(\pi+\phi_{\gamma 2,\min}\right) &= \vec{d}\left(\phi_{\gamma 2,\min}\right)\cdot\vec{d}\left(\phi_{\gamma 2,\max}\right) \\ &= d\left(\phi_{\gamma 2,\min}\right)d\left(\phi_{\gamma 2,\max}\right)Cos\,\alpha_{\left[\vec{d}\left(\phi_{\gamma 2,\min}\right)\measuredangle\vec{d}\left(\phi_{\gamma 2,\max}\right)\right],4^{th}\ iteration} \\ &= d_{(i=689)}\,d_{(i+730=1419)}Cos\,\alpha_{\left[\vec{d}\left(\phi_{\gamma 2,\min}\right)\measuredangle\vec{d}\left(\phi_{\gamma 2,\max}\right)\right],4^{th}\ iteration} \\ &= \left[\ell^2\left(\phi_{\gamma 2,\min}\right)\right]_{4^{th}\ iteration} - \left(\vec{a}\,Cos\,\phi_{\gamma 2,\min} + \vec{b}\,Sin\,\phi_{\gamma 2,\min}\right)^2\end{aligned} \quad (264)$$

In this case, utilizing (264) and $\boldsymbol{k}_1(\phi)$ (46) for $\left(\phi = \phi_{\gamma 2,\min} = -\phi_0\right)$ (139), which takes on the value $\boldsymbol{k}_{1,(i=689)}$ (175) based on the Excel sheet "Tan, A.P., Asli Pinar Tan Analysis Based on Earth-Sun distance (d) Landsat.xlsx"[13], and using $\left[\ell_{(i=689)}\right]_{4^{th}\ iteration}$ (262), the angle $\alpha_{\left[\vec{d}\left(\phi_{\gamma 2,\min}\right)\measuredangle\vec{d}\left(\phi_{\gamma 2,\max}\right)\right],4^{th}\ iteration}$ (265) between $\vec{d}\left(\phi_{\gamma 2,\min}\right)$ and $\vec{d}\left(\phi_{\gamma 2,\max}\right)$ is calculated, which turns out to be a value almost the same as that of $\alpha_{\left[\vec{d}\left(\phi_{\gamma 2,\min}\right)\measuredangle\vec{d}\left(\phi_{\gamma 2,\max}\right)\right],3^{rd}\ iteration}$ (245), close but not equal to $\pi\ radians = 180^\circ$.

$$\alpha_{\left[\vec{d}(\phi_{\gamma 2,\min})\measuredangle\vec{d}(\phi_{\gamma 2,\max})\right],4^{th}\ iteration} = Cos^{-1}\left(\frac{2\left[\ell_{(i=689)}\right]_{4^{th}\ iteration} - \boldsymbol{k}_{1,(i=689)}}{d_{(i=689)}\, d_{(i+730=1419)}}\right) \tag{265}$$

$$= \mathbf{3.13397114735832}\ radians = \mathbf{179^{\circ}.563}$$

Based on the definition of $\boldsymbol{k}_1(\phi)$ (46), we can find the value of $\left[b^2\right]_{4^{th}\ iteration}$ (266), making use of the already found values of $\boldsymbol{k}_1(-\phi_0)$ (175), $\left[2r_1 r_2 (1-Cos\beta)Cos\phi_0\right]$ (132), and $\left[\ell^2(-\phi_0)\right]_{4^{th}\ iteration}$ (262). This allows us to determine the magnitude $\left[b\right]_{4^{th}\ iteration}$ (267) of the semi-major axis vector $\vec{\boldsymbol{b}}$ (11) of the ellipse for the Sun-Earth system formed by their relative motions, which is the the semi-major axis vector based on the **ARTICLE 1** "Points on two circles in space mimic an ellipse with respect to a point" and (10) - (11), as $(Cos\phi_0 < 0)$ (129) and $(a^2 < b^2)$ (131). Based on (11) and (266), we also obtain the identity (268) at this point.

$$\left[b^2\right]_{4^{th}\ iteration} = \boldsymbol{k}_1(-\phi_0) - 2r_1 r_2 (1-Cos\beta)Cos\phi_0 - \left[\ell^2(-\phi_0)\right]_{4^{th}\ iteration} = \mathbf{1.00030220818578}\ au^2 \tag{266}$$

$$\left[b\right]_{4^{th}\ iteration} = \mathbf{1.00015109267839}\ au = \mathbf{149{,}620{,}473.842965}\ km \tag{267}$$

$$\left[b^2\right]_{4^{th}\ iteration} = \left[r_1^{\ 2}\right]_{4^{th}\ iteration} - 2\left[r_1\right]_{4^{th}\ iteration}\left[r_2\right]_{4^{th}\ iteration} Cos\,\phi_0 + \left[r_2^{\ 2}\right]_{4^{th}\ iteration} = \mathbf{1.00030220818578}\ au^2 \tag{268}$$

In this 4th iteration of calculations, our three equations (72), (141), and (11) are as restated in (269), (270), and (271), respectively, in the three variables r_1 (6), r_2 (7), and β (6), that we are set to solve for, where $\left[m_3\right]_{4^{th}\ iteration}$ (272) is given by (268).

$$2r_1 r_2 (1-Cos\beta) = \mathbf{0.000317144894687038}\ au^2 = m_1 \tag{269}$$

$$r_1\, Sin\,\beta = \pm\mathbf{0.397796274218557}\ au = m_2 \tag{270}$$

$$b^2 = r_1^{\ 2} - 2\ r_1 r_2 Cos\,\phi_0 + r_2^{\ 2} = m_3 \tag{271}$$

$$\left[m_3\right]_{4^{th}\ iteration} = \left[b^2\right]_{4^{th}\ iteration} = \mathbf{1.00030220818578}\ au^2 \tag{272}$$

Using the analysis in (187) - (197), and based on the values of m_1 (269), m_2 (270), and $\left[m_3\right]_{4^{th}\ iteration}$ (272), we would obtain the values for $\left[\mathbf{A}_\beta\right]_{4^{th}\ iteration}$ (273), $\mathbf{B}_\beta$ (274), and $\left[\mathbf{C}_\beta\right]_{4^{th}\ iteration}$ (275).

$$\left[\mathbf{A}_\beta\right]_{4^{th}\ iteration} = \frac{m_1^{\ 2}}{4m_2^{\ 2}} + \left[m_3\right]_{4^{th}\ iteration} = \mathbf{1.00030236708949}\ au^2 \tag{273}$$

$$\mathbf{B}_\beta = \frac{m_1^{\ 2}}{2m_2^{\ 2}} - m_1 Cos\,\phi_0 = \mathbf{0.00027799410952802}\ au^2 \tag{274}$$

$$\left[\mathbf{C}_\beta\right]_{4^{th}\ iteration} = \frac{m_1^{\ 2}}{4m_2^{\ 2}} + m_2^{\ 2} - m_1 Cos\,\phi_0 - \left[m_3\right]_{4^{th}\ iteration} = \mathbf{-0.841782497197793}\ au^2 \qquad (275)$$

Using (197), we can now solve for $\left[Cos\beta\right]_{4^{th}\ iteration}$ (276) to obtain its possible values from the quadratic equation in (192), or in (193), based on the calculated values of $\left[\mathbf{A}_\beta\right]_{4^{th}\ iteration}$ (273), $\mathbf{B}_\beta$ (274), and $\left[\mathbf{C}_\beta\right]_{4^{th}\ iteration}$ (275).

$$\left[Cos\beta\right]_{4^{th}\ iteration} = \frac{-\mathbf{B}_\beta \pm \sqrt{\mathbf{B}_\beta^{\ 2} - 4\left[\mathbf{A}_\beta\right]_{4^{th}\ iteration}\left[\mathbf{C}_\beta\right]_{4^{th}\ iteration}}}{2\left[\mathbf{A}_\beta\right]_{4^{th}\ iteration}} = \begin{cases} \mathbf{0.917209433589578} \\ \mathbf{-0.917487343668245} \end{cases} \qquad (276)$$

Based on the two possible solutions $\left[Cos\beta_1\right]_{4^{th}\ iteration}$ (277) and $\left[Cos\beta_2\right]_{4^{th}\ iteration}$ (277) for the value of $\left[Cos\beta\right]_{4^{th}\ iteration}$ (276), there are four possible values $\left[\beta_{1,1}\right]_{4^{th}\ iteration}$ (277), $\left[\beta_{1,2}\right]_{4^{th}\ iteration}$ (277), $\left[\beta_{2,1}\right]_{4^{th}\ iteration}$ (277), and $\left[\beta_{2,2}\right]_{4^{th}\ iteration}$ (277) that $\left[\beta\right]_{4^{th}\ iteration}$ (276) can take.

$$\left[Cos\beta\right]_{4^{th}\ iteration} = \left[Cos\beta_1\right]_{4^{th}\ iteration} = \mathbf{0.917209433589578}$$

$$\Rightarrow \begin{cases} \left[\beta\right]_{4^{th}\ iteration} = \left[\beta_{1,1}\right]_{4^{th}\ iteration} = \mathbf{0.409777643321792}\ radians = \mathbf{23^\circ.48} \\ \left[\beta\right]_{4^{th}\ iteration} = \left[\beta_{1,2}\right]_{4^{th}\ iteration} = \mathbf{-0.409777643321792}\ radians = \mathbf{-23^\circ.48} \end{cases}$$

(277)

$$\left[Cos\beta\right]_{4^{th}\ iteration} = \left[Cos\beta_2\right]_{4^{th}\ iteration} = \mathbf{-0.917487343668245}$$

$$\Rightarrow \begin{cases} \left[\beta\right]_{4^{th}\ iteration} = \left[\beta_{2,1}\right]_{4^{th}\ iteration} = \mathbf{2.73251312735455}\ radians = \mathbf{156^\circ.56} \\ \left[\beta\right]_{4^{th}\ iteration} = \left[\beta_{2,2}\right]_{4^{th}\ iteration} = \mathbf{-2.73251312735455}\ radians = \mathbf{-156^\circ.56} \end{cases}$$

For these four possible values (277) that $\left[\beta\right]_{4^{th}\ iteration}$ (276) can take, the possible values $\left[r_1\right]_{4^{th}\ iteration}$ (6) and $\hat{\boldsymbol{u}}_{2\perp}$ (97) can take based on (141) are listed in (278), and subsequently the possible values that $\left[r_2\right]_{4^{th}\ iteration}$ (7) can take based on (183) are also listed in (278).

$$\left[\beta_{1,1}\right]_{4^{th}\ iteration} = \mathbf{23^\circ.48} \Rightarrow Sin\left[\beta_{1,1}\right]_{4^{th}\ iteration} = \mathbf{0.398405390192307} \Rightarrow \hat{\boldsymbol{u}}_{2\perp} = \hat{z}$$

$$\left[\beta_{1,2}\right]_{4^{th}\ iteration} = \mathbf{-23^\circ.48} \Rightarrow Sin\left[\beta_{1,2}\right]_{4^{th}\ iteration} = \mathbf{-0.398405390192307} \Rightarrow \hat{\boldsymbol{u}}_{2\perp} = -\hat{z}$$

$$\Rightarrow \left[r_1\right]_{4^{th}\ iteration} = \left[r_{1,1}\right]_{4^{th}\ iteration} = \mathbf{0.998471115128596}\ au$$

$$\Rightarrow \left[r_2\right]_{4^{th}\ iteration} = \left[r_{2,1}\right]_{4^{th}\ iteration} = \mathbf{0.00191827722014718}\ au$$

(278)

$$\left[\beta_{2,1}\right]_{4^{th}\ iteration} = \mathbf{156^\circ.56} \Rightarrow Sin\left[\beta_{2,1}\right]_{4^{th}\ iteration} = \mathbf{0.397764973581873} \Rightarrow \hat{\boldsymbol{u}}_{2\perp} = \hat{z}$$

$$\left[\beta_{2,2}\right]_{4^{th}\ iteration} = \mathbf{-156^\circ.56} \Rightarrow Sin\left[\beta_{2,2}\right]_{4^{th}\ iteration} = \mathbf{-0.397764973581873} \Rightarrow \hat{\boldsymbol{u}}_{2\perp} = -\hat{z}$$

$$\Rightarrow \left[r_1\right]_{4^{th}\ iteration} = \left[r_{1,2}\right]_{4^{th}\ iteration} = \mathbf{1.00007869128446}\ au$$

$$\Rightarrow \left[r_2\right]_{4^{th}\ iteration} = \left[r_{2,2}\right]_{4^{th}\ iteration} = \mathbf{0.0000826915340950721}\ au$$

Calculating the value of $\left(r_1^2 - 2\,r_1\,r_2 Cos\,\phi_0 + r_2^{\,2}\right)$ (279) with both possible (r_1, r_2) (278) solution alternatives fo a validity check, we see that all match the expected $\left[b^2\right]_{4^{th}\ iteration}$ (268) value we have found before.

$$\left.\begin{array}{l}\left[r_1\right]_{4^{th}\ iteration} = \left[r_{1,1}\right]_{4^{th}\ iteration} = \mathbf{0.998471115128596}\ au \\ \left[r_2\right]_{4^{th}\ iteration} = \left[r_{2,1}\right]_{4^{th}\ iteration} = \mathbf{0.00191827722014718}\ au\end{array}\right\} \Rightarrow$$

$$\left[r_{1,1}^{\;2}\right]_{4^{th}\ iteration} - 2\left[r_{1,1}\right]_{4^{th}\ iteration}\left[r_{2,1}\right]_{4^{th}\ iteration} Cos\,\phi_0 + \left[r_{2,1}^{\;2}\right]_{4^{th}\ iteration}$$
$$= \mathbf{1.00030220818578}\ au^2 = \left[b^2\right]_{4^{th}\ iteration}$$

(279)

$$\left.\begin{array}{l}\left[r_1\right]_{4^{th}\ iteration} = \left[r_{1,2}\right]_{4^{th}\ iteration} = \mathbf{1.00007869128446}\ au \\ \left[r_2\right]_{4^{th}\ iteration} = \left[r_{2,2}\right]_{4^{th}\ iteration} = \mathbf{0.0000826915340950721}\ au\end{array}\right\} \Rightarrow$$

$$\left[r_{1,2}^{\;2}\right]_{4^{th}\ iteration} - 2\left[r_{1,2}\right]_{4^{th}\ iteration}\left[r_{2,2}\right]_{4^{th}\ iteration} Cos\,\phi_0 + \left[r_{2,2}^{\;2}\right]_{4^{th}\ iteration}$$
$$= \mathbf{1.00030220818578}\ au^2 = \left[b^2\right]_{4^{th}\ iteration}$$

The possible values of $\left[\ell_{(x,i=689)}\right]_{4^{th}\ iteration}$ and $\left[\ell_{(y,i=689)}\right]_{4^{th}\ iteration}$ are calculated in (280) - (281) based on (207) and (208), using the already calculated value of $\ell_z\left(\phi_{\gamma_{2,\min}}\right)$ (120) restated as $\ell_{(z,i=689)}$ (219), the solution alternatives we have found for $\left[\beta\right]_{4^{th}\ iteration}$ (277) and (r_1, r_2) (278), and the calculated values of $\vec{\boldsymbol{a}}\cdot\vec{\ell}_{(i=689)}$ (217) and $\vec{\boldsymbol{b}}\cdot\vec{\ell}_{(i=689)}$ (218) in the Excel sheet "Tan, A.P., Asli Pinar Tan Analysis Based on Earth-Sun distance (d) Landsat.xlsx"[13] corresponding to $\left(\ell_{(x,i=689)}, \ell_{(y,i=689)}, \ell_{(z,i=689)}\right)$ (177) at the Summer Solstice[5,6] on June 21.

$$\left.\begin{array}{l}\left.\begin{array}{l}\left[\beta_{1,1}\right]_{4^{th}\ iteration} = \mathbf{23^\circ.48} \\ \Rightarrow \hat{\boldsymbol{u}}_{2\perp} = \hat{z} \\ \Rightarrow \ell_{(z,i=689)} = \mathbf{0.00646733505568997}\ au\end{array}\right\} \\ \left.\begin{array}{l}\left[\beta_{1,2}\right]_{4^{th}\ iteration} = \mathbf{-23^\circ.48} \\ \Rightarrow \hat{\boldsymbol{u}}_{2\perp} = -\hat{z} \\ \Rightarrow \ell_{(z,i=689)} = \mathbf{-0.00646733505568997}\ au\end{array}\right\}\end{array}\right\} \left\{\begin{array}{l}\left[r_{1,1}\right]_{4^{th}\ iteration} = \mathbf{0.998471115128596}\ au \\ \left[r_{2,1}\right]_{4^{th}\ iteration} = \mathbf{0.00191827722014718}\ au \\ \left[\ell_{(x,i=689),1}\right]_{4^{th}\ iteration} = \mathbf{0.0149205191715268}\ au \\ \left[\ell_{(y,i=689),1}\right]_{4^{th}\ iteration} = \mathbf{0.00379474144551655}\ au\end{array}\right. \quad (280)$$

$$\left.\begin{array}{l}\left.\begin{array}{l}\left[\beta_{2,1}\right]_{4^{th}\ iteration} = \mathbf{156^\circ.56} \\ \Rightarrow \hat{\boldsymbol{u}}_{2\perp} = \hat{z} \\ \Rightarrow \ell_{(z,i=689)} = \mathbf{0.00646733505568997}\ au\end{array}\right\} \\ \left.\begin{array}{l}\left[\beta_{2,2}\right]_{4^{th}\ iteration} = \mathbf{-156^\circ.56} \\ \Rightarrow \hat{\boldsymbol{u}}_{2\perp} = -\hat{z} \\ \Rightarrow \ell_{(z,i=689)} = \mathbf{-0.00646733505568997}\ au\end{array}\right\}\end{array}\right\} \left\{\begin{array}{l}\left[r_{1,2}\right]_{4^{th}\ iteration} = \mathbf{1.00007869128446}\ au \\ \left[r_{2,2}\right]_{4^{th}\ iteration} = \mathbf{0.0000826915340950721}\ au \\ \left[\ell_{(x,i=689),2}\right]_{4^{th}\ iteration} = \mathbf{-0.0149168442957}\ au \\ \left[\ell_{(y,i=689),2}\right]_{4^{th}\ iteration} = \mathbf{0.00380916148831507}\ au\end{array}\right. \quad (281)$$

The resultant value of $\left[\ell_{(i=689)}\right]_{4^{th}\ iteration}$ (282), which is calculated based on (177), is the same and is $\mathbf{2,498,098.008400}\ km$ for all the possible solution values of $\left[\ell_{(x,i=689)}\right]_{4^{th}\ iteration}$ and $\left[\ell_{(y,i=689)}\right]_{4^{th}\ iteration}$ in (280) - (281), and the value of $\ell_z\left(\phi_{\gamma_{2,\min}}\right)$ (120) restated as $\ell_{(z,i=689)}$ (219).

$$\begin{aligned}\left[\ell_{(i=689)}\right]_{5^{th}\ iteration} &= \left[\ell\left(\phi_{\gamma_{2,\min}}\right)\right]_{5^{th}\ iteration} = \left[\ell\left(-\phi_0\right)\right]_{5^{th}\ iteration} \\ &= \mathbf{0.0166987537771169}\ au = \mathbf{2,498,098.008400}\ km\end{aligned} \quad (282)$$

As expressed in (283), the value we have found for $\left[\ell_{(i=689)}\right]_{5^{th}\ iteration}$ (282), which is $\mathbf{2,498,098.008400}\ km$, is sufficiently close in Sun-Earth distance scale and differs just about $\mathbf{0.001}\ m = \mathbf{1}\ mm$ from the value we have found earlier for $\left[\ell_{(i=689)}\right]_{4^{th}\ iteration}$ (262), which is $\mathbf{2,498,098.008399}\ km$ and was deviated by around $\mathbf{5}\ m$ from the value we have found earlier for $\left[\ell_{(i=689)}\right]_{3^{rd}\ iteration}$ (242), which is $\mathbf{2,498,098.003344}\ km$ and was deviated by around $\mathbf{18.1}\ km$ from the value we have found earlier for $\left[\ell_{(i=689)}\right]_{2^{nd}\ iteration}$ (222), which is $\mathbf{2,498,079.874210}\ km$ and was deviated by around $\mathbf{65,884.0}\ km$ from the value we have found earlier for $\left[\ell_{(i=689)}\right]_{1^{st}\ iteration}$ (179) value, which is $\mathbf{2,432,195.841362}\ km$, calculated based on an initial assumption that $\alpha_{\left[\vec{d}\left(\phi_{\gamma_{2,\min}}\right)\measuredangle\vec{d}\left(\phi_{\gamma_{2,\max}}\right)\right],1^{st}\ iteration} = 180^\circ$ (171).

$$\left[\ell_{(i=689)}\right]_{5^{th}\ iteration} = \mathbf{0.0166987537771169}\ au = \mathbf{2,498,098.008400}\ km$$

$$\neq \mathbf{0.0166987537771075}\ au = \mathbf{2,498,098.008399}\ km = \left[\ell_{(i=689)}\right]_{4^{th}\ iteration}$$

$$\neq \mathbf{0.0166987537433149}\ au = \mathbf{2,498,098.003344}\ km = \left[\ell_{(i=689)}\right]_{3^{rd}\ iteration} \quad (283)$$

$$\neq \mathbf{0.01669863255754}\ au = \mathbf{2,498,079.874210}\ km = \left[\ell_{(i=689)}\right]_{2^{nd}\ iteration}$$

$$\neq \mathbf{0.0162582250000018}\ au = \mathbf{2,432,195.841362}\ km = \left[\ell_{(i=689)}\right]_{1^{st}\ iteration}$$

As far as our analytical approximation is concerned, we have achieved a sufficiantly accurate result with $\left[\ell_{(i=689)}\right]_{5^{th}\ iteration}$ (282) in Sun-Earth distance scale (3). Therefore, we can end our iterative approach at this point and use the $\left[\ell\left(\phi_{\gamma_{2,\min}}\right)\right]_{5^{th}\ iteration} = \left[\ell\left(-\phi_0\right)\right]_{5^{th}\ iteration}$ (282) value for $\ell\left(\phi_{\gamma_{2,\min}}\right) = \ell\left(-\phi_0\right)$ (284), the $\left[b\right]_{4^{th}\ iteration}$ (267) value for the semi-major axis magnitude b (285) of the Sun-Earth system as a finding of this Article, thus $\left[b^2\right]_{4^{th}\ iteration}$ (266) value for b^2 (286), as well as the solution alternatives we have found for $\beta = \left[\beta\right]_{4^{th}\ iteration}$ (277) and $\left(r_1, r_2\right)$ (278) for β and $\left(r_1, r_2\right)$. We still have to determine *which* of the solution alternatives we have found for $\beta = \left[\beta\right]_{4^{th}\ iteration}$ (277) and $\left(r_1, r_2\right)$ (278) *is* the *actual* solution for β and $\left(r_1, r_2\right)$ values of the Sun-Earth system.

$$\begin{aligned} \ell\left(\phi_{\gamma_{2,\min}}\right) &= \ell\left(-\phi_0\right) = \ell_{(i=689)} \\ &= \left[\ell\left(\phi_{\gamma_{2,\min}}\right)\right]_{5^{th}\ iteration} = \left[\ell\left(-\phi_0\right)\right]_{5^{th}\ iteration} = \left[\ell_{(i=689)}\right]_{5^{th}\ iteration} \\ &= \mathbf{0.0166987537771169}\ au = \mathbf{2,498,098.008400}\ km \end{aligned} \quad (284)$$

$$b = \left[b\right]_{4^{th}\ iteration} = \mathbf{1.00015109267839}\ au = \mathbf{149,620,473.842965}\ km \quad (285)$$

$$b^2 = \left[b^2\right]_{4^{th}\ iteration} = \mathbf{1.00030220818578}\ au^2 \quad (286)$$

At this point we can also find the value of a^2 (287) using the values of b^2 (286) and $\left[2r_1 r_2\left(1 - Cos\beta\right)Cos\phi_0\right]$ (132). Subsequently, we are able to determine the magnitude a (288) of the semi-minor axis, as $\left(Cos\,\phi_0 < 0\right)$ (129) and $\left(a^2 < b^2\right)$ (131), vector $\vec{a}$ (10) of the ellipse for the Sun-Earth system formed by their relative motions, yet another finding of this Article.

$$a^2 = b^2 + 2r_1 r_2\left(1 - Cos\beta\right)Cos\phi_0 = \mathbf{1.00002453188367}\ au^2 \quad (287)$$

$$a = \mathbf{1.00001226586661}\ au = \mathbf{149,599,705.647527}\ km \quad (288)$$

Note that the difference between the semi-major axis magnitude b (285) and the semi-minor axis magnitude a (288) of the Sun-Earth system is about $\mathbf{20,768.2}\ km$ (289), which is relatively small on the Sun-Earth distance scale (3). Therefore, the ellipse for the Sun-Earth system formed by their relative motions reveals to be almost circular, as expected.

$$(b-a) = \mathbf{0.000138826811781456}\ au = \mathbf{20,768.195439}\ km \qquad (289)$$

We can also find the value of eccentricity[22] e (290) of the relative ellipse of the Sun-Earth system now, using found values of the focal[22,24] distance c (134) and semi-major[23] axis b (285), another one of the main findings of this Article.

$$e = \frac{c}{b} = \sqrt{\frac{2\, r_1 r_2 \left(1-Cos\beta\right)\left|Cos\,\phi_0\right|}{r_1^2 - 2 r_1 r_2 Cos\,\phi_0 + r_2^2}} = \mathbf{0.0166611047475858} \qquad (Eccentricity) \qquad (290)$$

Based the final value of $\ell\left(\phi_{\gamma_{2,\min}}\right) = \mathbf{0.0166987537771169}\ au$ (284), we can rewrite (169) as in (291) for $\left(\phi = \phi_{\gamma_{2,\min}} = -\phi_0\right)$ (139), to find the final value for $\alpha_{\left[\vec{d}\left(\phi_{\gamma_{2,\min}}\right)\measuredangle\vec{d}\left(\phi_{\gamma_{2,\max}}\right)\right]}$ (292).

$$\begin{aligned}\vec{d}\left(\phi_{\gamma_{2,\min}}\right)\cdot\vec{d}\left(\pi+\phi_{\gamma_{2,\min}}\right) &= \vec{d}\left(\phi_{\gamma_{2,\min}}\right)\cdot\vec{d}\left(\phi_{\gamma_{2,\max}}\right)\\ &= d\left(\phi_{\gamma_{2,\min}}\right)d\left(\phi_{\gamma_{2,\max}}\right)Cos\,\alpha_{\left[\vec{d}\left(\phi_{\gamma_{2,\min}}\right)\measuredangle\vec{d}\left(\phi_{\gamma_{2,\max}}\right)\right]}\\ &= d_{(i=689)}\, d_{(i+730=1419)} Cos\,\alpha_{\left[\vec{d}\left(\phi_{\gamma_{2,\min}}\right)\measuredangle\vec{d}\left(\phi_{\gamma_{2,\max}}\right)\right]}\\ &= \ell^2\left(\phi_{\gamma_{2,\min}}\right) - \left(\vec{a}\,Cos\,\phi_{\gamma_{2,\min}} + \vec{b}\,Sin\,\phi_{\gamma_{2,\min}}\right)^2\end{aligned} \qquad (291)$$

In this case, utilizing (291) and $\boldsymbol{k}_1(\phi)$ (46) for $\left(\phi = \phi_{\gamma_{2,\min}} = -\phi_0\right)$ (139), which takes on the value $\boldsymbol{k}_{1,(i=689)}$ (175) based on the Excel sheet "Tan, A.P., Asli Pinar Tan Analysis Based on Earth-Sun distance (d) Landsat.xlsx"[13], and using $\ell\left(\phi_{\gamma_{2,\min}}\right) = \ell\left(-\phi_0\right)$ (284), angle $\alpha_{\left[\vec{d}\left(\phi_{\gamma_{2,\min}}\right)\measuredangle\vec{d}\left(\phi_{\gamma_{2,\max}}\right)\right]}$ (292) between $\vec{d}\left(\phi_{\gamma_{2,\min}}\right)$ and $\vec{d}\left(\phi_{\gamma_{2,\max}}\right)$ is calculated, another one of the findings of this Article, which turns out to be a value close but not equal to $\pi\ radians = 180^\circ$, as expected.

$$\begin{aligned}\alpha_{\left[\vec{d}\left(\phi_{\gamma_{2,\min}}\right)\measuredangle\vec{d}\left(\phi_{\gamma_{2,\max}}\right)\right]} &= Cos^{-1}\left(\frac{2\ell^2\left(\phi_{\gamma_{2,\min}}\right) - \boldsymbol{k}_{1,(i=689)}}{d_{(i=689)}\, d_{(i+730=1419)}}\right)\\ &= \mathbf{3.13397114735826}\ radians = \mathbf{179^\circ.563}\end{aligned} \qquad (292)$$

We have now calculated $\ell^2(\phi)$ (9) for all ϕ, or equivalently $\ell_i^{\,2}$ (294), corresponding to each data point in the Sun-Earth system in the Excel sheet "Tan, A.P., Asli Pinar Tan Analysis Based on Earth-Sun distance (d) Landsat.xlsx"[13], using (293) obtained from $\boldsymbol{k}_1(\phi)$ (46), as well as the previously calculated b^2 (286) and the ϕ values for each data point in the Excel sheet "Tan, A.P., Asli Pinar Tan Analysis Based on Earth-Sun distance (d) Landsat.xlsx"[13], and applying the algorithm devised in (21) - (25).

$$\ell(\phi)^2 = \boldsymbol{k}_1(\phi) - b^2 - 2r_1 r_2\left(1-Cos\beta\right)Cos\phi\, Cos\left(\phi+\phi_0\right) \qquad (293)$$

$$\ell_i^{\,2} = \boldsymbol{k}_{1,i} - b^2 - 2r_1 r_2\left(1-Cos\beta\right)Cos\phi\, Cos\left(\phi+\phi_0\right) \qquad (294)$$

The obtained $\ell(\phi)$ (9) is plotted for all ϕ over a yearly Sun-Earth cycle in **Figure 19**, and it demonstrates similar behavior to that of **Figure 9** for $[\boldsymbol{k}_1(\phi)+\boldsymbol{k}_2(\phi)]$ (50), as expected. Working with the data in the Excel sheet "Tan, A.P., Asli Pinar Tan Analysis Based on Earth-Sun distance (d) Landsat.xlsx"[13], we observe that $\ell(\phi)$ (9) varies in a harmonic fashion between its maximum and minimum values expressed in (295) and (296), just as $[\boldsymbol{k}_1(\phi)+\boldsymbol{k}_2(\phi)]$ (50) varies in a harmonic fashion between its maximum and minimum values expressed in (79) and (80). $\left[\ell(\phi)\right]_{Max}$ (295) occurs on February 23 (DOY 54), May 26 (DOY 146), August 25 (DOY 237), and November 24 (DOY 328), whereas $\left[\ell(\phi)\right]_{Min}$ (296) occurs on January 1 (DOY 1), April 2 (DOY 92), July 2 (DOY 183), and October 1 (DOY 274). It is also worth noting here as an observation that the occurence dates of $\left\{\sqrt{[\boldsymbol{k}_3(\phi)]^2+[\boldsymbol{k}_4(\phi)]^2}\right\}_{Max}$ (68) and $\left[\ell(\phi)\right]_{Min}$ (296) coincide with each other. Based on the definition of $[\boldsymbol{k}_3(\phi)]^2+[\boldsymbol{k}_4(\phi)]^2$ (52), this observation allows us to conclude that the sum of squares of the projections of the constant $\vec{\boldsymbol{a}}$ (10) and $\vec{\boldsymbol{b}}$ (11) on varying $\vec{\ell}(\phi)$ (9), namely $\left[\vec{\boldsymbol{a}}\cdot\vec{\ell}(\phi)\right]^2+\left[\vec{\boldsymbol{b}}\cdot\vec{\ell}(\phi)\right]^2$, takes on its maximum value when the magnitude of $\ell(\phi)$ (9) is at its minimum value $\left[\ell(\phi)\right]_{Min}$ (296).

$$\left[\ell(\phi)\right]_{Max} = \mathbf{0.0168346184954446}\ au\ on\ 23\ February, 26\ May, 25\ August, 24\ November \quad (295)$$

$$\left[\ell(\phi)\right]_{Min} = \mathbf{0.0165381684638675}\ au \quad on \quad 1\ January, 2\ April, 2\ July, 1\ October \quad (296)$$

At this point, in order to determine which one of the four possible solution alternatives for (β, r_1, r_2) (278) that we have found for the Sun-Earth system, more specifically, which one of $(\beta_{1,1}, r_{1,1}, r_{2,1})$ (278), $(\beta_{1,2}, r_{1,1}, r_{2,1})$ (278), $(\beta_{2,1}, r_{1,2}, r_{2,2})$ (278), or $(\beta_{2,2}, r_{1,2}, r_{2,2})$ (278) is the valid solution for the Sun-Earth system, we will make use of the observation of what is known as Solar Analemma[17]. In astronomy, a solar analemma[17] is a diagram showing the position of the Sun in the sky, as seen from a fixed location on Earth at the same mean solar time, as that position varies over the course of the year. A solar analemma[17] diagram resembles the shape of the number "8", as in the two plotted observation examples seen in **Figure 20**[17] and **Figure 21**[17], and it is "oriented with the smaller loop appearing at the top of the larger loop" observed wherever on Earth in the Northern Hemisphere, and "oriented with the smaller loop appearing beneath the larger loop" observed wherever on Earth in the Southern Hemisphere, although this diagram may be differently tilted depending on the "time of day for observation" and "latitude of observation" on Earth. **Figure 20**[17] shows Solar Analemma[17] as observed on Earth as the position of the Sun is directly overhead, every 24 hours over one year. At the North Pole[29] of the Earth, the Solar Analemma[17] would be a completely upright "8" with the small loop at the top, and only the top half of it would be visible. At latitudes between the Arctic Circle and Equator in Northern Hemisphere, the entire Solar Analemma[17] would be visible, higher above the horizon as one goes further south, and continues to be an upright "8" with the small loop at the top at noon. When at the Equator, the Solar Analemma[17] is directly overhead. At latitudes between the Equator and Antarctic Circle in Southern Hemisphere, the Solar Analemma[17] would be inverted compared to the observation from Northern Hemisphere, this time the "8" with the small loop beneath the large loop in the sky. Once the Antarctic Circle is crossed towards to South Pole, the Solar Analemma[17] would be completely inverted compared to the observation from Northern Hemisphere and start to disappear partially, with only part of the larger loop visible at the South Pole. **Figure 21**[17] shows Solar Analemma[17] plotted as seen at 12:00 noon GMT from Greenwich Observatory (latitude $51^{\circ}.48$ North, longitude $0^{\circ}.0015$ West) during the year 2006, where the horizontal axis is the azimuth angle in degrees (180° is facing South), and the vertical axis is the altitude in degrees above horizon. The first day of each month is shown in black, and the Solstices[5] and Equinoxes[9] are shown in green.

Moreover, from a vantage point above the North Pole[29] of the Earth, the Earth appears to revolve[2] in a counterclockwise[27] direction in its relative motion around the Sun, and as seen from the Earth, the Sun appears to move[2] with respect to the other stars at a rate of about 1° eastward per solar day, therefore the Sun also appears to revolve[2] in a counterclockwise[27] direction in its relative motion with respect to the Earth. From the same vantage point, the Earth appears to rotate[2] also in a counterclockwise[27] direction about its axis, which is expected based on the "right hand rule"[18].

Considering the topology depicted in **Figure 2**, the Earth and the Sun's relative orbital[2] motion, and the four possible solution alternatives we have found for β (277), we can infer that the observed Solar Analemma[17] shape of the number "8" as in the two plotted observation examples seen in **Figure 20**[17] and **Figure 21**[17] can occur only for the cases when $\left(\frac{\pi}{2} \le \beta \le \frac{3\pi}{2}\right)$, and that the observed Solar Analemma[17] would have been closer to the number "0" shape in the case of $\left(-\frac{\pi}{2} \le \beta \le \frac{\pi}{2}\right)$, in the Sun-Earth system[2]. This analysis based on Solar Analemma[17] observation eliminates two of our possible β (277) solutions $\left(\beta = \beta_{1,1} = \mathbf{23^{\circ}.48}\right)$ (277) and $\left(\beta = \beta_{1,2} = \mathbf{-23^{\circ}.48}\right)$ (277), and leaves us with the other two of our possible valid β (297) solutions which are $\left(\beta = \beta_{2,1} = \mathbf{156^{\circ}.56}\right)$ (297) and $\left(\beta = \beta_{2,2} = \mathbf{-156^{\circ}.56}\right)$ (297) that meet the condition $\left(\frac{\pi}{2} \le \beta \le \frac{3\pi}{2}\right)$, with the associated $\left(r_1 = r_{1,2}, r_2 = r_{2,2}\right)$ (297) solution based on (278).

$$\left.\begin{aligned} \beta = \beta_{2,1} &= \mathbf{156^{\circ}.56} \Rightarrow Sin\left(\beta_{2,1}\right) = \mathbf{0.397764973581873} \Rightarrow \hat{\boldsymbol{u}}_{2\perp} = \hat{z} \\ \beta = \beta_{2,2} &= \mathbf{-156^{\circ}.56} \Rightarrow Sin\left(\beta_{2,1}\right) = \mathbf{-0.397764973581873} \Rightarrow \hat{\boldsymbol{u}}_{2\perp} = -\hat{z} \end{aligned}\right\}$$
$$\Rightarrow \quad r_1 = r_{1,2} = \mathbf{1.00007869128446}\,au \tag{297}$$
$$\Rightarrow \quad r_2 = r_{2,2} = \mathbf{0.0000826915340950721}\,au$$

Therefore, based on (297) and the data analysis in the Excel sheet "Tan, A.P., Asli Pinar Tan Analysis Based on Earth-Sun distance (d) Landsat.xlsx"[13], at this point we have solved for the values of r_1 (298) and r_2 (299), which are the radii of revolution of the Sun and the Earth around their individual circular orbits of revolution, respectively, based on our postulate in the **ARTICLE 1** "Points on two circles in space mimic an ellipse with respect to a point", which are among the most fundamental discoveries of this Article that we had set out to find.

$$r_1 = \mathbf{1.00007869128446}\,au = \mathbf{149{,}609{,}642.749}\,km \quad \left(radius\ of\ Sun's\ individual\ revolution\right) \tag{298}$$

$$r_2 = \mathbf{0.0000826915340950721}\ au = \mathbf{12,370.477}\ km \quad (radius\ of\ Earth's\ individual\ revolution) \quad (299)$$

We understand from (298) - (299) that it is the radius r_1 (298) of Sun's individual revolution which predominantly determines the distances between the Sun and the Earth over a yearly cycle, as well as the semi-major and semi-minor axes of the relative ellipse formed by the relative motions of the Sun and the Earth in their individual circular orbits, namely the magnitude b (285) of the semi-major axis vector $\bar{\boldsymbol{b}}$ (11) and the magnitude a (288) of the semi-minor axis vector $\bar{\boldsymbol{a}}$ (10).

Moreover, again considering the topology depicted in **Figure 2**, and as the Sun appears to move with respect to the other stars at a rate of about 1° eastward per solar day as seen from the Earth, at this point we can infer that the valid solution among $(\beta = \beta_{2,1} = \mathbf{156^\circ.56})$ (297) and $(\beta = \beta_{2,2} = -\mathbf{156^\circ.56})$ (297) that satisfies this motion, where the Sun appears to revolve[2] in a *counterclockwise*[27] direction in its relative motion with respect to the Earth from a vantage point above the North Pole[29] of the Earth, would have to be $(\beta = \beta_{2,2} = -\mathbf{156^\circ.56})$ (300), which finally also allows us to determine that $\hat{\boldsymbol{u}}_{2\perp} = -\hat{\boldsymbol{z}}$ (300) is the unit vector $\hat{\boldsymbol{u}}_{2\perp}$ (97) in the direction of self-rotation axis[30,7] of Earth, normal to $Circle_2$ of the Earth's revolution in its individual orbit, and also the direction of the North Pole[29] of the Earth in the configuration in **Figure 2**, which makes us understand that the Earth's yearly revolution around its own circular orbit is in a *clockwise*[27] direction as seen from above the North Pole[29] of the Earth, opposite to its daily *counterclockwise*[27] self rotation as seen from the same vantage point above the North Pole[29] of the Earth, and these findings in (300) are also among the most fundamental discoveries of this Article that we had set out to find.

This reversed direction of "the Earth's yearly revolution around its own circular orbit" and "the Earth's daily self rotation" as seen from a vantage point above its North Pole[29], hints a relation of this motion with the rule of Electromagnetism that is named as "Faraday's Law of Induction"[19], which states that "The Electromotive Force (EMF)[20] around a closed path is equal to the negative of the time rate of change of the Magnetic Flux[21] enclosed by the path." This implies that a charged particle moving in an external magnetic field with changing flux[21] moves in a direction to reverse the change in that magnetic flux[21]. The Earth may well be considered to behave like a charged particle obeying "Faraday's Law of Induction"[19], moving in space in an external magnetic field with changing flux[21] through the area of its individual circle of revolution, which may be the cause that triggers Earth's daily self rotation in the reverse direction, negative of the time rate of change of this flux[21] through the area of Earth's individual circle of revolution. In fact, all moving bodies in the Universe must be obeying "Faraday's Law of Induction"[19] in their motions, i.e. each one is possibly moving in space in external magnetic fields with changing flux[21] through the area of their individual circle of revolution while moving in *clockwise*[27] direction (Revolution Orbit "West" to Revolution Orbit "East"), which is the cause that triggers their self rotation in the reverse (*counterclockwise*[27]) direction (Self Rotation "West" to Self Rotation "East"), as seen from a vantage point above their North Pole[28,29], hence the two Easts and two Wests of bodies in the Universe, where we always accept the North Pole[28,29] of a body to be the pole rotating in counterclockwise[27] direction according to "right hand rule"[18]. This is an assertion we make in this Article.

$$\beta = \beta_{2,2} = \begin{cases} \mathbf{-156^\circ.56} \\ \mathbf{-2.73251312735455}\ radians \end{cases} \Rightarrow \begin{cases} Sin\beta = \mathbf{-0.397764973581873} \Rightarrow \hat{u}_{2\perp} = -\hat{z} \\ Cos\beta = \mathbf{-0.917487343668245} \end{cases} \quad (300)$$

$$(Sun - Earth\ system)$$

We eliminate the case of $(\beta = \beta_{2,1} = \mathbf{156^\circ.56})$ (297), because the topology depicted in **Figure 2** would imply a *clockwise*[27] motion of the Sun relative to the Earth, as seen from a vantage point above the North Pole[29] of the Earth, contradicting observational data.

Based on (141) and (300), we can now determine the actual value of $[r_1 Sin\beta]$ (301).

$$Sin\beta \le 0 \;\Rightarrow\; r_1\, Sin\beta = \mathbf{-0.397796274218557}\ au \qquad (Sun - Earth\ system) \quad (301)$$

At this point, having determined the values for ϕ_0 (138), r_1 (298), r_2 (299), and β (300) for the Sun-Earth system, as $(Cos\,\phi_0 < 0)$ (129) and $(a^2 < b^2)$ (131), we can determine the semi-minor[23] axis vector $\vec{a}$ (302) and the semi-major[23] axis vector $\vec{b}$ (303) of the ellipse for Sun-Earth system based on (10) - (11), also significant findings of this Article. As such, we again obtain $\vec{a}\cdot\vec{b}$ (304) for the Sun-Earth system from the Dot Product[14] of the semi-minor[23] axis vector $\vec{a}$ (302) with the semi-major[23] axis vector $\vec{b}$ (303) of the ellipse for Sun-Earth system. Comparing $\vec{a}\cdot\vec{b}$ (304) with $\vec{a}\cdot\vec{b}$ (143) we have found for the Sun-Earth system, we see that the result is consistent to 10 significant places after the decimal point in terms of astronomical units[4] (au) (3).

$$\vec{a} = \hat{x}\,120{,}168{,}467.040\ km - \hat{y}\,72{,}280{,}843.623\ km + \hat{z}\,52{,}106{,}536.351\ km \quad (302)$$

$$\vec{b} = -\hat{x}\,66{,}316{,}021.722\ km - \hat{y}\,131{,}002{,}922.948\ km - \hat{z}\,28{,}752{,}488.898\ km \quad (303)$$

$$\vec{a}\cdot\vec{b} \simeq \mathbf{1{,}714{,}509{,}335{,}021.425}\ km^2 \simeq \mathbf{0.0000766106293376090}\ au^2 \quad (304)$$

Based on (108) and (300), we can also determine the formula for $Sin[\gamma_{2,\Delta}(\phi)]$ (305).

$$\hat{u}_{2\perp} = -\hat{z} \;\Rightarrow\; Sin[\gamma_{2,\Delta}(\phi)] \simeq -\frac{2r_1 Sin\beta Cos(\phi+\phi_0)}{\left\{|\vec{d}(\phi)| + |\vec{d}(\pi+\phi)|\right\}} \quad (305)$$

As we have found the value of $[r_1 Sin\beta]$ (301), determined the formula for $Sin[\gamma_{2,\Delta}(\phi)]$ (305), and already calculated $Cos(\phi+\phi_0)$ for all ϕ of the Sun-Earth system, we now also calculate the values of $Sin[\gamma_{2,\Delta}(\phi)]$ (305) to an approximation for all ϕ in the Excel sheet "Tan, A.P., Asli Pinar Tan Analysis Based on Earth-Sun distance (d) Landsat.xlsx"[13].

Based on (107) and (300), we can now also finalize the identity for $\ell_z(\phi)$ (306).

$$\hat{u}_{2\perp} = -\hat{z} \;\Rightarrow\; \ell_z(\phi) = \ell_z(\phi+\pi) \simeq -\left[\frac{\left\{|\vec{d}(\phi)| - |\vec{d}(\pi+\phi)|\right\}}{2}\right] Sin[\gamma_{2,\Delta}(\phi)] \quad (306)$$

Having previously calculated $\vec{a}\cdot\vec{\ell}(\phi)$ and $\vec{b}\cdot\vec{\ell}(\phi)$ for all ϕ in the Excel sheet "Tan, A.P., Asli Pinar Tan Analysis Based on Earth-Sun distance (d) Landsat.xlsx"[13] using (158) and (159), as well as the value of $\ell_i^{\,2}$ (294) and $Sin[\gamma_{2,\Delta}(\phi)]$ (305) to an approximation for each data point, we now also calculate the values of $\left[\ell_x(\phi),\ell_y(\phi),\ell_z(\phi)\right]$ (9) for all ϕ in the Sun-Earth system, or equivalently the values of $\left(\ell_{x,i},\ell_{y,i},\ell_{z,i}\right)$ (177) for each data point in the Excel sheet "Tan, A.P., Asli Pinar Tan Analysis Based on Earth-Sun distance (d) Landsat.xlsx"[13], utilizing (306) and (209) - (213).

Furthermore, at this point we calculate the three components $\left[d_x(\phi),d_y(\phi),d_z(\phi)\right]$ (15) in Cartesian coordinates of $\vec{d}(\phi)$ (14), for all ϕ, in the Excel sheet "Tan, A.P., Asli Pinar Tan Analysis Based on Earth-Sun distance (d) Landsat.xlsx"[13], using found values of r_1 (298), r_2 (299), β (300), and the calculated values of $\left[\ell_x(\phi),\ell_y(\phi),\ell_z(\phi)\right]$ (9), $Cos\phi$, and $Cos(\phi+\phi_0)$ for all ϕ of Sun-Earth system.

Finally, to make a validity check, we also calculate the values of $d(\phi)$ (16) for all ϕ in the Excel sheet "Tan, A.P., Asli Pinar Tan Analysis Based on Earth-Sun distance (d) Landsat.xlsx"[13], using the values of $\left[d_x(\phi),d_y(\phi),d_z(\phi)\right]$ (15) calculated as described in the above paragraph, based on the values of the parameters we have found using our analytical approximation throughout this Article. The difference of the calculated value for $d(\phi)$ (16) and the actual value of $d(\phi)$ based on NASA Landsat observation data[3], for all ϕ, reveal that our analytical approximation in this Article has given us Sun-Earth distance results accurate to somewhere between five to eight decimal points in astronomical units[4] (au), for different ϕ.

It is important to note here that the inaccuracy in the calculated values of $d(\phi)$ (16) is totally due to the inaccuracy in the calculated values of $\left[\ell_x(\phi),\ell_y(\phi),\ell_z(\phi)\right]$ (9), based on the approximation we have made in (305) - (306).

Otherwise, our $Cos\phi$ and $Cos(\phi+\phi_0)$ calculation for all ϕ of Sun-Earth system is very reliable, and based on the iterative approach we have applied between (169) - (292), the values we have found for r_1 (298), r_2 (299), β (300) are very accurate, such that $\left[\ell_{(i=689)}\right]_{5^{th}\ iteration}$ (282) differs just about $\mathbf{0.001}\,m=\mathbf{1}\,mm$ from the value we have found earlier for $\left[\ell_{(i=689)}\right]_{4^{th}\ iteration}$ (262).

We had previously defined the unit vectors normal to $Circle_1$ of the Sun and $Circle_2$ of the Earth in the direction of their self-rotation axes[30,7] as $\hat{\boldsymbol{u}}_{1\perp}$ (97) and $\hat{\boldsymbol{u}}_{2\perp}$ (97), respectively, where the self-revolution axis and the self-rotation axis[30,7] of each body must be aligned along the normal of its equatorial[31] plane based on our assertion in the **ARTICLE 1** "Points on two circles in space mimic an ellipse with respect to a point", which are also the unit vectors pointing in the directions of the North Poles[28,29] of the Sun and the Earth along their rotational axes[7], respectively, and whose directions are determined according to 'right hand rule'[18] based on the direction of self-rotation of the body. Having determined the direction of $\hat{\boldsymbol{u}}_{2\perp}=-\hat{\boldsymbol{z}}$ (300), we still have to determine the direction of the unit vector $\hat{\boldsymbol{u}}_{1\perp}$ (97) normal to $Circle_1$ of the Sun's individual revolution in the direction of its self-rotation axis[30,7], i.e. the direction of the Sun's

North Pole[28,29] in the configuration in **Figure 2**, where the distance vector $\vec{d}(\phi)$ (15) is directed from the Earth to the Sun.

Based on the motion of the sunspots, the Sun rotates[10,11] about its axis once in about 25 Earth days at its equator, but more like in 27 Earth days at 35° from its equator. As an observer aligned with the North Pole[29] of the Earth sees, the sunspots on the Sun travel[10] from east to west, going across the Sun's disk in June and December, but their path dipping somewhat in March and September. This observation[10] implies that the Sun's self rotation is in *clockwise*[27] direction from a vantage point above the North Pole[29] of the Earth, meaning that the pole vector of the Sun that is angle-wise closer to the Earth's North Pole[29] should be the Sun's South Pole[28] vector, as the North Pole[28,29] is expected to be rotationg *counterclockwise*[27] based on the "right hand rule"[18]. Based on the configuration in **Figure 2** and the found $(\beta = -\mathbf{156^\circ.56})$ (300), from a vantage point above the North Pole[29] of the Earth, the Sun's motion in its individual circle of revolution is in *counterclockwise*[27] direction.

As a result of observation[10] and our analysis, from a vantage point above the North Pole[29] of the Earth, "the Sun's yearly revolution around its own circular orbit is in *counterclockwise*[27] direction" and "the Sun's self rotation is in in *clockwise*[27] direction". Equivalently, as seen from a vantage point above the North Pole[28,29] of the Sun, "the Sun's yearly revolution around its own circular orbit is in *clockwise*[27] direction" and "the Sun's self rotation is in in *counterclockwise*[27] direction". This is yet another fundamental discovery we make in this Article. This reversed direction of motion also indicates that the Sun obeys "Faraday's Law of Induction"[19], as we had asserted above for all moving bodies in the Universe.

Based on this discovery, as $\hat{\boldsymbol{u}}_{1\perp}$ (97) is the normal to $Circle_1$ of the Sun's individual revolution in the direction of its self-rotation axis[30,7], i.e. in the direction of its North Pole[28,29], $(-\hat{\boldsymbol{u}}_{1\perp})$ (97) must be the unit vector pointing in the direction of the Sun's South Pole[28]. As the Sun's South Pole[28] must be the one that appears to rotate[2] in a *clockwise*[27] direction from a vantage point above North Pole[28,29] vector $\hat{\boldsymbol{u}}_{2\perp} = -\hat{z}$ (300) of Earth, we can infer that $\left[(-\hat{\boldsymbol{u}}_{1\perp}) \cdot \hat{\boldsymbol{u}}_{2\perp} \geq 0\right]$ (307) must hold, leading to the information for the Sun-Earth system that the angle $\alpha_{\hat{\boldsymbol{u}}_{1\perp} \cdot \hat{\boldsymbol{u}}_{2\perp}}$ (307) between the North Pole[28,29] vectors $\hat{\boldsymbol{u}}_{1\perp}$ (97) and $(\hat{\boldsymbol{u}}_{2\perp} = -\hat{z})$ (300) of the Sun and the Earth, respectively, must obey $\left[\frac{\pi}{2} \leq \alpha_{\hat{\boldsymbol{u}}_{1\perp} \cdot \hat{\boldsymbol{u}}_{2\perp}} \leq \frac{3\pi}{2}\right]$ (307) and therefore $\left[Cos\left(\alpha_{\hat{\boldsymbol{u}}_{1\perp} \cdot \hat{\boldsymbol{u}}_{2\perp}}\right) \leq 0\right]$ (307). Based on (300) and (307), as $[Cos\,\beta < 0]$ (300), we can make the reasoning in (308) that for $[\hat{\boldsymbol{u}}_{1\perp} \cdot \hat{\boldsymbol{u}}_{2\perp} \leq 0]$ (307) to hold, the valid solution for the North Pole[28] vector $\hat{\boldsymbol{u}}_{1\perp}$ (97) of the Sun has to be $[\hat{\boldsymbol{u}}_{1\perp} = \hat{x} Sin\,\beta - \hat{z} Cos\,\beta]$ (308), another one of the fundamental findings of the Article.

$$(-\hat{u}_{1\perp})\cdot\hat{u}_{2\perp}\geq 0 \quad\Rightarrow\quad \hat{u}_{1\perp}\cdot\hat{u}_{2\perp}\leq 0 \quad\Rightarrow\quad \frac{\pi}{2}\leq\alpha_{\hat{u}_{1\perp}\cdot\hat{u}_{2\perp}}\leq\frac{3\pi}{2}$$

$$\begin{aligned}\Rightarrow\quad \hat{u}_{1\perp}\cdot\hat{u}_{2\perp} &= Cos\left(\alpha_{\hat{u}_{1\perp}\cdot\hat{u}_{2\perp}}\right)\leq 0\\ &=\pm(-\hat{x}Sin\beta+\hat{z}Cos\beta)\cdot(-\hat{z})\\ &=\pm\left[(-\hat{x}Sin\beta+\hat{z}Cos\beta)\cdot(-\hat{z})\right]\\ &=\pm(-Cos\beta)\end{aligned} \tag{307}$$

$$Cos\beta<0 \;\Rightarrow\; \hat{u}_{1\perp}\cdot\hat{u}_{2\perp}=Cos\beta\leq 0 \;\Rightarrow\; \hat{u}_{1\perp}=\hat{x}Sin\beta-\hat{z}Cos\beta \tag{308}$$

Moving on in our analysis, based on the definitions in (99) and (100), and also utilizing **(316)**, the cosine of the tilt[7,10,11] angle $\gamma_1(\phi)=\gamma_{Sun}(\phi)$ (98) of the Sun's North Pole[28] vector $\hat{u}_{1\perp}$ (308) along $\left[-\vec{d}(\phi)\right]$ (15) towards the Earth, and the cosine of the tilt[7] angle $\gamma_2(\phi)=\gamma_{Earth}(\phi)$ (98) of the Earth's North Pole[28,29] vector $\hat{u}_{2\perp}$ (300) along $\vec{d}(\phi)$ (15) towards the Sun, at each ϕ, are expressed in (309) and (311), respectively. Based on this topology, **(317)** - **(319)**, and **(322)**, it is further possible to define the tangent of $\gamma_{1,\Delta}(\phi)$ (98) as in (310).

$$Cos\left[\gamma_{Sun}(\phi)\right]=Cos\left[\gamma_1(\phi)\right]=\frac{\hat{u}_{1\perp}\cdot\left[-\vec{d}(\phi)\right]}{\left|\vec{d}(\phi)\right|}=\frac{\hat{u}_{1\perp}\cdot\left[-\vec{d}(\phi)\right]}{d(\phi)}=\hat{u}_{1\perp}\cdot\left[-\hat{d}(\phi)\right]=p_1(\phi) \tag{309}$$

$$Tan\left[\gamma_{1,\Delta}(\phi)\right]=\frac{p_1(\phi)}{\left|\vec{n}_1(\phi)\right|} \tag{310}$$

$$Cos\left[\gamma_{Earth}(\phi)\right]=Cos\left[\gamma_2(\phi)\right]=\frac{\hat{u}_{2\perp}\cdot\vec{d}(\phi)}{\left|\vec{d}(\phi)\right|}=\frac{\hat{u}_{2\perp}\cdot\vec{d}(\phi)}{d(\phi)}=\hat{u}_{2\perp}\cdot\hat{d}(\phi)=p_2(\phi) \tag{311}$$

To infer more insight, we have calculated the values of $Cos\left[\gamma_1(\phi)\right]$ (309), $Tan\left[\gamma_{1,\Delta}(\phi)\right]$ (310), and $Cos\left[\gamma_2(\phi)\right]$ (311) in the Excel sheet "Tan, A.P., Asli Pinar Tan Analysis Based on Earth-Sun distance (d) Landsat.xlsx"[13] for every ϕ, using the already calculated values of $\left[d_x(\phi),d_y(\phi),d_z(\phi)\right]$ (15), $\hat{u}_{1\perp}$ (308), $\hat{u}_{2\perp}$ (300), and β (300). We have also plotted these calculated values over a year, $\gamma_1(\phi)=\gamma_{Sun}(\phi)$ (98) in **Figure 22**, and $\gamma_2(\phi)=\gamma_{Earth}(\phi)$ (98) in **Figure 23**.

Checking the calculated data, we discover about Earth's tilt[7] that $\gamma_2(\phi)=90^\circ$ (312) on March 22 (Data Point $[i=322]$) and on September 20 (Data Point $[i=1052]$), $\gamma_2(\phi)=66^\circ.560=(90^\circ-23^\circ.440)$ (312) at the Summer Solstice[5] on June 21 - 22 (Data Points $[i=687-689]$), and $\gamma_2(\phi)=113^\circ.440=(90^\circ+23^\circ.440)$ (312) at the Winter Solstice[5] on December 21 (Data Points $[i=1417-1419]$). We also observe that the calculated $\gamma_2(\phi)$ (98) data and $\gamma_{2,\Delta}(\phi)$ (305) data of the Earth are consistent.

$$
\begin{aligned}
&\gamma_2(\phi)_{Data\ Point\,[i=322]} &&= \mathbf{90^\circ} && (March\ 22)\\
&\gamma_2(\phi)_{Data\ Point\,[i=687-689]} &&= \mathbf{66^\circ.560} = \left(90^\circ - 23^\circ.440\right) = \gamma_{2,\min} && (June\ 21-22)\\
&\gamma_2(\phi)_{Data\ Point\,[i=1052]} &&= \mathbf{90^\circ} && (September\ 20)\\
&\gamma_2(\phi)_{Data\ Point\,[i=1417-1419]} &&= \mathbf{113^\circ.440} = \left(90^\circ + 23^\circ.440\right) = \gamma_{2,\max} && (December\ 21)
\end{aligned}
\quad (312)
$$

The result in (313) is also consistent with (102) and the maximum tilt[7] of the Earth $\gamma_{Earth,\Delta Max}$ (4).

$$
\begin{aligned}
&\gamma_{2,\max} = \mathbf{113^\circ.440} = 90^\circ + 23^\circ.440 = 90^\circ + \gamma_{2,\Delta_1} && \Rightarrow && \gamma_{2,\Delta_1} = \mathbf{23^\circ.440} = \gamma_{Earth,\Delta Max}\\
&\gamma_{2,\min} = \ \mathbf{66^\circ.560} = 90^\circ - 23^\circ.440 = 90^\circ - \gamma_{2,\Delta_2} && \Rightarrow && \gamma_{2,\Delta_2} = \mathbf{23^\circ.440} = \gamma_{Earth,\Delta Max}
\end{aligned}
\quad (313)
$$

Along similar lines, checking the calculated data, we discover about the Sun's tilt[7,10,11] that $\gamma_1(\phi) = 90^\circ$ (314) on January 2 (Data Points $[i = 6-7]$) around minimum Sun-Earth distance in a yearly cycle, at the Summer Solstice[5] on June 20 - 21 (Data Points $[i = 683-685]$), on July 5 (Data Point $[i = 744]$) at maximum Sun-Earth distance in a yearly cycle, and at the Winter Solstice[5] on December 22 (Data Points $[i = 1421-1423]$), $\gamma_1(\phi) = 90^\circ.411 = \left(90^\circ + 0^\circ.411\right)$ (314) between March 26 - 30 (Data Points $[i = 340-356]$), and $\gamma_1(\phi) = 89^\circ.993 = \left(90^\circ - 0^\circ.007\right)$ (314) between June 26 - 29 (Data Points $[i = 707-720]$), $\gamma_1(\phi) = 90^\circ.408 = \left(90^\circ + 0^\circ.408\right)$ (314) between September 26 - 30 (Data Points $[i = 1074-1091]$), and $\gamma_1(\phi) = 89^\circ.996 = \left(90^\circ - 0^\circ.004\right)$ (314) between December 27 - 29 (Data Points $[i = 1441-1449]$). We also observe that the calculated $\gamma_1(\phi)$ (98) data and $\gamma_{1,\Delta}(\phi)$ (310) data of the Sun are consistent.

$$
\begin{aligned}
&\gamma_1(\phi)_{Data\ Points\,[i=6-7]} &&= \mathbf{90^\circ} && (January\ 2)\\
&\gamma_1(\phi)_{Data\ Points\,[i=340-356]} &&= \mathbf{90^\circ.411} = \left(90^\circ + 0^\circ.411\right) = \gamma_{1,\max 1} && (March\ 26-30)\\
&\gamma_1(\phi)_{Data\ Points\,[i=683-685]} &&= \mathbf{90^\circ} && (June\ 20-21)\\
&\gamma_1(\phi)_{Data\ Points\,[i=707-720]} &&= \mathbf{89^\circ.993} = \left(90^\circ - 0^\circ.007\right) = \gamma_{1,\min 1} && (June\ 26-29)\\
&\gamma_1(\phi)_{Data\ Points\,[i=744]} &&= \mathbf{90^\circ} && (July\ 5)\\
&\gamma_1(\phi)_{Data\ Points\,[i=1074-1091]} &&= \mathbf{90^\circ.408} = \left(90^\circ + 0^\circ.408\right) = \gamma_{1,\max 2} && (September\ 26-30)\\
&\gamma_1(\phi)_{Data\ Points\,[i=1421-1423]} &&= \mathbf{90^\circ} && (December\ 22)\\
&\gamma_1(\phi)_{Data\ Points\,[i=1441-1449]} &&= \mathbf{89^\circ.996} = \left(90^\circ - 0^\circ.004\right) = \gamma_{1,\min 2} && (December\ 27\ -\ 29\ )
\end{aligned}
\quad (314)
$$

Based on this calculation, we can make the analysis that the Sun's North Pole[28,29] vector is tilted[7] $\gamma_1(\phi)$ (98) around $\mathbf{90^\circ}$ (314) with respect to the Sun-Earth distance vector $\left[-\vec{\boldsymbol{d}}(\phi)\right]$ (15) throughout a yearly cycle, and it is exactly at $\mathbf{90^\circ}$ (314) with respect to the Sun-Earth distance vector $\left[-\vec{\boldsymbol{d}}(\phi)\right]$ (15) around the maximum and minimum Sun-Earth distance in a yearly cycle, as well as at the Summer and Winter Solstices[5], consistent with the observation[10,11] that in June and December, as viewed from Earth, the Sun's North and South Poles[28] do not to tip toward or away from Earth. Further, it is tilted *most towards* Earth twice a year, once a few days after the Spring Equinox and secondly a few days after the Autumn Equinox. Additionally, it is tilted *most away* from Earth twice a year, once between the Winter Solstice[5] and minimum Sun-Earth distance in a yearly cycle between $\left[\phi_{\gamma_{2,\max}} = \pi - \phi_0 = 331^{\circ}.11 = -28^{\circ}.89\right]$ (140) and $\left[\phi = 0^{\circ}\right]$, and secondly between the Summer Solstice[5] and maximum Sun-Earth distance in a yearly cycle between $\left[\phi_{\gamma_{2,\min}} = -\phi_0 = 151^{\circ}.11\right]$ (139) and $\left[\phi = 180^{\circ}\right]$, namely when the Sun and Earth are travelling in opposite directions towards each other, based on the topology in in **Figure 2**. Apparently, the maximum and minimum values that $\gamma_1(\phi)$ (98) take over a relative Sun-Earth cycle, namely $\gamma_{1,\max}$ (104) and $\gamma_{1,\min}$ (104), respectively, as well as the tilt[7] angles γ_{1,Δ_1} (104) and γ_{1,Δ_2} (104) take on more than a single value throughout a yearly cycle, that are expressed in terms of $\gamma_{1,\max 1}$ (315), $\gamma_{1,\max 2}$ (315), $\gamma_{1,\min 1}$ (315), $\gamma_{1,\min 2}$ (315), $\gamma_{1,\Delta_{1,1}}$ (315), $\gamma_{1,\Delta_{1,2}}$ (315), $\gamma_{1,\Delta_{2,1}}$ (315), and $\gamma_{1,\Delta_{2,2}}$ (315). These are also fundamental findings of this Article.

$$\begin{aligned}
\gamma_{1,\max 1} &= \mathbf{90^{\circ}.411} = 90^{\circ} + 0^{\circ}.411 = 90^{\circ} + \gamma_{1,\Delta_{1,1}} & \Rightarrow \quad & \gamma_{1,\Delta_{1,1}} = \mathbf{0^{\circ}.411} \\
\gamma_{1,\max 2} &= \mathbf{90^{\circ}.408} = 90^{\circ} + 0^{\circ}.408 = 90^{\circ} + \gamma_{1,\Delta_{1,2}} & \Rightarrow \quad & \gamma_{1,\Delta_{1,2}} = \mathbf{0^{\circ}.408} \\
\gamma_{1,\min 1} &= \mathbf{89^{\circ}.993} = 90^{\circ} - 0^{\circ}.007 = 90^{\circ} - \gamma_{1,\Delta_{2,1}} & \Rightarrow \quad & \gamma_{1,\Delta_{2,1}} = \mathbf{0^{\circ}.007} \\
\gamma_{1,\min 2} &= \mathbf{89^{\circ}.996} = 90^{\circ} - 0^{\circ}.004 = 90^{\circ} - \gamma_{1,\Delta_{2,2}} & \Rightarrow \quad & \gamma_{1,\Delta_{2,2}} = \mathbf{0^{\circ}.004}
\end{aligned} \quad (315)$$

According to observation[10,11], it is said that there is a day out of the year when the Sun's North Pole is said to tip most towards the Earth during September, a day out of the year when the Sun's North Pole is said to tip most away from the Earth during March, and that there are also two days during the year, in June and December, when the Sun's North and South Poles[28], as viewed from Earth, are not considered to tip toward or away from Earth. Based on our analysis above leading to the determination of Sun's North Pole[28,29] unit vector $\hat{\boldsymbol{u}}_{1\perp}$ (308) normal to $Circle_1$ of the Sun's individual revolution in the direction of its self-rotation axis[30,7], we have found out that it must be the Sun's South Pole[28,29] which appears to rotate[2] in a *clockwise*[27] direction from a vantage point above the North Pole[29] vector $\hat{\boldsymbol{u}}_{2\perp} = -\hat{\boldsymbol{z}}$ (300) of the Earth, and that $\left(-\hat{\boldsymbol{u}}_{1\perp}\right)$ (308) must be the unit vector pointing in the direction of the Sun's South Pole[28]. Thus, it must be the angles made by the Sun's South Pole[28] unit vector $\left(-\hat{\boldsymbol{u}}_{1\perp}\right)$ (308) which leads to the observation[10,11] stated above, i.e. the Sun's South Pole[28] vector must be observed to tip most towards the Earth during September, and the Sun's South Pole[28] vector must be observed to tip most away from the Earth during March.

To continue our analysis to obtain more insight into the topology of Sun-Earth system, we define the components $\bar{\boldsymbol{p}}_1(\phi)$ (318) and $\bar{\boldsymbol{n}}_1(\phi)$ (322) of the Sun's South Pole[28] vector $\left(-\hat{\boldsymbol{u}}_{1\perp}\right)$ (308) parallel and normal, respectively, to $\left[-\hat{\boldsymbol{d}}(\phi)\right]$ (316) from the Sun to the Earth, and the components $\bar{\boldsymbol{p}}_2(\phi)$ (320) and $\bar{\boldsymbol{n}}_2(\phi)$ (323) of the Earth's North Pole[29] vector $\hat{\boldsymbol{u}}_{2\perp}$ (300) parallel and normal, respectively, to $\vec{d}(\phi)$ (15) from the Earth to the Sun, defined as in (317) at each ϕ, where $\hat{d}(\phi)$ (316) is the unit vector in the direction of $\vec{d}(\phi)$ (15), which we obtain using its magnitude $|\vec{d}(\phi)| = d(\phi)$ (16). Magnitude of $\vec{p}_1(\phi)$ (318) in the direction of the unit vector $[-\hat{d}(\phi)]$ (316) from the Sun to the Earth is $p_1(\phi)$ (319), and the magnitude of $\vec{p}_2(\phi)$ (320) in the direction of the unit vector $\hat{d}(\phi)$ (316) from the Earth to the Sun is $p_2(\phi)$ (321). The unit vectors in the directions of $\vec{n}_1(\phi)$ (322) and $\vec{n}_2(\phi)$ (323) are $\hat{n}_1(\phi)$ (324) and $\hat{n}_2(\phi)$ (324), respectively, where $|\vec{n}_1(\phi)| = n_1(\phi)$ (324) is the magnitude of $\vec{n}_1(\phi)$ (322) in the direction of $\hat{n}_1(\phi)$ (324), and $|\vec{n}_2(\phi)| = n_2(\phi)$ (324) is the magnitude of $\vec{n}_2(\phi)$ (323) in the direction of $\hat{n}_2(\phi)$ (324). As our analysis has previously led to the conclusion that the Sun's self rotation is in *clockwise*[27] direction from a vantage point above the North Pole[29] of the Earth, meaning that the pole vector of the Sun that is angle-wise closer to the Earth's North Pole[29] should be the Sun's South Pole[28,29] vector, we can also infer the information that $[\hat{n}_1(\phi)\cdot\hat{n}_2(\phi) \geq 0]$ (324) must hold for all ϕ of the Sun-Earth system.

$$\hat{d}(\phi) = \frac{\vec{d}(\phi)}{|\vec{d}(\phi)|} = \frac{\vec{d}(\phi)}{d(\phi)} = \hat{x}\frac{d_x(\phi)}{d(\phi)} + \hat{y}\frac{d_y(\phi)}{d(\phi)} + \hat{z}\frac{d_z(\phi)}{d(\phi)} \qquad (\textit{unit vector from Earth to Sun}) \quad (316)$$

$$-\hat{u}_{1\perp} = \vec{p}_1(\phi) + \vec{n}_1(\phi) \quad ; \quad \hat{u}_{2\perp} = \vec{p}_2(\phi) + \vec{n}_2(\phi) \qquad (317)$$

$$\begin{aligned}\vec{p}_1(\phi) &= \left([-\hat{u}_{1\perp}]\cdot[-\hat{d}(\phi)]\right)[-\hat{d}(\phi)] = p_1(\phi)[-\hat{d}(\phi)] = -p_1(\phi)\hat{d}(\phi) = [(-\hat{u}_{1\perp})\cdot\hat{d}(\phi)]\hat{d}(\phi) \\ &= \left[\frac{d_x(\phi)}{d(\phi)} Sin\,\beta - \frac{d_z(\phi)}{d(\phi)} Cos\,\beta\right][-\hat{d}(\phi)] = \hat{x}\,p_{1x}(\phi) + \hat{y}p_{1y}(\phi) + \hat{z}\,p_{1z}(\phi)\end{aligned} \qquad (318)$$

$$\begin{aligned}&p_1(\phi) = [-\hat{u}_{1\perp}]\cdot[-\hat{d}(\phi)] = \frac{d_x(\phi)}{d(\phi)} Sin\,\beta - \frac{d_z(\phi)}{d(\phi)} Cos\,\beta \\ &\left(\textit{in the direction of unit vector } [-\hat{d}(\phi)] \textit{ from Sun to Earth}\right)\end{aligned} \qquad (319)$$

$$\begin{aligned}\vec{p}_2(\phi) &= [\hat{u}_{2\perp}\cdot\hat{d}(\phi)]\hat{d}(\phi) = p_2(\phi)\hat{d}(\phi) = -p_2(\phi)[-\hat{d}(\phi)] = \left[-\frac{d_z(\phi)}{d(\phi)}\right]\hat{d}(\phi) \\ &= -\hat{x}\frac{d_x(\phi)d_z(\phi)}{[d(\phi)]^2} - \hat{y}\frac{d_y(\phi)d_z(\phi)}{[d(\phi)]^2} - \hat{z}\frac{[d_z(\phi)]^2}{[d(\phi)]^2} = \hat{x}\,p_{2x}(\phi) + \hat{y}p_{2y}(\phi) + \hat{z}\,p_{2z}(\phi)\end{aligned} \qquad (320)$$

$$p_2(\phi) = \hat{u}_{2\perp}\cdot\hat{d}(\phi) = -\frac{d_z(\phi)}{d(\phi)} \qquad \left(\textit{in the direction of unit vector } \hat{d}(\phi) \textit{ from Earth to Sun}\right) \quad (321)$$

$$\vec{n}_1(\phi)=(-\hat{u}_{1\perp})-\vec{p}_1(\phi)=(-\hat{x}Sin\,\beta+\hat{z}Cos\,\beta)-\left[\frac{d_x(\phi)}{d(\phi)}Sin\,\beta-\frac{d_z(\phi)}{d(\phi)}Cos\,\beta\right]\left[-\hat{d}(\phi)\right] \qquad (322)$$

$$=\hat{x}n_{1x}(\phi)+\hat{y}n_{1y}(\phi)+\hat{z}n_{1z}(\phi)$$

$$\vec{n}_2(\phi)=\hat{u}_{2\perp}-\vec{p}_2(\phi)=\hat{x}\frac{d_x(\phi)d_z(\phi)}{\left[d(\phi)\right]^2}+\hat{y}\frac{d_y(\phi)d_z(\phi)}{\left[d(\phi)\right]^2}+\hat{z}\left(-1+\frac{\left[d_z(\phi)\right]^2}{\left[d(\phi)\right]^2}\right) \qquad (323)$$

$$=\hat{x}n_{2x}(\phi)+\hat{y}n_{2y}(\phi)+\hat{z}n_{2z}(\phi)$$

$$\hat{n}_1(\phi)=\frac{\vec{n}_1(\phi)}{\left|\vec{n}_1(\phi)\right|}=\frac{\vec{n}_1(\phi)}{n_1(\phi)} \quad ; \quad \hat{n}_2(\phi)=\frac{\vec{n}_2(\phi)}{\left|\vec{n}_2(\phi)\right|}=\frac{\vec{n}_2(\phi)}{n_2(\phi)} \quad ; \quad \hat{n}_1(\phi)\cdot\hat{n}_2(\phi)\geq 0 \qquad (324)$$

We have calculated the values of $\left[p_{1x}(\phi),p_{1y}(\phi),p_{1z}(\phi)\right]$ (318), $\left[n_{1x}(\phi),n_{1y}(\phi),n_{1z}(\phi)\right]$ (322), $\left[p_{2x}(\phi),p_{2y}(\phi),p_{2z}(\phi)\right]$ (320), $\left[n_{2x}(\phi),n_{2y}(\phi),n_{2z}(\phi)\right]$ (323), $p_1(\phi)$ (319), and $p_2(\phi)$ (321)in the Excel sheet "Tan, A.P., Asli Pinar Tan Analysis Based on Earth-Sun distance (d) Landsat.xlsx"[13] for all ϕ, using the values of $\left[d_x(\phi),d_y(\phi),d_z(\phi)\right]$ (15), $(-\hat{u}_{1\perp})$ (308), $\hat{u}_{2\perp}$ (300), and β (300) we have calculated and determined before.

At any point throughout a Sun-Earth yearly cycle, an observer on Earth aligned with the North Pole[29] vector $\hat{u}_{2\perp}$ (300) of the Earth would observe the relative angles that the Sun's South Pole[28] vector $(-\hat{u}_{1\perp})$ (308) makes, based on the observer's line of sight. What determines these relative angle observations from the point of view of the observer is specifically the relative directions of the parallel components $\vec{p}_1(\phi)$ (318) and $\vec{p}_2(\phi)$ (320), and normal components $\vec{n}_1(\phi)$ (322) and $\vec{n}_2(\phi)$ (323), of $(-\hat{u}_{1\perp})$ (308) and $\hat{u}_{2\perp}$ (300), respectively, to the line joining the Sun and the Earth along $\left[-\hat{d}(\phi)\right]$ (316).

With respect to the line of sight of an observer on Earth aligned with the North Pole[29] $\hat{u}_{2\perp}$ (300) vector of the Earth, we define the relative tilt vector[10,11] $\vec{T}(\phi)$ (325) of the Sun towards the Earth, in which case a *relative* tilt[10,11] angle $\alpha_{Sun,wobble}(\phi)$ (330) of the Sun towards the Earth would mean a positive wobble $\left\{Tan\left[\alpha_{Sun,wobble}(\phi)\right]\geq 0\right\}$ (330) at ϕ along the direction of $\left[-\hat{d}(\phi)\right]$ (316), different from the *individual* tilt[7,10,11] angle $\gamma_1(\phi)$ (98) of the Sun onto $\left[-\hat{d}(\phi)\right]$ (316).

The component of this relative tilt vector[10,11] $\vec{T}(\phi)$ (325) that is parallel to $\left[-\hat{d}(\phi)\right]$ (316) is $\Delta\vec{p}(\phi)$ (326), determined by the difference of $\vec{p}_1(\phi)$ (318) and $\vec{p}_2(\phi)$ (320), whose magnitude along $\left[-\hat{d}(\phi)\right]$ (316) from the Sun towards Earth is $\left|\Delta\vec{p}(\phi)\right|_{Sun\rightarrow Earth}$ (327). The component of this relative tilt vector[10,11] $\vec{T}(\phi)$ (325) that is normal to $\left[-\hat{d}(\phi)\right]$ (316) is $\Delta\vec{n}(\phi)$ (328), which is defined along the line of sight direction $\hat{\boldsymbol{n}}_2(\phi)$ (324) of the observer aligned with the North Pole[29] of Earth, and whose magnitude $\left|\Delta\vec{\boldsymbol{n}}(\phi)\right|_{Sun\to Earth}$ (329) is the projection on the line of sight direction $\hat{\boldsymbol{n}}_2(\phi)$ (324) of the observer, of the Sun's South Pole[28]'s normal $\vec{\boldsymbol{n}}_1(\phi)$ (322) to $\left[-\hat{\boldsymbol{d}}(\phi)\right]$ (316).

$$\vec{\boldsymbol{T}}(\phi)=\Delta\vec{\boldsymbol{p}}(\phi)+\Delta\vec{\boldsymbol{n}}(\phi) \tag{325}$$

$$\begin{aligned}\Delta\vec{\boldsymbol{p}}(\phi)=\vec{\boldsymbol{p}}_1(\phi)-\vec{\boldsymbol{p}}_2(\phi)&=\left\{\left[(-\hat{\boldsymbol{u}}_{1\perp})\cdot\hat{\boldsymbol{d}}(\phi)\right]-\left[\hat{\boldsymbol{u}}_{2\perp}\cdot\hat{\boldsymbol{d}}(\phi)\right]\right\}\hat{\boldsymbol{d}}(\phi)\\&=\left\{\left([-\hat{\boldsymbol{u}}_{1\perp}]\cdot\left[-\hat{\boldsymbol{d}}(\phi)\right]\right)-\left(\hat{\boldsymbol{u}}_{2\perp}\cdot\left[-\hat{\boldsymbol{d}}(\phi)\right]\right)\right\}\left[-\hat{\boldsymbol{d}}(\phi)\right]\\&=\left\{p_1(\phi)-\left[-p_2(\phi)\right]\right\}\left[-\hat{\boldsymbol{d}}(\phi)\right]\\&=\left\{p_1(\phi)+p_2(\phi)\right\}\left[-\hat{\boldsymbol{d}}(\phi)\right]\\&=\left|\Delta\vec{\boldsymbol{p}}(\phi)\right|_{Sun\to Earth}\left[-\hat{\boldsymbol{d}}(\phi)\right]\\&=\hat{\boldsymbol{x}}\,\Delta p_x(\phi)+\hat{\boldsymbol{y}}\Delta p_y(\phi)+\hat{\boldsymbol{z}}\,\Delta p_z(\phi)\end{aligned} \tag{326}$$

$$\left|\Delta\vec{\boldsymbol{p}}(\phi)\right|_{Sun\to Earth}=p_1(\phi)+p_2(\phi)\qquad\left(\textit{magnitude of } \Delta\vec{\boldsymbol{p}}(\phi) \textit{ in the direction of } \left[-\hat{\boldsymbol{d}}(\phi)\right]\right) \tag{327}$$

$$\Delta\vec{\boldsymbol{n}}(\phi)=\left[\vec{\boldsymbol{n}}_1(\phi)\cdot\hat{\boldsymbol{n}}_2(\phi)\right]\hat{\boldsymbol{n}}_2(\phi)=\left|\Delta\vec{\boldsymbol{n}}(\phi)\right|_{Sun\to Earth}\hat{\boldsymbol{n}}_2(\phi)=\hat{\boldsymbol{x}}\,\Delta n_x(\phi)+\hat{\boldsymbol{y}}\Delta n_y(\phi)+\hat{\boldsymbol{z}}\,\Delta n_z(\phi) \tag{328}$$

$$\left|\Delta\vec{\boldsymbol{n}}(\phi)\right|_{Sun\to Earth}=\vec{\boldsymbol{n}}_1(\phi)\cdot\hat{\boldsymbol{n}}_2(\phi)\qquad\left(\textit{magnitude of } \Delta\vec{\boldsymbol{n}}(\phi) \textit{ in the direction of } \hat{\boldsymbol{n}}_2(\phi)\right) \tag{329}$$

Using the parallel and normal compnents of this relative tilt vector[10,11] $\vec{\boldsymbol{T}}(\phi)$ (325) throughout a Sun-Earth yearly cycle, we define the relative angle that the vector $(-\hat{\boldsymbol{u}}_{1\perp})$ (308) seemingly makes towards or away from the Earth, with respect to the line of sight of an observer on Earth aligned with the North Pole[29] vector $\hat{\boldsymbol{u}}_{2\perp}$ (300) of Earth, as the Sun's "wobble" $\alpha_{Sun,wobble}(\phi)$ (330) angle at each ϕ, and also define $\alpha_{Sun,sideways}(\phi)$ (331) as the relative "sideways" swing angle that the vector $(-\hat{\boldsymbol{u}}_{1\perp})$ (308) seemingly makes with respect to the line of sight of an observer on Earth aligned with the North Pole[29] vector $\hat{\boldsymbol{u}}_{2\perp}$ (300) of the Earth.

$$\begin{aligned}Tan\left[\alpha_{Sun,wobble}(\phi)\right]&=\frac{\left|\Delta\vec{\boldsymbol{p}}(\phi)\right|_{Sun\to Earth}}{\left|\Delta\vec{\boldsymbol{n}}(\phi)\right|_{Sun\to Earth}}=\frac{\left\{p_1(\phi)+p_2(\phi)\right\}}{\left\{\vec{\boldsymbol{n}}_1(\phi)\cdot\hat{\boldsymbol{n}}_2(\phi)\right\}}\\&=\frac{[-\hat{\boldsymbol{u}}_{Sun\perp}]\cdot\left[-\hat{\boldsymbol{d}}(\phi)\right]+\hat{\boldsymbol{u}}_{Earth\perp}\cdot\hat{\boldsymbol{d}}(\phi)}{\left\{-\hat{\boldsymbol{u}}_{Sun\perp}-\left([-\hat{\boldsymbol{u}}_{Sun\perp}]\cdot\left[-\hat{\boldsymbol{d}}(\phi)\right]\right)\left[-\hat{\boldsymbol{d}}(\phi)\right]\right\}\cdot\left\{\dfrac{\hat{\boldsymbol{u}}_{Earth\perp}-\left[\hat{\boldsymbol{u}}_{Earth\perp}\cdot\hat{\boldsymbol{d}}(\phi)\right]\hat{\boldsymbol{d}}(\phi)}{\left|\hat{\boldsymbol{u}}_{Earth\perp}-\left[\hat{\boldsymbol{u}}_{Earth\perp}\cdot\hat{\boldsymbol{d}}(\phi)\right]\hat{\boldsymbol{d}}(\phi)\right|}\right\}}\end{aligned} \tag{330}$$

$$Cos\left[\alpha_{Sun,sideways}(\phi)\right]=\hat{\boldsymbol{n}}_1(\phi)\cdot\hat{\boldsymbol{n}}_2(\phi)=\frac{\left[\vec{\boldsymbol{n}}_1(\phi)\cdot\vec{\boldsymbol{n}}_2(\phi)\right]}{n_1(\phi)n_2(\phi)}$$

$$=\frac{\left\{-\hat{\boldsymbol{u}}_{Sun\perp}-\left(\left[-\hat{\boldsymbol{u}}_{Sun\perp}\right]\cdot\left[-\hat{\boldsymbol{d}}(\phi)\right]\right)\left[-\hat{\boldsymbol{d}}(\phi)\right]\right\}\cdot\left\{\hat{\boldsymbol{u}}_{Earth\perp}-\left[\hat{\boldsymbol{u}}_{Earth\perp}\cdot\hat{\boldsymbol{d}}(\phi)\right]\hat{\boldsymbol{d}}(\phi)\right\}}{\left|-\hat{\boldsymbol{u}}_{Sun\perp}-\left(\left[-\hat{\boldsymbol{u}}_{Sun\perp}\right]\cdot\left[-\hat{\boldsymbol{d}}(\phi)\right]\right)\left[-\hat{\boldsymbol{d}}(\phi)\right]\right|\left|\hat{\boldsymbol{u}}_{Earth\perp}-\left[\hat{\boldsymbol{u}}_{Earth\perp}\cdot\hat{\boldsymbol{d}}(\phi)\right]\hat{\boldsymbol{d}}(\phi)\right|} \quad (331)$$

Using the calculated values of $\left[p_{1x}(\phi),p_{1y}(\phi),p_{1z}(\phi)\right]$ (318), $\left[n_{1x}(\phi),n_{1y}(\phi),n_{1z}(\phi)\right]$ (322), $\left[p_{2x}(\phi),p_{2y}(\phi),p_{2z}(\phi)\right]$ (320), $\left[n_{2x}(\phi),n_{2y}(\phi),n_{2z}(\phi)\right]$ (323), $p_1(\phi)$ (319), and $p_2(\phi)$ (321) in the Excel sheet "Tan, A.P., Asli Pinar Tan Analysis Based on Earth-Sun distance (d) Landsat.xlsx"[13], we have also calculated the values of $\left[\Delta p_x(\phi),\Delta p_y(\phi),\Delta p_z(\phi)\right]$ (326), $\left|\Delta\vec{p}(\phi)\right|_{Sun\rightarrow Earth}$ (327), $\left[\Delta n_x(\phi),\Delta n_y(\phi),\Delta n_z(\phi)\right]$ (328), and $\left|\Delta\vec{n}(\phi)\right|_{Sun\rightarrow Earth}$ (329), as well as the values of the angles $\alpha_{Sun,wobble}(\phi)$ (330) and $\alpha_{Sun,sideways}(\phi)$ (331) for all ϕ. We have also plotted these calculated values over a year, $\alpha_{Sun,wobble}(\phi)$ (330) in **Figure 24**, and $\alpha_{Sun,sideways}(\phi)$ (331) in **Figure 25**.

$$\begin{aligned}
&\alpha_{Sun,wobble}(\phi)_{Data\ Point\,[i=245]} &&= \mathbf{-7^{\circ}.308} && (March\ 3)\\
&\alpha_{Sun,wobble}(\phi)_{Data\ Point\,[i=318]} &&= \mathbf{0^{\circ}.024} && (March\ 21)\\
&\alpha_{Sun,wobble}(\phi)_{Data\ Point\,[i=367]} &&= \mathbf{7^{\circ}.214} && (April\ 2)\\
&\alpha_{Sun,wobble}(\phi)_{Data\ Points\,[i=686-687]} &&= \mathbf{21^{\circ}.691} && (June\ 21)\\
&\alpha_{Sun,wobble}(\phi)_{Data\ Point\,[i=983]} &&= \mathbf{7^{\circ}.257} && (September\ 3)\\
&\alpha_{Sun,wobble}(\phi)_{Data\ Point\,[i=1056]} &&= \mathbf{0^{\circ}.002} && (September\ 21)\\
&\alpha_{Sun,wobble}(\phi)_{Data\ Point\,[i=1110]} &&= \mathbf{-7^{\circ}.262} && (October\ 5)\\
&\alpha_{Sun,wobble}(\phi)_{Data\ Points\,[i=1417-1419]} &&= \mathbf{-21^{\circ}.691} && (December\ 21)
\end{aligned} \quad (332)$$

The calculated values of $\alpha_{Sun,wobble}(\phi)$ (330) demonstrate that $\alpha_{Sun,wobble}(\phi)=-7^{\circ}.308$ (332) on March 3 (Data Point $[i=245]$) and that $\alpha_{Sun,wobble}(\phi)=-7^{\circ}.257$ (332) on September 3 (Data Point $[i=983]$), consistent with the observation[10,11] with our analysis result that the Sun's South Pole[28] is observed[10] to be tipping 7.25° (5) towards the Earth around the first week of September, and the Sun's South Pole[28] is observed[10] to be tipping 7.25° (5) away from the Earth around the first week of March. Hence, we have discovered and mathematically demonstrated in this Article that this $\alpha_{Sun,wobble}=7.25°$ (5) is *not* the maximum axial tilt $\gamma_{Sun,\Delta Max}$ (104) of the Sun's rotational pole with respect to the virtual ecliptic[8] plane along a yearly Sun-Earth cycle, but rather the angle that the Sun's South Pole[28] seemingly makes towards or away from the Earth during first weeks of September and March, respectively. This alignment of observation and our calculated results is another justification that our analysis in this Article is correct.

$$
\begin{aligned}
&\alpha_{Sun,sideways}(\phi)_{Data\ Points[i=317-320]} &&= \mathbf{23^\circ.439} && (March\ 21)\\
&\alpha_{Sun,sideways}(\phi)_{Data\ Point[i=687]} &&= \mathbf{0^\circ.092} && (June\ 21)\\
&\alpha_{Sun,sideways}(\phi)_{Data\ Points[i=1056-1059]} &&= \mathbf{23^\circ.439} && (September\ 21-22)\\
&\alpha_{Sun,wobble}(\phi)_{Data\ Point[i=1418]} &&= \mathbf{0^\circ.026} && (December\ 21)
\end{aligned}
\qquad (333)
$$

The variation of the calculated results for $\alpha_{Sun,sideways}(\phi)$ (331), including the values listed in (333), are consistent with the animation[12] of the sideways swing of the Sun's axis during the span of one year, as seen from the line of sight of an observer on Earth aligned with the North Pole[29] $\hat{\boldsymbol{u}}_{2\perp}$ (300) vector of the Earth, $\alpha_{Sun,sideways}(\phi)$ (331) swinging from $23^\circ.439$ (333) to the right side of the observer on March 21, to $23^\circ.439$ (333) to the left side of the observer on September 21-22. This alignment of observation and our results is yet another justification of our analysis and calculations in this Article.

Based on the results we have obtained in (312), (332), and (333) about the observed angles $\gamma_2(\phi)$ (98), $\alpha_{Sun,wobble}(\phi)$ (330), and $\alpha_{Sun,sideways}(\phi)$ (331), the Spring Equinox[9] must be around March 21-22 and the Autumn Equinox[9] must be around September 20-22.

CONCLUSIONS

In previous Articles, it is mathematically demonstrated and proven that bodies moving around different circles in space mimic vector-wise elliptical paths with respect to each other, whether they are moving with the *same*, *fixed different*, or *changing* angular velocities with respect to their own centers of revolution, and an analytical method is developed to calculate orbital parameters of two heavenly bodies moving with the *same* angular velocity in their own circular orbital revolutions. As the Sun and the Earth exhibit an almost fixed elliptic orbit around one another in a yearly[26] cycle of duration $(T_{Year} = T_{Sun} = T_{Earth})$ (334), based on the mathematical expectation that the Sun and the Earth may in reality be revolving around their individual circular orbits with the *same* angular velocity $(\omega = \omega_{Sun} = \omega_{Earth})$ (335) with respect to their individual centers of revolution, demonstrating this relative elliptical orbital behavior, in this Article, the individual orbital parameters of the Sun and Earth are calculated using that analytical method to an approximation, based on yearly[26] Sun-Earth distance data observed by NASA Landsat[3]. The obtained results strongly support that this expectation can be set forth as an assertion, that the Sun and the Earth may in reality be revolving around their individual circular orbits with the *same* angular velocity, demonstrating this relative elliptical orbital behavior.

$$T_{Year} = T_{Sun} = T_{Earth} = 365.256363004\ days \qquad (Sun\ and\ Earth's\ Orbital\ Period) \qquad (334)$$

$$\omega = \omega_{Sun} = \omega_{Earth} = \frac{2\pi}{T_{Year}} = \mathbf{0.017202124161519}\ radians/day \qquad (Sun\ \&\ Earth\ Angular\ Velocity) \qquad (335)$$

As a result of our analysis and calculations in this Article, we have obtained the following orbital parameter values for the Sun and the Earth.

$$\phi_0 = \mathbf{-2.63736997198951}\ rad = \mathbf{-151^\circ.11} \qquad (Phase\ difference\ for\ the\ Sun-Earth\ system) \qquad (336)$$

$$\beta = \mathbf{-156^\circ.56} = \mathbf{-2.73251312735455}\ radians \qquad (angle\ between\ Sun\ \&\ Earth\ revolution\ planes) \qquad (337)$$

$$r_1 = \mathbf{1.00007869128446}\ au = \mathbf{149,609,642.749}\ km \quad (radius\ of\ Sun's\ individual\ revolution) \tag{338}$$

$$r_2 = \mathbf{0.0000826915340950721}\ au = \mathbf{12,370.477}\ km \quad (radius\ of\ Earth's\ individual\ revolution) \tag{339}$$

$$\begin{gathered} a = \mathbf{1.00001226586661}\ au = \mathbf{149,599,705.647527}\ km \\ (Semi-minor\ axis\ magnitude\ of\ Sun-Earth\ system) \end{gathered} \tag{340}$$

$$\begin{gathered} b = \mathbf{1.00015109267839}\ au = \mathbf{149,620,473.842965}\ km \\ (Semi-major\ axis\ magnitude\ of\ Sun-Earth\ system) \end{gathered} \tag{341}$$

$$\begin{gathered} \vec{\boldsymbol{a}} = \hat{\boldsymbol{x}}120{,}168{,}467.040\ km - \hat{\boldsymbol{y}}\,72{,}280{,}843.623\ km + \hat{\boldsymbol{z}}\,52{,}106{,}536.351\ km \\ (Semi-minor\ axis\ vector\ of\ Sun-Earth\ system) \end{gathered} \tag{342}$$

$$\begin{gathered} \vec{\boldsymbol{b}} = -\hat{\boldsymbol{x}}\,66{,}316{,}021.722\ km - \hat{\boldsymbol{y}}131{,}002{,}922.948\ km - \hat{\boldsymbol{z}}\,28{,}752{,}488.898\ km \\ (Semi-major\ axis\ vector\ of\ Sun-Earth\ system) \end{gathered} \tag{343}$$

$$c = \mathbf{0.0166636221185271}\ au = \mathbf{2,492,842.4}\ km \quad (Focal\ distance\ of\ Sun-Earth\ system) \tag{344}$$

$$e = \frac{c}{b} = \mathbf{0.0166611047475858} \quad (Eccentricity) \tag{345}$$

$$\left[\ell(\phi)\right]_{Max} = \mathbf{0.0168346184954446}\ au\ on\ 23\ February, 26\ May, 25\ August, 24\ November \tag{346}$$

$$\left[\ell(\phi)\right]_{Min} = \mathbf{0.0165381684638675}\ au \quad on \quad 1\ January, 2\ April, 2\ July, 1\ October \tag{347}$$

$$\hat{\boldsymbol{u}}_{1\perp} = -\hat{\boldsymbol{x}}Sin\,\beta + \hat{\boldsymbol{z}}Cos\,\beta \quad (North\ Pole\ of\ Sun\ normal\ to\ Sun's\ circular\ revolution\ plane) \tag{348}$$

$$\boldsymbol{u}_{2\perp} = -\hat{\boldsymbol{z}} \quad (North\ Pole\ of\ Earth\ normal\ to\ Earth's\ circular\ revolution\ plane) \tag{349}$$

The most significant and interesting consequence of these orbital parameter values found for the Sun and the Earth is, the obtained results indicate that the Earth is *not* revolving around the Sun, but in fact the Sun is revolving around a larger individual circular orbit which encompasses the individual circular orbit of the Earth with an inclination, whose center of revolution is shifted compared to the center of revolution of the Sun. This revelation raises the need to reevaluate existing theories about the formation, expected motional behavior, and topologies of solar systems observed in the Universe in general as well as our Solar System, bringing new vision to the formation and motion structure of the bodies in the Universe. We can conclude that planets do *not* necessarily revolve around stars, and Earth is *not* revolving around the Sun.

Here are some other significant values for the Sun-Earth system, also obtained in this Article.

$$Sin\,\phi_0 = \mathbf{-0.483127005006538} \tag{350}$$

$$Cos\,\phi_0 = \mathbf{-0.875550282412959} \tag{351}$$

$$Sin\,\beta = \mathbf{-0.397764973581873} \tag{352}$$

$$Cos\,\beta = \mathbf{-0.917487343668245} \tag{353}$$

$$\phi_{Min\ Sun-Earth\ Distance} = \phi_{i=9} = \mathbf{-0.193236295254927}\ radians = \mathbf{-11^{\circ}.072} \quad on\ January\ 3 \tag{354}$$

$$\phi_{i=65} = \mathbf{0.00058673254949726}\ radians = \mathbf{0^{\circ}.034} \simeq 0 \qquad on\ January\ 17 \tag{355}$$

$$\phi_{Spring\ Equinox} = \mathbf{1.02676771671574}\ radians = \mathbf{58^\circ.829} \qquad on\ March\ 20 \quad (356)$$

$$\phi_{i=(65+365)} = \phi_{i=430} = \mathbf{1.57138305934439}\ radians = \mathbf{90^\circ.034} \simeq \frac{\pi}{2} \qquad on\ April\ 18 \quad (357)$$

$$\phi_{Summer\ Solstice} = \phi_{\gamma_{2,\min}} = -\phi_0 = \mathbf{2.63736997198951}\ rad = \mathbf{151^\circ.11} \qquad on\ June\ 21 \quad (358)$$

$$\phi_{Max\ Sun-Earth\ Distance} = \phi_{i=741} = \mathbf{2.94835635833487}\ radians = \mathbf{168^\circ.928} \qquad on\ July\ 5 \quad (359)$$

$$\phi_{i=(65+730)} = \phi_{i=795} = \mathbf{3.14217938613929}\ radians = \mathbf{180^\circ.034} \simeq \pi \qquad on\ July\ 18 \quad (360)$$

$$\phi_{Autumn\ Equinox} = \mathbf{-2.05488176708229}\ radians = \mathbf{242^\circ.264} = \mathbf{-117^\circ.736} \qquad on\ September\ 22 \quad (361)$$

$$\phi_{i=(65+1095)} = \phi_{i=1160} = \mathbf{-1.5702095942454}\ radians = \mathbf{-89^\circ.966} \simeq -\frac{\pi}{2} \qquad on\ October\ 17 \quad (362)$$

$$\phi_{Winter\ Solstice} = \phi_{\gamma_{2,\max}} = \pi - \phi_0 = \mathbf{5.7789626255793}\ rad = \mathbf{331^\circ.11} = \mathbf{-28^\circ.89} \qquad on\ December\ 21 \quad (363)$$

$$\vec{\boldsymbol{a}} \cdot \vec{\boldsymbol{b}} = -r_1 r_2 \left(1 - Cos\beta\right) Sin\phi_0 = \mathbf{0.00007661063}\ au^2 \qquad (364)$$

$$\left(b - a\right) = \mathbf{0.000138826811781456}\ au = \mathbf{20,768.195439}\ km \qquad (365)$$

$$a^2 = \mathbf{1.00002453188367}\ au^2 \quad \left(Semi-minor\ axis\ magnitude\ squared\right) \qquad (366)$$

$$b^2 = \mathbf{1.00030220818578}\ au^2 \quad \left(Semi-major\ axis\ magnitude\ squared\right) \qquad (367)$$

$$c^2 = \left|a^2 - b^2\right| = 2 r_1 r_2 \left(1 - Cos\beta\right)\left|Cos\phi_0\right| = \mathbf{0.000277676302109064}\ au^2 \quad \left(Eccentricity\ squared\right) \quad (368)$$

$$\ell\left(\phi_{\gamma_{2,\min}}\right) = \mathbf{0.0166987537771169}\ au = \mathbf{2,498,098.008400}\ km \qquad on\ June\ 21 \quad (369)$$

$$\ell\left(\phi_{\gamma_{2,\max}}\right) = \mathbf{0.0166987537771169}\ au = \mathbf{2,498,098.008400}\ km \qquad on\ December\ 21 \quad (370)$$

$$\alpha_{\left[\vec{d}\left(\phi_{\gamma_{2,\min}}\right) \measuredangle \vec{d}\left(\phi_{\gamma_{2,\max}}\right)\right]} = \mathbf{3.13397114735826}\ radians = \mathbf{179^\circ.563}\left[angle\ between\ \vec{\boldsymbol{d}}\left(\phi_{\gamma_{2,\min}}\right) \&\ \vec{\boldsymbol{d}}\left(\phi_{\gamma_{2,\max}}\right)\right] \quad (371)$$

$$\xi_{Max}\left(\phi\right) = \mathbf{2.96462117433723}\ radians = \mathbf{169^\circ.860} \qquad January\ 1,\ April\ 2,\ July\ 3,\ October\ 2 \quad (372)$$

$$\xi_{Min}\left(\phi\right) = \mathbf{2.8557213606605}\ radians = \mathbf{163^\circ.621} \qquad March\ 31,\ July\ 1,\ September\ 30,\ December\ 30 \quad (373)$$

$$\begin{aligned}
&\gamma_1\left(\phi\right) = \gamma_{Sun}\left(\phi\right) = \mathbf{90^\circ} && \left(January\ 2\right) \\
&\gamma_1\left(\phi\right) = \gamma_{Sun}\left(\phi\right) = \mathbf{90^\circ.411} = \left(90^\circ + 0^\circ.411\right) = \gamma_{1,\max 1} && \left(March\ 26-30\right) \\
&\gamma_1\left(\phi\right) = \gamma_{Sun}\left(\phi\right) = \mathbf{90^\circ} && \left(June\ 20-21\right) \\
&\gamma_1\left(\phi\right) = \gamma_{Sun}\left(\phi\right) = \mathbf{89^\circ.993} = \left(90^\circ - 0^\circ.007\right) = \gamma_{1,\min 1} && \left(June\ 26-29\right) \\
&\gamma_1\left(\phi\right) = \gamma_{Sun}\left(\phi\right) = \mathbf{90^\circ} && \left(July\ 5\right) \\
&\gamma_1\left(\phi\right) = \gamma_{Sun}\left(\phi\right) = \mathbf{90^\circ.408} = \left(90^\circ + 0^\circ.408\right) = \gamma_{1,\max 2} && \left(September\ 26-30\right) \\
&\gamma_1\left(\phi\right) = \gamma_{Sun}\left(\phi\right) = \mathbf{90^\circ} && \left(December\ 22\right) \\
&\gamma_1\left(\phi\right) = \gamma_{Sun}\left(\phi\right) = \mathbf{89^\circ.996} = \left(90^\circ - 0^\circ.004\right) = \gamma_{1,\min 2} && \left(December\ 27\ -\ 29\ \right)
\end{aligned} \qquad (374)$$

$$\gamma_{1,\max 1} = \mathbf{90^{\circ}.411} = 90^{\circ} + 0^{\circ}.411 = 90^{\circ} + \gamma_{1,\Delta_{1,1}} \quad \Rightarrow \quad \gamma_{1,\Delta_{1,1}} = \mathbf{0^{\circ}.411}$$
$$\gamma_{1,\max 2} = \mathbf{90^{\circ}.408} = 90^{\circ} + 0^{\circ}.408 = 90^{\circ} + \gamma_{1,\Delta_{1,2}} \quad \Rightarrow \quad \gamma_{1,\Delta_{1,2}} = \mathbf{0^{\circ}.408} \tag{375}$$
$$\gamma_{1,\min 1} = \mathbf{89^{\circ}.993} = 90^{\circ} - 0^{\circ}.007 = 90^{\circ} - \gamma_{1,\Delta_{2,1}} \quad \Rightarrow \quad \gamma_{1,\Delta_{2,1}} = \mathbf{0^{\circ}.007}$$
$$\gamma_{1,\min 2} = \mathbf{89^{\circ}.996} = 90^{\circ} - 0^{\circ}.004 = 90^{\circ} - \gamma_{1,\Delta_{2,2}} \quad \Rightarrow \quad \gamma_{1,\Delta_{2,2}} = \mathbf{0^{\circ}.004}$$

$$\gamma_2(\phi) = \gamma_{Earth}(\phi) = \mathbf{90^{\circ}} \quad (March\ 22)$$
$$\gamma_2(\phi) = \gamma_{Earth}(\phi) = \mathbf{66^{\circ}.560} = (90^{\circ} - 23^{\circ}.440) = \gamma_{2,\min} \quad (June\ 21-22) \tag{376}$$
$$\gamma_2(\phi) = \gamma_{Earth}(\phi) = \mathbf{90^{\circ}} \quad (September\ 20)$$
$$\gamma_2(\phi) = \gamma_{Earth}(\phi) = \mathbf{113^{\circ}.440} = (90^{\circ} + 23^{\circ}.440) = \gamma_{2,\max} \quad (December\ 21)$$

$$\gamma_{2,\max} = \mathbf{113^{\circ}.440} = 90^{\circ} + 23^{\circ}.440 = 90^{\circ} + \gamma_{2,\Delta_1} \quad \Rightarrow \quad \gamma_{2,\Delta_1} = \mathbf{23^{\circ}.440} = \gamma_{Earth,\Delta Max}$$
$$\gamma_{2,\min} = \mathbf{66^{\circ}.560} = 90^{\circ} - 23^{\circ}.440 = 90^{\circ} - \gamma_{2,\Delta_2} \quad \Rightarrow \quad \gamma_{2,\Delta_2} = \mathbf{23^{\circ}.440} = \gamma_{Earth,\Delta Max} \tag{377}$$

$$\alpha_{Sun,wobble}(\phi) = \mathbf{-7^{\circ}.308} \quad (March\ 3)$$
$$\alpha_{Sun,wobble}(\phi) = \mathbf{0^{\circ}.024} \quad (March\ 21)$$
$$\alpha_{Sun,wobble}(\phi) = \mathbf{7^{\circ}.214} \quad (April\ 2)$$
$$\alpha_{Sun,wobble}(\phi) = \mathbf{21^{\circ}.691} \quad (June\ 21)$$
$$\alpha_{Sun,wobble}(\phi) = \mathbf{7^{\circ}.257} \quad (September\ 3) \tag{378}$$
$$\alpha_{Sun,wobble}(\phi) = \mathbf{0^{\circ}.002} \quad (September\ 21)$$
$$\alpha_{Sun,wobble}(\phi) = \mathbf{-7^{\circ}.262} \quad (October\ 5)$$
$$\alpha_{Sun,wobble}(\phi) = \mathbf{-21^{\circ}.691} \quad (December\ 21)$$

$$\alpha_{Sun,sideways}(\phi)_{Data\ Points\,[i=317-320]} = \mathbf{23^{\circ}.439} \quad (March\ 21)$$
$$\alpha_{Sun,sideways}(\phi)_{Data\ Point\,[i=687]} = \mathbf{0^{\circ}.092} \quad (June\ 21)$$
$$\alpha_{Sun,sideways}(\phi)_{Data\ Points\,[i=1056-1059]} = \mathbf{23^{\circ}.439} \quad (September\ 21-22) \tag{379}$$
$$\alpha_{Sun,wobble}(\phi)_{Data\ Point\,[i=1418]} = \mathbf{0^{\circ}.026} \quad (December\ 21)$$

We have also reached the following conclusions, and made discoveries and observations about the facts listed below, for the Sun-Earth system, in this Article.

- The vector distance $\vec{\ell}(\phi)$ and its magnitude $\ell(\phi)$ (9) between centers of individual revolutions of Sun and Earth is *not constant* but rather *harmonically varying* over a year, and $\vec{\ell}(\phi)$ (9) repeats itself every quarter Sun-Earth cycle vector-wise as well as magnitude-wise (156), and is plotted over a yearly Sun-Earth cycle as in **Figure 19**.
- One oscillation of the magnitude $\ell(\phi)$ (9) of the distance between the centers of individual orbits of revolutions of the Earth and the Sun is about 12 times a year, which is apparently based on the impact of the relative Earth-Moon motion.

- The second oscillation of $\ell(\phi)$ (9) is seemingly based on the daily self-rotation of the Earth around its own axis.
- The possible cause of the third oscillation of $\ell(\phi)$ (9), which is about 4 times over a Sun-Earth year, is due to the impact of relative Sun-Mercury motion which has about four cycles over a Sun-Earth year. It seems that the Sun and Mercury form a twin or binary[25] system of bodies in space, in which Mercury behaves as the Sun's satellite, similar to Earth-Moon system, and implies the possibility that the individual circular orbits of revolution of the Sun and Mercury may also be revolving around each other.
- Further, our analysis also implies that it is very well possible that the relative motions of the other planets and their satellites, as well as other moving bodies, in the Solar system also impact the relative motion of the Sun and Earth in terms of additional minor oscillations of $\ell(\phi)$ (9) over longer periods than a Sun-Earth year, but their impact on Sun-Earth distance can be observed only over longer periods than a yearly cycle.
- It is worth noticing that ancient people did *not* select the first days of months in the Solar Calendar cycle based on Min-Max Sun-Earth distance dates (Jan 3 and July 5), or the dates of Solstices (June 21 and December 21)[5,6], or the Equinoxes (March 20 and September 23) [5,6], which slightly vary every year. Apparently, the selection of dates for the first days of months in the Solar Calendar by ancient people was not coincidental, but rather based on the phase ϕ_0 between individual revolutions of the Earth and the Sun in their own orbits, and also the aligned cycles of the harmonic $\ell(\phi)$ (9), the distance between centers of individual revolutions of the Sun and Earth, which subsequently has resulted in different duration months of 28, 30, and 31 days. Our ancestors possibly based these date selections on accurate Sun-Earth distance observations and the intermediate harmonic variations in it based on phase ϕ_0 and $\ell(\phi)$ (9), even if they were not aware of the existence of phase ϕ_0 or $\ell(\phi)$ (9).
- $\vec{\boldsymbol{a}}$ (342) based on (10) is the fixed semi-minor[23] axis vector and $\vec{\boldsymbol{b}}$ (343) based on (11) is the fixed semi-major[23] axis vector of the ellipse of the Sun-Earth system formed by their relative motions.
- The focal distance c (344) of the ellipse formed by the relative Sun-Earth motion is about $1/60$ of an astronomical unit[4] (au) (3).
- In its own cycle of individual revolution, the Sun is lagging with a fixed phase difference of $\left(\phi_0 = \mathbf{-2.63736997198951}\, rad = \mathbf{-151^{\circ}.11}\right)$ (138) behind Earth in its own cycle of individual revolution, or equivalently, the Earth in its own cycle of individual revolution is moving ahead of the Sun in its own cycle of individual revolution, by a phase difference of $\left(-\phi_0 = \mathbf{2.63736997198951}\, rad = \mathbf{151^{\circ}.11}\right)$.
- Phase $\phi\,[Degrees]$ against dates over a yearly Sun-Earth cycle is plotted in **Figure 14**.
- The difference between the semi-major axis magnitude b (285) and the semi-minor axis magnitude a (288) of the Sun-Earth system is about $\mathbf{20,768.2}\, km$ (289), which is relatively small on the Sun-Earth distance scale (3). Therefore, the ellipse formed by the relative motions of the Sun-Earth system reveals to be almost circular, as expected.

- We understand from (298) - (299) that it is the radius r_1 (298) of Sun's individual revolution which predominantly determines the distances between the Sun and the Earth over a yearly cycle, as well as the semi-major and semi-minor axes of the relative ellipse formed by the relative motions of the Sun and the Earth in their individual circular orbits, namely the magnitude b (285) of the semi-major axis vector $\vec{b}$ (11) and the magnitude a (288) of the semi-minor axis vector $\vec{a}$ (10).
- The plane of $Circle_1$, the circle of revolution of the Sun, is tilted by an angle β (300) with respect to the plane of $Circle_2$, the circle of revolution of the Earth.
- As an observer aligned with the North Pole[29] of the Earth sees, the sunspots on the Sun travel[10] from east to west, going across the Sun's disk in June and December, but their path dipping somewhat in March and September. This observation[10] implies that the Sun's self rotation is in *clockwise*[27] direction from a vantage point above the North Pole[29] of the Earth, meaning that the pole vector of the Sun that is angle-wise closer to the Earth's North Pole[29] should be the Sun's South Pole[28] vector, as the North Pole[28,29] is expected to be rotationg *counterclockwise*[27] based on the "right hand rule"[18].
- $\hat{u}_{1\perp} = \hat{x}Sin\beta - \hat{z}Cos\beta$ (308) is the unit vector $\hat{u}_{1\perp}$ (97) normal to $Circle_1$ of the Sun's revolution in its individual orbit in the direction of its self-rotation axis[30,7], and also the direction of the North Pole[28,29] of the Sun in the configuration in **Figure 2**, which ensures that the Sun's yearly revolution around its own circular orbit is in a *clockwise*[27] direction as seen from above the North Pole[28,29] of the Sun, reversed to its *counterclockwise*[27] self rotation direction as seen from the same vantage point above North Pole[28,29] of the Sun.
- $\hat{u}_{2\perp} = -\hat{z}$ (300) is the unit vector $\hat{u}_{2\perp}$ (97) normal to $Circle_2$ of the Earth's revolution in its individual orbit in the direction of self-rotation axis[30,7] of Earth, and also the direction of the North Pole[29] of the Earth in the configuration in **Figure 2**, which makes us understand that the Earth's yearly revolution around its own circular orbit is in a *clockwise*[27] direction as seen from above the North Pole[29] of the Earth, opposite to its daily *counterclockwise* self rotation as seen from the same vantage point above the North Pole[29] of the Earth.
- The calculated $\vec{a}\cdot\vec{\ell}(\phi)$ (158) and $\vec{b}\cdot\vec{\ell}(\phi)$ (159), namely the projections of the constant $\vec{a}$ (10) and $\vec{b}$ (11) on varying $\vec{\ell}(\phi)$ (9), have been plotted for all ϕ (136) over a yearly cycle in **Figure 15** and **Figure 16**, respectively. Both graphs demonstrate interesting behaviour overall. The results reveal a rippled graph in both cases, with $\vec{a}\cdot\vec{\ell}(\phi)$ (158) *negative* and $\vec{b}\cdot\vec{\ell}(\phi)$ (159) *positive* for all ϕ (136) over a yearly cycle, and both having sharp jumps quarterly around January 1 (DOY 1), April 2 (DOY 92), July 2 (DOY 183), and October 1 (DOY 274). This behavior seems to be due to the impact, on $\ell(\phi)$ (9), of the relative Sun-Mercury motion, which has about four cycles over a Sun-Earth year.
- Dates of occurrence of $\left[\ell(\phi)\right]_{Max}$ (295) are February 23 (DOY 54), May 26 (DOY 146), August 25 (DOY 237), and November 24 (DOY 328), whereas $\left[\ell(\phi)\right]_{Min}$ (296) occurs on January 1 (DOY 1), April 2 (DOY 92), July 2 (DOY 183), and October 1 (DOY 274). Sum of squares of the projections of the constant $\vec{a}$ (10) and $\vec{b}$ (11) on varying $\vec{\ell}(\phi)$ (9),

namely $\left[\vec{a}\cdot\vec{\ell}(\phi)\right]^2+\left[\vec{b}\cdot\vec{\ell}(\phi)\right]^2$, takes on its maximum value (68) when the magnitude $\ell(\phi)$ (9) is at its minimum value $\left[\ell(\phi)\right]_{Min}$ (296), and $\left[\vec{a}\cdot\vec{\ell}(\phi)\right]^2+\left[\vec{b}\cdot\vec{\ell}(\phi)\right]^2$ takes on its minimum value (69) about 5-6 days in advance of occurrence of $\left[\ell(\phi)\right]_{Max}$ (295).

- The values calculated for the virtual angles $\xi(\phi)$ (163) and $\left[\xi(\phi)-\phi\right]$ (166) are plotted over a yearly Sun-Earth cycle in **Figure 17** and **Figure 18**, respectively. **Figure 17** for $\xi(\phi)$ (163) exhibits symmetrical behavior compared to **Figure 15** and **Figure 16** of $\vec{a}\cdot\vec{\ell}(\phi)$ (158) and $\vec{b}\cdot\vec{\ell}(\phi)$ (159), respectively, as expected, as $Cos\left[\xi(\phi)\right]$ (163) and $Sin\left[\xi(\phi)\right]$ (163) values provide a measure of how much more or less is the projection of semi-minor axis vector $\vec{a}$ (10) on $\vec{\ell}(\phi)$ (9) with respect to the projection of semi-major axis vector $\vec{b}$ (11) on $\vec{\ell}(\phi)$ (9), for each phase ϕ (136) of the ellipse formed by the relative motions of the Sun-Earth system. The virtual angle $\xi(\phi)$ (163) has maximum and minimum values $\mathbf{169^{\circ}.860}$ (167) and $\mathbf{163^{\circ}.621}$ (168), with sharp jumps or value changes around December 31 (DOY 365), April 1 (DOY 91), July 2 (DOY 183), and October 1 (DOY 274), coinciding with the occurence dates of maximum value (68) of $\left[\vec{a}\cdot\vec{\ell}(\phi)\right]^2+\left[\vec{b}\cdot\vec{\ell}(\phi)\right]^2$ and minimum value $\left[\ell(\phi)\right]_{Min}$ (296) of $\ell(\phi)$ (9).
- The virtual angle $\xi(\phi)$ (163) provides a sense of how close or far is the direction of $\vec{d}(\phi)$ (13) to $\vec{\ell}(\phi)$ (9), at each ϕ (136) in a Sun-Earth cycle.
- Based on the obtained data and **Figure 17**, $\xi(\phi)$ (163) varies harmonically in a small angle range between $\mathbf{169^{\circ}.860}$ (167) and $\mathbf{163^{\circ}.621}$ (168). This important observation, as well as the definition of $\left[\xi(\phi)-\phi\right]$ (166) with (164), allows us to reach the understanding that throughout a year, $\vec{\ell}(\phi)$ (9) must be varying in a narrow angle range, which is closer to the sector of Earth's self-revolution when the Earth moves between $\phi=\mathbf{163^{\circ}.621}$ and $\phi=\mathbf{169^{\circ}.860}$, which approximately corresponds[13] to the sector of Earth's self-revolution between dates of July 2 and July 7 over a yearly[26] Sun-Earth cycle. The main factor influencing the existence of $\vec{\ell}(\phi)$ (9) vector, which is between centers of individual circular orbits of revolution of the Sun and the Earth, especially in this narrow direction range, may be the impact of the Milky Way[32]'s Galactic Center[33] located in the reverse direction, which is observed from Earth at the sector of the sky in the direction of constellation Sagittarius[34] in the same sector of the sky as the observed location of the Sun around Winter Solstice[5] on December 21. Note that as $\left(\beta\simeq\mathbf{-156^{\circ}.56}\right)$ (337), this sector of the sky where $\vec{\ell}(\phi)$ (9) exists, when the Earth moves between $\phi=\mathbf{163^{\circ}.621}$ and $\phi=\mathbf{169^{\circ}.860}$ in its individual circular orbit of revolution in the Sun-Earth configuration of **Figure 2**, is also one of the two times over a yearly[26] Sun-Earth cycle when the motion of Earth and Sun seemingly cross each other in *reverse* direction, in their motion within their individual circular orbits of revolution,

between the point when the phase $(\phi+\phi_0)$ of the Sun is $(\phi_{Summer\ Solstice}+\phi_0=0)$ (358) at the Summer Solstice[5] on June 21, and the point when the phase ϕ of Earth is $(\phi=\mathbf{180^\circ.034}\simeq\pi)$ (360) on July 18. Of the two times over a yearly[26] Sun-Earth cycle when the motion of the Earth and the Sun seemingly cross each other in *reverse* direction, in their motion within their individual circular orbits of revolution, the other one is between the point when phase $(\phi+\phi_0)$ of the Sun is $(\phi_{Winter\ Solstice}+\phi_0=\pi)$ (363) at the Winter Solstice[5] on December 21, and the point when phase ϕ of Earth is $(\phi=\mathbf{0^\circ.034}\simeq 0)$ (355) on January 17.

- Based on the obtained data and **Figure 18**, $\xi(\phi)$ (163) is equal to ϕ (136) on July 5 (DOY 186), and $\xi(\phi)$ (163) is equal to $(\pi+\phi)$ (136) on January 3 (DOY 3), which are the dates of maximum and minimum Sun-Earth distance over a year, respectively. This result is compliant with the expectation based on the definition of $[\xi(\phi)-\phi]$ (166), taking into consideration (164) and $d^2(\phi)$ (59). This also indicates that the direction of the distance vector $\vec{\boldsymbol{d}}(\phi)$ (13) from Earth to the Sun is most aligned with the direction of $\vec{\ell}(\phi)$ (9) on July 5, which is the date of maximum Sun-Earth distance over a year[26], and the direction of the distance vector $\vec{\boldsymbol{d}}(\phi)$ (13) from the Earth to the Sun is most aligned with the *reverse* direction of $\vec{\ell}(\phi)$ (9) on January 3, the date of minimum Sun-Earth distance over a year[26].
- The calculated values of $\gamma_2(\phi)=\gamma_{Earth}(\phi)$ (98) over a year are plotted in **Figure 23**.
- We discover about the Earth's North Pole[29] tilt[7], with respect to the distance $\vec{\boldsymbol{d}}(\phi)$ (15) vector directed from Earth to Sun, that $\gamma_{Earth}(\phi)=90^\circ$ (312) on March 22 and on September 20, $\gamma_{Earth}(\phi)=66^\circ.560=(90^\circ-23^\circ.440)$ (312) at the Summer Solstice[5] on June 21 – 22, and $\gamma_{Earth}(\phi)=113^\circ.440=(90^\circ+23^\circ.440)$ (312) at the Winter Solstice[5] on December 21.
- The calculated values of $\gamma_1(\phi)=\gamma_{Sun}(\phi)$ (98) over a year are plotted in **Figure 22**.
- We discover about the Sun's North Pole[28,29] tilt[7,10,11], with respect to the distance $[-\vec{\boldsymbol{d}}(\phi)]$ (15) vector directed from Sun to Earth, that $\gamma_{Sun}(\phi)=90^\circ$ (314) on January 2 around minimum Sun-Earth distance in a yearly cycle, at the Summer Solstice[5] on June 20 – 21, on July 5 at maximum Sun-Earth distance in a yearly cycle, and at the Winter Solstice[5] on December 22; $\gamma_{Sun}(\phi)=90^\circ.411=(90^\circ+0^\circ.411)$ (314) between dates of March 26 – 30, $\gamma_{Sun}(\phi)=89^\circ.993=(90^\circ-0^\circ.007)$ (314) between dates of June 26 - 29, $\gamma_{Sun}(\phi)=90^\circ.408=(90^\circ+0^\circ.408)$ (314) between dates of September 26 - 30, and $\gamma_{Sun}(\phi)=89^\circ.996=(90^\circ-0^\circ.004)$ (314) between dates of December 27 - 29.

- We can make the conclusion that the Sun's North Pole[28,29] vector is tilted[7] $\gamma_{Sun}(\phi)$ (98) around **90**$^{\circ}$ (314) with respect to the Sun-Earth distance vector $\left[-\vec{\boldsymbol{d}}(\phi)\right]$ (15) throughout a yearly cycle, and it is exactly at **90**$^{\circ}$ (314) with respect to the Sun-Earth distance vector $\left[-\vec{\boldsymbol{d}}(\phi)\right]$ (15) around the maximum and minimum Sun-Earth distance in a yearly cycle, as well as at the Summer and Winter Solstices[5], consistent with the observation[10,11] that in June and December, as viewed from Earth, the Sun's North and South Poles[28] do not to tip toward or away from Earth. Further, it is tilted *most towards* Earth twice a year, once a few days after the Spring Equinox and secondly a few days after the Autumn Equinox. Additionally, it is tilted *most away* from Earth twice a year, once between the Winter Solstice[5] and minimum Sun-Earth distance in a yearly cycle between $\left[\phi_{\gamma_{2,\max}} = \pi - \phi_0 = 331^{\circ}.11 = -28^{\circ}.89\right]$ (140) and $\left[\phi = -11^{\circ}.07\right]$ (144), and secondly between the Summer Solstice[5] and maximum Sun-Earth distance in a yearly cycle between $\left[\phi_{\gamma_{2,\min}} = -\phi_0 = 151^{\circ}.11\right]$ (139) and $\left[\phi = 168^{\circ}.93\right]$ (148), namely when the Sun and Earth are travelling in opposite directions towards each other, based on the topology in **Figure 2**.
- Maximum and minimum values that $\gamma_{Sun}(\phi)$ (98) takes over a relative Sun-Earth cycle, namely $\gamma_{1,\max}$ (104) and $\gamma_{1,\min}$ (104), respectively, as well as the tilt[7] angles γ_{1,Δ_1} (104) and γ_{1,Δ_2} (104) take on more than a single value throughout a yearly cycle, that are expressed in terms of $\gamma_{1,\max 1}$ (315), $\gamma_{1,\max 2}$ (315), $\gamma_{1,\min 1}$ (315), $\gamma_{1,\min 2}$ (315), $\gamma_{1,\Delta_{1,1}}$ (315), $\gamma_{1,\Delta_{1,2}}$ (315), $\gamma_{1,\Delta_{2,1}}$ (315), and $\gamma_{1,\Delta_{2,2}}$ (315).
- We define the relative angle that the Sun's South Pole unit vector $\left(-\hat{\boldsymbol{u}}_{1\perp}\right)$ (308) seemingly makes towards or away from the Earth, with respect to the line of sight of an observer on Earth aligned with the North Pole[29] unit vector $\hat{\boldsymbol{u}}_{2\perp}$ (300) of Earth, as $\alpha_{Sun,wobble}(\phi)$ (330), at each ϕ, and also define $\alpha_{Sun,sideways}(\phi)$ (331) as the relative "sideways" swing angle that the Sun's South Pole[28] unit vector $\left(-\hat{\boldsymbol{u}}_{1\perp}\right)$ (308) seemingly makes with respect to the line of sight of an observer on Earth aligned with the North Pole[29] vector $\hat{\boldsymbol{u}}_{2\perp}$ (300) of the Earth.
- The calculated values over a year for $\alpha_{Sun,wobble}(\phi)$ (330) are plotted in **Figure 24**, and the calculated values over a year for $\alpha_{Sun,sideways}(\phi)$ (331) are plotted in **Figure 25**.
- Calculated values of $\alpha_{Sun,wobble}(\phi)$ (330) demonstrate that $\alpha_{Sun,wobble}(\phi) = -7^{\circ}.308$ (332) on March 3 and that $\alpha_{Sun,wobble}(\phi) = 7^{\circ}.257$ (332) on September 3, reinforcing our analysis result that it must be South Pole[28] of the Sun which is observed[10] to be tipping 7.25° (5) towards the Earth around the first week of September, and the Sun's South Pole[28] is observed[10,11] to be tipping 7.25° (5) away from the Earth around the first week of March. Hence, we have discovered and mathematically demonstrated in this Article that this $\alpha_{Sun,wobble} = 7.25°$ (5) is *not* the maximum axial tilt $\gamma_{Sun,\Delta Max}$ (104) of the Sun's rotational pole with respect to the virtual ecliptic[8] plane along a yearly Sun-Earth cycle, but rather

the angle that the Sun's South Pole[28] seemingly makes towards or away from the Earth during first weeks of September and March, respectively. This alignment of observation and our calculated results is another justification that our analysis in this Article is correct.

- The variation of the calculated results for $\alpha_{Sun,sideways}(\phi)$ (331), including the values listed in (333), are consistent with the animation[12] of the sideways swing of the Sun's axis during the span of one year, as seen from the line of sight of an observer on Earth aligned with the North Pole[29] $\hat{\boldsymbol{u}}_{2\perp}$ (300) vector of the Earth, $\alpha_{Sun,sideways}(\phi)$ (331) swinging from $23^{\circ}.439$ (333) to the right side of the observer on March 21, to $23^{\circ}.439$ (333) to the left side of the observer on September 21-22. This alignment of observation and our results is yet another justification of our analysis and calculations in this Article.
- The results we have obtained for Earth's North Pole[29] tilt[7] (312), Sun's South Pole[28] "wobble" $\alpha_{Sun,wobble}(\phi)$ angle (332), and Sun's South Pole[28] relative "sideways" swing angle (333), all confirm that the Spring Equinox[9] must be around March 21-22 and the Autumn Equinox[9] must be around September 20-22.

Based on observed astronomical position data of heavenly objects in the Solar System and other planetary systems, all bodies in space seem to move in some kind of elliptical motion with respect to each other. As the results we have obtained in this Article for the Sun-Earth system support our assertion, we can extend our conclusion to the motions of all bodies in the Universe, stating that all bodies in the Universe move in their own circular orbits, thus demonstrating relative orbits of elliptical form, or in the broader sense, in the form of some conical function.

Moreover, the reversed direction of "Earth's yearly revolution around its own circular orbit" and "Earth's daily self rotation" as seen from a vantage point above its North Pole[29], hints a relation of this motion with the rule of Electromagnetism that is named as "Faraday's Law of Induction"[18], which states that "Electromotive Force (EMF)[20] around a closed path is equal to the negative of the time rate of change of the Magnetic Flux[21] enclosed by the path." This implies that a charged particle moving in an external magnetic field with changing flux[21] moves in a direction to reverse the change in that magnetic flux[21]. Earth may well be considered to behave like a charged particle obeying "Faraday's Law of Induction"[18], moving in space in an external magnetic field with changing flux[21] through the area of its individual circle of revolution, which may be the cause that triggers Earth's daily self rotation in the reverse direction, negative of the time rate of change of this flux[21] through the area of Earth's individual circle of revolution.

As a result of observation[10] and our analysis, from a vantage point above the North Pole[29] of the Earth, "the Sun's yearly revolution around its own circular orbit is in *counterclockwise*[27] direction" and "the Sun's self rotation is in in *clockwise*[27] direction". Equivalently, as seen from a vantage point above the North Pole[28,29] of the Sun, "the Sun's yearly revolution around its own circular orbit is in *clockwise*[27] direction" and "the Sun's self rotation is in in *counterclockwise*[27] direction". This reversed direction of motion also indicates that the Sun obeys "Faraday's Law of Induction"[19].

In fact, all moving bodies in space must obey "Faraday's Law of Induction"[19] in their motions, i.e. each one is possibly moving in space in external magnetic fields with changing flux[21] through the area of their individual circular orbit of revolution while moving in *clockwise*[27] direction (Revolution Orbit "West" to Revolution Orbit "East") as seen from a vantage point above their North Pole[28,29], which is the cause that triggers their self rotation in the reverse (*counterclockwise*[27]) direction (Self Rotation "West" to Self Rotation "East") as seen from a vantage point above their North Pole[28,29], hence the two Easts and two Wests of bodies in the Universe, where we always accept the North Pole[28,29] of a body to be the pole rotating in *counterclockwise*[27] direction according to "right hand rule"[18]. This is another very important assertion we make in this Article.

FIGURES

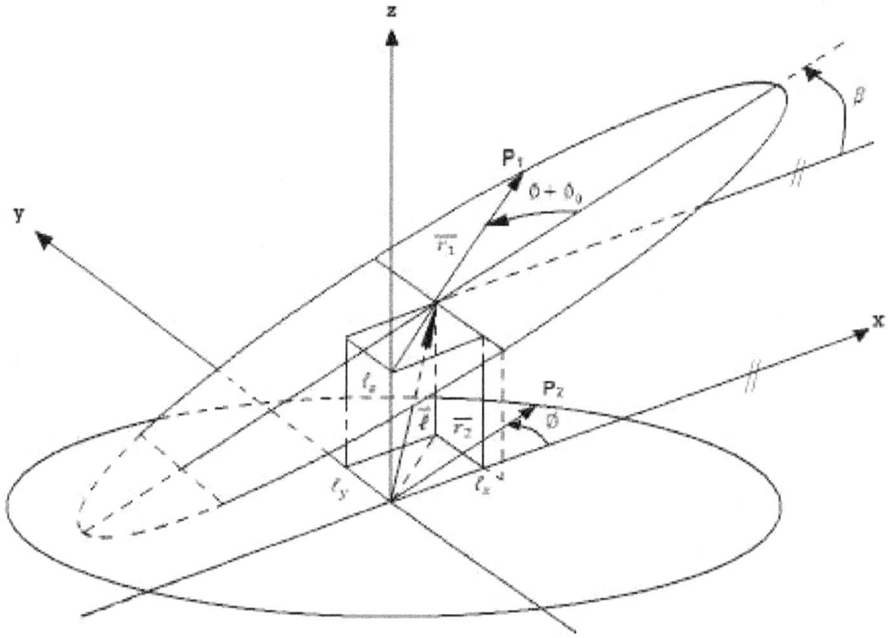

Figure 1 Point $\mathbf{P_1}$ for the Sun and $\mathbf{P_2}$ for the Earth moving on Two Circles in Space

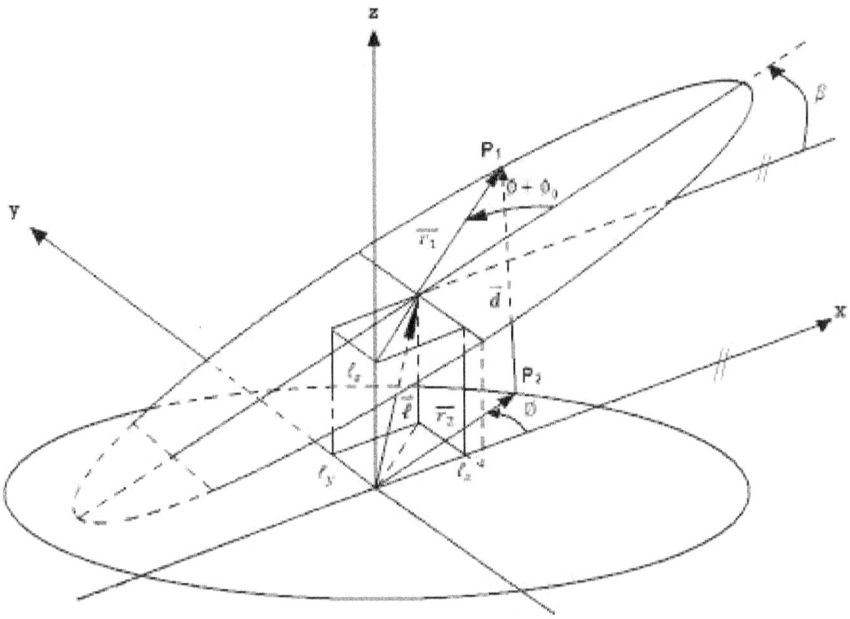

Figure 2 Distance Vector in Space Between Point $\mathbf{P}_1$ for the Sun and $\mathbf{P}_2$ for the Earth

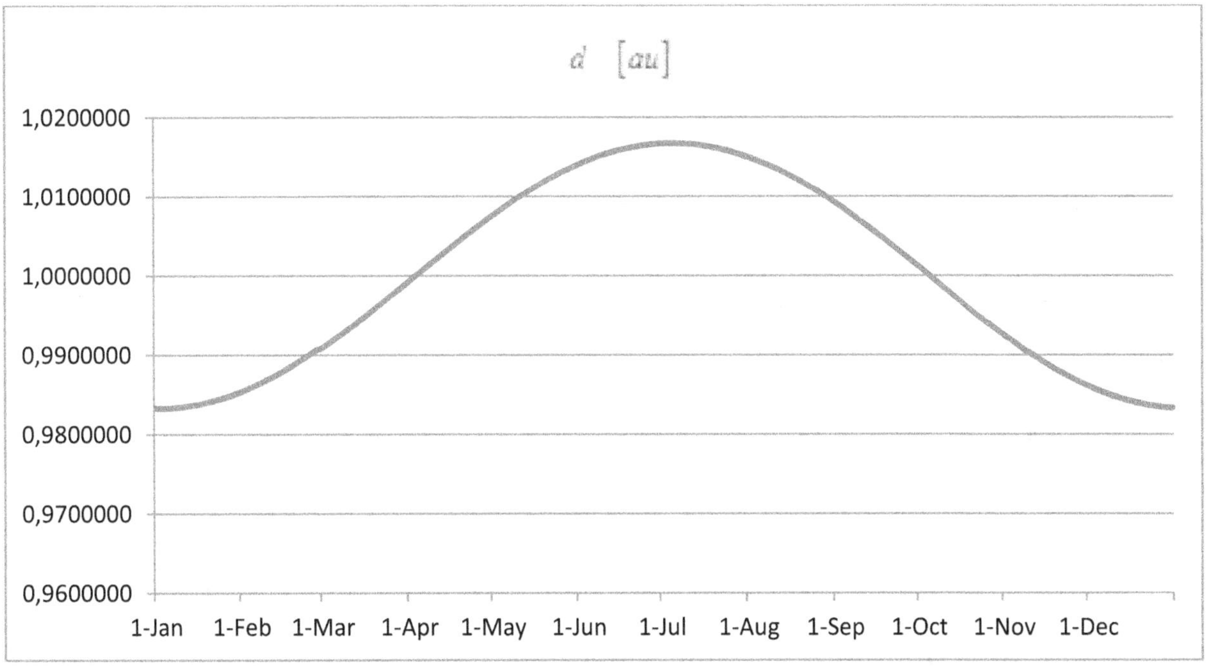

Figure 3 Sun-Earth distance (d) in astronomical units $[au]$ over a year

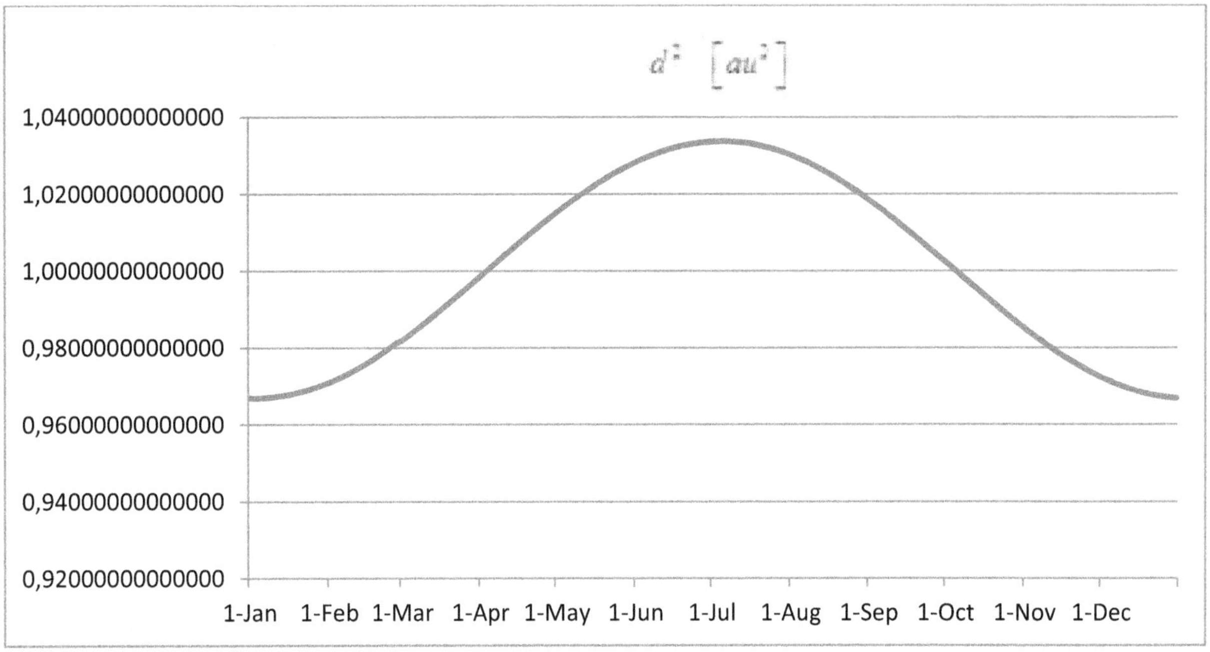

Figure 4 Sun-Earth distance squared $\left(d^2\right)$ in $\left[au^2\right]$ units over a year

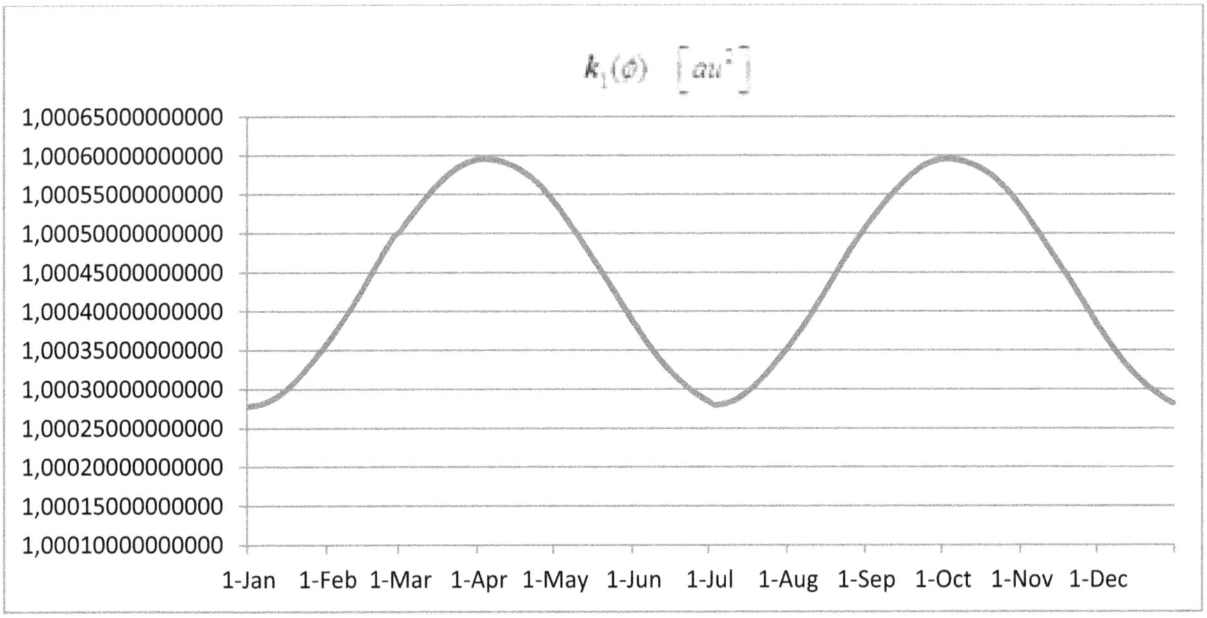

Figure 5 $k_1(\phi)$ in $\left[au^2\right]$ units for the Sun-Earth system over a year

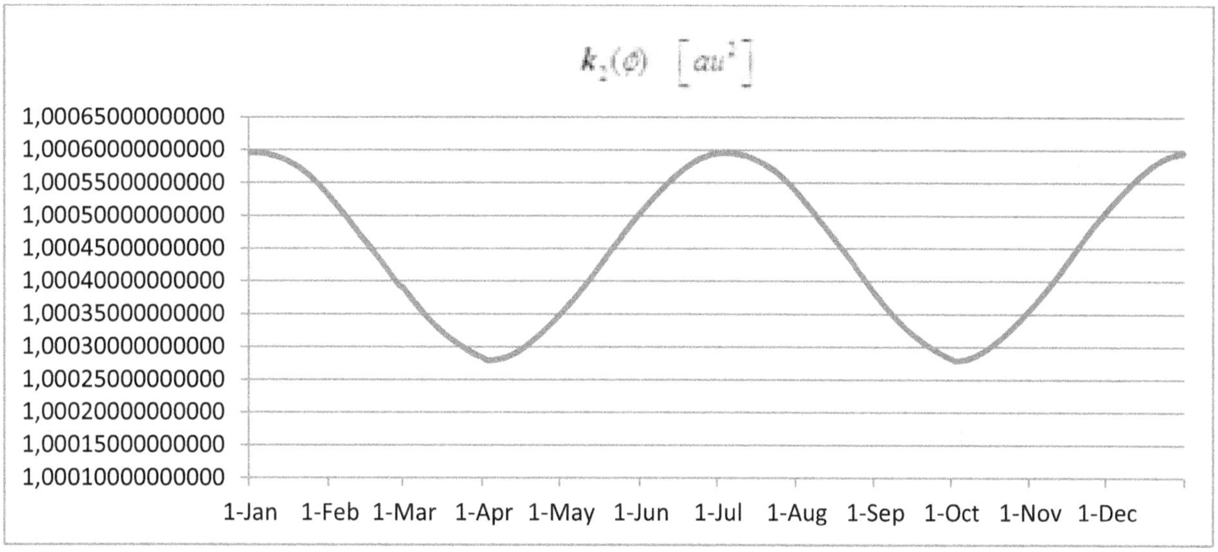

Figure 6 $k_2(\phi)$ in $\left[au^2\right]$ units for the Sun-Earth system over a year

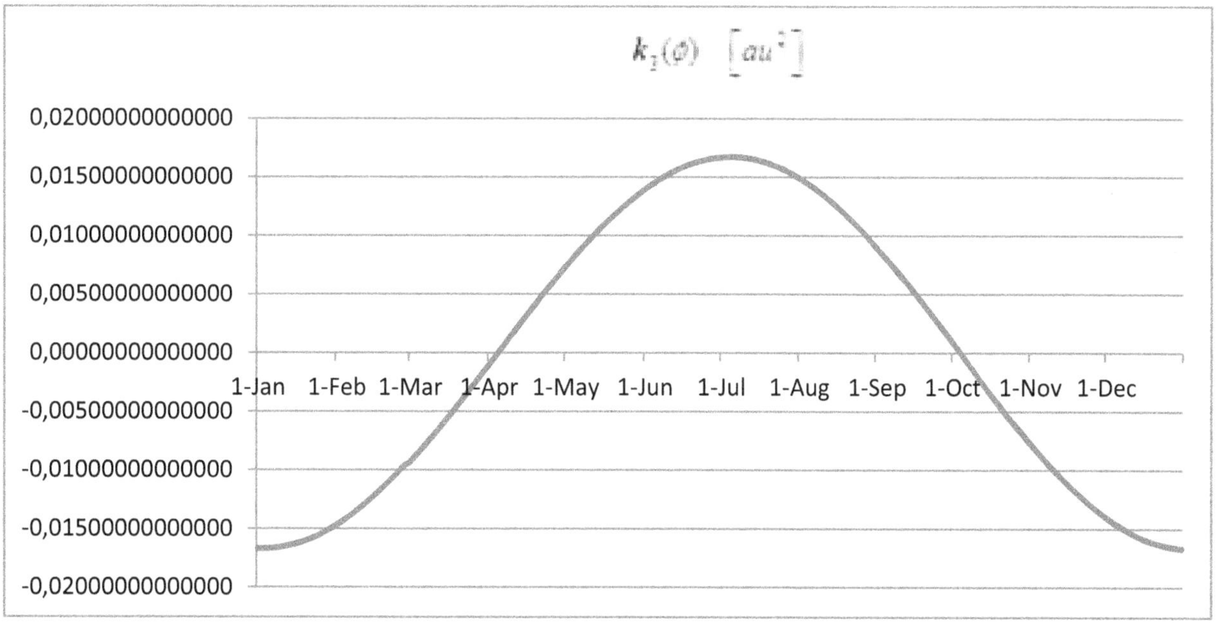

Figure 7 $k_3(\phi)$ in $\left[au^2\right]$ units for the Sun-Earth system over a year

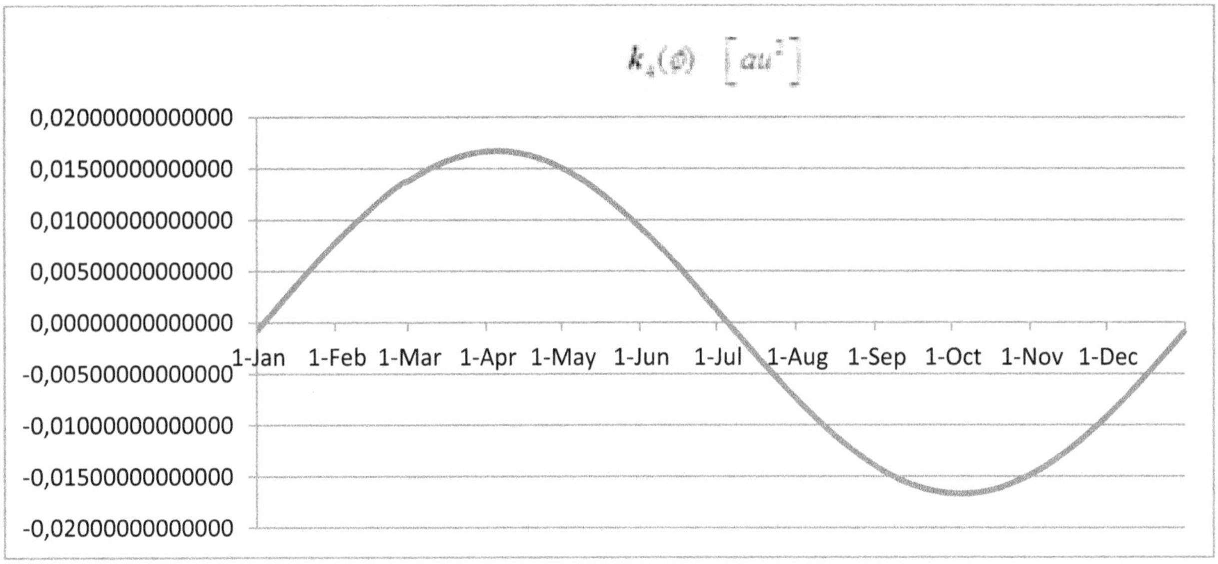

Figure 8 $k_4(\phi)$ in $\left[au^2\right]$ units for the Sun-Earth system over a year

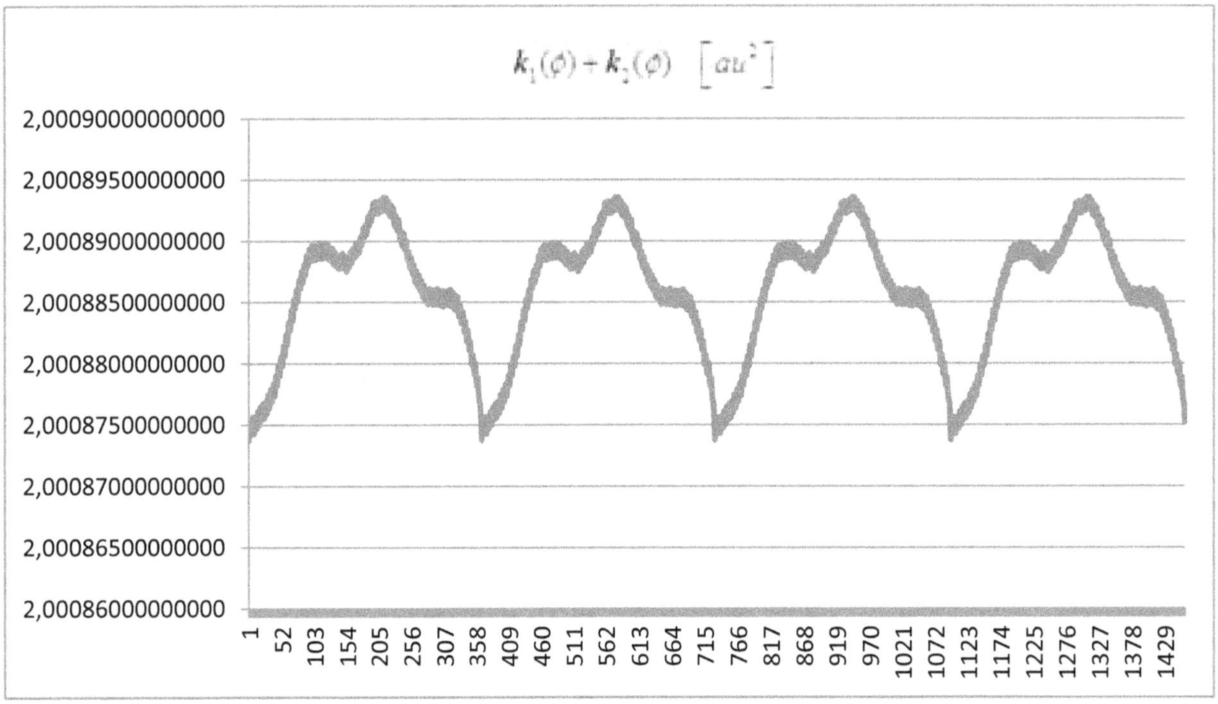

Figure 9 $k_1(\phi)+k_2(\phi)$ in $\left[au^2\right]$ units for the Sun-Earth system over a year

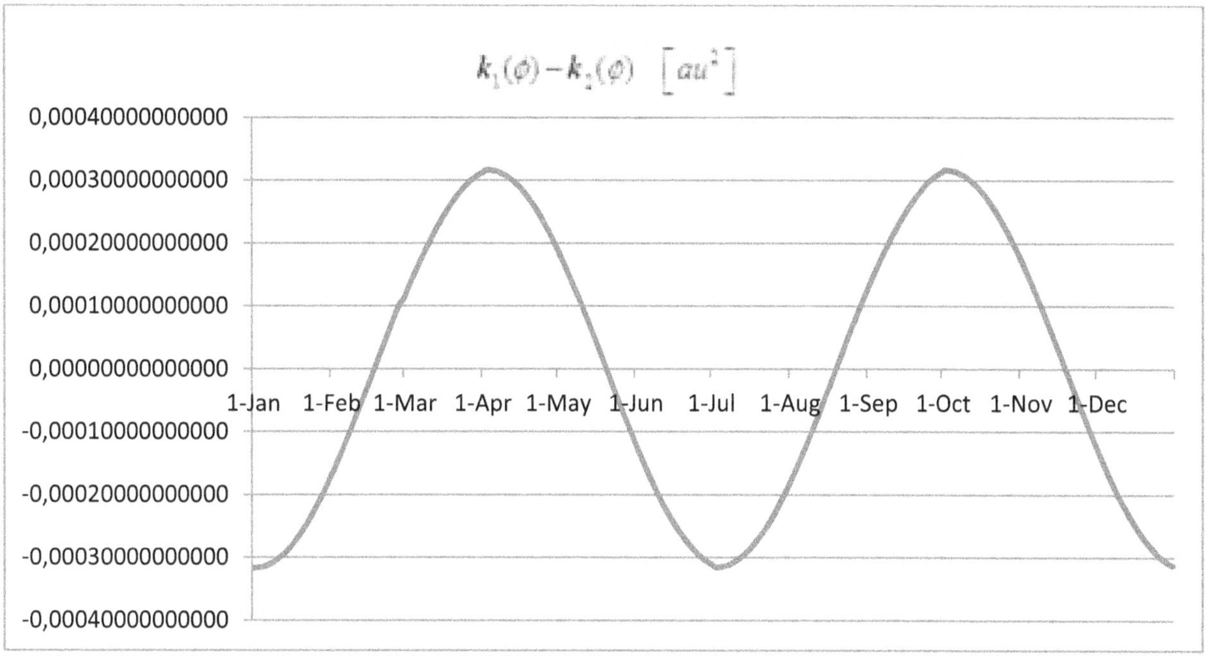

Figure 10 $k_1(\phi)-k_2(\phi)$ in $\left[au^2\right]$ units for the Sun-Earth system over a year

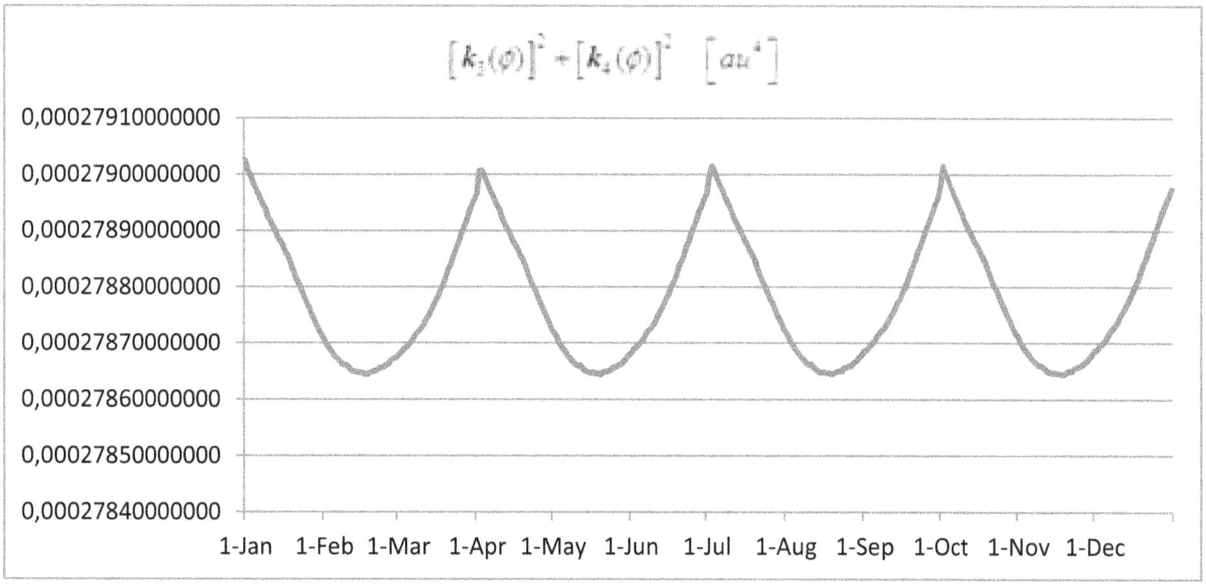

Figure 11 $[k_3(\phi)]^2+[k_4(\phi)]^2$ in $\left[au^4\right]$ units for the Sun-Earth system over a year

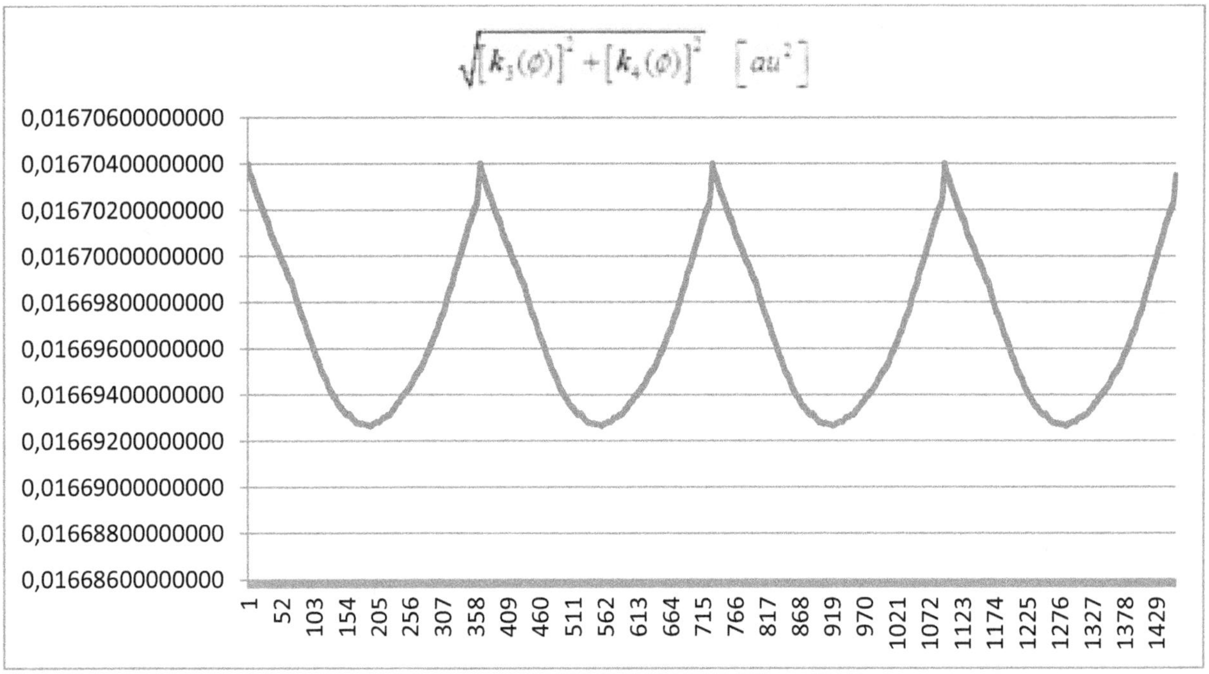

Figure 12 $\sqrt{[k_3(\phi)]^2+[k_4(\phi)]^2}$ in $\left[au^2\right]$ units for the Sun-Earth system over a year

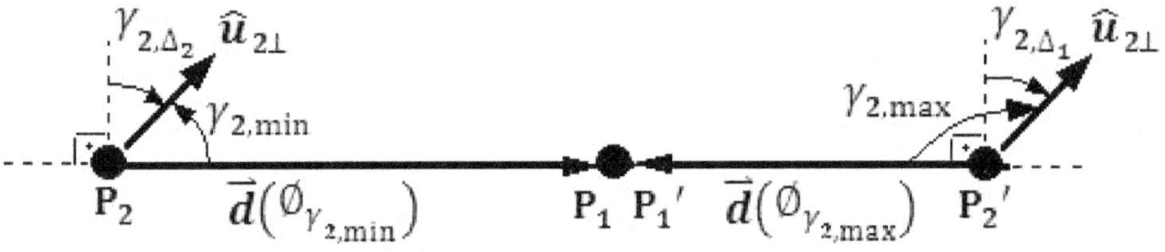

Figure 13 Simple geometric visualization of $\vec{d}(\phi)\cdot\hat{u}_{2\perp}$, $\gamma_{2,\max}$ and $\gamma_{2,\min}$

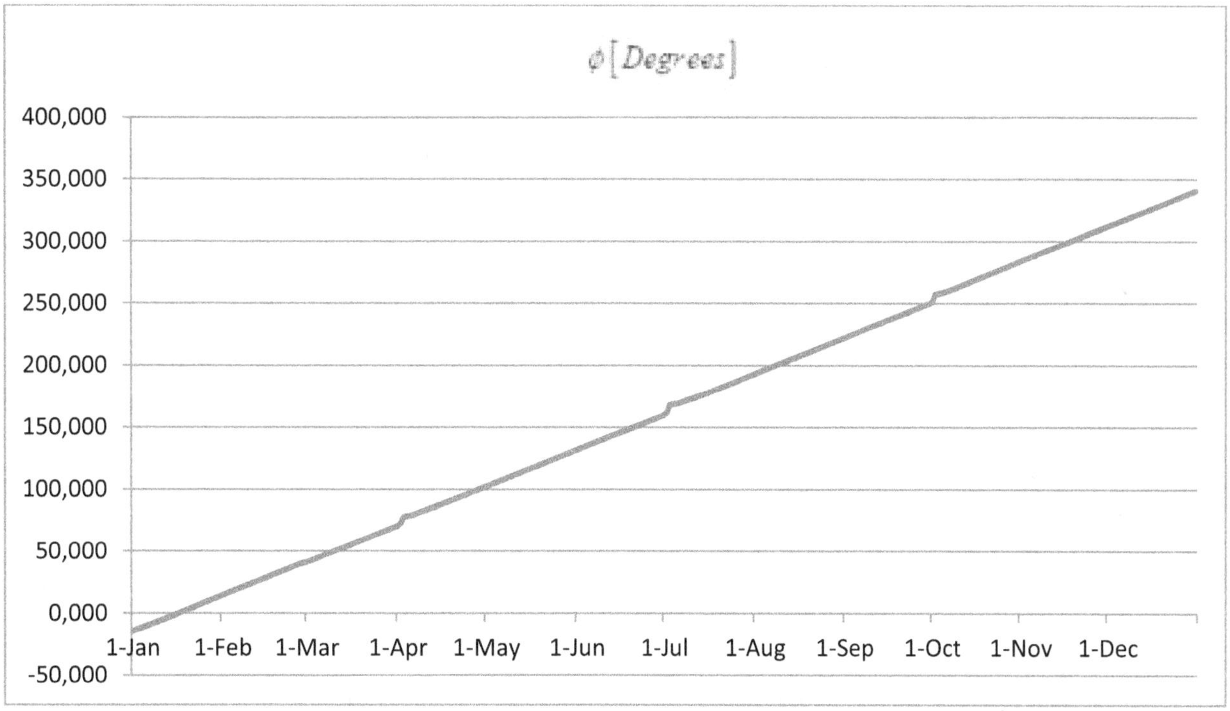

Figure 14 ϕ in $[Degrees]$ against dates for the Sun-Earth system over a year

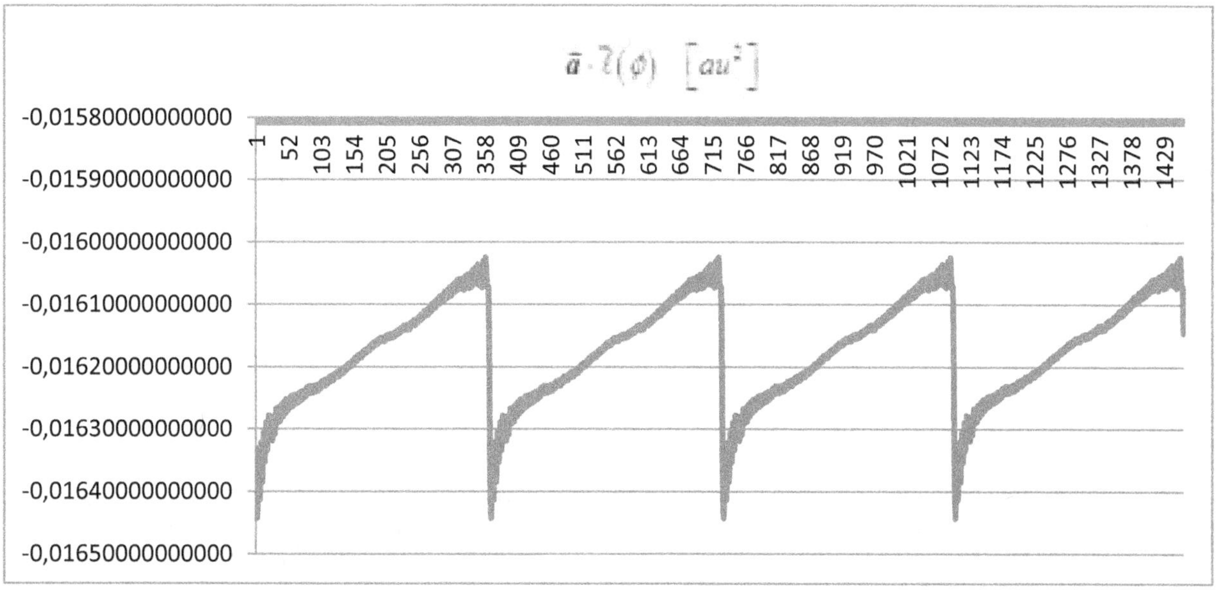

Figure 15 $\bar{a}\cdot\vec{\ell}(\phi)$ in $\left[au^2\right]$ units for the Sun-Earth system over a year

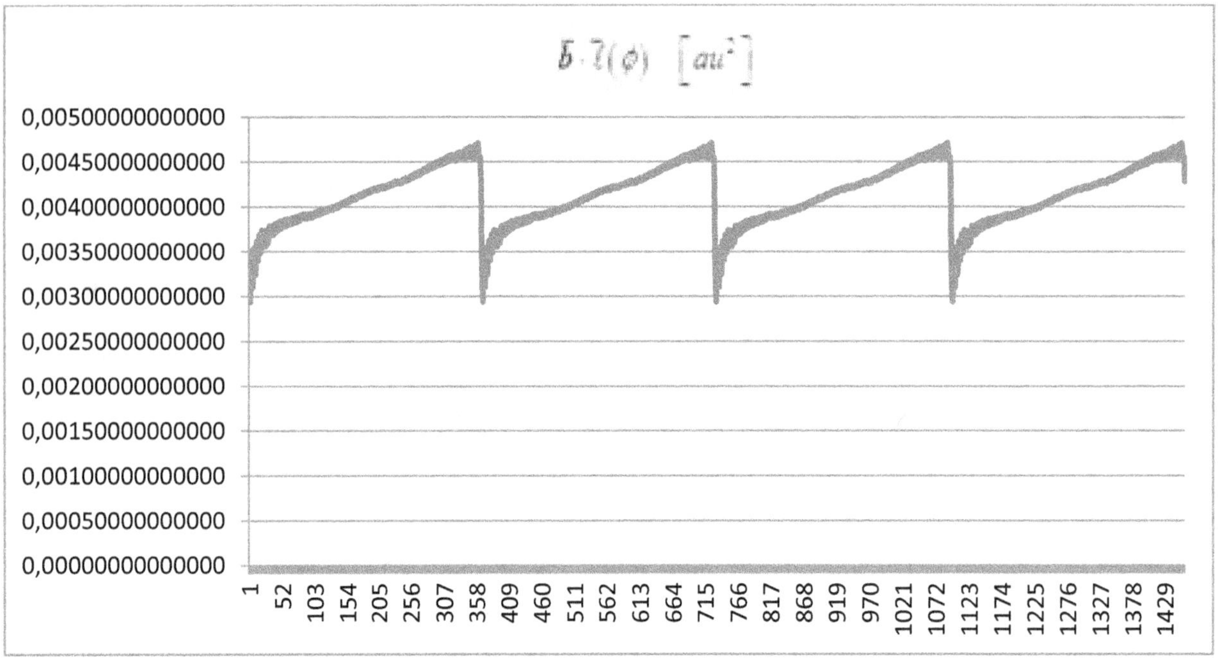

Figure 16 $\vec{b} \cdot \vec{\ell}(\phi)$ in $\left[au^2\right]$ units for the Sun-Earth system over a year

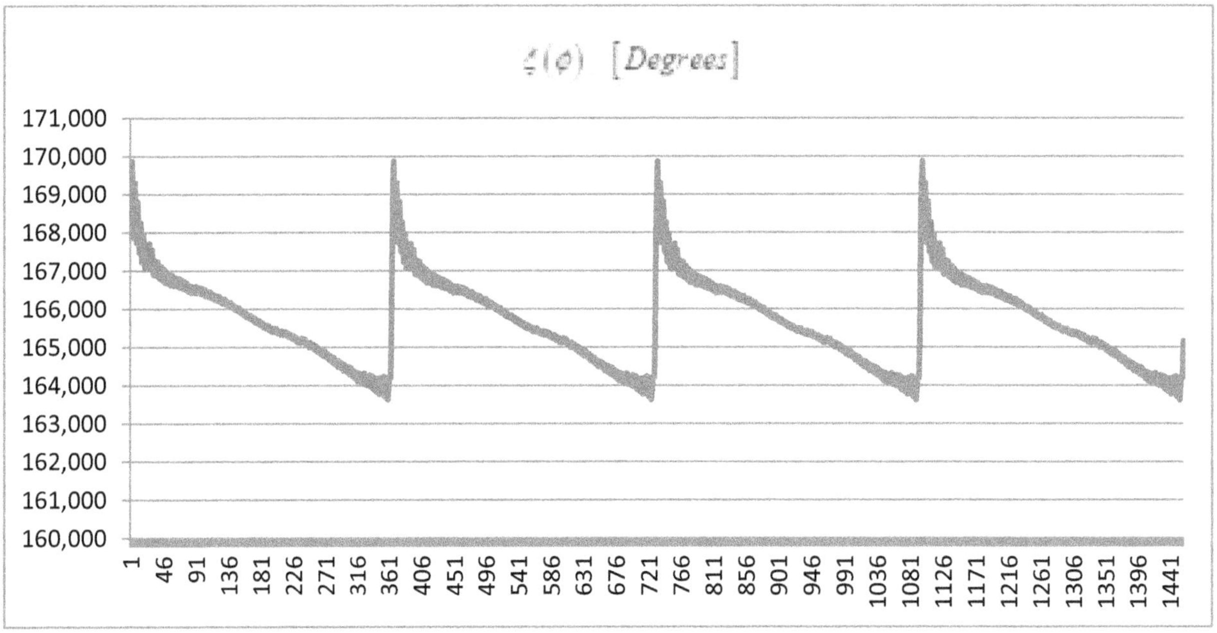

Figure 17 $\xi(\phi)$ in $[Degrees]$ for the Sun-Earth system over a year

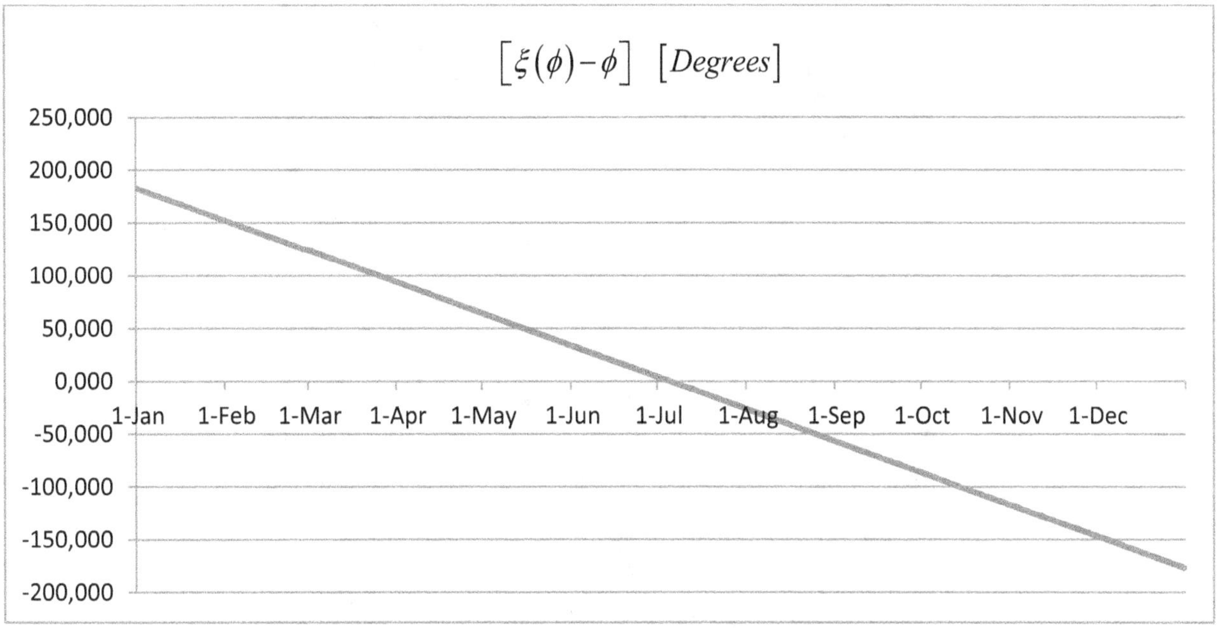

Figure 18 $[\xi(\phi)-\phi]$ in $[Degrees]$ for the Sun-Earth system over a year

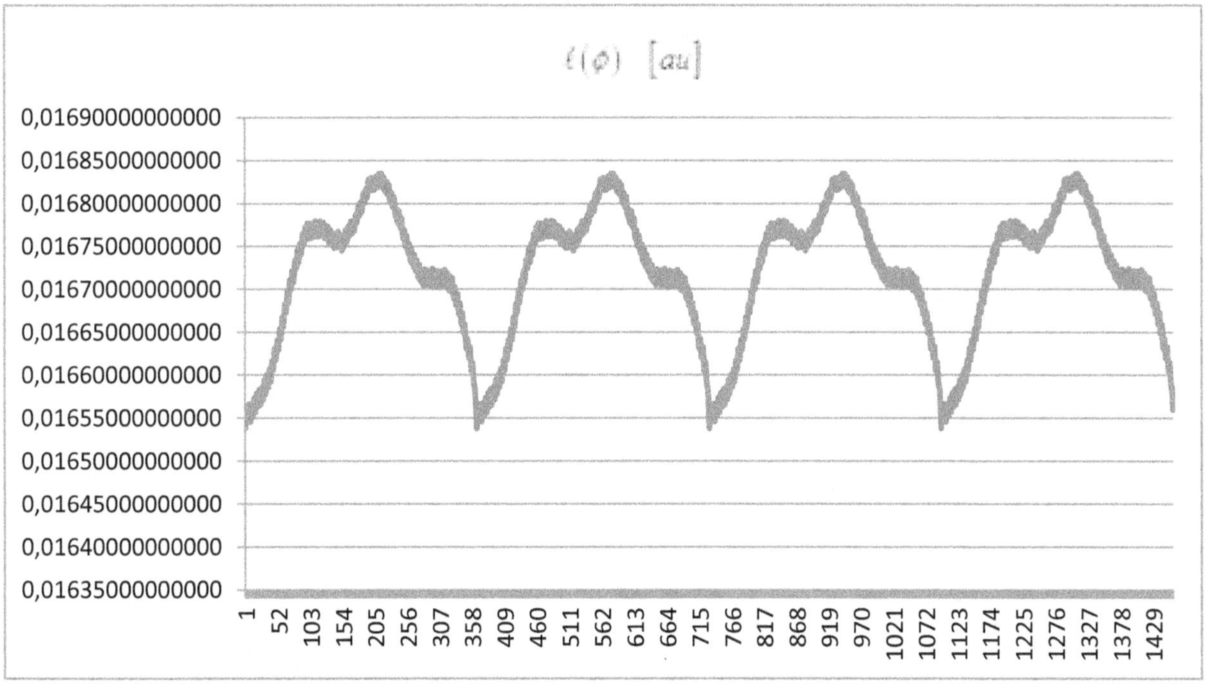

Figure 19 $\ell(\phi)$ in $[au]$ for the Sun-Earth system over a year

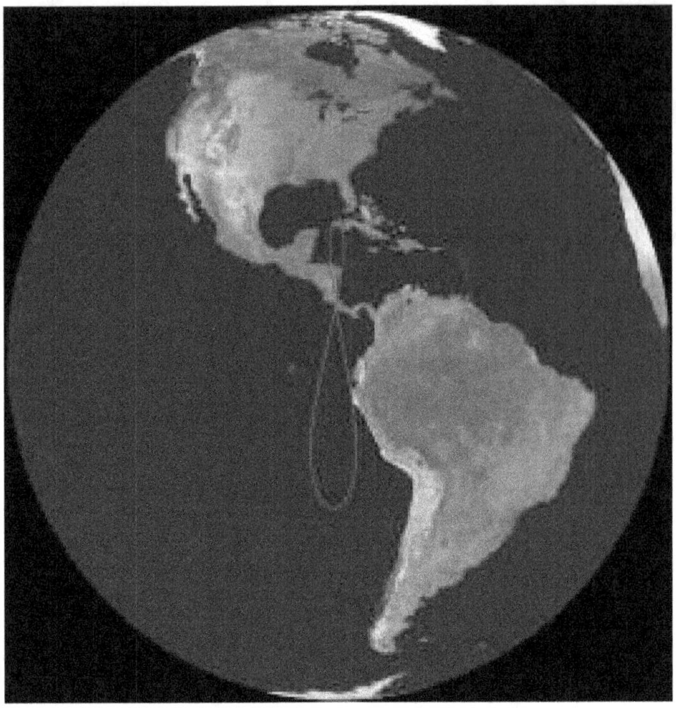

Figure 20 Analemma plotted on Earth over a year as the position of the Sun is directly overhead every 24 hours

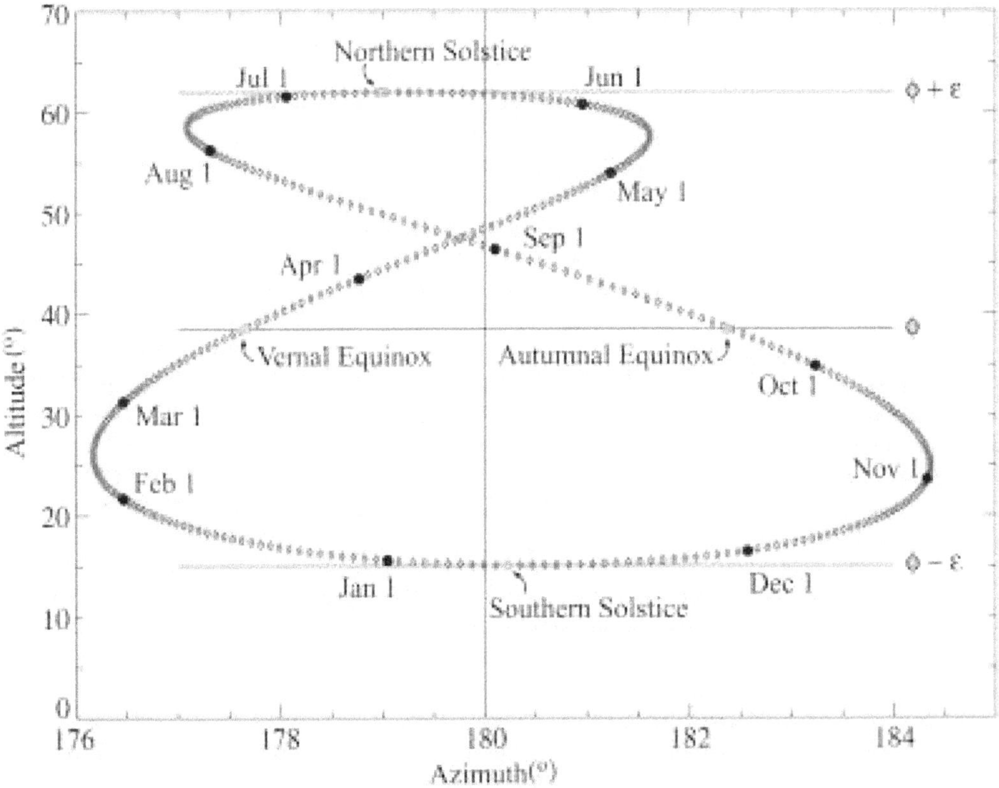

Figure 21 Analemma plotted over a year as seen at noon GMT from Greenwich Observatory

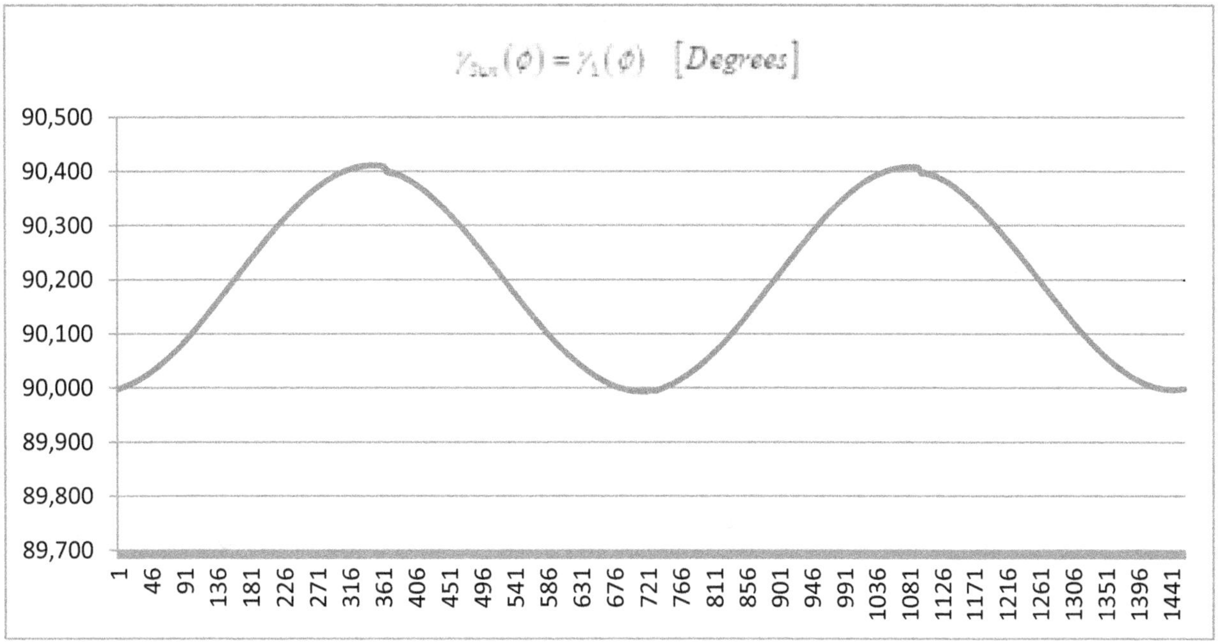

Figure 22 $\gamma_{Sun}(\phi) = \gamma_1(\phi)$ in $[Degrees]$ for the Sun-Earth system over a year

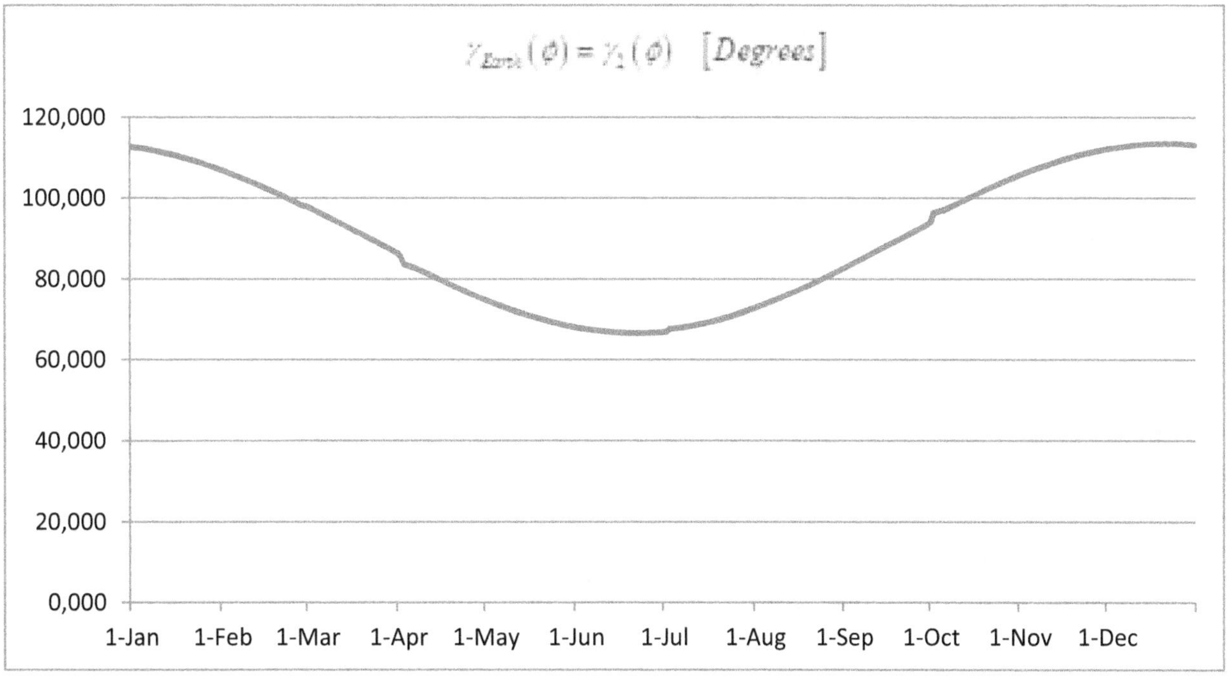

Figure 23 $\gamma_{Earth}(\phi) = \gamma_2(\phi)$ in $[Degrees]$ for the Sun-Earth system over a year

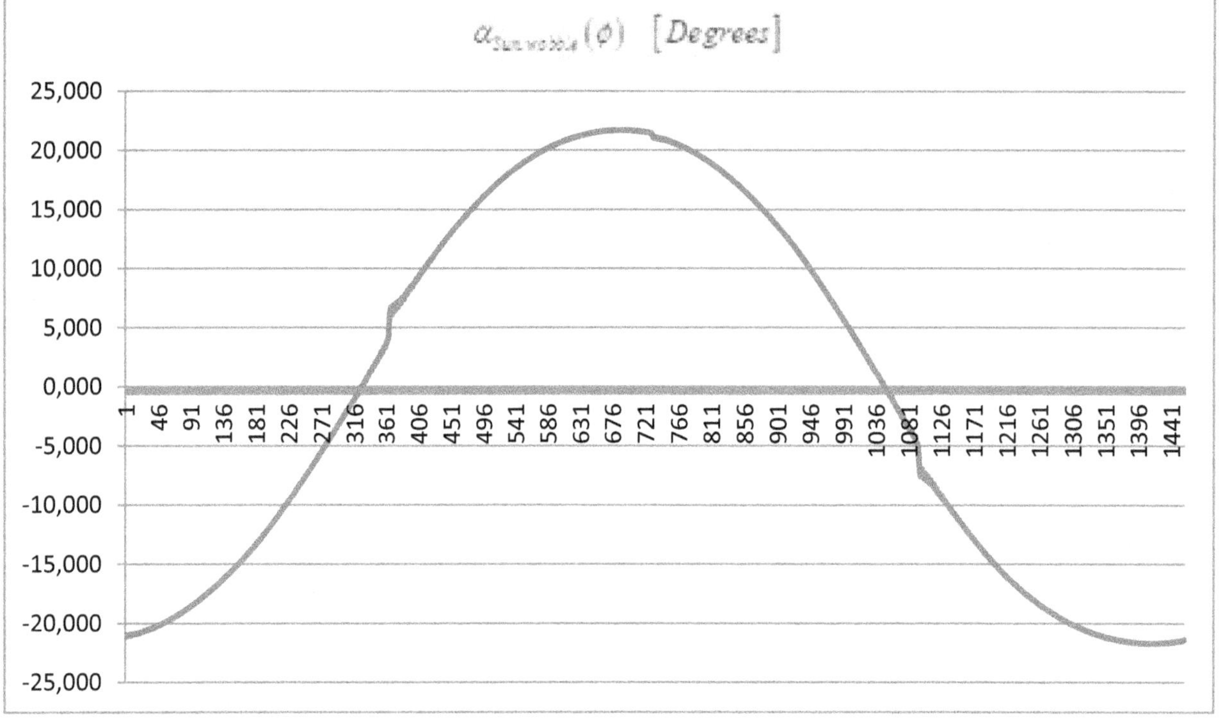

Figure 24 $\alpha_{Sun,wobble}(\phi)$ in $[Degrees]$ for the Sun-Earth system over a year

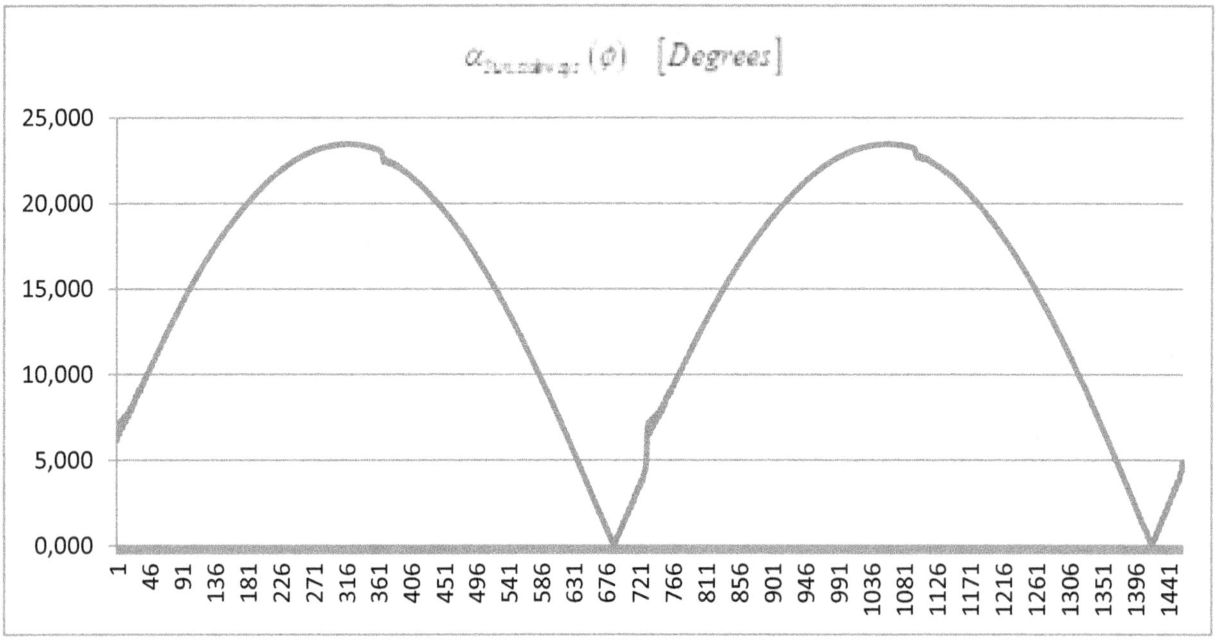

Figure 25 $\alpha_{Sun,sideways}(\phi)$ in $[Degrees]$ for the Sun-Earth system over a year

REFERENCES

References in this Article can be any Physics, Electromagnetics, and Calculus textbook, as the physics equations and mathematical identities used as a basis for the proof are all currently accepted theory in existing textbooks.

1. Halliday, D., Resnick, R., *Fundamentals of Physics (3rd Edition)*, John Wiley & Sons, 1988, ISBN 0-471-63735-1
2. Earth's orbit, Wikipedia, https://en.wikipedia.org/wiki/Earth%27s_orbit, (2020)
3. NASA *Landsat* Observation Data, http://landsathandbook.gsfc.nasa.gov/excel_docs/d.xls
4. Astronomical unit, Wikipedia, https://en.wikipedia.org/wiki/Astronomical_unit, (2020)
5. Solstice, Wikipedia, https://en.wikipedia.org/wiki/Solstice, (2020)
6. Equinox and solstice 2010-2020, Greenwich Mean Time, https://greenwichmeantime.com/longest-day/equinox-solstice-2010-2020/, (2020)
7. Axial tilt, Wikipedia, https://en.wikipedia.org/wiki/Axial_tilt, (2020)
8. Ecliptic, Wikipedia, https://en.wikipedia.org/wiki/Ecliptic, (2020)
9. Equinox, Wikipedia, https://en.wikipedia.org/wiki/Equinox, (2020)
10. June 2006: The Tilt of the Sun's Axis, Bruce McClure's Astronomy Page, http://www.idialstars.com/fipl.htm, (2020)
11. Zell, Holly, *Solar Rotation Varies by Latitude* (August 7, 2017), NASA, https://www.nasa.gov/mssion_pages/sunearth/science/solar-rotation.html, (2020)
12. The Sun's Tilt Thrughout The Year, YouTube, https://youtu.be/j44q2xvNePQ, @astroguyz , July 2, 2015
13. File: Tan, A. P., *Asli Pinar Tan Analysis Based on Earth-Sun distance (d) Landsat.xlsx*, https://www.dropbox.com/scl/fi/th5d3d5ur8d1mb60aitou/Asli-Pinar-Tan-Analysis-Based-on-Earth-Sun-distance-d-Landsat.xlsx?dl=0&rlkey=m3spxqxbxtnumrj2wvtg33l2v (2020)
14. Dot product, Wikipedia, https://en.wikipedia.org/wiki/Dot_product, (2020)
15. Planetary fact sheet – metric, NSSDCA, NASA Goddard Space Flight Center, https://nssdc.gsfc.nasa.gov/planetary/factsheet/, (2020)
16. Mercury (planet), Wikipedia, https://en.wikipedia.org/wiki/Mercury_(planet), (2020)
17. Analemma, Wikipedia, https://en.wikipedia.org/wiki/Analemma, (2020)
18. Right hand rule, Wikipedia, https://en.wikipedia.org/wiki/Right-hand_rule, (2020)
19. Faraday's law of induction, Wikipedia, https://en.wikipedia.org/wiki/Faraday%27s_law_of_induction, (2020)
20. Electromotive force, Wikipedia, https://en.wikipedia.org/wiki/Electromotive_force, (2020)
21. Magnetic flux, Wikipedia, https://en.wikipedia.org/wiki/Magnetic_flux, (2020)
22. Ellipse, Wikipedia, https://en.wikipedia.org/wiki/Ellipse, (2020)
23. Semi-major and semi-minor axes, Wikipedia, https://en.wikipedia.org/wiki/Semi-major_and_semi-minor_axes, (2020)
24. Focus (geometry), Wikipedia, https://en.wikipedia.org/wiki/Focus_(geometry), (2020)
25. Double planet, Wikipedia, https://en.wikipedia.org/wiki/Double_planet, (2020)
26. Year, Wikipedia, https://en.wikipedia.org/wiki/Year, (2020)
27. Clockwise, Wikipedia, https://en.wikipedia.org/wiki/Clockwise, (2020)
28. Geographical pole, Wikipedia, https://en.wikipedia.org/wiki/Geographical_pole, (2020)
29. North Pole, Wikipedia, https://en.wikipedia.org/wiki/North_Pole, (2020)

30. Rotation around a fixed axis, Wikipedia, http://en.wikipedia.org/wiki/Rotation_around_a_fixed_axis, (2020)
31. Equator, Wikipedia, https://en.wikipedia.org/wiki/Equator, (2020)
32. Milky Way, Wikipedia, https://en.wikipedia.org/wiki/Milky_Way, (2020)
33. Galactic Center, Wikipedia, https://en.wikipedia.org/wiki/Galactic_Center, (2020)
34. Sagittarius (constellation), Wikipedia, https://en.wikipedia.org/wiki/Sagittarius_(constellation), (2020)

ARTICLE 4

AN ANALYSIS OF THE CIRCULAR ORBITAL MOTION OF THE MOON AND THE EARTH

Author: Aslı Pınar Tan[IV]

SUMMARY

Based on observation, all bodies in space seem to move in some kind of elliptical motion with respect to each other. In a previous article, it is mathematically demonstrated and proven that the "distance between points on any two different circles in three-dimensional space" is equivalent to the "distance of points on a vector ellipse from another fixed or moving point", and that based on the elliptical variation of Earth-Moon distance[5] over their relative full cycle, with semi-major and semi-minor axis values also varying in a harmonic style, it is equivalently possible that the Earth and the Moon each revolve in a circular motion with fixed but different angular velocities around centers of their own individual circular orbits, where their orbital centers are displaced by a vector distance from each other. In this article, we analyze this possible motion of the Moon with respect to the Earth in more detail.

ARTICLE

According to Kepler's 1st Law[1,2,3], "the orbit of a planet with respect to the Sun is an ellipse, with the Sun at one of the two foci", which is an empirical rule concluded through near estimation based on the observation of astronomical position data measured over time in the Solar System. The orbit of the Moon[4,5] with respect to the Earth is also distinctly elliptical[6], but with a varying eccentricity[6] in a harmonic style as the Moon comes closer to and goes farther away from the Earth harmonically along a full cycle of this ellipse[6], again based on the observation of astronomical position data measured over time. In the **ARTICLE 1** "Points on two circles in space mimic an ellipse with respect to a point", it is mathematically demonstrated and proven that the "distance between points on any two different circles in three-dimensional space" is equivalent to the "distance of points on a vector ellipse from another fixed or moving point, similar to two-dimensional space". In other words, it is shown that moving points on two circles in space vector-wise mimic an elliptical path with respect to each other, whether they are moving with the *same* angular velocities, or *different* but *fixed* angular velocities, or even with *different* and *changing* angular velocities with respect to their own centers of revolution, virtually seeing each other as positioned at a instantaneously stationary point in space on their respective virtual ecliptic[29] plane. In the same **ARTICLE 1** "Points on two circles in space mimic an ellipse with respect to a point", it has also been demonstrated that, based on the astronomical position data measured over time, the variation of the Earth-Moon[4] distance over time t , as can be seen in **Figure 1**, exhibits a pattern of distance between points moving in their individual circular orbits, with ***fixed*** but ***different*** angular velocities around their own circles of revolution. Based on this fundamental observation, we make the assertion that the Earth and the Moon are possibly moving in different individual circular orbits, leading to the observation[19] that they are moving in an relative elliptical orbit with respect to each other, with harmonically varying semi-major[7] and semi-minor[7] axes. In this Article, we will investigate and further analyze the individual and

---

[IV] *ASLI PINAR TAN*
Linkedin Website: *https://www.linkedin.com./in/apinartan*

relative orbital motions of the Earth and the Moon, utilizing Earth-Moon distance[5] data with the given assumption.

Let us consider the Earth and the Moon to be a system of two different bodies moving around two different circular orbits in three-dimensional space, with the geometry of the system demonstrated as in **Figure 2** in Cartesian $(\hat{x}, \hat{y}, \hat{z})$ coordinates. We can randomly take $Circle_1$ to be the circle of revolution of the Moon, and $Circle_2$ in the selected horizontal plane to be the circle of revolution of the Earth. Note that the inclination angle β (5) between the planes of these two circles is also taken to be constant. Based on our previous analysis in the **ARTICLE 1** "Points on two circles in space mimic an ellipse with respect to a point", the Earth and the Moon obey a special relative motion case as expressed in (1) - (4), where the moving points $\mathbf{P}_1$ and $\mathbf{P}_2$, which represent the Moon and the Earth in their individual circular orbits, respectively, are phased with ***fixed*** but ***different*** (1) angular velocities ω_1 (2) and ω_2 (3), respectively, with respect to the centers of their own circles of revolution, their phase difference at time $(t = t_0 = 0)$ being $[\phi_0(t_0) = 0]$ (4), where the reference timestamp t_0 is taken to be a point in time when both of the moving points $\mathbf{P}_1$ (Moon) and $\mathbf{P}_2$ (Earth) are aligned in phase, with $\mathbf{P}_1$ having a phase of $[\phi_1(t_0 = 0) = 0]$ (2) and $\mathbf{P}_2$ having a phase of $[\phi_2(t_0) = 0]$ (3) in their respective orbits of revolution, in the configuration described in **Figure 2**.

$$\omega_1 \neq \omega_2 \qquad (\omega_1 \;\&\; \omega_2 \; constant) \tag{1}$$

$$\phi_1(t) = \omega_1 t = \phi_2(t) + \phi_0(t) = \phi(t) + \phi_0(t) = \phi + \phi_0 \;\; ; \;\; \phi_1(t_0 = 0) = 0 \qquad (Phase\ of\ \mathbf{P}_1) \tag{2}$$

$$\phi_2(t) = \omega_2 t = \phi(t) = \phi = \phi_{Earth-Moon} \;\; ; \;\; \phi_2(t_0 = 0) = 0 \qquad (Phase\ of\ \mathbf{P}_2) \tag{3}$$

$$\phi_0(t) = \phi_1(t) - \phi_2(t) = (\omega_1 - \omega_2)t \;\; ; \;\; \phi_0(t_0 = 0) = 0 \;\; ; \;\; (Phase\ difference\ of\ \mathbf{P}_1\ and\ \mathbf{P}_2) \tag{4}$$

In this case, at each $\phi(t)$ (3), based on the analysis in the **ARTICLE 1** "Points on two circles in space mimic an ellipse with respect to a point", we define the vector radii $\vec{r}_1[\phi(t)]$ (5) and $\vec{r}_2[\phi(t)]$ (6) of the two circles, the vector distance $\vec{\ell}[\phi(t)]$ (8) with magnitude $\ell[\phi(t)]$ (9) between the centers of these two circles, vector distance $\vec{d}[\phi(t)]$ (10) between the moving points $\mathbf{P}_1$ and $\mathbf{P}_2$ on the two respective circles, with Cartesian components in (11), as well as the semi-major[7] and semi-minor[7] axis vectors $\vec{a}(t)$ (12) and $\vec{b}(t)$ (13) of the vector ellipse[6], their Dot Product[9] $[\vec{a}(t) \cdot \vec{b}(t)]$ (16), their magnitudes squared $[a^2(t)]$ (14) and $[b^2(t)]$ (15), elliptic representation of the vector distance $\vec{d}[\phi(t)]$ (17) - (18), the elliptic coordinate vectors $\vec{\mathbf{X}}[\phi(t)]$ (19) - (20) and $\vec{\mathbf{Y}}[\phi(t)]$ (21) - (22), square of the scalar distance $d^2[\phi(t)]$ (25) between the moving points $\mathbf{P}_1$ and $\mathbf{P}_2$ on two respective circles, and square of the instantaneous focal[6,8] distance $[c^2(t)]$ (26) and instantaneous eccentricity[6] $e(t)$ (27) of the vector ellipse[6].

$$\vec{r}_1 = \vec{r}_1\left[\phi(t)\right] = \vec{r}_1(\omega_2 t)$$

$$= \left\{ \begin{array}{l} \hat{x}\, r_1 Cos\,\beta\, Cos\left[(\omega_1-\omega_2)t\right] + \hat{y}\, r_1 Sin\left[(\omega_1-\omega_2)t\right] + \\ \hat{z}\, r_1 Sin\,\beta\, Cos\left[(\omega_1-\omega_2)t\right] \end{array} \right\} Cos(\omega_2 t)$$

$$+ \left\{ \begin{array}{l} -\hat{x}\, r_1 Cos\,\beta\, Sin\left[(\omega_1-\omega_2)t\right] + \hat{y}\, r_1 Cos\left[(\omega_1-\omega_2)t\right] \\ -\hat{z}\, r_1 Sin\,\beta\, Sin\left[(\omega_1-\omega_2)t\right] \end{array} \right\} Sin(\omega_2 t) \quad (5)$$

$$= \hat{x}\, r_1 Cos\,\beta\, Cos(\omega_1 t) + \hat{y}\, r_1 Sin(\omega_1 t) + \hat{z}\, r_1 Sin\,\beta\, Cos(\omega_1 t) = \vec{r}_1(\omega_1 t)$$

$$\vec{r}_2 = \vec{r}_2\left[\phi(t)\right] = \vec{r}_2(\omega_2 t) = \hat{x}\, r_2\, Cos(\omega_2 t) + \hat{y}\, r_2\, Sin(\omega_2 t) \quad (6)$$

$$\vec{r}_1 \cdot \vec{r}_1 = r_1^2 \quad ; \quad \left|\vec{r}_1\right| = r_1 \quad ; \quad \vec{r}_2 \cdot \vec{r}_2 = r_2^2 \quad ; \quad \left|\vec{r}_2\right| = r_2 \quad (7)$$

$$\vec{\ell} = \vec{\ell}\left[\phi(t)\right] = \vec{\ell}(\omega_2 t) = \hat{x}\,\ell_x(\omega_2 t) + \hat{y}\,\ell_y(\omega_2 t) + \hat{z}\,\ell_z(\omega_2 t) \quad (8)$$

$$\ell\left[\phi(t)\right] = \ell(\omega_2 t) = \left|\vec{\ell}(\omega_2 t)\right| = \sqrt{\ell^2(\omega_2 t)} = \sqrt{\vec{\ell}(\omega_2 t)\cdot\vec{\ell}(\omega_2 t)} = \sqrt{\ell_x^{\,2}(\omega_2 t) + \ell_y^{\,2}(\omega_2 t) + \ell_z^{\,2}(\omega_2 t)} \quad (9)$$

$$\vec{d}\left[\phi(t)\right] = \vec{d}(\omega_2 t) = \hat{d}\left[\phi(t)\right] d\left[\phi(t)\right] = \vec{r}_1 - \vec{r}_2 + \vec{\ell}$$

$$= \left\{ \begin{array}{l} \hat{x}\left(r_1 Cos\,\beta\, Cos\left[(\omega_1-\omega_2)t\right] - r_2\right) + \hat{y}\, r_1 Sin\left[(\omega_1-\omega_2)t\right] + \\ \hat{z}\, r_1 Sin\,\beta\, Cos\left[(\omega_1-\omega_2)t\right] \end{array} \right\} Cos(\omega_2 t)$$

$$+ \left\{ \begin{array}{l} -\hat{x}\, r_1 Cos\,\beta\, Sin\left[(\omega_1-\omega_2)t\right] + \hat{y}\left(r_1 Cos\left[(\omega_1-\omega_2)t\right] - r_2\right) \\ -\hat{z}\, r_1 Sin\,\beta\, Sin\left[(\omega_1-\omega_2)t\right] \end{array} \right\} Sin(\omega_2 t) \quad (10)$$

$$+ \left\{ \hat{x}\,\ell_x(\omega_2 t) + \hat{y}\,\ell_y(\omega_2 t) + \hat{z}\,\ell_z(\omega_2 t) \right\}$$

$$\vec{d}(\phi) = \hat{x}\, d_x(\phi) + \hat{y}\, d_y(\phi) + \hat{z}\, d_z(\phi) \quad \Rightarrow \quad \left\{ \begin{array}{l} d_x(\phi) = r_1 Cos\,\beta\, Cos(\omega_1 t) - r_2 Cos(\omega_2 t) + \ell_x(\omega_2 t) \\ d_y(\phi) = r_1 Sin(\omega_1 t) - r_2 Sin(\omega_2 t) + \ell_y(\omega_2 t) \\ d_z(\phi) = r_1 Sin\beta\, Cos(\omega_1 t) + \ell_z(\omega_2 t) \end{array} \right. \quad (11)$$

$$\vec{a}(t) = \hat{x}\left\{ r_1 Cos\,\beta\, Cos\left[(\omega_1-\omega_2)t\right] - r_2 \right\} + \hat{y}\, r_1 Sin\left[(\omega_1-\omega_2)t\right] + \hat{z}\, r_1 Sin\,\beta\, Cos\left[(\omega_1-\omega_2)t\right] \quad (12)$$

$$\vec{b}(t) = -\hat{x}\, r_1 Cos\,\beta\, Sin\left[(\omega_1-\omega_2)t\right] + \hat{y}\left\{ r_1 Cos\left[(\omega_1-\omega_2)t\right] - r_2 \right\} - \hat{z}\, r_1 Sin\,\beta\, Sin\left[(\omega_1-\omega_2)t\right] \quad (13)$$

$$a^2(t) = \vec{a}(t)\cdot\vec{a}(t) = r_1^2 - 2 r_1 r_2\, Cos\beta\, Cos\left[(\omega_1-\omega_2)t\right] + r_2^2 \quad (14)$$

$$b^2(t) = \vec{b}(t)\cdot\vec{b}(t) = r_1^2 - 2 r_1 r_2\, Cos\left[(\omega_1-\omega_2)t\right] + r_2^2 \quad (15)$$

$$\vec{a}(t)\cdot\vec{b}(t) = r_1 r_2 (Cos\beta - 1) Sin\left[(\omega_1-\omega_2)t\right] \quad (16)$$

$$\vec{d}(\omega_2 t) = \vec{\mathbf{X}}(\omega_2 t) + \vec{\mathbf{Y}}(\omega_2 t) + \vec{\ell}(\omega_2 t) = \vec{r}_1(\omega_2 t) - \vec{r}_2(\omega_2 t) + \vec{\ell}(\omega_2 t) \quad (17)$$

$$\vec{r}_1 - \vec{r}_2 = \vec{r}_1(\omega_2 t) - \vec{r}_2(\omega_2 t) = \vec{\mathbf{X}}(\omega_2 t) + \vec{\mathbf{Y}}(\omega_2 t) = \vec{a}(t)Cos(\omega_2 t) + \vec{b}(t)Sin(\omega_2 t) \quad (18)$$

$$\vec{\mathbf{X}}[\phi(t)] = \vec{\mathbf{X}}(\omega_2 t) = \vec{a}(t)Cos(\omega_2 t) \quad ; \quad |\vec{\mathbf{X}}(\omega_2 t)| = X(\omega_2 t) \quad (19)$$

$$\vec{\mathbf{X}}(\omega_2 t)\cdot\vec{\mathbf{X}}(\omega_2 t) = X^2(\omega_2 t) = \vec{a}(t)\cdot\vec{a}(t)\,Cos^2(\omega_2 t) = a^2(t)Cos^2(\omega_2 t) \quad (20)$$

$$\vec{\mathbf{Y}}[\phi(t)] = \vec{\mathbf{Y}}(\omega_2 t) = \vec{b}(t)Sin(\omega_2 t) \quad ; \quad |\vec{\mathbf{Y}}(\omega_2 t)| = Y(\omega_2 t) \quad (21)$$

$$\vec{\mathbf{Y}}(\omega_2 t)\cdot\vec{\mathbf{Y}}(\omega_2 t) = Y^2(\omega_2 t) = \vec{b}(t)\cdot\vec{b}(t)\,Sin^2(\omega_2 t) = b^2(t)Sin^2(\omega_2 t) \quad (22)$$

$$d[\phi(t)] = |\vec{d}[\phi(t)]| = \sqrt{d^2[\phi(t)]} = \sqrt{\vec{d}[\phi(t)]\cdot\vec{d}[\phi(t)]} = \sqrt{[d_x(\phi)]^2 + [d_y(\phi)]^2 + [d_z(\phi)]^2} \quad (23)$$

$$\hat{d}[\phi(t)] = \frac{\vec{d}[\phi(t)]}{|\vec{d}[\phi(t)]|} = \frac{\vec{d}[\phi(t)]}{d[\phi(t)]} = \hat{x}\frac{d_x[\phi(t)]}{d(\phi)} + \hat{y}\frac{d_y[\phi(t)]}{d[\phi(t)]} + \hat{z}\frac{d_z[\phi(t)]}{d[\phi(t)]} \quad (24)$$

$$\begin{aligned} d^2[\phi(t)] = \vec{d}[\phi(t)]\cdot\vec{d}[\phi(t)] = d^2(\omega_2 t) = b^2(t) + 2r_1 r_2(1 - Cos\beta)Cos(\omega_1 t)Cos(\omega_2 t) \\ + 2\{\vec{a}(t)Cos(\omega_2 t) + \vec{b}(t)Sin(\omega_2 t)\}\cdot\vec{\ell}(\omega_2 t) + \ell^2(\omega_2 t) \end{aligned} \quad (25)$$

$$c^2(t) = |a^2(t) - b^2(t)| = 2\,r_1 r_2(1 - Cos\beta)|Cos[(\omega_1 - \omega_2)t]| \quad (Focal\ Distance\ Squared) \quad (26)$$

$$e(t) = \begin{cases} \dfrac{c(t)}{a(t)} = \sqrt{\dfrac{2\,r_1 r_2(1 - Cos\beta)|Cos[(\omega_1 - \omega_2)t]|}{r_1^2 - 2r_1 r_2 Cos\beta\,Cos[(\omega_1 - \omega_2)t] + r_2^2}} & if \quad a(t) > b(t) \\ \dfrac{c(t)}{b(t)} = \sqrt{\dfrac{2\,r_1 r_2(1 - Cos\beta)|Cos[(\omega_1 - \omega_2)t]|}{r_1^2 - 2r_1 r_2 Cos[(\omega_1 - \omega_2)t] + r_2^2}} & if \quad a(t) < b(t) \end{cases} \quad (Eccentricity) \quad (27)$$

Here, let us remember that the scalar distance $d[\phi(t)]$ (23) between $\mathbf{P}_1$ and $\mathbf{P}_2$ is calculated as the square root of $d^2[\phi(t)]$ (25), which in turn is calculated in terms of the Dot Product[9] $\{\vec{d}[\phi(t)]\cdot\vec{d}[\phi(t)]\}$ (25) of the vector distance $\vec{d}[\phi(t)]$ (10) with itself, where $\hat{d}[\phi(t)]$ (24) is the unit vector in the direction of $\vec{d}[\phi(t)]$ (10). The magnitudes of the vector radii $\vec{r}_1[\phi(t)]$ (5) and $\vec{r}_2[\phi(t)]$ (6) of the two circles are also r_1 (7) and r_2 (7), respectively.

According to the definitions of vectors $\vec{\mathbf{X}}[\phi(t)]$ (19) - (20), $\vec{\mathbf{Y}}[\phi(t)]$ (21) - (22), $\vec{a}(t)$ (12), and $\vec{b}(t)$ (13), as described in (12) - (22), the relation in (28) is valid and holds for all $\phi(t)$ (3).

$$\frac{\vec{\mathbf{X}}[\phi(t)]\cdot\vec{\mathbf{X}}[\phi(t)]}{\vec{a}(t)\cdot\vec{a}(t)} + \frac{\vec{\mathbf{Y}}[\phi(t)]\cdot\vec{\mathbf{Y}}[\phi(t)]}{\vec{b}(t)\cdot\vec{b}(t)} = \frac{X^2(\omega_2 t)}{a^2(t)} + \frac{Y^2(\omega_2 t)}{b^2(t)} = Cos^2[\phi(t)] + Sin^2[\phi(t)] = 1 \quad (28)$$

$$\boxed{\frac{\vec{\mathbf{X}}[\phi(t)]\cdot\vec{\mathbf{X}}[\phi(t)]}{\vec{a}(t)\cdot\vec{a}(t)} + \frac{\vec{\mathbf{Y}}[\phi(t)]\cdot\vec{\mathbf{Y}}[\phi(t)]}{\vec{b}(t)\cdot\vec{b}(t)} = 1} \quad (Definition\ of\ Vector\ Ellipse\ in\ 3-D) \quad (29)$$

$$\boxed{\frac{X^2(\omega_2 t)}{a^2(t)}+\frac{Y^2(\omega_2 t)}{b^2(t)}=1} \quad (Definition\ of\ Scalar\ Ellipse\ in\ 2-D) \qquad (30)$$

Therefore, the relation in (28) reveals the validity of (29) and (30) for the vector pair $\{\vec{\mathbf{X}}[\phi(t)],\vec{\mathbf{Y}}[\phi(t)]\}$ (19) - (22) and its magnitude pair $[X(\omega_2 t),Y(\omega_2 t)]$ (19) - (22), respectively. As (30) is the defining equation of an ellipse[6] in two dimensions, where $a(t)$ (14) is the semi-major[7] axis and $b(t)$ (15) is the semi-minor[7] axis of the ellipse[6] in the case when $[a(t)>b(t)]$[6,7] holds, and vice versa, with (30) reducing to the special case of a circle when $[a(t)=b(t)]$[6,7] holds, we can also claim that (29) indicates that the vector pair $\{\vec{\mathbf{X}}[\phi(t)],\vec{\mathbf{Y}}[\phi(t)]\}$ (19) - (22) defines points on a vector ellipse in three dimensions for the Earth-Moon system, as also stated in the **ARTICLE 1** "Points on two circles in space mimic an ellipse with respect to a point".

We know throughout a respective cycle of the points $\mathbf{P}_1$ and $\mathbf{P}_2$ moving around their own circles of revolution with ***fixed*** but ***different*** (1) angular velocities ω_1 (2) and ω_2 (3), respectively, that the semi-major[7] and semi-minor[7] axis vectors $\vec{a}(t)$ (12) and $\vec{b}(t)$ (13) of their relative elliptical motion are sinusoidal functions of time t, and that the $(\vec{r}_1-\vec{r}_2)$ (18) vector has a vector value of $[\vec{a}(t)Cos(\omega_2 t)+\vec{b}(t)Sin(\omega_2 t)]$ (18) at each phase value of $[\phi(t)=\omega_2 t]$ (3), moving in the plane formed by the $\vec{a}(t)$ (12) and $\vec{b}(t)$ (13) vectors, namely the $\vec{a}(t)-\vec{b}(t)$ plane, which is a <u>variable</u> moving plane in three dimensions for this case.

Based on (10) and depending on the values of ω_1 (2), ω_2 (3), and angular frequencies in the variation of $\vec{\ell}[\phi(t)]$ (8), $d^2[\phi(t)]$ (25) and therefore $d(\phi)$ (23) is expected to vary according to a sinusoid based on the <u>higher</u> of the angular frequencies ω_1 (2), ω_2 (3), $(\omega_1-\omega_2)$ (4), and angular frequencies in the variation of $\vec{\ell}[\phi(t)]$ (8), within a sinusoidal distance envelope varying according to the <u>smaller</u> of angular frequencies ω_1 (2), ω_2 (3), $(\omega_1-\omega_2)$ (4), and angular frequencies in the variation of $\vec{\ell}[\phi(t)]$ (8).

The curve[4] of the distance between the centers of Earth and Moon over 700 days, plotted in **Figure 1**, is an example of such a d (25)-curve as a function of time t. In this Article, in the configuration described in **Figure 2**, as we have randomly chosen $Circle_1$ to be the circle of revolution of the Moon, and $Circle_2$ in the selected horizontal plane to be the circle of revolution of the Earth, $(\omega_1=\omega_{Moon})$ (2) is the angular frequency of the Moon's circular orbit around its own center of revolution, and $(\omega_2=\omega_{Earth})$ (3) is the angular frequency of the Earth's circular orbit around its own center of revolution in this configuration. As we know the orbital period[10,11] T_{Moon} (31) of the Moon[11] and the Earth[12]'s yearly[13] orbital period[10,12] T_{Earth} (32), we can calculate the values of Moon[11] and Earth[12]'s angular velocities $(\omega_1=\omega_{Moon})$ (34) and $(\omega_2=\omega_{Earth})$ (35), respectively, as well as their angular velocity difference $(\omega_1 - \omega_2)$ (36). We can also calculate the synodic period[10,11] $T_{(Moon-Earth)}$ (38) of the Moon[11] to an approximation using the relation in (37). Therefore, there are approximately 13.3687466 orbital[10,11] (33) cycles of the Moon[11] in one orbital[10,12] cycle of the Earth[12], and there are approximately 12.3687463 (39) synodic[10,11] cycles of the Moon[11] in one orbital[10,12] cycle of the Earth[12].

$$T_{Moon} = 27.321661\ days = 27\ days\ 7\ hours\ 43\ minutes\ 11.5\ seconds \quad (Moon's\ Orbital\ Period) \quad (31)$$

$$T_{Earth} = 365.256363004\ days = 365\ days\ 6\ hours\ 9\ minutes\ 9.76\ seconds\ (Earth's\ Orbital\ Period) \quad (32)$$

$$\frac{T_{Earth}}{T_{Moon}} = \frac{365.256363004\ days}{27.321661\ days} = 13.3687466 \quad (33)$$

$$\omega_1 = \omega_{Moon} = \frac{2\pi}{T_{Moon}} = \mathbf{0.229970839151382}\ radians/day \quad (Moon's\ Angular\ Velocity) \quad (34)$$

$$\omega_2 = \omega_{Earth} = \frac{2\pi}{T_{Earth}} = \mathbf{0.017202124161519}\ radians/day \quad (Earth's\ Angular\ Velocity) \quad (35)$$

$$\omega_1 - \omega_2 = \omega_{Moon} - \omega_{Earth} = \mathbf{0.212768714989863}\ radians/day \quad (Angular\ Velocity\ Difference) \quad (36)$$

$$\omega_{Moon} - \omega_{Earth} = \frac{2\pi}{T_{(\omega_{Moon}-\omega_{Earth})}} \quad (Moon-Earth\ Angular\ Velocity\ Difference) \quad (37)$$

$$T_{(Moon-Earth)} = \frac{2\pi}{(\omega_{Moon} - \omega_{Earth})} = 29.530589\ days = 29\ days\ 12\ hours\ 44\ minutes\ 2.9\ seconds \quad (38)$$

$$(Moon's\ Synodic\ Period)$$

$$\frac{T_{Earth}}{T_{(Moon-Earth)}} = \frac{365.256363004\ days}{29.530589\ days} = 12.3687463 \quad (39)$$

Over a 2π cycle of $[\phi_0(t) = (\omega_1 - \omega_2)t]$ (4), squares of the semi-major[7] and semi-minor[7] axis magnitudes, namely $[a^2(t)]$ (14) and $[b^2(t)]$ (15), respectively, take their minimum $(a^2_{\min}, b^2_{\min})$ (40) and maximum $(a^2_{\max}, b^2_{\max})$ (40) values when $\{Cos[(\omega_1 - \omega_2)t] = \pm 1\}$ (40), in other words when points $\mathbf{P}_1$ (Moon) and $\mathbf{P}_2$ (Earth) moving around their own circles with ***fixed*** but ***different*** (1) angular velocities $(\omega_1 = \omega_{Moon})$ (2) and $(\omega_2 = \omega_{Earth})$ (3) are either aligned in phase $\{Cos[(\omega_1 - \omega_2)t] = +1\}$ (40) or opposite in phase $\{Cos[(\omega_1 - \omega_2)t] = -1\}$ (40).

$$
\begin{cases}
Cos\left[(\omega_1-\omega_2)t\right]=+1 \;\Rightarrow\; \begin{cases}
a^2(t)=r_1^2-2r_1r_2\,Cos\beta+r_2^2=\begin{cases} a_{\min}^2 & if\quad Cos\beta>0 \\ a_{\max}^2 & if\quad Cos\beta<0 \end{cases} \\
b^2(t)=r_1^2-2r_1r_2+r_2^2=(r_1-r_2)^2=b_{\min}^2 \\
\phi_0(t)=(\omega_1-\omega_2)t=2n\pi \qquad [n\ integer] \\
\mathbf{P}_1\ and\ \mathbf{P}_2\ are\ aligned\ in\ phase
\end{cases} \\
Cos\left[(\omega_1-\omega_2)t\right]=-1 \;\Rightarrow\; \begin{cases}
a^2(t)=r_1^2+2r_1r_2\,Cos\beta+r_2^2=\begin{cases} a_{\max}^2 & if\quad Cos\beta>0 \\ a_{\min}^2 & if\quad Cos\beta<0 \end{cases} \\
b^2(t)=r_1^2+2r_1r_2+r_2^2=(r_1+r_2)^2=b_{\max}^2 \\
\phi_0(t)=(\omega_1-\omega_2)t=(2n+1)\pi \qquad [n\ integer] \\
\mathbf{P}_1\ and\ \mathbf{P}_2\ are\ opposite\ in\ phase
\end{cases}
\end{cases}
\tag{40}
$$

Further to our analysis in (40), whenever $\{Cos\left[(\omega_1-\omega_2)t\right]=0\}$ (41) over a 2π cycle of $\left[\phi_0(t)=(\omega_1-\omega_2)t\right]$ (4), or if $(Cos\beta=1)$ (42) for any $\left[\phi_0(t)=(\omega_1-\omega_2)t\right]$ (4), $\left[a^2(t)\right]$ (14) and $\left[b^2(t)\right]$ (15) have equal values, as for magnitudes of radius vectors of a circle.

$$
Cos\left[(\omega_1-\omega_2)t\right]=0 \quad\Rightarrow\quad a^2(t)=b^2(t)=r_1^2+r_2^2 \tag{41}
$$

$$
Cos\beta=1 \;\Rightarrow\; \beta=0^\circ \;\Rightarrow\; a^2(t)=b^2(t)=r_1^2-2r_1r_2Cos\left[(\omega_1-\omega_2)t\right]+r_2^2 \tag{42}
$$

Inspecting $d^2\left[\phi(t)\right]$ (25) more closely, also utilizing $\left[b^2(t)\right]$ (15), we obtain (43) which is a superposed sinusoidal function of $(\omega_1 t)$ (2), $(\omega_2 t)$ (3), and $\left[(\omega_1-\omega_2)t\right]$ (4). As it can clearly be observed that the curve[4] of distance between the centers of Earth and Moon over 700 days plotted in **Figure 1** is such a d (25)-curve as a function of time t, we can utilize (43) to analyze Earth-Moon system in more detail, using the values of the Moon[11] and the Earth[12]'s angular frequencies $(\omega_1=\omega_{Moon})$ (34) and $(\omega_2=\omega_{Earth})$ (35), respectively, as well as their angular velocity difference $(\omega_1-\omega_2)$ (36), together with Earth-Moon distance[5] data provided in timeanddate.com website.

$$
\begin{aligned}
d^2\left[\phi(t)\right]=d^2(\omega_2 t)=r_1^2-2r_1r_2\,Cos\left[(\omega_1-\omega_2)t\right]+r_2^2+2r_1r_2(1-Cos\beta)Cos(\omega_1 t)Cos(\omega_2 t) \\
+2\left\{\vec{\boldsymbol{a}}(t)Cos(\omega_2 t)+\vec{\boldsymbol{b}}(t)Sin(\omega_2 t)\right\}\cdot\vec{\ell}(\omega_2 t)+\ell^2(\omega_2 t)
\end{aligned}
\tag{43}
$$

In **Table 1**, the minimum and maximum Earth-Moon distances[5] are listed, that occur between the dates of January 10, 2017 and January 9, 2021, based on the Earth-Moon distance[5] data provided in timeanddate.com website. In the same **Table 1**, the periods between consecutive minimum and maximum Earth-Moon distances[5] are also provided, as well as periods between consecutive minimums and periods between consecutive maximums of Earth-Moon distance[5], in terms of days and hours. Taking an average of the periods between consecutive minimums, and periods between consecutive maximums for the dates between January 10, 2017 and January 9, 2021 for the Earth-Moon system, we obtain an average period of 27 *days* 17 *hours* 23 *minutes* 22 *seconds* for the highest frequency variation sinusoid, which is very close to T_{Moon} (31) of the Moon[11] that is $27\ days\ 7\ hours\ 43\ minutes\ 11.5\ seconds$. Looking at **Figure 1**, as well as the data listed in **Table 1**, we observe that there are a little more than 13 cycles of a sinusoid in approximately 1 cycle of an envelope sinusoid, for the Earth-Moon system. As there are approximately 13.3687466 orbital[10,11] (33) cycles T_{Moon} (31) of the Moon[11] in one orbital[10,12] cycle T_{Earth} (32) of the Earth[12], and there are approximately 12.3687463 (39) synodic[10,11] cycles $T_{(Moon-Earth)}$ (38) of the Moon[11] in one orbital[10,12] cycle T_{Earth} (32) of the Earth[12], we can reach the conclusion that mainly the $(\omega_1 t)$ (2) component of $d^2\left[\phi(t)\right]$ (43) determines the frequency of the distance variation of Earth-Moon system within a distance envelope of angular velocity ω_2 (3), and that the $\left[Cos(\omega_1 t)Cos(\omega_2 t)\right]$ term determines the maximum and minimum values of $d^2\left[\phi(t)\right]$ (43) in its sinusoidal cycles, mainly the $Cos(\omega_1 t)$ term, which must be at dates when the Moon's phase $(\omega_1 t)$ (2) is a multiple of π, with a global maximum or minimum Earth-Moon distance[5] when the Earth's phase $(\omega_2 t)$ (3) is close to a multiple of π at the same time. In other words, the angular velocity of the Moon's circular orbit around its own center of revolution $(\omega_1 = \omega_{Moon})$ (34) determines the higher frequency distance variation of the Earth-Moon system, and the angular velocity of Earth's circular orbit around its own center of revolution $(\omega_2 = \omega_{Earth})$ (35) determines the frequency of Earth-Moon distance[5] envelope sinusoid, an important observation result in this Article which we will use in our further analysis.

Table 1 Minimum-Maximum Earth-Moon distances[5] in kilometers (km) between 2017-2021

Earth-Moon Distance[5] Data over Time								
Day	Date	Time (UTC)	Earth-Moon Distance[5] [km]	Notes	Difference from Previous Date [MIN-MAX] / [MAX-MIN]		Difference from Previous Date [MAX-MAX] / [MIN-MIN]	
					Days	+ Time (hrs)	Days	+ Time (hrs)
1	10.01.2017	06:01	363.238	MIN (Local)				
13	22.01.2017	00:13	404.914	**MAX (Local)**	11	18:12		
28	06.02.2017	14:02	368.816	MIN (Local)	15	13:49	27	08:01
40	18.02.2017	21:13	404.376	**MAX (Local)**	12	07:11	27	21:00
53	03.03.2017	07:33	369.062	MIN (Local)	12	10:20	24	17:31
68	**18.03.2017**	**17:25**	**404.650**	**MAX (Local)**	15	09:52	27	20:12
80	30.03.2017	12:32	363.854	MIN (Local)	11	19:07	27	04:59
96	15.04.2017	10:04	405.475	**MAX (Local)**	15	21:32	27	16:39
108	27.04.2017	16:14	359.327	MIN (Local)	12	06:10	28	03:42
123	12.05.2017	19:51	406.210	**MAX (Local)**	15	03:37	27	09:47
137	26.05.2017	01:20	357.208	MIN (Local)	13	05:29	28	09:06
150	08.06.2017	22:21	406.401	**MAX (Local)**	13	21:01	27	02:30

165	23.06.2017	10:50	357.937	MIN (Local)	14	12:29	28	09:30
178	06.07.2017	04:28	405.934	**MAX (Local)**	12	17:38	27	06:07
193	21.07.2017	17:11	361.236	MIN (Local)	15	12:43	28	06:21
205	02.08.2017	17:54	405.025	**MAX (Local)**	12	00:43	27	13:26
221	18.08.2017	13:19	366.121	MIN (Local)	15	19:25	27	20:08
233	30.08.2017	11:23	404.308	**MAX (Local)**	11	22:04	27	17:29
247	13.09.2017	16:06	369.860	MIN (Local)	14	04:43	26	02:47
261	27.09.2017	06:50	404.348	**MAX (Local)**	13	14:44	27	19:27
273	09.10.2017	05:54	366.855	MIN (Local)	11	23:04	25	13:48
289	25.10.2017	02:26	405.154	**MAX (Local)**	15	20:32	27	19:36
301	06.11.2017	00:10	361.438	MIN (Local)	11	21:44	27	18:16
316	21.11.2017	18:53	406.132	**MAX (Local)**	15	18:43	27	16:27
329	04.12.2017	08:45	357.492	MIN (Local)	12	13:52	28	08:35
344	**19.12.2017**	**01:25**	**406.603**	**MAX (Local)**	14	16:40	27	06:32
357	**01.01.2018**	**21:48**	**356.565**	**MIN (Global)**	13	20:23	28	13:03
371	15.01.2018	02:09	406.464	**MAX (Local)**	13	04:21	27	00:44
386	30.01.2018	09:56	358.993	MIN (Local)	15	07:47	28	12:08
398	11.02.2018	14:16	405.700	**MAX (Local)**	12	04:20	27	12:07
414	27.02.2018	14:39	363.932	MIN (Local)	16	00:23	28	04:43
426	11.03.2018	09:13	404.678	**MAX (Local)**	11	18:34	27	18:57
441	26.03.2018	17:16	369.106	MIN (Local)	15	08:03	27	02:37
454	08.04.2018	05:31	404.144	**MAX (Local)**	12	12:15	27	20:18
466	20.04.2018	14:41	368.714	MIN (Local)	12	09:10	24	21:25
482	06.05.2018	00:35	404.457	**MAX (Local)**	15	09:54	27	19:04
493	17.05.2018	21:04	363.776	MIN (Local)	11	20:29	27	06:23
509	02.06.2018	16:34	405.317	**MAX (Local)**	15	19:30	27	15:59
521	14.06.2018	23:52	359.503	MIN (Local)	12	07:18	28	02:48
537	30.06.2018	02:43	406.061	**MAX (Local)**	15	02:51	27	10:09
550	13.07.2018	08:24	357.431	MIN (Local)	13	05:41	28	08:32
564	27.07.2018	05:43	406.223	**MAX (Local)**	13	21:19	27	03:00
578	10.08.2018	18:06	358.078	MIN (Global)	14	12:23	28	09:42
591	23.08.2018	11:22	405.746	**MAX (Local)**	12	17:16	27	05:39
607	08.09.2018	01:19	361.351	MIN (Local)	15	13:57	28	07:13
619	**20.09.2018**	**00:53**	**404.876**	**MAX (Local)**	11	23:34	27	13:31
634	05.10.2018	22:26	366.392	MIN (Local)	15	21:33	27	21:07
646	17.10.2018	19:15	404.227	**MAX (Local)**	11	20:49	27	18:22
660	31.10.2018	20:23	370.204	MIN (Local)	14	01:08	25	21:57
674	14.11.2018	15:55	404.339	**MAX (Local)**	13	19:32	27	20:40
686	26.11.2018	12:12	366.620	MIN (Local)	11	20:17	25	15:49
702	12.12.2018	12:25	405.177	**MAX (Local)**	16	00:13	27	20:30
714	24.12.2018	09:48	361.061	MIN (Local)	11	21:23	27	21:36

730	09.01.2019	04:28	406.117	**MAX (Local)**	15	18:40	27	16:03
742	21.01.2019	19:59	357.342	MIN (Local)	12	15:31	28	10:11
757	05.02.2019	09:28	406.555	**MAX (Local)**	14	13:29	27	05:00
771	19.02.2019	09:02	356.761	MIN (Local)	13	23:34	28	13:03
784	04.03.2019	11:26	406.391	**MAX (Local)**	13	02:24	27	01:58
799	19.03.2019	19:47	359.377	MIN (Local)	15	08:21	28	10:45
812	01.04.2019	00:13	405.577	**MAX (Local)**	12	04:26	27	12:47
827	16.04.2019	22:04	364.205	MIN (Local)	15	21:51	28	02:17
839	28.04.2019	18:19	404.582	**MAX (Local)**	11	20:15	27	18:06
854	13.05.2019	21:52	369.009	MIN (Local)	15	03:33	26	23:48
867	26.05.2019	13:27	404.138	**MAX (Local)**	12	15:35	27	19:08
879	07.06.2019	23:15	368.504	MIN (Local)	12	09:48	25	01:23
895	**23.06.2019**	**07:49**	**404.548**	**MAX (Local)**	15	08:34	27	18:22
907	05.07.2019	05:00	363.726	MIN (Local)	11	21:11	27	05:45
922	20.07.2019	23:58	405.481	**MAX (Local)**	15	18:58	27	16:09
935	02.08.2019	07:11	359.398	MIN (Local)	12	07:13	28	02:11
950	17.08.2019	10:49	406.244	**MAX (Local)**	15	03:38	27	10:51
963	30.08.2019	15:53	357.176	MIN (Local)	13	05:04	28	08:42
977	13.09.2019	13:32	406.377	**MAX (Local)**	13	21:39	27	02:43
992	28.09.2019	02:24	357.802	MIN (Local)	14	12:52	28	10:31
1004	10.10.2019	18:28	405.899	**MAX (Local)**	12	16:04	27	04:56
1020	26.10.2019	10:38	361.311	MIN (Local)	15	16:10	28	08:14
1032	07.11.2019	08:35	405.058	**MAX (Local)**	11	21:57	27	14:07
1048	23.11.2019	07:40	366.716	MIN (Local)	15	23:05	27	21:02
1060	05.12.2019	04:08	404.446	**MAX (Local)**	11	20:28	27	19:33
1073	18.12.2019	20:25	370.265	MIN (Local)	13	16:17	25	12:45
1088	02.01.2020	01:30	404.580	**MAX (Local)**	14	05:05	27	21:22
1099	13.01.2020	20:20	365.958	MIN (Local)	11	18:50	25	23:55
1115	29.01.2020	21:26	405.393	**MAX (Local)**	16	01:06	27	19:56
1127	10.02.2020	20:27	360.461	MIN (Local)	11	23:01	28	00:07
1143	26.02.2020	11:34	406.278	**MAX (Local)**	15	15:07	27	14:08
1156	10.03.2020	06:30	357.122	MIN (Local)	12	18:56	28	10:03
1170	**24.03.2020**	**15:23**	**406.692**	MAX (Global)	14	08:53	27	03:49
1184	07.04.2020	18:08	356.906	MIN (Local)	14	02:45	28	11:38
1197	20.04.2020	19:00	406.462	**MAX (Local)**	13	00:52	27	03:37
1213	06.05.2020	03:03	359.654	MIN (Local)	15	08:03	28	08:55
1225	18.05.2020	07:44	405.583	**MAX (Local)**	12	04:41	27	12:44
1241	03.06.2020	03:38	364.366	MIN (Local)	15	19:54	28	00:35
1253	15.06.2020	00:56	404.595	**MAX (Local)**	11	21:18	27	17:12
1268	30.06.2020	02:12	368.598	MIN (Local)	15	01:16	26	22:34
1280	12.07.2020	19:26	404.199	**MAX (Local)**	12	17:14	27	18:30

1293	25.07.2020	05:01	368.361	MIN (Local)	12	09:35	25	02:49
1308	09.08.2020	13:50	404.659	**MAX (Local)**	15	08:49	27	18:24
1320	21.08.2020	10:55	363.513	MIN (Local)	11	21:05	27	05:54
1336	06.09.2020	06:29	405.607	**MAX (Local)**	15	19:34	27	16:39
1348	18.09.2020	13:48	359.082	MIN (Local)	12	07:19	28	02:53
1363	03.10.2020	17:22	406.321	**MAX (Local)**	15	03:34	27	10:53
1376	16.10.2020	23:46	356.912	MIN (Local)	13	06:24	28	09:58
1390	30.10.2020	18:45	406.394	**MAX (Local)**	13	18:59	27	01:23
1405	14.11.2020	11:42	357.837	MIN (Local)	14	16:57	28	11:56
1418	27.11.2020	00:28	405.894	**MAX (Local)**	12	12:46	27	05:43
1433	12.12.2020	20:41	361.773	MIN (Local)	15	20:13	28	08:59
1445	24.12.2020	16:31	405.012	**MAX (Local)**	11	19:50	27	16:03
1461	09.01.2021	15:36	367.387	MIN (Local)	15	23:05	27	18:55
						AVERAGE	27	17:32:22

Using $\vec{\ell}[\phi(t)]$ (8), $\vec{\boldsymbol{a}}(t)$ (12), and $\vec{\boldsymbol{b}}(t)$ (13), $d^2[\phi(t)]$ (43) can be expressed in a more detailed way as in $d^2[\phi(t)]$ (44) to visualize all angular velocity components involved.

$$\begin{aligned} d^2[\phi(t)] &= r_1^2 - 2r_1 r_2 Cos[(\omega_1-\omega_2)t] + r_2^2 + 2r_1 r_2 (1-Cos\beta)Cos(\omega_1 t)Cos(\omega_2 t) + \ell^2(\omega_2 t) \\ &+2\begin{Bmatrix} \{r_1 Cos\beta\, Cos[(\omega_1-\omega_2)t] - r_2\}\ell_x(\omega_2 t) \\ +r_1 Sin[(\omega_1-\omega_2)t]\ell_y(\omega_2 t) \\ +r_1 Sin\beta\, Cos[(\omega_1-\omega_2)t]\ell_z(\omega_2 t) \end{Bmatrix} Cos(\omega_2 t) \\ &+2\begin{Bmatrix} -r_1 Cos\beta\, Sin[(\omega_1-\omega_2)t]\ell_x(\omega_2 t) \\ +\{r_1 Cos[(\omega_1-\omega_2)t] - r_2\}\ell_y(\omega_2 t) \\ -r_1 Sin\beta\, Sin[(\omega_1-\omega_2)t]\ell_z(\omega_2 t) \end{Bmatrix} Sin(\omega_2 t) \end{aligned} \quad (44)$$

The variation frequencies of $\vec{\boldsymbol{a}}(t)$ (12) and $\vec{\boldsymbol{b}}(t)$ (13) are based on the angular velocity difference $(\omega_1-\omega_2)$ (36), that is, it depends on the synodic period[10,11] $T_{(Moon-Earth)}$ (38) of the Earth-Moon[11] system. However, looking at **Figure 1**, as well as the data listed in **Table 1**, we do not observe an effective frequency impact of the synodic motion[10,11] of the Earth-Moon[11] system at every 2π cycle of $[(\omega_1-\omega_2)t]$ (36) on the Earth-Moon distance[5] based on $d^2[\phi(t)]$ (44). Therefore, the Dot Product[9] of the vector distance $\vec{\ell}[\phi(t)]$ (8) between the centers of individual circular revolutions of the Earth and the Moon, with $\vec{\boldsymbol{a}}(t)$ (12) and $\vec{\boldsymbol{b}}(t)$ (13), does not seem to have too much impact on the Earth-Moon distance[5] variation. As a result, we can infer, to an approximation, that either $\vec{\ell}[\phi(t)]$ (8) has a very small magnitude variation $\ell[\phi(t)]$ (9) such that $\vec{\ell}[\phi(t)]$ (8) is almost a constant vector over time t compared to the individual motion frequencies of the Earth and the Moon, and/or $\vec{\ell}\left[\phi(t)\right]$ (8) itself also varies with the same frequency and angular velocity $\left(\omega_1 = \omega_{Moon}\right)$ (34) of the Moon, which is the dominant variation frequency of Earth-Moon distance. We shall investigate the details of this subject in this Article. Note that in the **ARTICLE 3** "Analytical calculation of earth and sun orbital parameters from distance data", the magnitude r_2 (7) of the vector radius $\vec{r}_2\left[\phi(t)\right]$ (6) of the Earth in its individual circular orbit of revolution in space is already found to be r_2 (45), the magnitude of the radius of individual circular orbit of revolution of the Sun is found to be as r_{Sun} (46), and the inclination angle of the plane of the individual circular revolution of the Sun with respect to the plane of the individual circular revolution of the Earth is also found as β_{Sun} (47), as well as the North Pole vector of the Sun (48) and the North Pole vector of the Earth (49) in the configuration of **Figure 2**, whereas $\phi_{0,Earth-Sun}$ (50) is the constant phase difference between the Sun and the Earth in their individual orbits of revolution throughout a year.

$$r_2 = r_{Earth} = \mathbf{0.0000826915340950721}\, au = \mathbf{12,370.477}\, km \;\left(Earth\ individual\ revolution\ radius\right) \quad (45)$$

$$r_{Sun} = \mathbf{1.00007869128446}\, au = \mathbf{149,609,642.749}\, km \;\left(radius\ of\ Sun's\ individual\ revolution\right) \quad (46)$$

$$\beta_{Sun} = \mathbf{-156^{\circ}.56} = \mathbf{-2.73251312735455}\, radians \left(angle\ between\ Sun\,\&\,Earth\ revolution\ planes\right) \quad (47)$$

$$\hat{\boldsymbol{u}}_{Sun\perp} = -\hat{\boldsymbol{x}} Sin\beta_{Sun} + \hat{\boldsymbol{z}} Cos\beta_{Sun} \quad \left(North\ Pole\ of\ Sun\ normal\ to\ Sun's\ circular\ revolution\ plane\right) \quad (48)$$

$$\hat{\boldsymbol{u}}_{Earth\perp} = -\hat{\boldsymbol{z}} \quad \left(North\ Pole\ of\ Earth\ normal\ to\ Earth's\ circular\ revolution\ plane\right) \quad (49)$$

$$\phi_{0,Earth-Sun} = \mathbf{-2.63736997198951}\, rad = \mathbf{-151^{\circ}.11} \;\left(Phase\ difference\ for\ Sun - Earth\ system\right) \quad (50)$$

The observed trajectory of the Moon[11,19] around the Sun throughout a year is quite consistent and is somewhat as in **Figure 3**, exaggerated for purposes of visualization and not to scale, as it moves around the Earth with approximately 13.3687466 orbital[10,11] (33) cycles T_{Moon} (31) of the Moon[11] in one orbital[10,12] cycle T_{Earth} (32) of the Earth[12]. Moon phases[11,18] occur over every synodic period[10,11] $T_{(Moon-Earth)}$ (38) of the Moon[11], observed[30] as sketched in **Figure 4**. Therefore, we can conclude that the Moon[11]'s Orbit[19] is quite fixed over a Lunar[11,33] Month, and moves together with Earth[12] in its individual circular orbital[10,12] cycle of period T_{Earth} (32), leading to the Moon[11]'s observed relative motion around the Sun over a year as in **Figure 3**. As a result, we can infer that the Earth-Moon distance[5,11] $\vec{d}\left[\phi(t)\right]$ (10) consists of two main components that can be expressed as $\vec{d}\left[\phi(t)\right]$ (51), one component that is based on $\vec{r}_2\left[\phi(t)\right]$ (6) due to the individual circular orbit of revolution of Earth[12] in its yearly cycle of angular velocity $\left(\omega_2 = \omega_{Earth}\right)$ (35), and another component $\vec{d}_{MoonOrbit}\left[\phi(t)\right]$ (51) that is due to Moon[11] Orbit[19], which in turn is based on $\vec{r}_1\left[\phi(t)\right]$ (5) due to the individual circular orbit of revolution of the Moon[11] in its lunar cycle of angular velocity $\left(\omega_1 = \omega_{Moon}\right)$ (34), as well as the vector distance $\vec{\ell}\left[\phi(t)\right]$ (8) between centers of individual circular orbits of revolution of Earth and Moon. This is another fundamental analysis result set forth in this Article.

$$\vec{d}\left[\phi(t)\right]=\vec{d}_{MoonOrbit}\left[\phi(t)\right]-\vec{r}_2\left[\phi(t)\right] \quad ; \quad \vec{d}_{MoonOrbit}\left[\phi(t)\right]=\vec{r}_1\left[\phi(t)\right]+\vec{\ell}\left[\phi(t)\right] \quad (51)$$

Therefore, the above analysis and $\vec{d}_{MoonOrbit}\left[\phi(t)\right]$ (51) tells us that "individual circular orbit of revolution of the Moon with radius vector $\vec{r}_1\left[\phi(t)\right]$ (5)" as a whole is seemingly revolving around the Earth with an instantanous vector radius of $\vec{\ell}\left[\phi(t)\right]$ (8) at each $\phi(t)$ (3), during every orbital period[10,11] T_{Moon} (31) of the Moon[11], or vice versa, so that the Moon and the Earth move in a relative trajectory in which the Moon passes between the Sun and the Earth once every Lunar[11,33] Month and the Earth passes between the Sun and Moon once every Lunar[11,33] Month, as visualized in **Figure 3**, and the Moon phases[11,18] occur over every synodic period[10,11] $T_{(Moon-Earth)}$ (38) of the Moon[11], observed[30] as sketched in **Figure 4**. In this sense, the Earth and the Moon must be forming a twin or binary[20] system of bodies in space, whose individual circular orbits of revolution also revolve around each other with an instantaneous vector radius of $\vec{\ell}\left[\phi(t)\right]$ (8) at each $\phi(t)$ (3), which is an already observed[19] phenomenon. This Earth-Moon motion topology assertion, such that the "individual circular orbit of revolution of the Moon with radius vector $\vec{r}_1\left[\phi(t)\right]$ (5)" as a whole is seemingly revolving around the Earth with an instantanous vector radius of $\vec{\ell}\left[\phi(t)\right]$ (8) at each $\phi(t)$ (3), is also a significant finding of this Article, which is compliant with observation[19], supporting our findings and assertions in this Article about the relative motions of the Earth and the Moon.

We also know that the Moon[11] is in synchronous[19] rotation with Earth, and thus always shows the same side to Earth, and that due to libration[19,21,22] there is a wobble or wavering of the Moon perceived by Earth-bound observers throughout a lunar[11,19,33] month of period T_{Moon} (31), and slightly more than half, which is about 59%, of the total lunar surface can be viewed[11] from Earth. In light of our conclusion above about Earth-Moon motion topology, we can infer that the lunar[11] monthly libration[21,22] of the Moon observed from Earth is mainly due to the individual circular orbit of revolution of the Moon around a radius of $\vec{r}_1\left[\phi(t)\right]$ (5), and the more-than-half view[11] of the lunar surface from Earth is apparently based on the relative view angle of the Earth-bound observers, based on *both* the direction of $\vec{\ell}\left[\phi(t)\right]$ (8) vector and the position of the Moon in its individual circular orbit of revolution around a vector radius of $\vec{r}_1\left[\phi(t)\right]$ (5), on that date throughout a cycle, which is another fundamental analysis result of this Article. We further infer that the self-rotation of the Moon[11] has the same period T_{Moon} (31) and angular velocity $(\omega_1=\omega_{Moon})$ (34) as its individual circular orbit of revolution.

The Moon[11] is observed to move in counterclockwise[35] direction as seen from a vantage point from above the North Pole[23,34] of the Earth throughout a lunar[11,33] month of period T_{Moon} (31). Therefore, based on our analysis above, the Moon[11] Orbit[19] vector $\vec{d}_{MoonOrbit}\left[\phi(t)\right]$ (51) directed from Earth towards the Moon must be moving in counterclockwise[35] direction as seen from a vantage point from above the North Pole[23,34] of the Earth, in its plane of motion. Further, based on our topology analysis above as well as observation[19], in order to maintain its synchronous[19] rotation with Earth, the Moon[11]'s self-rotation must also be in counterclockwise[35] direction from a vantage point from above the North Pole[23,34] of the Earth.

Additionally, the libration[19,21,22] of the Moon perceived by an Earth-bound observer aligned with the North Pole[23,34] direction of the Earth is observed[21,22] seemingly clockwise[35] throughout a lunar[11,33] month of period T_{Moon} (31). Therefore, we also can also infer that the Moon's individual circular orbit of revolution around radius $\vec{r}_1[\phi(t)]$ (5) is in clockwise[35] direction from a vantage point from above the North Pole[23,34] of the Earth, yet another finding of this Article.

Based on the assertion in the **ARTICLE 3** "Analytical calculation of earth and sun orbital parameters from distance data", all moving bodies in the Universe must be obeying "Faraday's Law of Induction"[24] in their motions, i.e. each one is possibly moving in space in external magnetic fields with changing flux[25] through the area of their individual circle of revolution, which is the cause that triggers their self rotation in the reverse direction, as seen from a vantage point above their North Pole[23,26,34], where we always accept the North Pole[23,26,34] of a body to be the pole rotating in counterclockwise[35] direction according to "right hand rule"[27]. As seen from a vantage point from above the North Pole[23,34] of the Earth, the Moon[11]'s observed *clockwise*[35] motion in its individual circular orbit of revolution around vector radius $\vec{r}_1[\phi(t)]$ (5) and its *counterclockwise*[35] self-rotation synchronous[19] with Earth are in alignment with this assertion. This is a very important finding of this Article, reinforcing our assertions with observation.

To better understand the behavior of $\vec{\ell}[\phi(t)]$ (8) between centers of individual circular orbits of revolution of the Earth and the Moon, which we assert is a determining component of the Moon[11] Orbit[19] based on $\vec{d}_{MoonOrbit}[\phi(t)]$ (51), we need to check observation. Based on observation[19], the shape of Moon[11]'s Orbit[19] is pretty much fixed in its plane of motion over a lunar[11,19,33] month of period T_{Moon} (31), but the *orientation* of the Moon[11]'s Orbit[19] with respect to the Earth and Sun is *not* fixed in space, and moves over time. As the Moon[11]'s motion in its individual circular orbit of revolution around vector radius $\vec{r}_1[\phi(t)]$ (5) has an angular velocity of $(\omega_1 = \omega_{Moon})$ (34), the observation[19] that the shape of Moon[11]'s Orbit[19] is pretty much fixed in its plane of motion over a lunar[11,19,33] month of period T_{Moon} (31) leads us to determine, based on $\vec{d}_{MoonOrbit}[\phi(t)]$ (51), that the $\ell[\phi(t)]$ (9) magnitude cycle of $\vec{\ell}[\phi(t)]$ (8) over a lunar[11,19,33] month of period T_{Moon} (31) must also be pretty much fixed in its plane of motion, and must also be changing with an angular velocity of $(\omega_1 = \omega_{Moon})$ (34) with a phase shift of $[-\phi_\ell(t)]$ (52) over time t, as described in $\ell[\omega_1 t - \phi_\ell(t)]$ (52). The nature of $[-\phi_\ell(t)]$ (52) is to be determined based on observation. Apart from this lunar[11,19,33] monthly variation of Moon[11]'s motion in its individual circular orbit of revolution around vector radius $\vec{r}_1[\phi(t)]$ (5) and the lunar[11,19,33] monthly $\ell[\omega_1 t - \phi_\ell(t)]$ (52) almost-fixed magnitude cycle of $\vec{\ell}[\phi(t)]$ (8), we are now able to consider any other variations in the orientation of Moon[11]'s Orbit[19] in space over different periods, i.e. with angular velocities other than $(\omega_1 = \omega_{Moon})$ (34), to be due to *only* orientation variations in $\vec{\ell}[\phi(t)]$ (8), based on $\vec{d}_{MoonOrbit}[\phi(t)]$ (51).

$$\ell\left[\phi(t)\right]=\ell\left[\omega_1 t-\phi_\ell(t)\right]=\left|\vec{\ell}\left[\omega_1 t-\phi_\ell(t)\right]\right|$$
$$=\sqrt{\ell_x^2\left[\omega_1 t-\phi_\ell(t)\right]+\ell_y^2\left[\omega_1 t-\phi_\ell(t)\right]+\ell_z^2\left[\omega_1 t-\phi_\ell(t)\right]} \quad (52)$$

One orbital precession of the Moon[11]'s Orbit[19] is called the apsidal[32] precession, and is the rotation of the Moon[11]'s Orbit[19] within its orbital plane, i.e. the axes of its apparent elliptical orbit change direction. The Moon[11]'s major axis[7] – the longest diameter of the orbit, joining its nearest and farthest points, the perigee[36] and apogee[36], respectively – makes[19] one revolution every 8.85 Earth[12] years, or 3,232.6054 days, as it rotates slowly in the same direction as the self-rotation of the Moon[11] itself, which is in *counterclockwise*[35] direction from a vantage point from above the North Pole[23,34] $\left(\hat{\boldsymbol{u}}_{Earth\perp}=-\hat{z}\right)$ (49) of the Earth[12] in the configuration of **Figure 2**. Considering that the Moon[11] Orbit[19]'s major axis[7] revolves to come back to the same position in 8.85 Earth[12] years, or 3,232.6054 days, in which case the position of the perigee[36] and apogee[36] are reversed with respect to the so called ecliptic[29] plane, the complete period for the perigee[36] and apogee[36] to come back to the same position with respect to the so called ecliptic[29] plane, namely one full apsidal[32] precession cycle, must be double this time, that is 17.7 Earth[12] years, or 6,465.2108 days. Therefore, we can expect $\vec{\boldsymbol{d}}_{MoonOrbit}\left[\phi(t)\right]$ (51) to have an apsidal variation over a period of $T_{A,\vec{\boldsymbol{d}}_{MoonOrbit}\left[\phi(t)\right]}$ (53), with angular velocity $\omega_{A,\vec{\boldsymbol{d}}_{MoonOrbit}\left[\phi(t)\right]}$ (54). Based on our analysis result above, as the Moon[11]'s motion in its individual circular orbit of revolution around the vector radius $\vec{\boldsymbol{r}}_1\left[\phi(t)\right]$ (5) has an angular velocity of $\left(\omega_1=\omega_{Moon}\right)$ (34), and as $\vec{\ell}\left[\phi(t)\right]$ (8) has an almost fixed magnitude cycle of $\ell\left[\omega_1 t-\phi_\ell(t)\right]$ (52) over a lunar[11,19,33] month of period T_{Moon} (31), any other motion variation of $\vec{\boldsymbol{d}}_{MoonOrbit}\left[\phi(t)\right]$ (51) with an angular velocity different from $\left(\omega_1=\omega_{Moon}\right)$ (34), such as this observed apsidal[32] precession of $\vec{\boldsymbol{d}}_{MoonOrbit}\left[\phi(t)\right]$ (51) with angular velocity $\omega_{A,\vec{\boldsymbol{d}}_{MoonOrbit}\left[\phi(t)\right]}$ (54), must be due to an orientation variation of $\vec{\ell}\left[\phi(t)\right]$ (8). As a result, we can infer, to an approximation, that the variation in the orientation of the $\vec{\ell}\left[\phi(t)\right]$ (8) cycle, or $\ell\left[\omega_1 t-\phi_\ell(t)\right]$ (52) cycle, is very small over a couple of consecutive lunar[11,19,33] months of period T_{Moon} (31), compared to the individual motion frequencies of the Earth and the Moon, another important analysis result of this Article. Further, the *negative* $\left[-\phi_\ell(t)\right]$ (52) phase shift in $\ell\left[\omega_1 t-\phi_\ell(t)\right]$ (52) lunar[11,19,33] orbit magnitude cycle orientation we have chosen is in alignment with the observed *counterclockwise*[35] revolution of Moon[11] Orbit[19]'s major axis[7] from a vantage point from above North Pole[23,34] $\left(\hat{\boldsymbol{u}}_{Earth\perp}=-\hat{z}\right)$ (49) of the Earth[12] in the configuration of **Figure 2**. Following along the lines of the assertion in the **ARTICLE 3** "Analytical calculation of earth and sun orbital parameters from distance data" that all moving bodies in the Universe must be obeying "Faraday's Law of Induction"[24] in their motions, we also assert that each revolution in space, which is in external magnetic fields with changing flux[25] through the area of this revolution, triggers a rotation in the reverse direction. Hence, as Moon[11] and Earth[12] revolve in *clockwise*[35] directions in their individual circular orbits of revolution, as seen from a vantage point from above North Pole[23,34] $\left(\hat{\boldsymbol{u}}_{Earth\perp}=-\hat{z}\right)$ (49) of the Earth[12] in the configuration of

Figure 2, with angular velocities $(\omega_1 = \omega_{Moon})$ (34) and $(\omega_2 = \omega_{Earth})$ (35), respectively, the observed *counterclockwise*[35] revolution of Moon[11] Orbit[19]'s major axis[7], or equivalently $\ell\left[\omega_1 t - \phi_\ell(t)\right]$ (52) lunar[11,19,33] orbit magnitude cycle orientation, from a vantage point from above North Pole[23,34] $(\hat{\boldsymbol{u}}_{Earth\perp} = -\hat{z})$ (49) of Earth[12] in the configuration of **Figure 2**, is also in alignment this expectation that all bodies in the Universe must be obeying "Faraday's Law of Induction"[24] in their motions, and even extending the scope of our previous assertion such that the harmonic motion of the distance vector $\vec{\ell}(\phi)$ (8) between centers of individual circular orbits of revolutions of twin or binary[20] system of bodies in space are triggered to move in reverse direction to their revolution directions in their individual circular orbits, because they obey "Faraday's Law of Induction"[24] as macro-scale charged "system of particles" moving in external magnetic fields with changing flux[25] through the area of their "twin or binary[20] system" motion planes.

$$T_{A,\vec{d}_{MoonOrbit}[\phi(t)]} \simeq 17.7\ Earth\ years \simeq 6{,}465.2108\ days\ \left(\vec{d}_{MoonOrbit}\left[\phi(t)\right]\ Apsidal\ Variation\ Period\right) \quad (53)$$

$$\omega_{A,\vec{d}_{MoonOrbit}[\phi(t)]} = \frac{2\pi}{T_{A,\vec{d}_{MoonOrbit}[\phi(t)]}} \simeq \mathbf{0.000971845389353675}\ radians\,/\,day \quad (54)$$

$$\left(\vec{d}_{MoonOrbit}\left[\phi(t)\right]\ Apsidal\ Variation\ Angular\ Velocity\right)$$

Another observation[19] is based on the nodes or points at which the Moon[11]'s Orbit[19] crosses the so called ecliptic[29] plane. The line of nodes, the intersection between the two respective planes, namely the orbital plane of the Moon[11] and the so called ecliptic[29] plane, has a retrograde motion for an observer on Earth[12], such that it rotates westward along the ecliptic[29], or equivalently in a *clockwise*[35] direction from a vantage point from above the North Pole[23,34] $(\hat{\boldsymbol{u}}_{Earth\perp} = -\hat{z})$ (49) of the Earth[12] in the configuration of **Figure 2**, with a period of 18.6 years (60) or $-19^\circ.3549$ (59) per year. The draconic year[40] $T_{Draconic\ Year}$ (55) is the time taken for the Sun, as seen from Earth, to complete one revolution with respect to the same lunar[11] node, namely a point at which the Moon[11]'s Orbit[19] crosses the so called ecliptic[29] plane, and has an observed period of about 346.6 days (55). The draconic[40] year is associated with eclipses; Lunar and Solar Eclipses are observed[19] to occur when the nodes align with the Sun, roughly every 173.3 days. The Moon[11] crosses the same node every 27.2122 days (56), an interval called the Draconic[19,33] Month $T_{Draconic\ Month}$ (56), which we expect, according to our accepted Moon-Earth topology, to be the synodic[38] period $T_{\left(Moon-\vec{\ell}[\phi(t)]\right)}$ (57) of the motion of the Moon[11] in its individual circular orbit of revolution around radius $\vec{r}_1\left[\phi(t)\right]$ (5) with an angular velocity of $(\omega_1 = \omega_{Moon})$ (34), with respect to the nodal orientation variation motion of $\vec{\ell}\left[\phi(t)\right]$ (8) with an angular velocity $\omega_{N,\vec{\ell}[\phi(t)]}$ (59). We reach this conclusion based on our analysis result above which states that, as the Moon[11]'s motion in its individual circular orbit of revolution around the vector radius $\vec{r}_1\left[\phi(t)\right]$ (5) has an angular velocity of $(\omega_1 = \omega_{Moon})$ (34), any other motion variation of $\vec{d}_{MoonOrbit}\left[\phi(t)\right]$ (51) with an angular velocity different from $(\omega_1 = \omega_{Moon})$ (34), such as this observed rotation of line of nodes of the Moon[11]'s Orbit[19] with angular velocity $\omega_{N,\vec{\ell}[\phi(t)]}$ (59), must be entirely due to an orientation variation of $\vec{\ell}\left[\phi(t)\right]$ (8). This synodic[38] period $T_{\left(Moon-\vec{\ell}[\phi(t)]\right)}$ (57) is a little less than the orbital period[10,11] T_{Moon} (31) of Moon[11], from which we calculate the corresponding synodic[38] angular velocity $\omega_{\left(Moon-\vec{\ell}[\phi(t)]\right)}$ (58). It is significantly worth noting that $\omega_{\left(Moon-\vec{\ell}[\phi(t)]\right)}$ (58) leads to a phase difference of $-152^{\circ}.1$ (58) per year in the synodic[38] motion of nodal orientation variation motion of $\vec{\ell}\left[\phi(t)\right]$ (8) with respect to the motion of the Moon[11] in its individual circular orbit of revolution around radius $\vec{r}_1\left[\phi(t)\right]$ (5), which is very close to the value $-151^{\circ}.11$ (50) of the constant phase difference $\phi_{0,Earth-Sun}$ (50) of the Sun in its individual circular orbit of revolution throughout a year relative to Earth's phase in its own individual circular orbit of revolution, based on our findings in the **ARTICLE 3** "Analytical calculation of earth and sun orbital parameters from distance data" This points out an alignment of the nodal orientation variation motion of $\vec{\ell}\left[\phi(t)\right]$ (8) with the yearly motion of the Sun, a significant finding we have now made in this Article, also supported by the observation[11,19] that the Moon[11] orbits closer to the so called ecliptic[29] plane than the equatorial plane of Earth, which apparently must be due to orientation of $\vec{\ell}\left[\phi(t)\right]$ (8). Utilizing $\left(\omega_1 = \omega_{Moon}\right)$ (34) and $\omega_{\left(Moon-\vec{\ell}[\phi(t)]\right)}$ (58), at this point we are able to calculate the value of the angular velocity $\omega_{N,\vec{\ell}[\phi(t)]}$ (59) of nodal orientation variation motion of $\vec{\ell}\left[\phi(t)\right]$ (8), which turns out to be $-19^{\circ}.3549$ (59), from which we calculate the value of the period $T_{N,\vec{\ell}[\phi(t)]}$ (60) of the nodal orientation variation motion of $\vec{\ell}\left[\phi(t)\right]$ (8), that happens to be 18.59 years (60), both consistent with the observation[19] that the line of nodes has a retrograde motion for an observer on Earth[12], such that it rotates westward along the ecliptic[29], or equivalently in *clockwise*[35] direction, with a period of 18.6 years (60) or $-19^{\circ}.3549$ (59) per year from a vantage point from above the North Pole[23,34] $\left(\hat{\boldsymbol{u}}_{Earth\perp} = -\hat{z}\right)$ (49) of the Earth[12] in the configuration of **Figure 2**. This is a reassuring result that indicates we are on the right track in our Earth-Moon topology analysis. Subsequently, we can infer that the synodic[38] motion of Earth[12] with respect to the nodal orientation variation of Moon[11] Orbit[19] $\vec{\boldsymbol{d}}_{MoonOrbit}\left[\phi(t)\right]$ (51), or equivalently the nodal orientation variation motion of $\vec{\ell}\left[\phi(t)\right]$ (8) based on our analysis above, has a nodal synodic[38] angular velocity $\omega_{\left(Earth-\vec{\ell}[\phi(t)]\right)}$ (61), which can be computed using the values of $\left(\omega_2 = \omega_{Earth}\right)$ (35) and $\omega_{N,\vec{\ell}[\phi(t)]}$ (59), from which the nodal synodic[38] cycle period $T_{\left(Earth-\vec{\ell}[\phi(t)]\right)}$ (62) is also calculated, yielding a value of 346.6 days (55). This is also a result confirming our Earth-Moon topology analysis, such that the observed[19] draconic year[40] $T_{Draconic\ Year}$ (55) turns out to be the period $T_{\left(Earth-\vec{\ell}[\phi(t)]\right)}$ (62) of the synodic[38] motion of Earth[12] with respect to the nodal orientation variation motion of $\vec{\ell}\left[\phi(t)\right]$ (8) in Moon[11] Orbit[19] $\vec{\boldsymbol{d}}_{MoonOrbit}\left[\phi(t)\right]$ (51).

$$T_{Draconic\,Year} = 346.620075883\,days = 346\,days\,14\,hours\,52\,min.s\,54\,seconds \quad (Draconic\,Year) \quad (55)$$

$$T_{Draconic\,Month} \simeq 27.2122247\,days \qquad (Moon's\,Draconic\,Month) \qquad (56)$$

$$T_{(Moon-\vec{\ell}[\phi(t)])} = \frac{2\pi}{\omega_{(Moon-\vec{\ell}[\phi(t)])}} = T_{Draconic\,Month} \simeq \mathbf{27.2122247}\,days \quad (Moon-\vec{\ell}[\phi(t)]\,Synodic\,Period) \quad (57)$$

$$\omega_{(Moon-\vec{\ell}[\phi(t)])} = \frac{2\pi}{T_{(Moon-\vec{\ell}[\phi(t)])}} = \omega_{Moon} - \omega_{N,\vec{\ell}[\phi(t)]} \quad (Moon-\vec{\ell}[\phi(t)]\,Synodic\,Angular\,Velocity)$$
$$\simeq \mathbf{0.2308956866}\,radians/day \simeq \mathbf{2.65470974}\,radians/year \simeq \mathbf{152^{\circ}.1}/year \quad (58)$$

$$\omega_{N,\vec{\ell}[\phi(t)]} = \frac{2\pi}{T_{N,\vec{\ell}[\phi(t)]}} = \omega_{Moon} - \omega_{(Moon-\vec{\ell}[\phi(t)])} \simeq -\mathbf{0.3378}\,rad.s/year \simeq -\mathbf{19^{\circ}.3549}/year \quad (59)$$
$$(Angular\,Velocity\,of\,\vec{\ell}[\phi(t)]\,Nodal\,Orientation\,Variation)$$

$$T_{N,\vec{\ell}[\phi(t)]} = \frac{2\pi}{\left|\omega_{N,\vec{\ell}[\phi(t)]}\right|} \simeq \mathbf{18.599}\,years \quad (Nodal\,Orientation\,Variation\,Period\,of\,\vec{\ell}[\phi(t)]) \quad (60)$$

$$\omega_{(Earth-\vec{\ell}[\phi(t)])} = \frac{2\pi}{T_{(Earth-\vec{\ell}[\phi(t)])}} = \omega_{Earth} - \omega_{N,\vec{\ell}[\phi(t)]} \simeq \mathbf{6.621}\,rad.s/year \simeq \mathbf{379^{\circ}.3549}/year \quad (61)$$
$$(Earth-\vec{\ell}[\phi(t)]\,Synodic\,Angular\,Velocity)$$

$$T_{(Earth-\vec{\ell}[\phi(t)])} = \frac{2\pi}{\omega_{(Earth-\vec{\ell}[\phi(t)])}} = T_{Draconic\,Year} \simeq \mathbf{0.94898}\,years \simeq \mathbf{346.620}\,days \quad (62)$$
$$(Earth-\vec{\ell}[\phi(t)]\,Synodic\,Period)$$

$$T_{N,\vec{\ell}[\phi(t)]} \times T_{(Earth-\vec{\ell}[\phi(t)])}/year \simeq \mathbf{17.651}\,years \simeq T_{A,\vec{d}_{MoonOrbit}[\phi(t)]} \quad (63)$$

Another significant outcome (63) we obtain here as a result of our analysis so far is that, the period of nodal orientation variation cycle $T_{N,\vec{\ell}[\phi(t)]}$ (60) of $\vec{\ell}[\phi(t)]$ (8), multiplied by the synodic[38] cycle $T_{(Earth-\vec{\ell}[\phi(t)])}$ (62) of Earth[12] per year with respect to the nodal orientation variation motion of $\vec{\ell}[\phi(t)]$ (8), yields a result which is about the same as the apsidal[32] variation period $T_{A,\vec{d}_{MoonOrbit}[\phi(t)]}$ (53) of $\vec{d}_{MoonOrbit}[\phi(t)]$ (51) mentioned above. The slight difference is most probably due to the effect of the orientation of $\vec{r}_1[\phi(t)]$ (5) relative to $\vec{\ell}[\phi(t)]$ (8) at the end of a $\left[T_{N,\vec{\ell}[\phi(t)]} \times T_{(Earth-\vec{\ell}[\phi(t)])}\right]$ (63) cycle, leading to a number of additional days for reaching a local maximum magnitude for $\vec{d}_{MoonOrbit}[\phi(t)]$ (51), in determining the apsidal[32] variation period $T_{A,\vec{d}_{MoonOrbit}[\phi(t)]}$ (53). As a result, we can conclude that the observed[19] apsidal[32] precession of lunar[11,19,33] orbit magnitude $\ell\left[\omega_1 t-\phi_\ell\left(t\right)\right]$ (52) in *counterclockwise*[35] direction, and the observed[19] rotation of the line of nodes of the Moon[11] Orbit[19] in *clockwise*[35] direction, from a vantage point from above North Pole[23,34] $\left(\hat{\boldsymbol{u}}_{Earth\perp}=-\hat{z}\right)$ (49) of the Earth[12] in the configuration of **Figure 2**, are both phenomena related to the same orientation variation motion of $\vec{\ell}\left[\phi(t)\right]$ (8) as part of $\vec{\boldsymbol{d}}_{MoonOrbit}\left[\phi(t)\right]$ (51). We thus understand that $\phi_\ell\left(t\right)$ (64) in $\ell\left[\omega_1 t-\phi_\ell\left(t\right)\right]$ (52) is mainly a function $f\left(\omega_{Earth}t,\omega_{Moon}t,\omega_{N,\vec{\ell}[\phi(t)]}t\right)$ (64) of $\left(\omega_{Earth}t\right)$ (35), $\left(\omega_{Moon}t\right)$ (34), and $\left(\omega_{N,\vec{\ell}[\phi(t)]}t\right)$ (59), along the lines of the relations through (53) - (63). Further, following along the lines of the assertion in the **ARTICLE 3** "Analytical calculation of earth and sun orbital parameters from distance data" that all moving bodies in the Universe must be obeying "Faraday's Law of Induction"[24] in their motions, we can also assert that each revolution in space, which is in external magnetic fields with changing flux[25] through the area of this revolution, triggers a rotation in the reverse direction. The observed[19] apsidal[32] precession of lunar[11,19,33] orbit magnitude $\ell\left[\omega_1 t-\phi_\ell\left(t\right)\right]$ (52) in *counterclockwise*[35] direction, and the observed[19] rotation of the line of nodes of the Moon[11] Orbit[19] in *clockwise*[35] direction, from a vantage point from above North Pole[23,34] $\left(\hat{\boldsymbol{u}}_{Earth\perp}=-\hat{z}\right)$ (49) of the Earth[12] in the configuration of **Figure 2**, are a motion pair in alignment with this assertion. The conclusions in this paragraph are among the most significant findings we make as a result of our analysis in this Article.

$$\phi_\ell\left(t\right)=f\left(\omega_{Earth}t,\omega_{Moon}t,\omega_{N,\vec{\ell}[\phi(t)]}t\right) \tag{64}$$

Let us now investigate the angles involved in the Sun-Earth-Moon motion topology. The observed[11,19] topology of relative angles between poles[26,23] and planes of Moon, Earth, and Sun are sketched in **Figure 5** to provide a visual idea. The Moon[11] is observed[11,19] to orbit closer to the so called ecliptic[29] plane than the equatorial plane of Earth. The observed axial tilt[11,19,28] angle of the Moon[11]'s North Pole[26] vector with respect to the so called observed ecliptic[29] plane, or lunar[11,19] obliquity to ecliptic[29], namely $\alpha_{Moon\measuredangle Ecliptic}$ (65), is around $1^\circ.5424$; whereas the observed axial tilt[11,28] angle of the Moon[11]'s North Pole[26] vector with respect to Earth[12]'s equatorial plane, namely $\alpha_{Moon\measuredangle Earth\ Equator}$ (66), is said to be about 24°. On the other hand, the observed mean axial tilt[11,28] angle of the Moon[11]'s North Pole[26] vector with respect to the Moon's observed approximate plane of orbit around the Earth, or mean lunar[11,19] obliquity, namely $\alpha_{Moon\measuredangle Moon\ Orbit}$ (67), is about $6^\circ.687$. Further, the maximum angle that the Moon[11]'s observed approximate plane of orbit around the Earth makes with respect to the so called ecliptic[29] plane, or mean lunar[11,19] orbital inclination, namely $\alpha_{Moon\ Orbit\measuredangle Ecliptic}$ (68), is around $5^\circ.09'$ or equivalently $5^\circ.1446$. Additionally, the observed axial tilt[3,28] angle of the Earth[12], or Earth[12] obliquity, or equivalently the angle that the so called ecliptic[29] plane makes with respect to the Earth[12]'s equatorial plane, namely $\alpha_{Ecliptic\measuredangle Earth\ Equator}$ (69), is approximately $23^\circ.27'$ or equivalently $23^\circ.44$, which sums up to 180° with the found β_{Sun} (47) found in the **ARTICLE 3** "Analytical calculation of earth and sun orbital parameters from distance data"

$$\alpha_{Moon\measuredangle Ecliptic}\simeq\alpha_{Moon\measuredangle Sun\ Orbital\ Plane}\simeq 1^\circ.5424 \tag{65}$$

$$\alpha_{Moon \measuredangle Earth\ Equator} = \alpha_{Moon \measuredangle Earth\ Orbital\ Plane} \simeq 24^\circ \tag{66}$$

$$\alpha_{Moon \measuredangle Moon\ Orbit} = \alpha_{Moon \measuredangle \vec{\ell}[\phi(t)]-plane} \simeq 6^\circ.687 \tag{67}$$

$$\alpha_{Moon\ Orbit \measuredangle Ecliptic} \simeq \alpha_{\vec{\ell}[\phi(t)]-plane \measuredangle Sun\ Orbital\ Plane} \simeq 5^\circ.09' \simeq 5^\circ.1446 \simeq 6^\circ.687 - 1^\circ.5424 \tag{68}$$

$$\alpha_{Ecliptic \measuredangle Earth\ Equator} \simeq \alpha_{Sun\ Orbital\ Plane \measuredangle Earth\ Orbital\ Plane} \simeq 23^\circ.27' \simeq 23^\circ.44 \simeq 180^\circ - 156^\circ.56 = 180^\circ + \beta_{Sun} \tag{69}$$

As also expressed a little above, based on our findings and assertions about the actual topology of motions of the Sun, Earth, and Moon in the **ARTICLE 1** "Points on two circles in space mimic an ellipse with respect to a point" and "Analytical calculation of earth and sun orbital parameters from distance data", as well as in this Article so far:

- The equatorial plane of the Earth is in fact the plane of individual circular orbit of revolution of Earth, which is the horizontal plane of $Circle_2$ in **Figure 2**;
- The Moon's observed plane of orbit around the Earth should be the plane of motion of $\vec{d}_{MoonOrbit}\left[\phi(t)\right]$ (51). However, based on our Moon[11] Orbit[19] analysis so far, and our assertion in the **ARTICLE 1** "Points on two circles in space mimic an ellipse with respect to a point" that the self-rotation axis and the normal of the individual orbit of revolution of every spatial body is the same, the angles made by the Moon[11]'s North Pole[26] vector determine angles made by the plane of individual circular orbit of revolution of the Moon[11] around $\vec{r}_1\left[\phi(t)\right]$ (5), and thus the angles said to be made by the Moon's observed[19] plane of orbit around the Earth should be the angles made by the plane of motion of the vector distance $\vec{\ell}\left[\phi(t)\right]$ (8), which is between the centers of individual circular orbits of revolution of the Earth and the Moon, such that the individual circular orbit of revolution of the Moon around $\vec{r}_1\left[\phi(t)\right]$ (5) in turn revolves as a whole around the Earth with a radius of $\vec{\ell}\left[\phi(t)\right]$ (8) at each $\phi(t)$ (3) through every orbital period[10,11] T_{Moon} (31) of the Moon[11], or vice versa;
- The so called observed ecliptic[29] plane of the Earth in its motion around the Sun is in fact not really a plane, but the approximate angles said to be observed with respect to the ecliptic[29] are angles observed with respect to the plane of the distance vector between the Earth and the Sun around the year, and in fact the so called observed ecliptic[29] plane can approximately be considered to be the plane of individual circular orbit of revolution of the Sun, as $\left(\mathbf{12,370.477}\ km \simeq r_2 \ll r_{Sun} \simeq \mathbf{149,609,642.749}\ km\right)$ (45) - (46).

In light of all this information, the above information tells us that the observed axial tilt[11,28] angle of the Moon[11]'s North Pole[26] vector with respect to the normal of the plane of motion of the vector distance $\vec{\ell}\left[\phi(t)\right]$ (8), which is between centers of individual circular orbits of revolution of Earth and Moon, namely $\alpha_{Moon \measuredangle \vec{\ell}[\phi(t)]-plane}$ (67), would be about $6^\circ.687$; and the observed axial tilt[11,28] angle of the Moon[11]'s North Pole[26] vector approximately with respect to the normal of the plane of individual circular orbit of revolution of the Sun, namely $\alpha_{Moon \measuredangle Sun\ Orbital\ Plane}$ (65), would be around $1^\circ.5424$; and the observed angle between the individual circular orbits of revolution of the Earth and the Sun, namely $\alpha_{Sun\ Orbital\ Plane \measuredangle Earth\ Orbital\ Plane}$ (69), is $23^\circ.44$. In light of this, the observed angle that the plane of motion of the vector distance $\vec{\ell}\left[\phi(t)\right]$ (8), which is between centers of individual circular orbits of revolution of Earth and Moon, makes approximately with respect to the normal of the plane of individual circular orbit of revolution of the Sun, namely $\alpha_{\vec{\ell}[\phi(t)]-plane \measuredangle Sun\,Orbital\,Plane}$ (68), would be around $5^{\circ}.1446$.

Further, the observed axial tilt[11,28] angle of the Moon[11]'s North Pole[26] vector with respect to Earth's plane of individual circular orbit of revolution, namely $\alpha_{Moon \measuredangle Earth\,Orbital\,Plane}$ (66), would be about 24°, which is expected to reveal the value of the inclination angle $\left(\beta = \beta_{Moon}\right)$ (5) of the plane of the Moon's individual orbit $\left(Circle_1\right)$ with respect to the plane of the Earth's individual circular orbit of revolution $\left(Circle_2\right)$ in **Figure 2**, also based on our assertion in the **ARTICLE 1** "Points on two circles in space mimic an ellipse with respect to a point" that the self-rotation axis and the normal of the individual orbit of revolution of every spatial body is the same. The direction of Earth's North Pole[23,26] vector is $\left(\hat{\boldsymbol{u}}_{Earth\perp} = -\hat{z}\right)$ (49) in the Earth-Sun the configuration of **Figure 2**, which is the same in the Earth-Moon configuration of **Figure 2**, as the plane of individual circular orbit of revolution of the Earth is chosen to be the horizontal $\hat{x}-\hat{y}$ plane in both the Earth-Moon topology and the Earth-Sun topology of the configuration in **Figure 2**. Based on our analysis above, as Moon[11]'s self-rotation must be in counterclockwise[35] direction from a vantage point from above the North Pole[23,34] $\left(\hat{\boldsymbol{u}}_{Earth\perp} = -\hat{z}\right)$ (49) of the Earth in order to maintain its synchronous[19] rotation with Earth, we can determine that the direction of Moon[11]'s North Pole[26] vector should be $\left(\hat{\boldsymbol{u}}_{Moon\perp} = \hat{x}Sin\beta - \hat{z}Cos\beta\right)$ (70) in the Earth-Moon configuration of **Figure 2**, with $\left(-\dfrac{\pi}{2} \leq \beta \leq \dfrac{\pi}{2}\right)$ (71). As the Moon[11] is observed[11,19] to orbit close to the so called ecliptic[29] plane, or equivalently close to the plane of individual circular orbit of revolution of the Sun based on our analysis above, we can conclude that $\left(\beta = \beta_{Moon} \simeq +24^{\circ}\right)$ (72) based on (66) and (69), which is a value we had set out to find, another fundamental finding of this Article.

$$\hat{\boldsymbol{u}}_{Moon\perp} = \hat{x}Sin\beta - \hat{z}Cos\beta \quad \left(North\ Pole\ of\ Moon\ normal\ to\ Moon's\ circular\ revolution\ plane\right) \quad (70)$$

$$-\frac{\pi}{2} \leq \beta \leq \frac{\pi}{2} \quad (71)$$

$$\beta = \beta_{Moon} = +\alpha_{Moon \measuredangle Earth\,Equator} = +\alpha_{Moon \measuredangle Earth\,Orbital\,Plane} \simeq +24^{\circ} \quad (72)$$

Every 18.6 years (60), the angle between the Moon[11]'s Orbit[19] and Earth[12]'s equator reaches a maximum $\alpha_{Moon\,Orbit \measuredangle Earth\,Equator,Max}$ (73) of $28^{\circ}.36'$ or equivalently $28^{\circ}.5846$, the sum of Earth[12]'s equatorial tilt $\alpha_{Ecliptic \measuredangle Earth\,Equator}$ (69), or equivalently $\alpha_{Sun\,Orbital\,Plane \measuredangle Earth\,Orbital\,Plane}$ (69), which is $23^{\circ}.27'$ or equivalently $23^{\circ}.44$, and the Moon[11]'s orbital inclination $\alpha_{Moon\,Orbit \measuredangle Ecliptic}$ (68) to the so called ecliptic[29], or equivalently $\alpha_{\vec{\ell}[\phi(t)]-plane \measuredangle Sun\,Orbital\,Plane}$ (68), which is $5^{\circ}.09'$ or equivalently $5^{\circ}.1446$. This is called a major lunar[11] standstill[11,41]. Conversely, 9.3 years later, the angle between the Moon[11]'s Orbit[19] and the Earth[12]'s equator reaches a minimum

$\alpha_{Moon\ Orbit \measuredangle Earth\ Equator, Min}$ (74) of $18^{\circ}.20'$ or equivalently $18^{\circ}.33$, approximately near $18^{\circ}.2954$, which is the Moon[11]'s orbital inclination $\alpha_{Moon\ Orbit \measuredangle Ecliptic}$ (68) to the so called ecliptic[29], or equivalently $\alpha_{\vec{\ell}[\phi(t)]-plane \measuredangle Sun\ Orbital\ Plane}$ (68), subtracted from Earth[12]'s equatorial tilt $\alpha_{Ecliptic \measuredangle Earth\ Equator}$ (69), or equivalently $\alpha_{Sun\ Orbital\ Plane \measuredangle Earth\ Orbital\ Plane}$ (69). This is called a minor lunar[11] standstill[11,41], which last occurred in October 2015. This variation of the angle $\alpha_{Moon\ Orbit \measuredangle Earth\ Equator}$ (73) - (75) between the Moon[11]'s Orbit[19] and Earth[12]'s equator over a 18.6 year cycle is apparently related to the observed[19] rotation of the line of nodes of the Moon[11] Orbit[19] over a 18.6 year (60) cycle, which we have previously concluded to be related to the orientation variation motion of $\vec{\ell}\left[\phi(t)\right]$ (8) as part of $\vec{\boldsymbol{d}}_{MoonOrbit}\left[\phi(t)\right]$ (51), with a period $T_{N,\vec{\ell}[\phi(t)]}$ (60) of 18.6 years. Thus, we conclude here that the angle $\alpha_{Moon\ Orbit \measuredangle Earth\ Equator}$ (73) - (75) observed[19] to be between the Moon[11]'s Orbit[19] and Earth[12]'s equator is equivalently the angle $\alpha_{\vec{\ell}[\phi(t)]-plane \measuredangle Earth\ Orbital\ Plane}$ (73) - (75) between the plane of motion of $\vec{\ell}\left[\phi(t)\right]$ (8), which is the vector between the centers of individual circular orbits of revolution of the Earth and the Moon, and Earth[12]'s equator. This is another fundamental conclusion of our analysis in this Article.

$$\begin{aligned}\alpha_{Moon\ Orbit \measuredangle Earth\ Equator, Max} &= \alpha_{\vec{\ell}[\phi(t)]-plane \measuredangle Earth\ Orbital\ Plane, Max} \simeq 5^{\circ}.1446 + 23^{\circ}.44 \simeq 28^{\circ}.5846 \simeq 28^{\circ}.36' \\ &= \alpha_{\vec{\ell}[\phi(t)]-plane \measuredangle Sun\ Orbital\ Plane} + \alpha_{Sun\ Orbital\ Plane \measuredangle Earth\ Orbital\ Plane}\end{aligned} \quad (73)$$

$$\begin{aligned}\alpha_{Moon\ Orbit \measuredangle Earth\ Equator, Min} &= \alpha_{\vec{\ell}[\phi(t)]-plane \measuredangle Earth\ Orbital\ Plane, Min} \simeq -5^{\circ}.1446 + 23^{\circ}.44 \simeq 18^{\circ}.2954 \simeq 18^{\circ}.20' \\ &= -\alpha_{\vec{\ell}[\phi(t)]-plane \measuredangle Sun\ Orbital\ Plane} + \alpha_{Sun\ Orbital\ Plane \measuredangle Earth\ Orbital\ Plane}\end{aligned} \quad (74)$$

$$\alpha_{Moon\ Orbit \measuredangle Earth\ Equator, Min} \leq \alpha_{Moon\ Orbit \measuredangle Earth\ Equator} \leq \alpha_{Moon\ Orbit \measuredangle Earth\ Equator, Max} \quad (over\ 18.6\ years\ cycle) \quad (75)$$

The resultant expected overall orientation variation motion of $\vec{\ell}\left[\phi(t)\right]$ (8) as part of $\vec{\boldsymbol{d}}_{MoonOrbit}\left[\phi(t)\right]$ (51), with a period $T_{N,\vec{\ell}[\phi(t)]}$ (60) of 18.6 years, is depicted with Sun-Earth Cartesian $(\hat{\boldsymbol{x}}, \hat{\boldsymbol{y}}, \hat{\boldsymbol{z}})$ coordinates, roughly in **Figure 6** at $(t = t_0)$ and $(t = t_0 + 18.6\, years)$ corresponding to a major lunar[11] standstill[11,41], in **Figure 7** at $(t = t_0 + 4.65\, years)$ end of 1st Quarter Cycle, in **Figure 8** at $(t = t_0 + 9.3\, years)$ end of 1 Half Cycle at a minor lunar[11] standstill[11,41], and in **Figure 9** at $(t = t_0 + 13.95\, years)$ end of 3rd Quarter Cycle, not to scale and with shape of lunar[11,19,33] monthly cycle of $\vec{\ell}\left[\phi(t)\right]$ (8) not necessarily accurate, to enable visualization of the motion of $\vec{\ell}\left[\phi(t)\right]$ (8) cycle, leading to the combined perception by an observer[19] on Earth[12], of Moon[11] Orbit[19]'s apsidal[32] precession, rotation of the line of nodes of Moon[11] Orbit[19] over a 18.6 year (60) cycle, and the variation of $\alpha_{Moon\ Orbit \measuredangle Earth\ Equator}$ (73) - (75) angle between the Moon[11]'s Orbit[19] and the Earth[12]'s equator, or equivalently $\alpha_{\vec{\ell}[\phi(t)]-plane \measuredangle Earth\ Orbital\ Plane}$ (73) - (75), over a 18.6 year (60) cycle.

These are conclusions we make based on observation[19] and our Sun-Earth-Moon motion topology analysis in the **ARTICLE 1** "Points on two circles in space mimic an ellipse with respect to a point" and **ARTICLE 3** "Analytical calculation of earth and sun orbital parameters from distance data", as well as in this Article so far.

As we have concluded in our analysis above that it is the angular frequency of the Moon's circular orbit around its own center of revolution $(\omega_1 = \omega_{Moon})$ (34) which determines the angular velocity of the higher frequency distance variation of the Earth-Moon system, and that mainly the $Cos(\omega_1 t)$ term determines the maximum and minimum values of $d^2[\phi(t)]$ (44) in its sinusoidal cycles, let us determine $d^2[\phi(t)]$ (79) for local maximum Earth-Moon distances[5] at dates when the Moon's phase $(\omega_1 t)$ (76) is an even multiple of π, with $(\omega_2 t)$ (77) and $(\omega_1 - \omega_2)t$ (78), and $d^2[\phi(t)]$ (83) for local minimum Earth-Moon distances[5] at dates when the Moon's phase $(\omega_1 t)$ (80) is an odd multiple of π, with $(\omega_2 t)$ (81) and $(\omega_1 - \omega_2)t$ (82). Note here that $(\omega_1 = \omega_{Moon})$ (34) and $(\omega_2 = \omega_{Earth})$ (35) are the values of the Moon[11] and the Earth[12]'s angular frequencies around their individual circular orbits of revolution, respectively, their angular velocity difference being $(\omega_1 - \omega_2)$ (36), where we have $(\omega_{Moon} T_{Moon} = 2\pi \; radians)$ (34), and we make use of trigonometric identities, as well as utilizing the definition of the magnitude $\ell[\phi(t)]$ (9) of the vector distance $\vec{\ell}[\phi(t)]$ (8), which is between the centers of individual circular orbits of revolution of the Earth and the Moon.

$$t = 2n\frac{T_{Moon}}{2} \quad \Rightarrow \quad \omega_1 t = n\omega_{Moon} T_{Moon} = n \times 2\pi \; radians \quad (Moon's \; Phase) \quad ; \quad n \; integer \quad (76)$$

$$t = 2n\frac{T_{Moon}}{2} \quad \Rightarrow \quad \omega_2 t = \omega_{Earth} n T_{Moon} = \frac{\omega_2}{\omega_1} n \times 2\pi \; radians \quad (Earth's \; Phase) \quad ; \quad n \; integer \quad (77)$$

$$t = 2n\frac{T_{Moon}}{2} \quad \Rightarrow \quad (\omega_1 - \omega_2)t = \left(1 - \frac{\omega_2}{\omega_1}\right) n \times 2\pi \; radians = -\frac{\omega_2}{\omega_1} 2n\pi \; radians \quad ; \quad n \; integer \quad (78)$$

$$
\begin{aligned}
d^2\left[\phi(t)\right] = d^2\left(\frac{\omega_2}{\omega_1}2n\pi\right) &= r_1^2 - 2r_1 r_2 Cos\beta Cos\left(\frac{\omega_2}{\omega_1}2n\pi\right) + r_2^2 + \ell^2\left(\frac{\omega_2}{\omega_1}2n\pi\right) \\
&+ 2\left\{r_1 Cos\,\beta\ell_x\left(\frac{\omega_2}{\omega_1}2n\pi\right) + r_1 Sin\,\beta\,\ell_z\left(\frac{\omega_2}{\omega_1}2n\pi\right)\right\} \\
&- 2r_2 Cos\left(\frac{\omega_2}{\omega_1}2n\pi\right)\ell_x\left(\frac{\omega_2}{\omega_1}2n\pi\right) - 2r_2 Sin\left(\frac{\omega_2}{\omega_1}2n\pi\right)\ell_y\left(\frac{\omega_2}{\omega_1}2n\pi\right) \\
&= \left\{r_1^2 Cos^2\beta + 2r_1 Cos\,\beta\ell_x\left(\frac{\omega_2}{\omega_1}2n\pi\right) + \ell_x^2\left(\frac{\omega_2}{\omega_1}2n\pi\right)\right\} \\
&+ \left\{r_1^2 Sin^2\beta + 2r_1 Sin\,\beta\,\ell_z\left(\frac{\omega_2}{\omega_1}2n\pi\right) + \ell_z^2\left(\frac{\omega_2}{\omega_1}2n\pi\right)\right\} \\
&- 2r_2 Cos\left(\frac{\omega_2}{\omega_1}2n\pi\right)\left[r_1 Cos\,\beta + \ell_x\left(\frac{\omega_2}{\omega_1}2n\pi\right)\right] + r_2^2 Cos^2\left(\frac{\omega_2}{\omega_1}2n\pi\right) \\
&+ r_2^2 Sin^2\left(\frac{\omega_2}{\omega_1}2n\pi\right) - 2r_2 Sin\left(\frac{\omega_2}{\omega_1}2n\pi\right)\ell_y\left(\frac{\omega_2}{\omega_1}2n\pi\right) + \ell_y^2\left(\frac{\omega_2}{\omega_1}2n\pi\right) \\
&= \left[r_1 Cos\,\beta + \ell_x\left(\frac{\omega_2}{\omega_1}2n\pi\right) - r_2 Cos\left(\frac{\omega_2}{\omega_1}2n\pi\right)\right]^2 \\
&+ \left[r_1 Sin\,\beta + \ell_z\left(\frac{\omega_2}{\omega_1}2n\pi\right)\right]^2 + \left[r_2 Sin\left(\frac{\omega_2}{\omega_1}2n\pi\right) - \ell_y\left(\frac{\omega_2}{\omega_1}2n\pi\right)\right]^2
\end{aligned}
\qquad (79)
$$

$$
t = (2n+1)\frac{T_{Moon}}{2} \Rightarrow \omega_1 t = \frac{(2n+1)}{2}\omega_{Moon}T_{Moon} = (2n+1)\times\pi \text{ radians } (Moon's\ \ Phase);\ n \text{ integer} \quad (80)
$$

$$
t = (2n+1)\frac{T_{Moon}}{2} \Rightarrow \omega_2 t = \omega_{Earth}(2n+1)\frac{T_{Moon}}{2} = \frac{\omega_2}{\omega_1}(2n+1)\times\pi \text{ radians } (Earth\,Phase);n \text{ integer} \quad (81)
$$

$$
t = (2n+1)\frac{T_{Moon}}{2} \Rightarrow (\omega_1 - \omega_2)t = \left(1 - \frac{\omega_2}{\omega_1}\right)(2n+1)\times\pi \text{ radians} = \pi - \frac{\omega_2}{\omega_1}(2n+1)\times\pi\ \ ;\ \ n \text{ int} \quad (82)
$$

$$
\begin{aligned}
d^2\left[\phi(t)\right] &= d^2\left[\frac{\omega_2}{\omega_1}(2n+1)\pi\right] = r_1^2 + 2r_1 r_2 Cos\beta\, Cos\left[\frac{\omega_2}{\omega_1}(2n+1)\pi\right] + r_2^2 + \ell^2\left[\frac{\omega_2}{\omega_1}(2n+1)\times\pi\right] \\
&\quad -2\left\{r_1 Cos\beta\,\ell_x\left[\frac{\omega_2}{\omega_1}(2n+1)\times\pi\right] + r_1 Sin\beta\,\ell_z\left[\frac{\omega_2}{\omega_1}(2n+1)\times\pi\right]\right\} \\
&\quad -2r_2 Cos\left[\frac{\omega_2}{\omega_1}(2n+1)\times\pi\right]\ell_x\left[\frac{\omega_2}{\omega_1}(2n+1)\times\pi\right] \\
&\quad -2r_2 Sin\left[\frac{\omega_2}{\omega_1}(2n+1)\times\pi\right]\ell_y\left[\frac{\omega_2}{\omega_1}(2n+1)\times\pi\right] \\
&= \left\{r_1^2 Cos^2\beta - 2r_1 Cos\beta\,\ell_x\left[\frac{\omega_2}{\omega_1}(2n+1)\times\pi\right] + \ell_x^2\left[\frac{\omega_2}{\omega_1}(2n+1)\times\pi\right]\right\} \\
&\quad +\left\{r_1^2 Sin^2\beta - 2r_1 Sin\beta\,\ell_z\left[\frac{\omega_2}{\omega_1}(2n+1)\times\pi\right] + \ell_z^2\left[\frac{\omega_2}{\omega_1}(2n+1)\times\pi\right]\right\} \\
&\quad +2r_2 Cos\left[\frac{\omega_2}{\omega_1}(2n+1)\times\pi\right]\left\{r_1 Cos\beta - \ell_x\left[\frac{\omega_2}{\omega_1}(2n+1)\times\pi\right]\right\} \\
&\quad +r_2^2 Cos^2\left[\frac{\omega_2}{\omega_1}(2n+1)\times\pi\right] + r_2^2 Sin^2\left[\frac{\omega_2}{\omega_1}(2n+1)\times\pi\right] \\
&\quad -2r_2 Sin\left[\frac{\omega_2}{\omega_1}(2n+1)\times\pi\right]\ell_y\left[\frac{\omega_2}{\omega_1}(2n+1)\times\pi\right] + \ell_y^2\left[\frac{\omega_2}{\omega_1}(2n+1)\times\pi\right] \\
&= \left\{r_1 Cos\beta - \ell_x\left[\frac{\omega_2}{\omega_1}(2n+1)\times\pi\right] + r_2 Cos\left[\frac{\omega_2}{\omega_1}(2n+1)\times\pi\right]\right\}^2 \qquad (83) \\
&\quad +\left\{r_1 Sin\beta - \ell_z\left[\frac{\omega_2}{\omega_1}(2n+1)\times\pi\right]\right\}^2 \\
&\quad +\left\{r_2 Sin\left[\frac{\omega_2}{\omega_1}(2n+1)\times\pi\right] - \ell_y\left[\frac{\omega_2}{\omega_1}(2n+1)\times\pi\right]\right\}^2
\end{aligned}
$$

We can actually distinguish the Cartesian components (11) of $\vec{\boldsymbol{d}}\left[\phi(t)\right]$ (10) in $d^2\left[\phi(t)\right]$ (79) and $d^2\left[\phi(t)\right]$ (83), as in (84) and (85), respectively.

$$\vec{\boldsymbol{d}}\left(\frac{\omega_2}{\omega_1}2n\pi\right)=\hat{\boldsymbol{x}}\,d_x\left(\frac{\omega_2}{\omega_1}2n\pi\right)+\hat{\boldsymbol{y}}\,d_y\left(\frac{\omega_2}{\omega_1}2n\pi\right)+\hat{\boldsymbol{z}}\,d_z\left(\frac{\omega_2}{\omega_1}2n\pi\right)$$

$$\Rightarrow\begin{cases}d_x\left(\frac{\omega_2}{\omega_1}2n\pi\right)=r_1Cos\,\beta-r_2Cos\left(\frac{\omega_2}{\omega_1}2n\pi\right)+\ell_x\left(\frac{\omega_2}{\omega_1}2n\pi\right)\\ d_y\left(\frac{\omega_2}{\omega_1}2n\pi\right)=-r_2Sin\left(\frac{\omega_2}{\omega_1}2n\pi\right)+\ell_y\left(\frac{\omega_2}{\omega_1}2n\pi\right)\\ d_z\left(\frac{\omega_2}{\omega_1}2n\pi\right)=r_1Sin\,\beta+\ell_z\left(\frac{\omega_2}{\omega_1}2n\pi\right)\end{cases}\tag{84}$$

$$\vec{\boldsymbol{d}}\left[\frac{\omega_2}{\omega_1}(2n+1)\times\pi\right]=\hat{\boldsymbol{x}}\,d_x\left[\frac{\omega_2}{\omega_1}(2n+1)\times\pi\right]+\hat{\boldsymbol{y}}\,d_y\left[\frac{\omega_2}{\omega_1}(2n+1)\times\pi\right]+\hat{\boldsymbol{z}}\,d_z\left[\frac{\omega_2}{\omega_1}(2n+1)\times\pi\right]$$

$$\Rightarrow\begin{cases}d_x\left[\frac{\omega_2}{\omega_1}(2n+1)\times\pi\right]=-r_1Cos\,\beta-r_2Cos\left[\frac{\omega_2}{\omega_1}(2n+1)\times\pi\right]+\ell_x\left[\frac{\omega_2}{\omega_1}(2n+1)\times\pi\right]\\ d_y\left[\frac{\omega_2}{\omega_1}(2n+1)\times\pi\right]=-r_2Sin\left[\frac{\omega_2}{\omega_1}(2n+1)\times\pi\right]+\ell_y\left[\frac{\omega_2}{\omega_1}(2n+1)\times\pi\right]\\ d_z\left[\frac{\omega_2}{\omega_1}(2n+1)\times\pi\right]=-r_1Sin\,\beta+\ell_z\left[\frac{\omega_2}{\omega_1}(2n+1)\times\pi\right]\end{cases}\tag{85}$$

Checking the data in **Table 1**, between the dates of January 10, 2017 and January 9, 2021, minimum Earth-Moon distance[5] is observed as $356,565$ km occurring on January 1, 2018, and maximum Earth-Moon distance[5] is observed as $406,692$ km occurring on March 24, 2020. Considering our previous analysis that there are approximately 13.3687466 orbital[10,11] (33) cycles of a distance variation sinusoid in approximately 1 cycle of a distance envelope sinusoid for the Earth-Moon system, which is also compliant with the observation in **Figure 1** and the data listed in **Table 1**, we can infer as in (86) - (90) that once in every 3-year cycles (86) of the Earth, which is approximately 40 cycles (86) of the Moon, both the Moon (88) and Earth (89) moving around their own circles of revolution must be passing from the same aligned phase of a multiple of 2π (76) - (77) with n being a multiple of 40 (86) around the same date, after their initial phase alignment of $\left[\phi_0(t_0)=0\right]$ (4) at time $(t=t_0=0)$ with $(n\simeq 0)$, in the configuration described in **Figure 2**. Thus, on these dates, the various *Sine* and *Cosine* values in $d^2\left[\phi(t)\right]$ (79) take on their values in (91). Based on our previous analysis result, it is the $\left[Cos(\omega_1 t)Cos(\omega_2 t)\right]$ term which determines the maximum and minimum values of $d^2\left[\phi(t)\right]$ (44) in its sinusoidal cycles, mainly the $Cos(\omega_1 t)$ term, which must be at dates when the Moon phase $(\omega_1 t)$ (76) is a multiple of π, with a global maximum or minimum Earth-Moon distance[5] when the Earth phase $(\omega_2 t)$ (77) is close to a multiple of π at the same time. On these dates when time t is a multiple of $(t=3T_{Earth}\simeq 40T_{Moon})$ (86), Moon's phase $(\omega_1 t)$ (76) is an even multiple of π, with $(\omega_2 t)$

(77) also an even multiple of π, and the $\left[Cos(\omega_1 t)Cos(\omega_2 t)\right]$ term, and therefore $d^2\left[\phi(t)\right]$ (79), must be taking on its global maximum value within a 3-year cycle of the Earth. We can thus conclude to an approximation that within the 3-year cycle of Earth after the initial Earth-Moon phase alignment of $\left[\phi_0(t_0)=0\right]$ (4) at time $(t=t_0=0)$ with $(n\simeq 0)$ between the dates of January 10, 2017 and January 9, 2021, the global maximum Earth-Moon distance[5] observed as 406,692 km (92) occurring on March 24, 2020 (87) corresponds to this $\left[Cos(\omega_1 t)Cos(\omega_2 t)\simeq 1\right]$ (91) value in $d^2\left[\phi(t)\right]$ (79), on which date the phase of the Moon (88) and the phase of the Earth (89) are aligned as a multiple of 2π as in (90).

$$T_{Earth}\times 3=\left(T_{Moon}\times 13.3687466\right)\times 3\simeq T_{Moon}\times 40 \quad\Rightarrow\quad t=t_0+\frac{\omega_2}{\omega_1}nT_{Earth}\simeq t_0+nT_{Moon}$$
$$\left(n\ is\ a\ multiple\ of\ 40 \quad \& \quad \frac{\omega_2}{\omega_1}n\ is\ a\ multiple\ of\ 3 \quad \& \quad t_0=0\right) \tag{86}$$

$$n\simeq 40 \quad \& \quad t_0=0 \quad\Rightarrow\quad t=3T_{Earth}\simeq 40T_{Moon} \qquad \left(March\ 24,\ 2020\right) \tag{87}$$

$$t=3T_{Earth}\simeq 40T_{Moon} \quad\Rightarrow\quad \omega_1 t\simeq 40\,\omega_{Moon}T_{Moon}=40\times 2\pi\ radians \quad \left(Moon's\ Phase\right) \tag{88}$$

$$t=3T_{Earth}\simeq 40T_{Moon} \quad\Rightarrow\quad \omega_2 t=3\,\omega_{Earth}T_{Earth}=3\times 2\pi\ radians \quad \left(Earth's\ Phase\right) \tag{89}$$

$$t=3T_{Earth}\simeq 40T_{Moon} \quad\Rightarrow\quad \left(\omega_1-\omega_2\right)t\simeq\left(40\,\omega_{Moon}T_{Moon}-3\,\omega_{Earth}T_{Earth}\right)=37\times 2\pi\ radians \tag{90}$$

$$t=3T_{Earth}\simeq 40T_{Moon} \quad\Rightarrow\quad \begin{cases} Cos(\omega_1 t)=Cos(\omega_2 t)=Cos\left[(\omega_1-\omega_2)t\right]\simeq 1 \\ Sin(\omega_1 t)=Sin(\omega_2 t)=Sin\left[(\omega_1-\omega_2)t\right]\simeq 0 \end{cases} \tag{91}$$

Subsequently, $d^2\left[\phi(t)\right]$ (79) at time $(t=3T_{Earth}\simeq 40T_{Moon})$ (87) is $d^2(6\pi)$ (92) with $(n\simeq 40)$.

$$\begin{aligned} d^2\left[\phi(t=3T_{Earth})\right] &= d^2(6\pi)=(406{,}692\ km)^2=165{,}398{,}382{,}864\ km^2 \qquad \left(March\ 24,\ 2020\right) \\ &= r_1^2-2r_1r_2\,Cos\beta+r_2^2+\ell^2(6\pi)+2\left[(r_1Cos\beta-r_2)\ell_x(6\pi)+r_1Sin\beta\,\ell_z(6\pi)\right] \\ &= \left[r_1Cos\beta+\ell_x(6\pi)-r_2\right]^2+\left[r_1Sin\beta+\ell_z(6\pi)\right]^2+\ell_y^2(6\pi) \end{aligned} \tag{92}$$

Again checking the data in **Table 1**, between the dates of January 10, 2017 and January 9, 2021, the date corresponding to the initial Earth-Moon phase alignment of $\left[\phi_0(t_0)=0\right]$ (4) at time $(t=t_0=0)$ with $(n\simeq 0)$ must be approximately 40 Moon cycles (86) before the occurrence of this 3-year cycle (87) global maximum Earth-Moon distance[5] observed as 406,692 km (92) on March 24, 2020 (92), which is observed to be 1102 days before, on March 18, 2017 (93), when the Earth-Moon distance[5] observed is 404,650 km (93), different in value from each other by 2,042 *km*, significantly smaller than the given Earth-Moon distances[5] observed. Calculating $d^2\left[\phi(t)\right]$ (79) at time $(t=t_0=0)$ with $(n\simeq 0)$, which is $d^2(0)$ (93), and comparing, we see it would have been approximately the same as $d^2(6\pi)$ (92) at $(t=3T_{Earth}\simeq 40T_{Moon})$ (87) with $(n\simeq 40)$ if $\vec{\ell}\left[\phi(t)\right]$ (8) were a constant vector over time t. However, they have different values, and this is an indication that $\vec{\ell}\left[\phi(t)\right]$ (8) is in fact variable over time t, even if it has a very small magnitude variation $\ell\left[\phi(t)\right]$ (9) compared to Earth-Moon distances[5] observed. This finding is also in accordance with our expectation of $\vec{\ell}\left[\phi(t)\right]$ (8) orientation variation over time, observed as apsidal[32] precession and motion of line of nodes[19], based on our previous analysis results above.

$$\begin{aligned} d^2\left[\phi(t=t_0=0)\right] &= d^2(0) = (404{,}650\,km)^2 = 163{,}741{,}622{,}500\,km^2 \qquad (March\ 18,\ 2017) \\ &= r_1^2 - 2r_1 r_2 Cos\beta + r_2^2 + \ell^2(0) + 2\left[(r_1 Cos\beta - r_2)\ell_x(0) + r_1 Sin\beta\,\ell_z(0)\right] \qquad (93) \\ &= \left[r_1 Cos\beta + \ell_x(0) - r_2\right]^2 + \left[r_1 Sin\beta + \ell_z(0)\right]^2 + \ell_y^2(0) \end{aligned}$$

Based on our analysis in (86) - (93) above, halfway (94) to every 3-year cycle (86) of Earth after its phase alignment with Moon in the configuration described in **Figure 2**, which is approximately 20 cycles (94) of the Moon after and before the 3-year cycle global maximum Earth-Moon distances[5], such as midway between March 18, 2017 (93) and March 24, 2020 (92), the Moon (95) and the Earth (96) moving around their own circles of revolution must be passing through an opposite phase (97), with Moon at a multiple of 2π (95) and Earth at an odd multiple of π (96). As we have previously assessed that the $\left[Cos(\omega_1 t)Cos(\omega_2 t)\right]$ term determines the maximum and minimum values of $d^2\left[\phi(t)\right]$ (44) in its sinusoidal cycles, mainly the $Cos(\omega_1 t)$ term, which must be at dates when the Moon phase $(\omega_1 t)$ (76) is a multiple of π, with a global maximum or minimum Earth-Moon distance[5] when the Earth phase $(\omega_2 t)$ (77) is close to a multiple of π at the same time, we can approximately determine that the date halfway (94) of the 3-year cycle (86) of Earth between global maximum Earth-Moon distances[5] on March 18, 2017 (93) and March 24, 2020 (92) from **Table 1**, which is observed to occur 551 days after and before these dates, on September 20, 2018 (99), when the Earth-Moon distance[5] is a local maximum and is 404,876 km (99), 20 cycles (95) of Moon after March 18, 2017 (93) and before March 24, 2020 (92). On this date, the *Sine* and *Cosine* values in $d^2\left[\phi(t)\right]$ (79) have values as in (98).

$$T_{Earth} \times 1.5 = (T_{Moon} \times 13.3687466) \times 1.5 \simeq T_{Moon} \times 20 \quad \Rightarrow \quad t = 1.5\,T_{Earth} \simeq 20\,T_{Moon} \qquad (94)$$

$$t = 1.5\,T_{Earth} \simeq 20\,T_{Moon} \;\Rightarrow\; \omega_1 t \simeq 20\,\omega_{Moon} T_{Moon} = 20 \times 2\pi\ radians \quad (Moon's\ Phase) \qquad (95)$$

$$t = 1.5\,T_{Earth} \simeq 20\,T_{Moon} \;\Rightarrow\; \omega_2 t = 1.5\,\omega_{Earth} T_{Earth} = 1.5 \times 2\pi\ radians \quad (Earth's\ Phase) \qquad (96)$$

$$t = 1.5\,T_{Earth} \simeq 20\,T_{Moon} \;\Rightarrow\; (\omega_1 - \omega_2)t \simeq (20\,\omega_{Moon} T_{Moon} - 1.5\,\omega_{Earth} T_{Earth}) = 37\pi\ radians \qquad (97)$$

$$t = 1.5\,T_{Earth} \simeq 20\,T_{Moon} \;\Rightarrow\; \begin{cases} Cos(\omega_1 t) \simeq 1 \\ Cos(\omega_2 t) = Cos\left[(\omega_1 - \omega_2)t\right] \simeq -1 \\ Sin(\omega_1 t) = Sin(\omega_2 t) = Sin\left[(\omega_1 - \omega_2)t\right] \simeq 0 \end{cases} \qquad (98)$$

Subsequently, $d^2\left[\phi(t)\right]$ (79) at $(t = 1.5\,T_{Earth} \simeq 20\,T_{Moon})$ (94) is $d^2(3\pi)$ (99) with $(n \simeq 20)$.

$$
\begin{aligned}
d^2\left[\phi\left(t=1.5\,T_{Earth}\right)\right] &= d^2(3\pi) = (404{,}876\ km)^2 = 163{,}924{,}575{,}376\ km^2 \\
&= r_1^2 + 2 r_1 r_2\, Cos\beta + r_2^2 + \ell^2(3\pi) \qquad \qquad (September\ 20,\ 2018) \\
&+ 2\left[\left(r_1 Cos\,\beta + r_2\right)\ell_x(3\pi) + r_1 Sin\,\beta\,\ell_z(3\pi)\right] \\
&= \left[r_1 Cos\,\beta + \ell_x(3\pi) + r_2\right]^2 + \left[r_1 Sin\,\beta + \ell_z(3\pi)\right]^2 + \ell_y^2(3\pi)
\end{aligned} \tag{99}
$$

Along similar lines that we have obtained (94) - (99) from (86) - (93) above, one fourth (100) of the way to every 3-year cycle (86) of Earth after its phase alignment with Moon in the configuration described in **Figure 2**, which is approximately 10 Moon (101) cycles after and 30 Moon (101) cycles before the 3-year global maximum Earth-Moon distances[5], the Earth moving around its own circle of revolution must be passing through a phase of $\left(\omega_2 t = \frac{3\pi}{2}\right)$ (102), with Moon at a phase of a multiple of 2π (101). As we have previously assessed that the $\left[Cos(\omega_1 t)Cos(\omega_2 t)\right]$ term determines the maximum and minimum values of $d^2\left[\phi(t)\right]$ (44) in its sinusoidal cycles, mainly the $Cos(\omega_1 t)$ term, which must be at dates when the Moon's phase $(\omega_1 t)$ (101) is a multiple of π, we can approximately determine the date one fourth (100) of the way between the 3-year global maximum Earth-Moon distances[5] on March 18, 2017 (93) and March 24, 2020 (92) from **Table 1**, 10 Moon (101) cycles after March 18, 2017 (93) and 30 Moon (101) cycles before March 24, 2020 (92). As such, looking at **Table 1**, we observe this date to be December 19, 2017 (105), when the Earth-Moon distance[5] is a local maximum and is 406,603 km (105), 276 days after March 18, 2017 (93) and 275 days before September 20, 2018 (99), which is 30 cycles of the Moon before March 24, 2020 (92). On this date, the various *Sine* and *Cosine* values in $d^2\left[\phi(t)\right]$ (79) have values as in (104). Based on the data in **Table 1**, the global minimum Earth-Moon distance[5] between the dates of January 10, 2017 and January 9, 2021, observed as 356,565 km on January 1, 2018, is also just half a Moon cycle (31) after December 19, 2017 (105).

$$
T_{Earth} \times 0.75 = \left(T_{Moon} \times 13.3687466\right) \times 0.75 \simeq T_{Moon} \times 10 \quad \Rightarrow \quad t = 0.75\,T_{Earth} \simeq 10\,T_{Moon} \tag{100}
$$

$$
t = 0.75\,T_{Earth} \simeq 10\,T_{Moon} \quad \Rightarrow \quad \omega_1 t \simeq 10\,\omega_{Moon} T_{Moon} = 10 \times 2\pi\ radians \quad (Moon's\ \ Phase) \tag{101}
$$

$$
t = 0.75\,T_{Earth} \simeq 10\,T_{Moon} \Rightarrow \omega_2 t = 0.75\,\omega_{Earth} T_{Earth} = 0.75 \times 2\pi = \frac{3\pi}{2}\ radians \quad (Earth's\ \ Phase) \tag{102}
$$

$$
t = 0.75\,T_{Earth} \simeq 10\,T_{Moon} \Rightarrow (\omega_1 - \omega_2)t \simeq \left(10\,\omega_{Moon} T_{Moon} - 0.75\,\omega_{Earth} T_{Earth}\right) = \left[(9 \times 2\pi) + \frac{\pi}{2}\right] radians \tag{103}
$$

$$
t = 0.75\,T_{Earth} \simeq 10\,T_{Moon} \quad \Rightarrow \quad \begin{cases} Cos(\omega_1 t) = Sin\left[(\omega_1 - \omega_2)t\right] \simeq 1 \\ Sin(\omega_2 t) \simeq -1 \\ Sin(\omega_1 t) = Cos(\omega_2 t) = Cos\left[(\omega_1 - \omega_2)t\right] \simeq 0 \end{cases} \tag{104}
$$

Subsequently, $d^2\left[\phi(t)\right]$ (79) at $\left(t = 0.75\,T_{Earth} \simeq 10\,T_{Moon}\right)$ (100) is $d^2\left(\frac{3\pi}{2}\right)$ (105) with $(n \simeq 10)$.

$$d^2\left[\phi\left(t=0.75\,T_{Earth}\right)\right]=d^2\left(\frac{3\pi}{2}\right)=\left(406{,}603\ km\right)^2=165{,}325{,}999{,}609\ km^2$$

$$=r_1^{\,2}+r_2^{\,2}+\ell^2\left(\frac{3\pi}{2}\right) \qquad \left(December\ 19,\ 2017\right)$$

$$+2\left[r_1 Cos\,\beta\,\ell_x\left(\frac{3\pi}{2}\right)+r_2\ell_y\left(\frac{3\pi}{2}\right)+r_1 Sin\,\beta\,\ell_z\left(\frac{3\pi}{2}\right)\right] \qquad (105)$$

$$=\left[r_1 Cos\,\beta+\ell_x\left(\frac{3\pi}{2}\right)\right]^2+\left[r_1 Sin\,\beta+\ell_z\left(\frac{3\pi}{2}\right)\right]^2+\left[r_2+\ell_y\left(\frac{3\pi}{2}\right)\right]^2$$

Similarly, as we have obtained (100) - (105) from (86) - (93) above, three fourths (106) of the way to every 3-year cycle (86) of Earth after its phase alignment with Moon in the configuration described in **Figure 2**, which is approximately 30 Moon (101) cycles after and 10 Moon (101) cycles before the 3-year global maximum Earth-Moon distances[5], the Earth moving around its own circle of revolution must be passing through a phase of $\left(\omega_2 t=\frac{9\pi}{2}\right)$ (108) with the Moon at a phase of multiple of 2π (107). As we have previously assessed that the $\left[Cos\left(\omega_1 t\right)Cos\left(\omega_2 t\right)\right]$ term determines the maximum and minimum values of $d^2\left[\phi(t)\right]$ (44) in its sinusoidal cycles, mainly the $Cos\left(\omega_1 t\right)$ term, which must be at dates when the Moon's phase $\left(\omega_1 t\right)$ (107) is a multiple of π, we can approximately determine that the date three fourths (106) of the way between the 3-year global maximum Earth-Moon distances[5] on March 18, 2017 (93) and March 24, 2020 (92) from **Table 1**, 30 Moon (101) cycles after March 18, 2017 (93) and 10 Moon (101) cycles before March 24, 2020 (92). As such, checking from **Table 1**, this date is observed to be June 23, 2019 (111), when the Earth-Moon distance[5] is a local maximum at a value of 404,548 km (111), 276 days after September 20, 2018 (99) and 275 days before March 24, 2020 (92), which is 30 cycles of the Moon after March 18, 2017 (93). On this date, the various *Sine* and *Cosine* values in $d^2\left[\phi(t)\right]$ (79) have values as in (110).

$$T_{Earth}\times 2.25=\left(T_{Moon}\times 13.3687466\right)\times 2.25\simeq T_{Moon}\times 30 \quad\Rightarrow\quad t=2.25\,T_{Earth}\simeq 30\,T_{Moon} \quad (106)$$

$$t=2.25\,T_{Earth}\simeq 30\,T_{Moon} \;\Rightarrow\; \omega_1 t\simeq 30\,\omega_{Moon}T_{Moon}=30\times 2\pi\ radians \quad \left(Moon's\ Phase\right) \quad (107)$$

$$t=2.25\,T_{Earth}\simeq 30\,T_{Moon} \Rightarrow \omega_2 t=2.25\,\omega_{Earth}T_{Earth}=2.25\times 2\pi=\frac{9\pi}{2}\ radians \quad \left(Earth's\ Phase\right) \quad (108)$$

$$t=2.25\,T_{Earth}\simeq 30\,T_{Moon}\Rightarrow\left(\omega_1-\omega_2\right)t\simeq\left(30\,\omega_{Moon}T_{Moon}-2.25\,\omega_{Earth}T_{Earth}\right)=\left[\left(28\times 2\pi\right)-\frac{\pi}{2}\right]rad.s \quad (109)$$

$$t=2.25\,T_{Earth}\simeq 30\,T_{Moon} \quad\Rightarrow\quad \begin{cases} Cos\left(\omega_1 t\right)=Sin\left(\omega_2 t\right)\simeq 1 \\ Sin\left[\left(\omega_1-\omega_2\right)t\right]\simeq -1 \\ Sin\left(\omega_1 t\right)=Cos\left(\omega_2 t\right)=Cos\left[\left(\omega_1-\omega_2\right)t\right]\simeq 0 \end{cases} \quad (110)$$

As such, $d^2\left[\phi(t)\right]$ (79) at $\left(t=2.25T_{Earth}\simeq 30T_{Moon}\right)$ (106) is $d^2\left(\frac{9\pi}{2}\right)$ (111) with $(n\simeq 30)$.

$$
\begin{aligned}
d^2\left[\phi\left(t=2.25T_{Earth}\right)\right] &= d^2\left(\frac{9\pi}{2}\right)=\left(404{,}548\ km\right)^2=163{,}659{,}084{,}304\ km^2 \\
&= r_1^2+r_2^2+\ell^2\left(\frac{9\pi}{2}\right) \qquad \left(June\ 23,\ 2019\right) \\
&+2\left[r_1 Cos\,\beta\,\ell_x\left(\frac{9\pi}{2}\right)-r_2\ell_y\left(\frac{9\pi}{2}\right)+r_1 Sin\,\beta\,\ell_z\left(\frac{9\pi}{2}\right)\right] \qquad (111) \\
&= \left[r_1 Cos\,\beta+\ell_x\left(\frac{9\pi}{2}\right)\right]^2+\left[r_1 Sin\,\beta+\ell_z\left(\frac{9\pi}{2}\right)\right]^2+\left[r_2-\ell_y\left(\frac{9\pi}{2}\right)\right]^2
\end{aligned}
$$

At this point, we can rewrite $d^2(0)$ (93), $d^2\left(\frac{3\pi}{2}\right)$ (105), $d^2(3\pi)$ (99), $d^2\left(\frac{9\pi}{2}\right)$ (111), and $d^2(6\pi)$ (92) as in (112) - (116), to visualize better the Earth-Moon distance[5] behavior relative to their phases in their individual circular orbits of revolution, and make further analysis to analytically be able to find an approximate value for r_1 (7).

$$
\begin{aligned}
d^2\left[\phi\left(t=t_0=0\right)\right] &= d^2(0)=\left(404{,}650\ km\right)^2=163{,}741{,}622{,}500\ km^2 \qquad \left(March\ 18,\ 2017\right) \\
&= r_1^2-2r_1r_2\,Cos\beta+r_2^2+\ell^2(0)+2\left[\left(r_1 Cos\,\beta-r_2\right)\ell_x(0)+r_1 Sin\,\beta\,\ell_z(0)\right] \qquad (112) \\
&= \left[r_1 Cos\,\beta+\ell_x(0)-r_2\right]^2+\left[r_1 Sin\,\beta+\ell_z(0)\right]^2+\ell_y^2(0)
\end{aligned}
$$

$$
\begin{aligned}
d^2\left[\phi\left(t=0.75T_{Earth}\right)\right] &= d^2\left(\frac{3\pi}{2}\right)=\left(406{,}603\ km\right)^2=165{,}325{,}999{,}609\ km^2 \\
&= r_1^2+r_2^2+\ell^2\left(\frac{3\pi}{2}\right) \qquad \left(December\ 19,\ 2017\right) \\
&+2\left[r_1 Cos\,\beta\,\ell_x\left(\frac{3\pi}{2}\right)+r_2\ell_y\left(\frac{3\pi}{2}\right)+r_1 Sin\,\beta\,\ell_z\left(\frac{3\pi}{2}\right)\right] \qquad (113) \\
&= \left[r_1 Cos\,\beta+\ell_x\left(\frac{3\pi}{2}\right)\right]^2+\left[r_1 Sin\,\beta+\ell_z\left(\frac{3\pi}{2}\right)\right]^2+\left[r_2+\ell_y\left(\frac{3\pi}{2}\right)\right]^2
\end{aligned}
$$

$$
\begin{aligned}
d^2\left[\phi\left(t=1.5T_{Earth}\right)\right] &= d^2(3\pi)=\left(404{,}876\ km\right)^2=163{,}924{,}575{,}376\ km^2 \\
&= r_1^2+2r_1r_2\,Cos\beta+r_2^2+\ell^2(3\pi) \qquad \left(September\ 20,\ 2018\right) \\
&+2\left[\left(r_1 Cos\,\beta+r_2\right)\ell_x(3\pi)+r_1 Sin\,\beta\,\ell_z(3\pi)\right] \qquad (114) \\
&= \left[r_1 Cos\,\beta+\ell_x(3\pi)+r_2\right]^2+\left[r_1 Sin\,\beta+\ell_z(3\pi)\right]^2+\ell_y^2(3\pi)
\end{aligned}
$$

$$d^2\left[\phi\left(t=2.25T_{Earth}\right)\right]=d^2\left(\frac{9\pi}{2}\right)=\left(404,548\ km\right)^2=163,659,084,304\ km^2$$

$$=r_1^2+r_2^2+\ell^2\left(\frac{9\pi}{2}\right) \qquad (June\ 23,\ 2019)$$

$$+2\left[r_1 Cos\,\beta\ell_x\left(\frac{9\pi}{2}\right)-r_2\ell_y\left(\frac{9\pi}{2}\right)+r_1 Sin\,\beta\,\ell_z\left(\frac{9\pi}{2}\right)\right] \qquad (115)$$

$$=\left[r_1 Cos\,\beta+\ell_x\left(\frac{9\pi}{2}\right)\right]^2+\left[r_1 Sin\,\beta+\ell_z\left(\frac{9\pi}{2}\right)\right]^2+\left[r_2-\ell_y\left(\frac{9\pi}{2}\right)\right]^2$$

$$d^2\left[\phi\left(t=3T_{Earth}\right)\right]=d^2\left(6\pi\right)=\left(406,692\ km\right)^2=165,398,382,864\ km^2 \qquad (March\ 24,\ 2020)$$

$$=r_1^2-2r_1r_2\,Cos\beta+r_2^2+\ell^2\left(6\pi\right)+2\left[\left(r_1 Cos\,\beta-r_2\right)\ell_x\left(6\pi\right)+r_1 Sin\,\beta\,\ell_z\left(6\pi\right)\right] \qquad (116)$$

$$=\left[r_1 Cos\,\beta+\ell_x\left(6\pi\right)-r_2\right]^2+\left[r_1 Sin\,\beta+\ell_z\left(6\pi\right)\right]^2+\ell_y^2\left(6\pi\right)$$

Based on our analysis above for the Earth-Moon configuration described in **Figure 2**, we have made the assessment that around every 10 cycles of the Moon (31), the Earth has a phase change of $\left(\phi\simeq\frac{3\pi}{2}\right)$ (112) - (116) in its individual circular orbit of revolution, with a phase of a multiple of 6π on March 18, 2017 (93) and March 24, 2020 (92). More specifically, in its individual circular orbit of revolution, the Earth passes through phase $\left(\phi\simeq 0\right)$ (112) in its motion relative to Moon on March 18, 2017 (93), and the Earth passes through phase $\left(\phi\simeq\frac{3\pi}{2}\right)$ (113) in its motion relative to Moon on December 19, 2017 (105), and Earth passes through phase $\left(\phi\simeq 3\pi\right)$ (114) in its motion relative to the Moon on September 20, 2018 (99), and the Earth passes through phase $\left(\phi\simeq\frac{9\pi}{2}\right)$ (115) in its motion relative to the Moon on June 23, 2019 (111), and the Earth passes through phase $\left(\phi\simeq 6\pi\right)$ (116) in its motion relative to the Moon on March 24, 2020 (92). Looking at the dates of phases $\left(\phi\simeq 0\right)$ (112), $\left(\phi\simeq\frac{3\pi}{2}\right)$ (113), $\left(\phi\simeq 3\pi\right)$ (114), $\left(\phi\simeq\frac{9\pi}{2}\right)$ (115), and $\left(\phi\simeq 6\pi\right)$ (116) of Earth in its individual circular orbit of revolution within a 3-year cycle with respect to the Moon in the configuration described in **Figure 2**, which is approximately 40 cycles (87) of the Moon, we observe that the dates of these $\left(\phi\simeq\frac{3\pi}{2}\right)$ multiples of phase for the Earth-Moon system occur around equinoxes[14,16] and solstices[15,16] of the Earth with respect to the Sun during their yearly relative cycles. This is one of the major findings of this Article.

Further, along the lines of our analysis in (86) - (116), we can conclude that based on (86), (3), and (34) - (35), around every 1 cycle of the Moon (31), the Earth has a phase change of $\left(\phi\simeq\frac{3\pi}{20}\right)$ (117) in its individual circular orbit of revolution, another finding of this Article.

$$T_{Earth}\times 3 \simeq T_{Moon}\times 40 \quad \& \quad t \simeq T_{Moon} = \frac{3}{40}T_{Earth}$$
$$\Rightarrow \phi(t)=\phi=\phi_{Earth-Moon}=\omega_2 t=\omega_2 T_{Moon}=\omega_2\frac{2\pi}{\omega_1}=\frac{2\pi}{T_{Earth}}T_{Moon}\simeq\frac{3\pi}{20} \qquad (117)$$

In the **ARTICLE 3** "Analytical calculation of earth and sun orbital parameters from distance data", for the Earth-Sun system, it is discovered that the Earth passes through its $(\phi_{Earth-Sun}=0)$ (118) phase point every year around January 17 with respect to its motion relative to the Sun, and the phase $\phi_{Earth-Sun}$ of the Earth in its individual circular orbit of revolution in space with respect to its motion relative to the Sun takes the values listed in (118) - (130), also based on the data calculated in the Excel sheet "Tan, A.P., Asli Pinar Tan Analysis Based on Earth-Sun distance (d) Landsat.xlsx"[17].

$$\phi_{Earth-Sun}=\mathbf{0.00058673254949726}\,radians=\mathbf{0^\circ.034}\simeq 0 \qquad on\ January\ 17 \qquad (118)$$

$$\phi_{Earth-Sun}=\mathbf{0.9943139612468}\,radians=\mathbf{56^\circ.970} \qquad on\ March\ 18 \qquad (119)$$

$$\phi_{Spring\ Equinox}=\mathbf{1.02676771671574}\,radians=\mathbf{58^\circ.829} \qquad on\ March\ 20 \qquad (120)$$

$$\phi_{Earth-Sun}=\mathbf{1.09345173465532}\,radians=\mathbf{62^\circ.650} \qquad on\ March\ 24 \qquad (121)$$

$$\phi_{Earth-Sun}=\mathbf{1.57138305934439}\,radians=\mathbf{90^\circ.034}\simeq\frac{\pi}{2} \qquad on\ April\ 18 \qquad (122)$$

$$\phi_{Earth-Sun}=-\phi_{0,Earth-Sun}=\mathbf{2.63736997198951}\,rad=\mathbf{151^\circ.11} \qquad on\ June\ 21 \qquad (123)$$

$$\phi_{Earth-Sun}=\mathbf{2.65857971564607}\,rad=\mathbf{152^\circ.325} \qquad on\ June\ 23 \qquad (124)$$

$$\phi_{Earth-Sun}=\mathbf{3.14217938613929}\,radians=\mathbf{180^\circ.034}\simeq\pi \qquad on\ July\ 18 \qquad (125)$$

$$\phi_{Earth-Sun}=\mathbf{-2.08889119201738}\,radians=\mathbf{240^\circ.315}=\mathbf{-119^\circ.625} \qquad on\ September\ 20 \quad (126)$$

$$\phi_{Autumn\ Equinox}=\mathbf{-2.05488176708229}\,radians=\mathbf{242^\circ.264}=\mathbf{-117^\circ.736} \qquad on\ September\ 22 \quad (127)$$

$$\phi_{Earth-Sun}=\mathbf{-1.5702095942454}\,radians=\mathbf{-89^\circ.966}\simeq-\frac{\pi}{2} \qquad on\ October\ 17 \qquad (128)$$

$$\phi_{Earth-Sun}=\mathbf{5.74535336837779}\,rad=\mathbf{329^\circ.184}=\mathbf{-30^\circ.816} \qquad on\ December\ 19 \quad (129)$$

$$\phi_{Earth-Sun}=\pi-\phi_{0,Earth-Sun}=\mathbf{5.7789626255793}\,rad=\mathbf{331^\circ.11}=\mathbf{-28^\circ.89} \quad on\ December\ 21 \quad (130)$$

Based on (112) and (119), as well as (116) and (121), we can infer that whenever the Earth and the Moon moving around their individual circular orbits of revolution pass from the same aligned phase of a multiple of 2π around the same date every 3-year cycles of the Earth, which is approximately every 40 cycles (87) of the Moon, after their initial phase alignment of $[\phi_0(t_0)=0]$ (4) at time $(t=t_0=0)$, in the configuration described in **Figure 2**, the Earth passes through its Spring Equinox[14,16] with respect to the Sun around March 20 (120), one of the dates around which Solar and Lunar Eclipses[11,19] occur due to Sun-Earth-Moon motion topology. Therefore, based on the configuration in **Figure 2**, whenever the phase of the Earth in its individual circular orbit of revolution is around $(\phi=\phi_{Earth-Moon}\simeq 0^\circ)$ (112) with respect to its motion relative to the Moon, the phase of the Earth in its individual circular orbit of revolution is around $\left(\phi_{Earth-Sun} \simeq 60^{\circ}\right)$ (119) - (121) with respect to its motion relative to the Sun. Therefore, the $\left(\phi_{Earth-Moon} \simeq 0^{\circ}\right)$ (112) phase of Earth in its Earth-Moon configuration of **Figure 2** corresponds to the $\left(\phi_{Earth-Sun} \simeq 60^{\circ}\right)$ (119) - (121) phase of Earth in its Earth-Sun configuration of **Figure 2**, revealing that the $\left(\phi = \phi_{Earth-Moon} = 0^{\circ}\right)$ (112) plane of Earth-Moon configuration and the $\left(\phi = \phi_{Earth-Sun} = 0^{\circ}\right)$ (118) plane Earth-Sun configuration are around 60° (131) apart on the Earth's individual plane of revolution. This is another fundamental finding of this Article.

Similarly, whenever the phase of the Earth in its individual circular orbit of revolution is around $\left(\phi = \phi_{Earth-Moon} \simeq 3\pi = 180^{\circ}\right)$ (114), or an odd multiple of 3π in the more general sense, in its motion relative to Moon, the phase of the Earth in its individual circular orbit of revolution is around $\left(\phi_{Earth-Sun} \simeq 240^{\circ} = -120^{\circ}\right)$ (126) with respect to its motion relative to the Sun, and Earth passes through its Autumn Equinox[14,16] with respect to the Sun around September 20 (126) and September 22 (127), another one of the dates around which Solar and Lunar Eclipses[11,19] occur due to motion topology of Sun-Earth-Moon system. Moreover, whenever the phase of the Earth in its individual circular orbit of revolution is around $\left(\phi = \phi_{Earth-Moon} \simeq \frac{9\pi}{2} = 90^{\circ}\right)$ (115) in its motion relative to Moon, the phase of the Earth in its individual circular orbit of revolution is around $\left(\phi_{Earth-Sun} = -\phi_{0,Earth-Sun} \simeq 151^{\circ}\right)$ (123) - (124) with respect to its motion relative to the Sun, and Earth passes through its Summer Solstice[15,16] with respect to the Sun. Further, whenever the phase of the Earth in its individual circular orbit of revolution is around $\left(\phi = \phi_{Earth-Moon} \simeq \frac{3\pi}{2} = -90^{\circ}\right)$ (113) in its motion relative to Moon, the phase of the Earth in its individual circular orbit of revolution is around $\left(\phi_{Earth-Sun} \simeq \pi - \phi_{0,Earth-Sun} \simeq -30^{\circ}\right)$ (129) - (130) with respect to its motion relative to the Sun, and Earth passes through its Winter Solstice[15,16] with respect to the Sun. With the orbital period[10,11] T_{Moon} (31) of the Moon[11] and the Earth[12]'s yearly[13] orbital period[10,12] T_{Earth} (32), which is the same as the orbital period[10,12] of Sun, all consistently constant, and thus having constant values for the Moon[11] and Earth[12]'s angular frequencies $\left(\omega_1 = \omega_{Moon}\right)$ (34) and $\left(\omega_2 = \omega_{Earth}\right)$ (35), respectively, we naturally expect a constant difference between the phase value of the Earth relative to the Earth-Sun configuration and the phase value of the Earth relative to the Earth-Moon configuration at any date, within Earth's individual circular orbit of revolution. As a result of these observations (112) - (130), we reach the overall conclusion that, in its individual circular orbit of revolution, the phase $\phi_{Earth-Sun}$ (131) of the Earth in its motion relative to the Earth- Sun configuration at any date is around 60° ahead of its phase $\phi_{Earth-Moon}$ (3) for its motion relative to the Earth-Moon configuration. This is yet another fundamental finding of this Article.

$$\phi_{Earth-Sun} \simeq \phi_{Earth-Moon} + 60^{\circ} \qquad (at\ any\ date\ of\ the\ year) \qquad (131)$$

Our finding in the in the **ARTICLE 3** "Analytical calculation of earth and sun orbital parameters from distance data" that in its own cycle of individual revolution, the Sun is lagging with a phase difference of $\left(-\phi_{Earth-Sun} = \phi_{0,Earth-Sun} \simeq -151^{\circ}\right)$ (123) - (124) behind Earth in its own cycle of individual revolution, or equivalently, the Earth in its own cycle of individual revolution is moving ahead of the Sun in its own cycle of individual revolution, by a phase difference of $\left(\phi_{Earth-Sun} = -\phi_{0,Earth-Sun} \simeq 151^{\circ}\right)$ (123) - (124), as well as our finding in this Article that in its individual circular orbit of revolution, the phase $\phi_{Earth-Sun}$ (131) of the Earth in its motion relative to the Earth- Sun configuration at any date is around 60° ahead of its phase $\phi_{Earth-Moon}$ (3) for its motion relative to the Earth-Moon configuration, reveal and exhibit an alignment of angles in the Sun-Earth motion topology and Earth-Moon motion topology. This alignment was also discovered a little above in our analysis, where it was found that $\omega_{\left(Moon-\vec{\ell}[\phi(t)]\right)}$ (58) leads to a phase difference of $-152^{\circ}.1$ (58) per year in the synodic[38] motion of the nodal orientation variation motion of $\vec{\ell}\left[\phi(t)\right]$ (8) with respect to the motion of the Moon[11] in its individual circular orbit of revolution around radius $\vec{r}_1\left[\phi(t)\right]$ (5), which is very close to $-151^{\circ}.11$ (50) of the constant phase difference $\phi_{0,Earth-Sun}$ (50) of the Sun in its individual circular orbit of revolution throughout a year relative to Earth's phase in its own individual circular orbit of revolution, based on our findings in the **ARTICLE 3** "Analytical calculation of earth and sun orbital parameters from distance data", which points out an alignment of the nodal orientation variation motion of $\vec{\ell}\left[\phi(t)\right]$ (8) with the yearly motion of the Sun, also supported by the observation[11,19] that the Moon[11] orbits closer to the so called ecliptic[29] plane than the equatorial plane of Earth, which apparently must be due to orientation of $\vec{\ell}\left[\phi(t)\right]$ (8). This is another important analysis result we have obtained in this Article.

As mentioned a little above, the last minor lunar[11] standstill[11,41] occurred[42] in October 2015, within the 18.6 year (60) cycle of the angle between the Moon[11]'s Orbit[19] and Earth[12]'s equator. The angle between the Moon[11]'s Orbit[19] and Earth[12]'s equator was observed[42] to reach a minimum of $18^{\circ}.134$ south of the equator on September 21, 2015, namely at the southern standstill[42], followed by a minimum of $18^{\circ}.140$ north of the equator on October 3, 2015, namely at the northern standstill[42]. Checking the Sun-Earth-Moon topology we have set forth as a result of our previous analysis in **Figure 8** at $\left(t = t_0 + 9.3\,years\right)$ end of 1 Half Cycle at a minor lunar[11] standstill[11,41], and considering that according to our findings in the **ARTICLE 3** "Analytical calculation of earth and sun orbital parameters from distance data" the North Pole[23,34] of the Earth[12] in the configuration of **Figure 2** is directed as $\left(\hat{\boldsymbol{u}}_{Earth\perp} = -\hat{z}\right)$ (49), any angle $\alpha_{Moon\ Orbit\ \measuredangle\ Earth\ Equator}$ (73) - (75), or equivalently any angle $\alpha_{\vec{\ell}[\phi(t)]-plane\ \measuredangle\ Earth\ Orbital\ Plane}$ (73) - (75) between the plane of motion of the vector $\vec{\ell}\left[\phi(t)\right]$ (8) and Earth[12]'s equator, to the south of equator must take on a value $\left(0 \leq \alpha_{\vec{\ell}[\phi(t)]-plane\ \measuredangle\ Earth\ Orbital\ Plane} \leq 180^{\circ}\right)$, and to the north of equator must take on a value $\left(180^{\circ} \leq \alpha_{\vec{\ell}[\phi(t)]-plane\ \measuredangle\ Earth\ Orbital\ Plane} \leq 360^{\circ}\right)$ in the configuration of **Figure 2**.

Note that we had previously made the analysis that the so called observed ecliptic[29] plane of the

Earth in its motion around the Sun is in fact not really a plane, but the approximate angles said to be observed with respect to the ecliptic[29] are angles observed with respect to the plane of the distance vector between the Earth and the Sun around the year, and in fact the so called observed ecliptic[29] plane can approximately be considered to be the plane of individual circular orbit of revolution of the Sun, as $\left(\mathbf{12,370.477}\,km \simeq r_2 \ll r_{Sun} \simeq \mathbf{149,609,642.749}\,km\right)$ (45) - (46). Therefore, based on the configuration in **Figure 8** depicted with Sun-Earth Cartesian $(\hat{x},\hat{y},\hat{z})$ coordinates, where $\phi_{Earth-Sun}$ (131) of the Earth in its motion relative to the Earth- Sun configuration at any date is around 60° ahead of its phase $\phi_{Earth-Moon}$ (3), we reason that $\alpha_{\vec{\ell}[\phi(t)]-plane \measuredangle Earth\ Equator,Min,North}$ (133) must be $\left(-18^\circ.140\right)$ at the northern standstill[42] on October 3, 2015, followed by $\alpha_{\vec{\ell}[\phi(t)]-plane \measuredangle Earth\ Equator,Min,South}$ (132) of $161^\circ.866$ at the southern standstill[42] on September 21, 2015, with respect to the $\hat{x}-\hat{y}$ plane in Moon-Earth configuration of **Figure 2**, also supported by the observation[11,19] that Moon[11] orbits closer to the so called ecliptic[29] plane, which apparently must be due to orientation of $\vec{\ell}\left[\phi(t)\right]$ (8) as we had concluded before.

$$\alpha_{\vec{\ell}[\phi(t)]-plane \measuredangle Earth\ Orbital\ Plane,Min,South} \simeq 180^\circ - 18^\circ.134 = 161^\circ.866 \qquad \left(September\ 21,\ 2015\right) \quad (132)$$

$$\alpha_{\vec{\ell}[\phi(t)]-plane \measuredangle Earth\ Orbital\ Plane,Min,North} \simeq -18^\circ.140 \qquad \left(October\ 3,\ 2015\right) \quad (133)$$

Table 2 Min-Max Earth-Moon distances[5] in kilometers (km) around lunar[11] standstill[11,41] in 2015

Earth-Moon Distance Data					
Date	Time (UTC)	Earth-Moon Distance [km]	Approximate Earth Phase $(\phi=\omega_2 t)$ [radians]	Approximate Moon Phase $(\omega_1 t)$ [radians]	Notes
18.08.2015	02:32	405.848	-3.150π (135)	-42π (140)	MAX (Local)
30.08.2015	15:21	358.290	-3.075π (136)	-41π (149)	MIN (Local)
14.09.2015	11:27	**406.464**	-3.000π (137)	-40π (143)	**MAX (Yearly)**
28.09.2015	01:45	**356.877**	-2.925π (138)	-39π (152)	**MIN (Yearly)**
11.10.2015	13:17	406.388	-2.850π (139)	-38π (146)	MAX (Local)

Based on Earth-Moon distance[5] data provided in timeanddate.com website, as also presented in **Table 2**, the yearly maximum and minimum Earth-Moon distances[5] of 2015 are around these dates, in fact at a quarter lunar[11,33] month before and after the southern standstill[42] on September 21, 2015, as the yearly maximum of 406,464 km is on September 14, 2015 and the yearly minimum of 356,877 km is on September 28, 2015. Symmetrically around these yearly extrema, also observed[5] are local maximum Earth-Moon distances[5] of 405,848 km on August 18, 2015 and 406,388 km on October 11, 2015, with a local minimum Earth-Moon distance[5] of 358,290 km on August 30, 2015.

Again based on observation[19], the shape of Moon[11]'s Orbit[19] is pretty much fixed in its plane of motion over a lunar[11,19,33] month of period T_{Moon} (31), and the variation over $T_{N,\bar{\ell}[\phi(t)]}$ (60), or an 18.6 year (60) cycle of the angle $\alpha_{\bar{\ell}[\phi(t)]-plane\,\measuredangle\,Earth\,Orbital\,Plane}$ (73) - (75) between the plane of motion of $\bar{\ell}\left[\phi(t)\right]$ (8) and the Earth[12]'s equator must be relatively small over consecutive lunar[11,19,33] months of period T_{Moon} (31), as $\left(T_{N,\bar{\ell}[\phi(t)]} \gg T_{Moon}\right)$ (134).

$$T_{N,\bar{\ell}[\phi(t)]} \simeq 18.599\,years \quad \& \quad T_{Moon} \simeq 27.321661\,days \;\Rightarrow\; T_{N,\bar{\ell}[\phi(t)]} \gg T_{Moon} \qquad (134)$$

The southern standstill[42] which is on September 21, 2015 is midway between the yearly maximum on September 14, 2015 and the yearly minimum on September 28, 2015, at which dates the phase $(\omega_1 t)$ (2) of the Moon[11,19] in its individual circular orbit of revolution must be a multiple of 2π with $\bar{r}_1$ (5) in $\hat{x}-\hat{z}$ plane and an odd multiple of π with $\bar{r}_1$ (5) in $(-\hat{x})-(-\hat{z})$ plane, respectively, in the Moon-Earth configuration of **Figure 2**. Similarly, the northern standstill[42] on October 3, 2015 is about midway between the yearly minimum on September 28, 2015 and local maximum on October 11, 2015, at which dates the phase $(\omega_1 t)$ (2) of the Moon[11,19] in its individual circular orbit of revolution must be an odd multiple of π with $\bar{r}_1$ (5) in $(-\hat{x})-(-\hat{z})$ plane and a multiple of 2π with $\bar{r}_1$ (5) in $\hat{x}-\hat{z}$ plane, respectively, according to our previous analysis. Also as part of our previous analysis in this Article, we had determined that the Moon's individual circular orbit of revolution around radius $\bar{r}_1\left[\phi(t)\right]$ (5) must be in clockwise[35] direction from a vantage point from above the North Pole[23,34] of the Earth[12], which is directed as $\left(\hat{\boldsymbol{u}}_{Earth\perp} = -\hat{z}\right)$ (49) in the configuration of **Figure 2**. In light of all this information, we can reason that at the southern standstill[42] on September 21, 2015, the phase $(\omega_1 t)$ (2) of the Moon[11,19] in its individual circular orbit of revolution must have a value equivalent to $\frac{\pi}{2}$ radians, with $\bar{r}_1$ (5) in the $(+\hat{y})$ direction in the Moon-Earth configuration of **Figure 2**, midway between September 14, 2015, where the phase $(\omega_1 t)$ (2) of the Moon[11,19] is an a multiple of 2π, and September 28, 2015, where the phase $(\omega_1 t)$ (2) of the Moon[11,19] is an odd multiple of π. Along the same lines of reasoning, at the northern standstill[42] on October 3, 2015, the phase $(\omega_1 t)$ (2) of the Moon[11,19] in its individual circular orbit of revolution must have a value equivalent to $\left(-\frac{\pi}{2}\right)$ radians, with $\bar{r}_1$ (5) in the $(-\hat{y})$ direction in the Moon-Earth configuration of **Figure 2**, midway between September 28, 2015, where phase $(\omega_1 t)$ (2) of the Moon[11,19] is an odd multiple of π, and October 11, 2015, where phase $(\omega_1 t)$ (2) of the Moon[11,19] is a multiple of 2π. Subsequently, also based on the Sun-Earth-Moon configuration of **Figure 8**, we can make an intelligent inference that observed[42] minimum angles $\alpha_{\bar{\ell}[\phi(t)]-plane\,\measuredangle\,Earth\,Equator,Min,South}$ (132) and $\alpha_{\bar{\ell}[\phi(t)]-plane\,\measuredangle\,Earth\,Equator,Min,North}$ (133) between the plane of motion of vector $\bar{\ell}\left[\phi(t)\right]$ (8) and the

Earth[12]'s equator at the southern standstill[42] on September 21, 2015 and at northern standstill[42] on October 3, 2015, respectively, must have been observed especially on these dates during the 2015 standstill[42] lunar[11,19,33] month, because the $\vec{r}_1$ (5) vector at the $\left(\pm\frac{\pi}{2}\right)$ radian values of phase $(\omega_1 t)$ (2) of the Moon[11,19] must be perpendicular $(\perp)$ to the axis in the plane of motion of vector $\vec{\ell}\left[\phi(t)\right]$ (8) that makes the minimum angles (132) - (133) with Earth[12]'s equatorial plane. That is, the observed[42] minimum angles $\alpha_{\vec{\ell}[\phi(t)]-plane\,\measuredangle\,Earth\,Equator,Min,South}$ (132) and $\alpha_{\vec{\ell}[\phi(t)]-plane\,\measuredangle\,Earth\,Equator,Min,North}$ (133) between the plane of motion of vector $\vec{\ell}\left[\phi(t)\right]$ (8) and the Earth[12]'s equator at the southern standstill[42] on September 21, 2015 and at northern standstill[42] on October 3, 2015, respectively, must be the angles that vector $\vec{\ell}\left[\phi(t)\right]$ (8) makes with Earth[12]'s equatorial plane on the dates where the phase $(\omega_1 t)$ (2) of the Moon[11,19] in its individual circular orbit of revolution is a multiple of 2π and an odd multiple of π, respectively, such as the ones in **Table 2**.

Again based on the Earth-Moon distance[5] data provided in timeanddate.com, the yearly maximum Earth-Moon distance[5] of 406,464 km on September 14, 2015 occurs[5] 20 Moon cycles (31) before the occurrence of the 3-year cycle (93) global maximum Earth-Moon distance[5] observed on March 18, 2017 (93) at time $(t = t_0 = 0)$. Therefore, along the same lines of our reasoning above in (86) - (116), and based on (117), on September 14, 2015 the Earth must have a phase of approximately $(\phi \simeq -3\pi)$ (137) in its individual circular orbit of revolution. Similarly, the local maximum Earth-Moon distances[5] of 405,848 km on August 18, 2015 and 406,388 km on October 11, 2015 occur[5] 21 and 19 Moon cycles (31), respectively, before the occurrence of the 3-year cycle (93) global maximum Earth-Moon distance[5] observed on March 18, 2017 (93) at time $(t = t_0 = 0)$. Hence, on August 18, 2015, Earth must have a phase of approximately $(\phi \simeq -3.15\pi)$ (135), and on October 11, 2015, Earth must have a phase of approximately $(\phi \simeq -2.85\pi)$ (139), in its individual circular orbit of revolution. With similar reasoning, based on the Earth-Moon distance[5] data provided in timeanddate.com, the yearly minimum Earth-Moon distance[5] of 356,877 km on September 28, 2015 occurs[5] 19.5 Moon cycles (31), and the local minimum of 358,290 km on August 30, 2015 occurs[5] 20.5 Moon cycles (31), before occurrence of the 3-year cycle (93) global maximum Earth-Moon distance[5] observed on March 18, 2017 (93) at time $(t = t_0 = 0)$. As a result, along the same lines of our reasoning above in (86) - (116), and based on (117), on September 28, 2015 and August 30, 2015, Earth must have phases of approximately $(\phi \simeq -2.925\pi)$ (138) and $(\phi \simeq -3.075\pi)$ (136), respectively, in its individual circular orbit of revolution.

$$T_{Earth} \times 3 \simeq T_{Moon} \times 40 \quad \& \quad t \simeq -21 \times T_{Moon} \simeq -21 \times \frac{3}{40} T_{Earth} \qquad (August\ 18,\ 2015)$$
$$\Rightarrow \quad \phi(t) = \phi = \phi_{Earth-Moon} = \omega_2 t \simeq -\frac{2\pi}{T_{Earth}} \times 21 \times \frac{3}{40} T_{Earth} \simeq -\frac{126\pi}{40} = -3.15\pi\ radians \qquad (135)$$

$$T_{Earth}\times 3\simeq T_{Moon}\times 40 \quad \& \quad t\simeq -20.5\times T_{Moon}\simeq -\frac{41}{2}\times\frac{3}{40}T_{Earth} \qquad (August\ 30,\ 2015)$$
$$\Rightarrow \quad \phi(t)=\phi=\phi_{Earth-Moon}=\omega_2 t\simeq -\frac{2\pi}{T_{Earth}}\times\frac{41}{2}\times\frac{3}{40}T_{Earth}\simeq -\frac{123\pi}{40}=-3.075\pi\ radians \qquad (136)$$

$$T_{Earth}\times 3\simeq T_{Moon}\times 40 \quad \& \quad t\simeq -20\times T_{Moon}\simeq -20\times\frac{3}{40}T_{Earth} \qquad (September\ 14,\ 2015)$$
$$\Rightarrow \quad \phi(t)=\phi=\phi_{Earth-Moon}=\omega_2 t\simeq -\frac{2\pi}{T_{Earth}}\times 20\times\frac{3}{40}T_{Earth}\simeq -\frac{120\pi}{40}=-3\pi\ radians \qquad (137)$$

$$T_{Earth}\times 3\simeq T_{Moon}\times 40 \quad \& \quad t\simeq -19.5\times T_{Moon}\simeq -\frac{39}{2}\times\frac{3}{40}T_{Earth} \qquad (September\ 28,\ 2015)$$
$$\Rightarrow \quad \phi(t)=\phi=\phi_{Earth-Moon}=\omega_2 t\simeq -\frac{2\pi}{T_{Earth}}\times\frac{39}{2}\times\frac{3}{40}T_{Earth}\simeq -\frac{117\pi}{40}=-2.925\pi\ radians \qquad (138)$$

$$T_{Earth}\times 3\simeq T_{Moon}\times 40 \quad \& \quad t\simeq -19\times T_{Moon}\simeq -19\times\frac{3}{40}T_{Earth} \qquad (October\ 11,\ 2015)$$
$$\Rightarrow \quad \phi(t)=\phi=\phi_{Earth-Moon}=\omega_2 t\simeq -\frac{2\pi}{T_{Earth}}\times 19\times\frac{3}{40}T_{Earth}\simeq -\frac{114\pi}{40}=-2.85\pi\ radians \qquad (139)$$

Using the found phase $\left[\phi(t)=\omega_2 t\right]$ (3) values in (135) - (139) that Earth must have in its individual circular orbit of revolution on the respective dates, based on $d^2\left[\phi(t)\right]$ (44) and $\vec{d}\left[\phi(t)\right]$ (10) - (11), we determine $d^2\left(-\frac{126\pi}{40}\right)$ (142) on August 18, 2015, $d^2\left(-\frac{123\pi}{40}\right)$ (151) on August 30, 2015, $d^2(-3\pi)$ (145) on September 14, 2015, $d^2\left(-\frac{117\pi}{40}\right)$ (154) on September 28, 2015, and $d^2\left(-\frac{114\pi}{40}\right)$ (148) on October 11, 2015. The phase $(\omega_1 t)$ (2) of the Moon and the Moon-Earth phase difference $\left[(\omega_1-\omega_2)t\right]$ (4) associated with the given Earth phase $\left[\phi(t)=\omega_2 t\right]$ (3) values are, (140) – (141) with (135), (149) - (150) with (136), (143) - (144) with (137), (152) - (153) with (138), and (146) - (147) with (139). Note here that in these calculations, $(\omega_1=\omega_{Moon})$ (34) and $(\omega_2=\omega_{Earth})$ (35) are the values of the Moon[11] and the Earth[12]'s angular frequencies around their individual circular orbits of revolution, respectively, their angular velocity difference being $(\omega_1-\omega_2)$ (36), where we have $(\omega_{Moon}T_{Moon}=2\pi\ radians)$ (34), and we make use of trigonometric identities, as well as utilizing the definition of the magnitude $\ell\left[\phi(t)\right]$ (9) of the vector distance $\vec{\ell}\left[\phi(t)\right]$ (8), which is between the centers of individual circular orbits of revolution of the Earth and the Moon.

$$t\simeq -21\times T_{Moon} \quad \Rightarrow \quad \omega_1 t\simeq -21\omega_{Moon}T_{Moon}\simeq -21\times 2\pi\ radians \qquad (August\ 18,\ 2015)\ (140)$$

$$t \simeq -21\times T_{Moon} \Rightarrow (\omega_1-\omega_2)t \simeq (-42+3.15\pi)\pi \; radians = -38.85\pi \; radians \quad (August\ 18,\ 2015) \quad (141)$$

$$d^2\left(-\frac{126\pi}{40}\right) = (405{,}848\,km)^2 = 164{,}712{,}599{,}104\,km^2 \qquad (August\ 18,\ 2015)$$

$$= r_1^2 - 2r_1r_2 Cos\beta Cos\left(\frac{34\pi}{40}\right) + r_2^2 + \ell^2\left(-\frac{126\pi}{40}\right)$$

$$+2\left\{\begin{matrix}\left[r_1 Cos\beta - r_2 Cos\left(\frac{34\pi}{40}\right)\right]\ell_x\left(-\frac{126\pi}{40}\right) - r_2 Sin\left(\frac{34\pi}{40}\right)\ell_y\left(-\frac{126\pi}{40}\right)\\ + r_1 Sin\beta\,\ell_z\left(-\frac{126\pi}{40}\right)\end{matrix}\right\} \quad (142)$$

$$= \left[r_1 Cos\beta - r_2 Cos\left(\frac{34\pi}{40}\right) + \ell_x\left(-\frac{126\pi}{40}\right)\right]^2 + \left[-r_2 Sin\left(\frac{34\pi}{40}\right) + \ell_y\left(-\frac{126\pi}{40}\right)\right]^2$$

$$+\left[r_1 Sin\beta + \ell_z\left(-\frac{126\pi}{40}\right)\right]^2$$

$$t \simeq -20\times T_{Moon} \quad \Rightarrow \quad \omega_1 t \simeq -20\omega_{Moon}T_{Moon} \simeq -20\times 2\pi \; radians \quad (September\ 14,\ 2015) \quad (143)$$

$$t \simeq -20\times T_{Moon} \quad \Rightarrow \quad (\omega_1-\omega_2)t \simeq (-40+3)\pi \; radians = -37\pi \; radians \quad (September\ 14,\ 2015) \quad (144)$$

$$d^2(-3\pi) = (406{,}464\,km)^2 = 165{,}212{,}983{,}296\,km^2 \qquad (September\ 14,\ 2015)$$

$$= r_1^2 + 2r_1r_2\,Cos\beta + r_2^2 + \ell^2(-3\pi) + 2\left[(r_1 Cos\beta + r_2)\ell_x(-3\pi) + r_1 Sin\beta\,\ell_z(-3\pi)\right] \quad (145)$$

$$= \left[r_1 Cos\beta + r_2 + \ell_x(-3\pi)\right]^2 + \ell_y^2(-3\pi) + \left[r_1 Sin\beta + \ell_z(-3\pi)\right]^2$$

$$t \simeq -19\times T_{Moon} \quad \Rightarrow \quad \omega_1 t \simeq -19\omega_{Moon}T_{Moon} \simeq -19\times 2\pi \; radians \quad (October\ 11,\ 2015) \quad (146)$$

$$t \simeq -19\times T_{Moon} \Rightarrow (\omega_1-\omega_2)t \simeq (-38+2.85)\pi \; radians = -35.15\pi \; radians \quad (October\ 11,\ 2015) \quad (147)$$

$$d^2\left(-\frac{114\pi}{40}\right) = (406{,}388\,km)^2 = 165{,}151{,}206{,}544\,km^2 \qquad (October\ 11,\ 2015)$$

$$= r_1^2 - 2r_1r_2 Cos\beta Cos\left(\frac{34\pi}{40}\right) + r_2^2 + \ell^2\left(-\frac{114\pi}{40}\right)$$

$$+2\left\{\begin{matrix}\left[r_1 Cos\beta - r_2 Cos\left(\frac{34\pi}{40}\right)\right]\ell_x\left(-\frac{114\pi}{40}\right)\\ + r_2 Sin\left(\frac{34\pi}{40}\right)\ell_y\left(-\frac{114\pi}{40}\right) + r_1 Sin\beta\,\ell_z\left(-\frac{114\pi}{40}\right)\end{matrix}\right\} \quad (148)$$

$$= \left[r_1 Cos\beta - r_2 Cos\left(\frac{34\pi}{40}\right) + \ell_x\left(-\frac{114\pi}{40}\right)\right]^2 + \left[r_2 Sin\left(\frac{34\pi}{40}\right) + \ell_y\left(-\frac{114\pi}{40}\right)\right]^2$$

$$+\left[r_1 Sin\beta + \ell_z\left(-\frac{114\pi}{40}\right)\right]^2$$

$$t \simeq -20.5 \times T_{Moon} \quad \Rightarrow \quad \omega_1 t \simeq -\frac{41}{2}\omega_{Moon}T_{Moon} = -41\pi \ radians \qquad \left(August\ 30,\ 2015\right) \ (149)$$

$$t \simeq -20.5 \times T_{Moon} \Rightarrow \left(\omega_1 - \omega_2\right)t \simeq \left(-41 + 3.075\right)\pi = -37.925\pi \ radians \quad \left(August\ 30,\ 2015\right) \ (150)$$

$$d^2\left(-\frac{123\pi}{40}\right) = \left(358,290\ km\right)^2 = 128,371,724,100\ km^2 \qquad \left(August\ 30,\ 2015\right)$$

$$\begin{aligned} &= r_1^2 + 2r_1r_2Cos\beta\,Cos\left(\frac{37\pi}{40}\right) + r_2^2 + \ell^2\left(-\frac{123\pi}{40}\right) \\ &-2\begin{Bmatrix} \left[r_1Cos\,\beta + r_2Cos\left(\frac{37\pi}{40}\right)\right]\ell_x\left(-\frac{123\pi}{40}\right) \\ +r_2Sin\left(\frac{37\pi}{40}\right)\ell_y\left(-\frac{123\pi}{40}\right) + r_1Sin\,\beta\,\ell_z\left(-\frac{123\pi}{40}\right) \end{Bmatrix} \\ &= \left[-r_1Cos\,\beta - r_2Cos\left(\frac{37\pi}{40}\right) + \ell_x\left(-\frac{123\pi}{40}\right)\right]^2 + \left[-r_2Sin\left(\frac{37\pi}{40}\right) + \ell_y\left(-\frac{123\pi}{40}\right)\right]^2 \\ &+ \left[-r_1Sin\beta + \ell_z\left(-\frac{123\pi}{40}\right)\right]^2 \end{aligned} \qquad (151)$$

$$t \simeq -19.5 \times T_{Moon} \quad \Rightarrow \quad \omega_1 t \simeq -\frac{39}{2}\omega_{Moon}T_{Moon} = -39\pi \ radians \quad \left(September\ 28,\ 2015\right) \ (152)$$

$$t \simeq -19.5 \times T_{Moon} \Rightarrow \left(\omega_1 - \omega_2\right)t \simeq \left(-39 + 2.925\right)\pi = -36.075\pi \ radians \quad \left(September\ 28,\ 2015\right) \ (153)$$

$$d^2\left(-\frac{117\pi}{40}\right) = \left(356,877\ km\right)^2 = 127,361,193,129\ km^2 \qquad \left(September\ 28,\ 2015\right)$$

$$\begin{aligned} &= r_1^2 + 2r_1r_2\,Cos\beta Cos\left(\frac{37\pi}{40}\right) + r_2^2 + \ell^2\left(-\frac{117\pi}{40}\right) \\ &-2\begin{Bmatrix} \left[r_1Cos\,\beta + r_2Cos\left(\frac{37\pi}{40}\right)\right]\ell_x\left(-\frac{117\pi}{40}\right) \\ -r_2Sin\left(\frac{37\pi}{40}\right)\ell_y\left(-\frac{117\pi}{40}\right) + r_1Sin\,\beta\,\ell_z\left(-\frac{117\pi}{40}\right) \end{Bmatrix} \\ &= \left[-r_1Cos\,\beta - r_2Cos\left(\frac{37\pi}{40}\right) + \ell_x\left(-\frac{117\pi}{40}\right)\right]^2 + \left[r_2Sin\left(\frac{37\pi}{40}\right) + \ell_y\left(-\frac{117\pi}{40}\right)\right]^2 \\ &+ \left[-r_1Sin\beta + \ell_z\left(-\frac{117\pi}{40}\right)\right]^2 \end{aligned} \qquad (154)$$

As we have previously stated, based on observation[19], the shape of Moon[11]'s Orbit[19] is pretty much fixed in its plane of motion over a lunar[11,19,33] month of period T_{Moon} (31), but the *orientation* of the Moon[11]'s Orbit[19] with respect to the Earth and Sun is *not* fixed in space, and moves over time, over longer periods such as $T_{N,\vec{\ell}[\phi(t)]}$ (60) that happens to be 18.59 years (60), which we have assessed to be due to the nodal orientation variation motion of $\vec{\ell}\left[\phi(t)\right]$ (8). As Moon[11]'s motion in its individual circular orbit of revolution around vector radius $\vec{r}_1\left[\phi(t)\right]$ (5) has an angular velocity of $(\omega_1 = \omega_{Moon})$ (34), the observation[19] that the shape of Moon[11]'s Orbit[19] is pretty much fixed in its plane of motion over a lunar[11,19,33] month of period T_{Moon} (31) has led us to determine, based on $\vec{\boldsymbol{d}}_{MoonOrbit}\left[\phi(t)\right]$ (51), that $\ell\left[\phi(t)\right]$ (9) magnitude cycle of $\vec{\ell}\left[\phi(t)\right]$ (8) over a lunar[11,19,33] month of period T_{Moon} (31) must also be pretty much fixed in its plane of motion, and the direction of this almost constant $\ell\left[\phi(t)\right]$ (9) magnitude cycle changing with angular velocity of $(\omega_1 = \omega_{Moon})$ (34), or equivalently the orientation of $\vec{\ell}\left[\phi(t)\right]$ (8), must be varying very little over consecutive lunar[11,19,33] months of period T_{Moon} (31), as $\left(T_{N,\vec{\ell}[\phi(t)]} \gg T_{Moon}\right)$ (134). Therefore, for our purpose of analytically finding an approximate value for r_1 (7), for the dates corresponding to yearly and local maximum Earth-Moon distances[5] around the last minor lunar[11] standstill[11,41] which occurred[42] in October 2015, we can take the variation between $\vec{\ell}\left(-\frac{126\pi}{40}\right)$ (142) on August 18, 2015, $\vec{\ell}(-3\pi)$ (145) on September 14, 2015, and $\vec{\ell}\left(-\frac{114\pi}{40}\right)$ (148) on October 11, 2015, to be sufficiently negligible (155), and take $\vec{\ell}\left[\phi(t)\right]$ (8) vector values to be approximately the same on these dates, as $\vec{\ell}_{2\pi}$ (156), with a magnitude of $\ell_{2\pi}$ (157), where the phase $(\omega_1 t)$ (2) of the Moon is a multiple of 2π as stated in (140), (143), and (146), respectively. Similarly, for the dates corresponding to yearly and local minimum Earth-Moon distances[5] around the last minor lunar[11] standstill[11,41] occurring[42] in October 2015, we can take the variation between $\vec{\ell}\left(-\frac{123\pi}{40}\right)$ (151) on August 30, 2015 and $\vec{\ell}\left(-\frac{117\pi}{40}\right)$ (154) on September 28, 2015, to be sufficiently negligible (155), and take the $\vec{\ell}\left[\phi(t)\right]$ (8) vector values to be approximately the same on these dates, as $\vec{\ell}_{\pi}$ (158), with a magnitude of ℓ_{π} (159), where the phase $(\omega_1 t)$ (2) of the Moon is an odd multiple of π as stated in (149) and (152), respectively.

$$T_{N,\vec{\ell}[\phi(t)]} \gg T_{Moon} \Rightarrow \vec{\ell}_{2\pi} \simeq \vec{\ell}\left(-\frac{126\pi}{40}\right) \simeq \vec{\ell}(-3\pi) \simeq \vec{\ell}\left(-\frac{114\pi}{40}\right) \,\&\, \vec{\ell}_{\pi} \simeq \vec{\ell}\left(-\frac{123\pi}{40}\right) \simeq \vec{\ell}\left(-\frac{117\pi}{40}\right) \quad (155)$$

$$\begin{aligned} \vec{\ell}_{2\pi} = \hat{\boldsymbol{x}}\ell_{x,2\pi} + \hat{\boldsymbol{y}}\ell_{y,2\pi} + \hat{\boldsymbol{z}}\ell_{z,2\pi} &\simeq \vec{\ell}\left(-\frac{126\pi}{40}\right) \simeq \hat{\boldsymbol{x}}\ell_x\left(-\frac{126\pi}{40}\right) + \hat{\boldsymbol{y}}\ell_y\left(-\frac{126\pi}{40}\right) + \hat{\boldsymbol{z}}\ell_z\left(-\frac{126\pi}{40}\right) \\ &\simeq \vec{\ell}(-3\pi) \simeq \hat{\boldsymbol{x}}\ell_x(-3\pi) + \hat{\boldsymbol{y}}\ell_y(-3\pi) + \hat{\boldsymbol{z}}\ell_z(-3\pi) \quad (156) \\ &\simeq \vec{\ell}\left(-\frac{114\pi}{40}\right) \simeq \hat{\boldsymbol{x}}\ell_x\left(-\frac{114\pi}{40}\right) + \hat{\boldsymbol{y}}\ell_y\left(-\frac{114\pi}{40}\right) + \hat{\boldsymbol{z}}\ell_z\left(-\frac{114\pi}{40}\right) \end{aligned}$$

$$\ell_{2\pi} = \left|\vec{\ell}_{2\pi}\right| = \sqrt{\ell_{2\pi}{}^2} = \sqrt{\vec{\ell}_{2\pi}\cdot\vec{\ell}_{2\pi}} = \sqrt{\ell_{x,2\pi}{}^2 + \ell_{y,2\pi}{}^2 + \ell_{z,2\pi}{}^2} \simeq \ell\left(-\frac{126\pi}{40}\right) \simeq \ell(-3\pi) \simeq \ell\left(-\frac{114\pi}{40}\right) \quad (157)$$

$$\begin{aligned}\vec{\ell}_{\pi} = \hat{\boldsymbol{x}}\,\ell_{x,\pi} + \hat{\boldsymbol{y}}\,\ell_{y,\pi} + \hat{\boldsymbol{z}}\ell_{z,\pi} &\simeq \vec{\ell}\left(-\frac{123\pi}{40}\right) \simeq \hat{\boldsymbol{x}}\,\ell_x\left(-\frac{123\pi}{40}\right) + \hat{\boldsymbol{y}}\,\ell_y\left(-\frac{123\pi}{40}\right) + \hat{\boldsymbol{z}}\left(-\frac{123\pi}{40}\right)\\ &\simeq \vec{\ell}\left(-\frac{117\pi}{40}\right) \simeq \hat{\boldsymbol{x}}\,\ell_x\left(-\frac{117\pi}{40}\right) + \hat{\boldsymbol{y}}\,\ell_y\left(-\frac{117\pi}{40}\right) + \hat{\boldsymbol{z}}\left(-\frac{117\pi}{40}\right)\end{aligned} \quad (158)$$

$$\ell_{\pi} = \left|\vec{\ell}_{\pi}\right| = \sqrt{\ell_{\pi}{}^2} = \sqrt{\vec{\ell}_{\pi}\cdot\vec{\ell}_{\pi}} = \sqrt{\ell_{x,\pi}{}^2 + \ell_{y,\pi}{}^2 + \ell_{z,\pi}{}^2} \simeq \ell\left(-\frac{123\pi}{40}\right) \simeq \ell\left(-\frac{117\pi}{40}\right) \quad (159)$$

Based on the approximation in (155) - (159), $d^2\left(-\frac{126\pi}{40}\right)$ (142) on August 18, 2015 can be restated as $d^2\left(-\frac{126\pi}{40}\right)$ (160), $d^2\left(-\frac{123\pi}{40}\right)$ (151) on August 30, 2015 can be restated as $d^2\left(-\frac{123\pi}{40}\right)$ (163), $d^2(-3\pi)$ (145) on September 14, 2015 can be restated as $d^2(-3\pi)$ (161), $d^2\left(-\frac{117\pi}{40}\right)$ (154) on September 28, 2015 can be restated as $d^2\left(-\frac{117\pi}{40}\right)$ (164), and $d^2\left(-\frac{114\pi}{40}\right)$ (148) on October 11, 2015 can be restated as $d^2\left(-\frac{114\pi}{40}\right)$ (162).

$$\begin{aligned} d^2\left(-\frac{126\pi}{40}\right) &= (405{,}848\ km)^2 = 164{,}712{,}599{,}104\ km^2 \qquad (August\ 18,\ 2015)\\ &= r_1^2 - 2r_1 r_2 Cos\beta Cos\left(\frac{34\pi}{40}\right) + r_2^2 + \ell_{2\pi}{}^2\\ &\quad + 2\left\{\left[r_1 Cos\,\beta - r_2 Cos\left(\frac{34\pi}{40}\right)\right]\ell_{x,2\pi} - r_2 Sin\left(\frac{34\pi}{40}\right)\ell_{y,2\pi} + r_1 Sin\,\beta\,\ell_{z,2\pi}\right\}\\ &= \left[r_1 Cos\,\beta - r_2 Cos\left(\frac{34\pi}{40}\right) + \ell_{x,2\pi}\right]^2 + \left[-r_2 Sin\left(\frac{34\pi}{40}\right) + \ell_{y,2\pi}\right]^2 + \left(r_1 Sin\beta + \ell_{z,2\pi}\right)^2 \end{aligned} \quad (160)$$

$$\begin{aligned} d^2(-3\pi) &= (406{,}464\ km)^2 = 165{,}212{,}983{,}296\ km^2 \qquad (September\ 14,\ 2015)\\ &= r_1^2 + 2r_1 r_2\,Cos\beta + r_2^2 + \ell_{2\pi}{}^2 + 2\left[\left(r_1 Cos\,\beta + r_2\right)\ell_{x,2\pi} + r_1 Sin\,\beta\,\ell_{z,2\pi}\right]\\ &= \left(r_1 Cos\,\beta + r_2 + \ell_{x,2\pi}\right)^2 + \ell_{y,2\pi}{}^2 + \left(r_1 Sin\beta + \ell_{z,2\pi}\right)^2 \end{aligned} \quad (161)$$

$$d^2\left(-\frac{114\pi}{40}\right) = (406,388\ km)^2 = 165,151,206,544\ km^2 \qquad (October\ 11,\ 2015)$$

$$= r_1^2 - 2r_1 r_2 Cos\beta Cos\left(\frac{34\pi}{40}\right) + r_2^2 + \ell_{2\pi}^2$$

53

$$+2\left\{\left[r_1 Cos\,\beta - r_2 Cos\left(\frac{34\pi}{40}\right)\right]\ell_{x,2\pi} + r_2 Sin\left(\frac{34\pi}{40}\right)\ell_{y,2\pi} + r_1 Sin\,\beta\,\ell_{z,2\pi}\right\} \qquad (162)$$

$$= \left[r_1 Cos\,\beta - r_2 Cos\left(\frac{34\pi}{40}\right) + \ell_{x,2\pi}\right]^2 + \left[r_2 Sin\left(\frac{34\pi}{40}\right) + \ell_{y,2\pi}\right]^2 + \left(r_1 Sin\beta + \ell_{z,2\pi}\right)^2$$

$$d^2\left(-\frac{123\pi}{40}\right) = (358,290\ km)^2 = 128,371,724,100\ km^2 \qquad (August\ 30,\ 2015)$$

$$= r_1^2 + 2r_1 r_2 Cos\beta\, Cos\left(\frac{37\pi}{40}\right) + r_2^2 + \ell_{\pi}^2$$

$$-2\left\{\left[r_1 Cos\,\beta + r_2 Cos\left(\frac{37\pi}{40}\right)\right]\ell_{x,\pi} + r_2 Sin\left(\frac{37\pi}{40}\right)\ell_{y,\pi} + r_1 Sin\,\beta\,\ell_{z,\pi}\right\} \qquad (163)$$

$$= \left[-r_1 Cos\,\beta - r_2 Cos\left(\frac{37\pi}{40}\right) + \ell_{x,\pi}\right]^2 + \left[-r_2 Sin\left(\frac{37\pi}{40}\right) + \ell_{y,\pi}\right]^2 + \left(-r_1 Sin\beta + \ell_{z,\pi}\right)^2$$

$$d^2\left(-\frac{117\pi}{40}\right) = (356,877\ km)^2 = 127,361,193,129\ km^2 \qquad (September\ 28,\ 2015)$$

$$= r_1^2 + 2r_1 r_2\, Cos\beta Cos\left(\frac{37\pi}{40}\right) + r_2^2 + \ell_{\pi}^2$$

$$-2\left\{\left[r_1 Cos\,\beta + r_2 Cos\left(\frac{37\pi}{40}\right)\right]\ell_{x,\pi} - r_2 Sin\left(\frac{37\pi}{40}\right)\ell_{y,\pi} + r_1 Sin\,\beta\,\ell_{z,\pi}\right\} \qquad (164)$$

$$= \left[-r_1 Cos\,\beta - r_2 Cos\left(\frac{37\pi}{40}\right) + \ell_{x,\pi}\right]^2 + \left[r_2 Sin\left(\frac{37\pi}{40}\right) + \ell_{y,\pi}\right]^2 + \left(-r_1 Sin\beta + \ell_{z,\pi}\right)^2$$

Subtracting (160) from (162), and using the value of r_2 (45), we obtain the value of $\ell_{y,2\pi}$ (165).

$$\ell_{y,2\pi} \simeq \frac{d^2\left(-\frac{114\pi}{40}\right) - d^2\left(-\frac{126\pi}{40}\right)}{4r_2 Sin\left(\frac{34\pi}{40}\right)} \simeq \frac{(406,388\ km)^2 - (405,848\ km)^2}{4r_2 Sin\left(\frac{34\pi}{40}\right)} \simeq \mathbf{19,524.629}\ km \qquad (165)$$

Subtracting (164) from (163), and using the value of r_2 (45), we obtain the value of $\ell_{y,\pi}$ (166).

$$\ell_{y,\pi} \simeq -\frac{d^2\left(-\frac{123\pi}{40}\right) - d^2\left(-\frac{117\pi}{40}\right)}{4r_2 Sin\left(\frac{37\pi}{40}\right)} \simeq -\frac{\left[(358,290\ km)^2 - (356,877\ km)^2\right]}{4r_2 Sin\left(\frac{37\pi}{40}\right)} \simeq -\mathbf{87,481.846}\ km \qquad (166)$$

To meet our purpose of analytically finding an approximate value for r_1 (7), we can make use of our previous inference that the observed[42] minimum angles $\alpha_{\vec{\ell}[\phi(t)]-plane\measuredangle Earth\,Equator,Min,South}$ (132) and $\alpha_{\vec{\ell}[\phi(t)]-plane\measuredangle Earth\,Equator,Min,North}$ (133) between the plane of motion of vector $\vec{\ell}\left[\phi(t)\right]$ (8) and the Earth[12]'s equator at the southern standstill[42] on September 21, 2015 and at northern standstill[42] on October 3, 2015, respectively, must be the angles that vector $\vec{\ell}\left[\phi(t)\right]$ (8) makes with Earth[12]'s equatorial plane on the dates where the phase $(\omega_1 t)$ (2) of the Moon[11,19] in its individual circular orbit of revolution is an odd multiple of π and a multiple of 2π, such as the ones in **Table 2**, in the Moon-Earth configuration of **Figure 2**. So, based on the Sun-Earth-Moon configuration of **Figure 8**, and Moon-Earth configuration of **Figure 2**, as $\left(\ell_{y,2\pi}>0\right)$ (165) we can use $\alpha_{\vec{\ell}[\phi(t)]-plane\measuredangle Earth\,Equator,Min,North}$ (133) as an approximation for the angle between $\vec{\ell}_{2\pi}$ (156) on September 14, 2015, August 18, 2015, and October 11, 2015, namely the dates corresponding to yearly and local maximum Earth-Moon distances[5] around the minor standstill[42] in 2015; and as $\left(\ell_{y,\pi}<0\right)$ (166), we can use $\alpha_{\vec{\ell}[\phi(t)]-plane\measuredangle Earth\,Equator,Min,South}$ (132) as an approximation for the angle between $\vec{\ell}_{\pi}$ (158) on September 28, 2015 and on August 30, 2015, namely the dates corresponding to yearly and local minimum Earth-Moon distances[5] around the minor standstill[42] in 2015, based on data in **Table 2**. This would yield the approximate equations in (167) - (168) for $\vec{\ell}_{2\pi}$ (156) and $\vec{\ell}_{\pi}$ (158), respectively, in the Moon-Earth configuration of **Figure 2**.

$$\ell_{z,2\pi}\simeq\sqrt{\ell_{x,2\pi}^2+\ell_{y,2\pi}^2}\,tan\left(\alpha_{\vec{\ell}[\phi(t)]-plane\measuredangle Earth\,Equator,Min,North}\right)\simeq\sqrt{\ell_{x,2\pi}^2+\ell_{y,2\pi}^2}\,tan\left(-18^\circ.140\right)\quad(167)$$

$$\ell_{z,\pi}\simeq\sqrt{\ell_{x,\pi}^2+\ell_{y,\pi}^2}\,tan\left(\alpha_{\vec{\ell}[\phi(t)]-plane\measuredangle Earth\,Equator,Min,South}\right)\simeq\sqrt{\ell_{x,\pi}^2+\ell_{y,\pi}^2}\,tan\left(161^\circ.866\right)\quad(168)$$

Subtracting $\left[-r_2Sin\left(\frac{34\pi}{40}\right)+\ell_{y,2\pi}\right]^2$ from $d^2\left(-\frac{126\pi}{40}\right)$ in (160), or equivalently subtracting $\left[r_2Sin\left(\frac{34\pi}{40}\right)+\ell_{y,2\pi}\right]^2$ from $d^2\left(-\frac{114\pi}{40}\right)$ in (162), we obtain the result in (169), using the values of r_2 (45) and $\ell_{y,2\pi}$ (165).

$$\begin{aligned}d^2\left(-\frac{126\pi}{40}\right)-\left[-r_2Sin\left(\frac{34\pi}{40}\right)+\ell_{y,2\pi}\right]^2&=d^2\left(-\frac{114\pi}{40}\right)-\left[r_2Sin\left(\frac{34\pi}{40}\right)+\ell_{y,2\pi}\right]^2\\&=\left[r_1Cos\beta-r_2Cos\left(\frac{34\pi}{40}\right)+\ell_{x,2\pi}\right]^2+\left(r_1Sin\beta+\ell_{z,2\pi}\right)^2\quad(169)\\&\simeq164,519,151,339.886\,km^2\end{aligned}$$

Subtracting $\left(\ell_{y,2\pi}{}^2\right)$ from $d^2\left(-3\pi\right)$ in (161), we also obtain the result in (170), using the value of $\ell_{y,2\pi}$ (165).

$$d^2\left(-3\pi\right)-\ell_{y,2\pi}{}^2=\left(r_1Cos\beta+r_2+\ell_{x,2\pi}\right)^2+\left(r_1Sin\beta+\ell_{z,2\pi}\right)^2\simeq164,831,772,155.616\,km^2\quad(170)$$

This time subtracting (169) from (170) and reorganizing, we obtain the value for $\left(r_1 Cos\,\beta + \ell_{x,2\pi}\right)$ (171), using the value of r_2 (45).

$$\left(r_1 Cos\,\beta + \ell_{x,2\pi}\right) \simeq \frac{164,831,772,155.616\ km^2 - 164,519,151,339.886\ km^2 - r_2^{\,2} Sin^2\left(\frac{34\pi}{40}\right)}{2r_2\left[1 + Cos\left(\frac{34\pi}{40}\right)\right]} \qquad (171)$$

$$\simeq \mathbf{104,235.038}\ km$$

At this point, replacing the values for $\left(r_1 Cos\,\beta + \ell_{x,2\pi}\right)$ (171) and r_2 (45) in (170) and reorganizing, we also obtain the value of $\left(r_1 Sin\beta + \ell_{z,2\pi}\right)$ (172).

$$\left(r_1 Sin\beta + \ell_{z,2\pi}\right) \simeq \sqrt{164,831,772,155.616\ km^2 - \left[\left(r_1 Cos\,\beta + \ell_{x,2\pi}\right) + r_2\right]^2} \simeq \pm 388,889.349\ km \qquad (172)$$

Moving further, reorganizing (167) and replacing the value for the $\ell_{y,2\pi}$ (165) term yields (173).

$$\ell^2_{z,2\pi} = 0.107337\,\ell^2_{x,2\pi} + 40,918,063.372970\ km^2 \qquad (173)$$

Extracting $\ell_{x,2\pi}$ (156) from (171) and $\ell_{z,2\pi}$ (156) from (172) and placing in (173) gives us (174).

$$\begin{aligned}\left(\pm 388,889.349\ km - r_1 Sin\beta\right)^2 &= 0.107337\left(104,235.038\ km - r_1 Cos\,\beta\right)^2 \\ &\quad + 40,918,063.372970\ km^2\end{aligned} \qquad (174)$$

Placing the value for $\left(\beta = \beta_{Moon}\right)$ (72) in (174) and reorganizing, we would obtain the quadratic equations (175) in r_1 (7), based on the possible values of (172), whose possible solutions are given in (176).

$$\begin{cases}\left(r_{1,1} Sin\beta + \ell_{z,2\pi,1}\right) \simeq +388,889.349\ km \\ \quad \Rightarrow\ 0.075854953872858\, r_{1,1}^{\ 2} - 295,909.097\, r_{1,1} + 150,027,797,422.383 = 0 \\ \left(r_{1,2} Sin\beta + \ell_{z,2\pi,2}\right) \simeq -388,889.349\ km \\ \quad \Rightarrow\ 0.075854953872858\, r_{1,2}^{\ 2} + 336,793.096\, r_{1,2} + 150,027,797,422.383 = 0\end{cases} \qquad (175)$$

$$r_{1,1} \simeq \frac{295{,}909.097 \pm 205{,}038.502}{0.1517}\,km$$
$$\Rightarrow \begin{cases} r_{1,1,1} \simeq \dfrac{295{,}909.097+205{,}038.502}{0.1517}\,km \simeq 3{,}302{,}009.781\,km \Rightarrow \textit{POSSIBLY VALID} \\ r_{1,1,2} \simeq \dfrac{295{,}909.097-205{,}038.502}{0.1517}\,km \simeq 598{,}976.011\,km \Rightarrow \textit{POSSIBLY VALID} \end{cases}$$
$$r_{1,2} \simeq \frac{-336{,}793.096 \pm 260{,}591.986}{0.1517}\,km \quad (176)$$
$$\Rightarrow \begin{cases} r_{1,2,1} \simeq \dfrac{-336{,}793.096+260{,}591.986}{0.1517}\,km \simeq -502{,}281.700\,km < 0 \Rightarrow \textit{NOT VALID} \\ r_{1,2,2} \simeq \dfrac{-336{,}793.096-260{,}591.986}{0.1517}\,km \simeq -3{,}937{,}680.086\,km < 0 \Rightarrow \textit{NOT VALID} \end{cases}$$

Looking at the possible valid solutions of r_1 (7) in (176) based on the two quadratic equations in (175), we understand that the valid solution for $(r_1 Sin\beta + \ell_{z,2\pi})$ (172) is $(r_1 Sin\beta + \ell_{z,2\pi})$ (177).

$$(r_1 Sin\beta + \ell_{z,2\pi}) \simeq \sqrt{164{,}831{,}772{,}155.616\,km^2 - \left[(r_1 Cos\,\beta + \ell_{x,2\pi}) + r_2\right]^2} \simeq \mathbf{388{,}889.349}\,km \quad (177)$$

Based on $(r_1 Sin\beta + \ell_{z,2\pi})$ (177) and (176), and using the value of $(\beta = \beta_{Moon})$ (72), we also find the possible solutions of $\ell_{z,2\pi}$ (156) in (178).

$$\ell_{z,2\pi} \simeq 388{,}889.349\,km - r_1 Sin\beta \Rightarrow \begin{cases} \ell_{z,2\pi,1} \simeq 388{,}889.349\,km - r_{1,1,1} Sin\beta \simeq -954{,}159.024\,km \\ \ell_{z,2\pi,2} \simeq 388{,}889.349\,km - r_{1,1,2} Sin\beta \simeq 145{,}263.857\,km \end{cases} \quad (178)$$

Further, based on $(r_1 Cos\,\beta + \ell_{x,2\pi})$ (171) and (176), and also using the value of $(\beta = \beta_{Moon})$ (72), we find the possible solutions of $\ell_{x,2\pi}$ (156) in (179).

$$\ell_{x,2\pi} \simeq 104{,}235.038\,km - r_1 Cos\beta$$
$$\Rightarrow \begin{cases} \ell_{x,2\pi,1} \simeq 104{,}235.038\,km - r_{1,1,1} Cos\beta \simeq -2{,}912{,}300.998\,km \\ \ell_{x,2\pi,2} \simeq 104{,}235.038\,km - r_{1,1,2} Cos\beta \simeq -442{,}956.776\,km \end{cases} \quad (179)$$

Continuing our analysis, by subtracting $\left[-r_2 Sin\left(\frac{37\pi}{40}\right) + \ell_{y,\pi}\right]^2$ from $d^2\left(-\frac{123\pi}{40}\right)$ in (163), or equivalently subtracting $\left[r_2 Sin\left(\frac{37\pi}{40}\right) + \ell_{y,\pi}\right]^2$ from $d^2\left(-\frac{117\pi}{40}\right)$ in (164), we obtain the result in (180), using the values of r_2 (45) and $\ell_{y,\pi}$ (166).

$$d^2\left(-\frac{123\pi}{40}\right)-\left[-r_2 Sin\left(\frac{37\pi}{40}\right)+\ell_{y,\pi}\right]^2 = d^2\left(-\frac{117\pi}{40}\right)-\left[r_2 Sin\left(\frac{37\pi}{40}\right)+\ell_{y,\pi}\right]^2$$

$$=\left[-r_1 Cos\,\beta - r_2 Cos\left(\frac{37\pi}{40}\right)+\ell_{x,\pi}\right]^2+\left(-r_1 Sin\beta+\ell_{z,\pi}\right)^2$$

$$=r_1^2+2r_1 r_2 Cos\beta\, Cos\left(\frac{37\pi}{40}\right)+r_2^2 Cos^2\left(\frac{37\pi}{40}\right)+\ell_{x,\pi}^{\ 2}+\ell_{z,\pi}^{\ 2} \tag{180}$$

$$-2\left\{\left[r_1 Cos\,\beta+r_2 Cos\left(\frac{37\pi}{40}\right)\right]\ell_{x,\pi}+r_1 Sin\,\beta\,\ell_{z,\pi}\right\}$$

$$\simeq 120{,}205{,}045{,}712.590\ km^2$$

Further, reorganizing (168) and replacing the value for the $\ell_{y,\pi}$ (166) term yields (181).

$$\ell_{z,\pi}^2 = 0.107261\,\ell_{x,\pi}^2 + 820{,}876{,}620.705158\ km^2 \tag{181}$$

Placing (181) in (180), we obtain the quadratic equation (182) in $\ell_{x,\pi}$ (158).

$$1.107261\,\ell_{x,\pi}^{\ 2}-2\left[r_1 Cos\,\beta+r_2 Cos\left(\frac{37\pi}{40}\right)\right]\ell_{x,\pi}$$
$$+\left\{r_1^2+2r_1 r_2 Cos\beta\, Cos\left(\frac{37\pi}{40}\right)+r_2^2 Cos^2\left(\frac{37\pi}{40}\right)-2r_1 Sin\,\beta\,\ell_{z,\pi}-119{,}384{,}169{,}091.885\ km^2\right\}\simeq 0 \tag{182}$$

Based on the observation[42] of approximate symmetry of $\alpha_{\vec{\ell}[\phi(t)]-plane\,\measuredangle\,Earth\,Equator,Min,South}$ (132) and $\alpha_{\vec{\ell}[\phi(t)]-plane\,\measuredangle\,Earth\,Equator,Min,North}$ (133) angles between the plane of motion of vector $\vec{\ell}\left[\phi(t)\right]$ (8) and Earth[12]'s equator at 2015 southern and northern standstills[42], we can expect the angle between $\ell_{y,2\pi}$ (165) and $\sqrt{\ell_{x,2\pi}^2+\ell_{y,2\pi}^2}$ (167) to be about the same as the angle between $\ell_{y,\pi}$ (166) and $\sqrt{\ell_{x,\pi}^2+\ell_{y,\pi}^2}$ (168), which can be expressed as in (183) by equating the cosines of these angles, and utilizing (167) - (168). Utilizing this approximation in (183), we can calculate the possible values for $\ell_{z,\pi}$ (158) as in (184), using the values of $\ell_{y,2\pi}$ (165), $\ell_{y,\pi}$ (166), and the possible two solutions of $\ell_{z,2\pi}$ (156) in (178).

$$\frac{\ell_{y,\pi}}{\sqrt{\ell_{x,\pi}^2+\ell_{y,\pi}^2}}\simeq\frac{\ell_{y,2\pi}}{\sqrt{\ell_{x,2\pi}^2+\ell_{y,2\pi}^2}} \quad\Rightarrow\quad \frac{\ell_{y,\pi}}{\ell_{z,2\pi}}tan\left(-18^\circ.140\right)\simeq\frac{\ell_{y,2\pi}}{\ell_{z,\pi}}tan\left(161^\circ.866\right) \tag{183}$$

$$\frac{\ell_{z,\pi}}{\ell_{y,\pi} tan(-18^\circ.140)} \simeq \frac{\ell_{z,2\pi}}{\ell_{y,2\pi} tan(161^\circ.866)}$$

$$\Rightarrow \begin{cases} \ell_{z,\pi,1} \simeq \ell_{z,2\pi,1} \dfrac{\ell_{y,\pi} tan(-18^\circ.140)}{\ell_{y,2\pi} tan(161^\circ.866)} \simeq 4{,}276{,}708.457\ km \\ \ell_{z,\pi,2} \simeq \ell_{z,2\pi,2} \dfrac{\ell_{y,\pi} tan(-18^\circ.140)}{\ell_{y,2\pi} tan(161^\circ.866)} \simeq -651{,}098.140\ km \end{cases} \quad (184)$$

Placing the possible solution values of $\ell_{z,\pi}$ (184) in the $[-2r_1 Sin\beta \ell_{z,\pi}]$ term of (182), the quadratic equation (182) in $\ell_{x,\pi}$ (158) becomes either one of the two possible quadratic equations (185) and (186) in $\ell_{x,\pi}$ (158), based on the two possible solutions for r_1 (7) in (176).

$$1.107261 \ell_{x,\pi,1}^{2} - 2\left[r_{1,1,1} Cos\beta + r_2 Cos\left(\frac{37\pi}{40}\right)\right]\ell_{x,\pi,1} + \begin{Bmatrix} r_{1,1,1}^{2} + 2 r_{1,1,1} r_2 Cos\beta Cos\left(\dfrac{37\pi}{40}\right) \\ + r_2^{2} Cos^{2}\left(\dfrac{37\pi}{40}\right) \\ -11{,}607{,}036{,}844{,}818.271\ km^2 \end{Bmatrix} \simeq 0 \quad (185)$$

$$1.107261 \ell_{x,\pi,2}^{2} - 2\left[r_{1,1,2} Cos\beta + r_2 Cos\left(\frac{37\pi}{40}\right)\right]\ell_{x,\pi,2} + \begin{Bmatrix} r_{1,1,2}^{2} + 2 r_{1,1,2} r_2 Cos\beta Cos\left(\dfrac{37\pi}{40}\right) \\ + r_2^{2} Cos^{2}\left(\dfrac{37\pi}{40}\right) \\ +197{,}864{,}040{,}617.741\ km^2 \end{Bmatrix} \simeq 0 \quad (186)$$

Using the values of $(\beta = \beta_{Moon})$ (72) and r_2 (45), (185) - (186) become (187) - (188), respectively. Subsequently, the possible solutions for $\ell_{x,\pi}$ (158) as a result of these possible quadratic equations (187) - (188) can be found in (189).

$$1.107261 \ell_{x,\pi,1}^{2} - (6{,}009{,}014.714\ km)\ell_{x,\pi,1} - (776{,}193{,}452{,}924.193\ km^2) \simeq 0 \quad (187)$$

$$1.107261 \ell_{x,\pi,2}^{2} - (1{,}070{,}326.269\ km)\ell_{x,\pi,2} + (543{,}617{,}001{,}606.835\ km^2) \simeq 0 \quad (188)$$

$$\ell_{x,\pi,1} \simeq \frac{6{,}009{,}014.714 \pm 6{,}288{,}565.251}{2.2145} km$$

$$\Rightarrow \begin{cases} \ell_{x,\pi,1,1} \simeq \dfrac{6{,}009{,}014.714 + 6{,}288{,}565.251}{2.2145} km \simeq 5{,}553{,}153.005\, km \Rightarrow POSSIBLY\ VALID \\ \ell_{x,\pi,1,2} \simeq \dfrac{6{,}009{,}014.714 - 6{,}288{,}565.251}{2.2145} km \simeq -126{,}235.154\, km \Rightarrow POSSIBLY\ VALID \end{cases} \quad (189)$$

$$\ell_{x,\pi,2} \simeq \frac{1{,}070{,}326.269 \pm \sqrt{-1{,}262{,}105{,}388{,}246.855}}{2.2145} km \Rightarrow NOT\ VALID \Rightarrow r_{1,1,2}\ NOT\ VALID$$

As a result of (189), because inside of the square root term is negative for the possible $\ell_{x,\pi}$ (158) solution involving $r_{1,1,2}$ (176), we figure that the valid solution for r_1 (7) in (176) is $(r_1 = r_{Moon} = r_{1,1,1})$ (191), which is the value of the radius of Moon's individual circular orbit of revolution, according to our claim in the **ARTICLE 1** "Points on two circles in space mimic an ellipse with respect to a point" that all bodies moving in relative elliptical orbits with respect to each other must be revolving around an individual circular orbit themselves. This is the most significant result we had set out to find, as well as the most fundamental finding in this Article. $(r_1 = r_{Moon})$ (191) is also expressed in astronomical units[43] (au) (190) for comparison.

$$1\ astronomical\ unit\ (au) = 149{,}597{,}870.700\ km\ (exactly) \quad (190)$$

$$r_1 = r_{Moon} = r_{1,1,1} \simeq \mathbf{3{,}302{,}009.781}\ km \simeq \mathbf{0.0220725720604161}\ au \quad (radius\ of\ Moon's\ individual\ revolution) \quad (191)$$

Even if we have made approximations in the analytical calculation, the result we have obtained is important in comparing the order of magnitude of the radii of individual circular orbits of revolution of the Earth, the Sun, and the Moon, as $(r_{Earth} \ll r_{Moon} \ll r_{Sun})$ (192), based on our findings of r_1 (191), r_2 (45), and r_{Sun} (46).

$$(r_2 = r_{Earth} \simeq \mathbf{12{,}370.477}\ km) \ll (r_1 = r_{Moon} \simeq \mathbf{3{,}302{,}009.781}\ km) \ll (r_{Sun} \simeq \mathbf{149{,}609{,}642.749}\ km) \quad (192)$$

As we have now found out that the valid solution alternative for r_1 (7) in (176) is $(r_1 = r_{Moon} = r_{1,1,1})$ (191), we are now able to determine in turn the valid solutions for $\ell_{z,2\pi}$ (156) in (178), $\ell_{x,2\pi}$ (156) in (179), and $\ell_{z,\pi}$ (158) in (184) from the ones that are based on the $(r_1 = r_{Moon} = r_{1,1,1})$ (191) solution in (176), which are $(\ell_{z,2\pi} \simeq \ell_{z,2\pi,1})$ (193), $(\ell_{x,2\pi} \simeq \ell_{x,2\pi,1})$ (194), and $(\ell_{z,\pi} \simeq \ell_{z,\pi,1})$ (195), respectively.

$$\ell_{z,2\pi} \simeq \ell_{z,2\pi,1} \simeq \mathbf{-954{,}159.024}\ km \quad (193)$$

$$\ell_{x,2\pi} \simeq \ell_{x,2\pi,1} \simeq \mathbf{-2{,}912{,}300.998}\ km \quad (194)$$

$$\ell_{z,\pi} \simeq \ell_{z,\pi,1} \simeq \mathbf{4{,}276{,}708.457}\ km \quad (195)$$

Using the value we have found for $(\ell_{z,\pi} \simeq \ell_{z,\pi,1})$ (195) based on approximations, the relation in (168) yields the value in (196) for $\ell_{x,\pi}$ (158), to test against the possible values in (189).

Although both of the possible solution alternative values found in (189) are different from our test value in (196) for $\ell_{x,\pi}$ (158), the possible solution value for $\ell_{x,\pi}$ (158) in (189) that is closer in terms of order of magnitude to the test value in (196), as well as the value that ensures symmetry of sign with respect to $\ell_{x,2\pi}$ (194) for $\ell_{x,\pi}$ (158) is $\left(\ell_{x,\pi} \simeq \ell_{x,\pi,1,1}\right)$ (197).

$$\ell_{x,\pi(Test)} \simeq 13{,}058{,}067.031\,km \tag{196}$$

$$\ell_{x,\pi} \simeq \ell_{x,\pi,1,1} \simeq \mathbf{5{,}553{,}153.005}\,km \tag{197}$$

Note that these are all approximate values obtained with our analytical simplifications based on observation, but sufficient to provide an idea about Earth-Moon topology. Looking at the signs of the found Cartesian $(\hat{x},\hat{y},\hat{z})$ coordinates of $\vec{\ell}_{2\pi}$ (156) and $\vec{\ell}_{\pi}$ (158) in the Earth-Moon system configuration of **Figure 2**, one directly notices that these vectors are somewhat in a reverse direction relative to the vector radius $\vec{r}_1$ (5) of the individual circular orbit of the Moon, as $\vec{r}_1$ (5) is in $(+\hat{x})$-direction in the configuration of **Figure 2** when $(\omega_1 t)$ (2) of the Moon[11,19] in its individual circular orbit of revolution is a multiple of 2π, and $\vec{r}_1$ (5) is in $(-\hat{x})$-direction in the configuration of **Figure 2** when $(\omega_1 t)$ (2) of the Moon[11,19] in its individual circular orbit of revolution is an odd multiple of π, based on our previous analysis. This is another significant finding of this Article regarding the relative motion of $\vec{\ell}\left[\phi(t)\right]$ (8), namely the vector distance between the centers of individual circular orbits of revolution of the Earth and the Moon, and $\vec{r}_1\left[\phi(t)\right]$ (5), namely the vector radius of the Moon's individual circular orbit of revolution.

Finally, using the found approximate values of $\ell_{x,2\pi}$ (194), $\ell_{y,2\pi}$ (165), and $\ell_{z,2\pi}$ (193), we find the value of $\ell_{2\pi}$ (198) from (157). Similarly, using the found approximate values of $\ell_{x,\pi}$ (197), $\ell_{y,\pi}$ (166), and $\ell_{z,\pi}$ (195), we find the value of ℓ_{π} (199) from (159).

$$\ell_{2\pi} = \sqrt{\ell_{x,2\pi}^{\;2} + \ell_{y,2\pi}^{\;2} + \ell_{z,2\pi}^{\;2}} \simeq \mathbf{3{,}064{,}685.589}\,km \tag{198}$$

$$\ell_{\pi} = \sqrt{\ell_{x,\pi}^{\;2} + \ell_{y,\pi}^{\;2} + \ell_{z,\pi}^{\;2}} \simeq \mathbf{7{,}009{,}664.513}\,km \tag{199}$$

This finding of $\ell_{2\pi}$ (198) and ℓ_{π} (199), even if an approximate values, provides us the insight that the magnitude $\ell\left[\phi(t)\right]$ (9) of $\vec{\ell}\left[\phi(t)\right]$ (8), which is the vector between the centers of individual circular orbits of revolution of the Earth and the Moon, is on a similar order of magnitude as $\left(r_1 = r_{Moon}\right)$ (191), the radius of individual circular orbit of revolution of the Moon. This is another important finding of this Article.

Having found the value of $\left(r_1 = r_{Moon} = r_{1,1,1}\right)$ (191) approximately, which is the value of the radius of Moon's individual circular orbit of revolution that we had set out to find, according to our claim in the **ARTICLE 1** "Points on two circles in space mimic an ellipse with respect to a point" that all bodies moving in relative elliptical orbits with respect to each other must be revolving around an individual circular orbit themselves, we can now calculate the values of the expected time(t)-varying semi-major[7] and semi-minor[7] axis vectors $\vec{a}(t)$ (12) and $\vec{b}(t)$ (13) of the relative vector ellipse[6] between Earth and Moon in (200) and (201), respectively, with their

Dot Product[9] $\left[\vec{a}(t)\cdot\vec{b}(t)\right]$ (16) in (202), as well as their magnitudes squared $\left[a^2(t)\right]$ (14) in (203) and $\left[b^2(t)\right]$ (15) in (204), with square of the instantaneous focal[6,8] distance $\left[c^2(t)\right]$ (26) and instantaneous eccentricity[6] $e(t)$ (27) of the vector ellipse[6] in (205) and (206), respectively, in the elliptic representation (29) - (30) of the vector distance $\vec{d}\left[\phi(t)\right]$ (17) - (18), using values of $(\beta = \beta_{Moon})$ (72), $(r_2 = r_{Earth})$ (45), $(\omega_1 = \omega_{Moon})$ (34), $(\omega_2 = \omega_{Earth})$ (35), and $(\omega_1 - \omega_2)$ (36).

$$\begin{aligned}\vec{a}(t) \simeq\ & \hat{x}\left[3{,}016{,}536.037\, Cos(0.212768714989863t) - 12{,}370.477\right] km \\ & + \hat{y}\, 3{,}302{,}009.781\, Sin(0.212768714989863t)\ km \\ & + \hat{z}\, 1{,}343{,}048.374\, Cos(0.212768714989863t)\ km\end{aligned} \tag{200}$$

$$\begin{aligned}\vec{b}(t) =\ & -\hat{x}\, 3{,}016{,}536.037\, Sin(0.212768714989863t)\ km \\ & + \hat{y}\left[3{,}302{,}009.781\, Cos(0.212768714989863t) - 12{,}370.477\right] km \\ & - \hat{z}\, 1{,}343{,}048.374\, Sin(0.212768714989863t)\ km\end{aligned} \tag{201}$$

$$\vec{a}(t)\cdot\vec{b}(t) = -3{,}531{,}446{,}390.263\, Sin(0.212768714989863t)\ km^2 \tag{202}$$

$$\begin{aligned}a^2(t) = \vec{a}(t)\cdot\vec{a}(t) =\ & 10{,}903{,}421{,}623{,}251.049\ km^2 \\ & - 74{,}631{,}979{,}321.481\, Cos(0.212768714989863t)\ km^2\end{aligned} \tag{203}$$

$$\begin{aligned}b^2(t) = \vec{b}(t)\cdot\vec{b}(t) =\ & 10{,}903{,}421{,}623{,}251.049\ km^2 \\ & - 81{,}694{,}872{,}102.007\, Cos(0.212768714989863t)\ km^2\end{aligned} \tag{204}$$

$$c^2(t) = \left|a^2(t) - b^2(t)\right| = 7{,}062{,}892{,}780.525\left|Cos(0.212768714989863t)\right|\ km^2 \quad (Focal\ Distance\ Squared) \tag{205}$$

$$e(t) = \begin{cases} \dfrac{c(t)}{a(t)} = \sqrt{\dfrac{7{,}062{,}892{,}780.525\left|Cos(0.212768714989863t)\right|}{10{,}903{,}421{,}623{,}251.049 - 74{,}631{,}979{,}321.481\, Cos(0.212768714989863t)}} & , \ if\ a(t) > b(t) \\ \dfrac{c(t)}{b(t)} = \sqrt{\dfrac{7{,}062{,}892{,}780.525\left|Cos(0.212768714989863t)\right|}{10{,}903{,}421{,}623{,}251.049 - 81{,}694{,}872{,}102.007\, Cos(0.212768714989863t)}} & , \ if\ a(t) < b(t) \end{cases} \quad (Eccentricity) \tag{206}$$

Note here that in (200) - (206), the time (t) must be measured in $[days]$ from a date that coincides with a $(t = t_0 = 0)$ when the Earth-Moon distance[5] is a maximum in the month of March once in every 3-year cycles (86) of the Earth, which is approximately 40 cycles (86) of the Moon, after March 18, 2017 (93) or March 24, 2020 (92), when both the Moon (88) and the

Earth (89) moving around their own circles of revolution must be passing from the same aligned phase of a multiple of 2π (76) - (77) with n being a multiple of 40 (86) around the same date. Furthermore, based on the values we have found for $(\beta = \beta_{Moon})$ (72) and $(r_1 = r_{Moon})$ (191) to an approximation, as well as the found value of $(r_2 = r_{Earth})$ (45) in the **ARTICLE 3** "Analytical calculation of earth and sun orbital parameters from distance data", the expected minimum (a_{min}^2, b_{min}^2) (40) and maximum (a_{max}^2, b_{max}^2) (40) values for squares of the semi-major[7] and semi-minor[7] axis magnitudes, namely for $[a^2(t)]$ (14) and $[b^2(t)]$ (15), over a 2π cycle of $[\phi_0(t) = (\omega_1 - \omega_2)t]$ (4) become as in (207) for the Earth-Moon system.

$$
\begin{aligned}
a_{max}^2 &= r_1^2 + 2r_1 r_2 Cos\beta + r_2^2 \simeq \mathbf{10{,}978{,}053{,}602{,}572.530599}\ km^2 &\Rightarrow\ & a_{max} \simeq \mathbf{3{,}313{,}314.595}\ km \\
a_{min}^2 &= r_1^2 - 2r_1 r_2 Cos\beta + r_2^2 \simeq \mathbf{10{,}828{,}789{,}643{,}929.567938}\ km^2 &\Rightarrow\ & a_{min} \simeq \mathbf{3{,}290{,}712.635}\ km \\
b_{max}^2 &= (r_1 + r_2)^2 \simeq \mathbf{10{,}985{,}116{,}495{,}353.055828}\ km^2 &\Rightarrow\ & b_{max} = r_1 + r_2 \simeq \mathbf{3{,}314{,}380.258}\ km \\
b_{min}^2 &= (r_1 - r_2)^2 \simeq \mathbf{10{,}821{,}726{,}751{,}149.042710}\ km^2 &\Rightarrow\ & b_{min} = r_1 - r_2 \simeq \mathbf{3{,}289{,}639.304}\ km
\end{aligned}
\tag{207}
$$

Further to our analysis in (40), whenever $\{Cos[(\omega_1 - \omega_2)t] = 0\}$ (41) over a 2π cycle of $[\phi_0(t) = (\omega_1 - \omega_2)t]$ (4), $[a^2(t)]$ (14) and $[b^2(t)]$ (15) have equal values, as for magnitudes of radius vectors of a circle, which is approximately the value in (208) based on the found values of $(r_1 = r_{Moon})$ (191) and $(r_2 = r_{Earth})$ (45), which is a value close to $(r_1 = r_{Moon})$ (191).

$$
\begin{aligned}
Cos[(\omega_1 - \omega_2)t] = 0 \ &\Rightarrow\ a^2(t) = b^2(t) = r_1^2 + r_2^2 \simeq \mathbf{10{,}903{,}421{,}623{,}251.049269}\ km^2 \\
&\Rightarrow\ a(t) = b(t) \simeq \mathbf{3{,}302{,}032.953}\ km
\end{aligned}
\tag{208}
$$

With these important parameter values found, which are the final important findings of this Article, we wrap up our Sun-Earth-Moon system analysis so far.

CONCLUSIONS

Based on observation, all bodies in space seem to move in some kind of elliptical motion with respect to each other. In a previous article, it was mathematically demonstrated and proven that the "distance between points on any two different circles in three-dimensional space" is equivalent to the "distance of points on a vector ellipse from another fixed or moving point", and that based on the elliptical variation of Earth-Moon distance[5] over their relative full cycle, with semi-major and semi-minor axis values also varying in a harmonic style, we have made the assertion that it is equivalently possible for the Earth and the Moon to be each revolving in a circular motion with fixed but different angular velocities around centers of their own individual circular orbits, where their orbital centers are displaced by a vector distance from each other. In this Article, we have analyzed the individual and relative orbital motions of the Earth and the Moon in detail, utilizing Earth-Moon distance[5] and observation[19] data with the given assumption.

Based on the analysis demonstrated above, and our calculations in this Article, we have obtained the following orbital parameter values for the Moon and the Earth, also relative to the Sun, to an approximation, in the Earth-Moon configuration of **Figure 2**.

$$r_1 = r_{Moon} \simeq \mathbf{3{,}302{,}009.781}\ km \simeq \mathbf{0.0220725720604161}\ au \quad (radius\ of\ Moon's\ revolution\ orbit) \tag{209}$$

$$\beta = \beta_{Moon} \simeq 24^\circ \qquad (angle\ between\ Moon\ \&\ Earth\ revolution\ planes) \tag{210}$$

$$\hat{\boldsymbol{u}}_{Moon\perp} = \hat{\boldsymbol{x}}Sin\beta - \hat{\boldsymbol{z}}Cos\beta \quad (North\ Pole\ of\ Moon\ normal\ to\ Moon's\ circular\ revolution\ plane) \tag{211}$$

$$\phi_{Earth-Sun} \simeq \phi_{Earth-Moon} + 60^\circ \qquad (at\ any\ date\ of\ the\ year) \tag{212}$$

According to our claim in the **ARTICLE 1** "Points on two circles in space mimic an ellipse with respect to a point" that all bodies moving in relative elliptical orbits with respect to each other must be revolving around an individual circular orbit themselves, the values of the expected time (t)-varying semi-major[7] and semi-minor[7] axis vectors $\vec{\boldsymbol{a}}(t)$ (213) and $\vec{\boldsymbol{b}}(t)$ (214) of the relative vector ellipse[6] between Earth and Moon, their Dot Product[9] $\left[\vec{\boldsymbol{a}}(t)\cdot\vec{\boldsymbol{b}}(t)\right]$ (215), their magnitudes squared $\left[a^2(t)\right]$ (216) and $\left[b^2(t)\right]$ (217), respectively, with square of the instantaneous focal[6,8] distance $\left[c^2(t)\right]$ (218) and instantaneous eccentricity[6] $e(t)$ (219) of the vector ellipse[6], respectively, in the elliptic representation (29) - (30) of the vector distance $\vec{\boldsymbol{d}}\left[\phi(t)\right]$ (17) - (18), have been found approximately using the found approximate values of $(r_1 = r_{Moon})$ (209), $(\beta = \beta_{Moon})$ (210), $(r_2 = r_{Earth})$ (45), $(\omega_1 = \omega_{Moon})$ (34), $(\omega_2 = \omega_{Earth})$ (35), and $(\omega_1 - \omega_2)$ (36).

$$\begin{aligned}\vec{\boldsymbol{a}}(t) \simeq\ & \hat{\boldsymbol{x}}\left[3{,}016{,}536.037\,Cos(0.212768714989863t) - 12{,}370.477\right] km \\ & + \hat{\boldsymbol{y}}\,3{,}302{,}009.781\,Sin(0.212768714989863t)\ km \\ & + \hat{\boldsymbol{z}}\,1{,}343{,}048.374\,Cos(0.212768714989863t)\ km\end{aligned} \tag{213}$$

$$\begin{aligned}\vec{\boldsymbol{b}}(t) =\ & -\hat{\boldsymbol{x}}\,3{,}016{,}536.037\,Sin(0.212768714989863t)\ km \\ & + \hat{\boldsymbol{y}}\left[3{,}302{,}009.781\,Cos(0.212768714989863t) - 12{,}370.477\right] km \\ & - \hat{\boldsymbol{z}}\,1{,}343{,}048.374\,Sin(0.212768714989863t)\ km\end{aligned} \tag{214}$$

$$\vec{\boldsymbol{a}}(t)\cdot\vec{\boldsymbol{b}}(t) = -3{,}531{,}446{,}390.263\,Sin(0.212768714989863t)\ km^2 \tag{215}$$

$$\begin{aligned}a^2(t) = \vec{\boldsymbol{a}}(t)\cdot\vec{\boldsymbol{a}}(t) =\ & 10{,}903{,}421{,}623{,}251.049\ km^2 \\ & - 74{,}631{,}979{,}321.481\,Cos(0.212768714989863t)\ km^2\end{aligned} \tag{216}$$

$$\begin{aligned}b^2(t) = \vec{\boldsymbol{b}}(t)\cdot\vec{\boldsymbol{b}}(t) =\ & 10{,}903{,}421{,}623{,}251.049\ km^2 \\ & - 81{,}694{,}872{,}102.007\,Cos(0.212768714989863t)\ km^2\end{aligned} \tag{217}$$

$$\begin{aligned}c^2(t) = \left|a^2(t) - b^2(t)\right| = 7{,}062{,}892{,}780.525\left|Cos(0.212768714989863t)\right| km^2 \\ (Focal\ Distance\ Squared)\end{aligned} \tag{218}$$

$$e(t)=\begin{cases}\frac{c(t)}{a(t)}=\sqrt{\frac{7{,}062{,}892{,}780.525\left|Cos(0.212768714989863t)\right|}{10{,}903{,}421{,}623{,}251.049-74{,}631{,}979{,}321.481Cos(0.212768714989863t)}} & ,\ if\ a(t)>b(t)\\ \frac{c(t)}{b(t)}=\sqrt{\frac{7{,}062{,}892{,}780.525\left|Cos(0.212768714989863t)\right|}{10{,}903{,}421{,}623{,}251.049-81{,}694{,}872{,}102.007Cos(0.212768714989863t)}} & ,\ if\ a(t)<b(t)\end{cases} \quad (219)$$

$$(Eccentricity)$$

Note here that in (213) - (219), the time (t) must be measured in $[days]$ from a date that coincides with a $(t=t_0=0)$, which is when the Earth-Moon distance[5] is a maximum in the month of March once in every 3-year cycles (86) of the Earth, which is approximately once every 40 cycles (86) of the Moon, such as March 18, 2017 (93) or March 24, 2020 (92), when both the Moon (88) and the Earth (89) moving around their own circles of revolution must be passing from the same aligned phase of a multiple of 2π (76) - (77) with n being a multiple of 40 (86) around the same date.

Furthermore, based on the values we have found for $(\beta=\beta_{Moon})$ (210) and $(r_1=r_{Moon})$ (209) to an approximation, as well as the found value of $(r_2=r_{Earth})$ (45) in the **ARTICLE 3** "Analytical calculation of earth and sun orbital parameters from distance data", the expected maximum and minimum $(a_{\max},a_{\min},b_{\max},b_{\min})$ (220) values for the semi-major[7] and semi-minor[7] axis magnitudes and their squares, over a 2π cycle of $[\phi_0(t)=(\omega_1-\omega_2)t]$ (4) are also found for the Earth-Moon system.

$$\begin{aligned} a_{\max}^2 &= r_1^2+2r_1r_2Cos\beta+r_2^2 \simeq \mathbf{10{,}978{,}053{,}602{,}572.530599}\ km^2 &&\Rightarrow\ a_{\max}\simeq \mathbf{3{,}313{,}314.595}\ km\\ a_{\min}^2 &= r_1^2-2r_1r_2Cos\beta+r_2^2 \simeq \mathbf{10{,}828{,}789{,}643{,}929.567938}\ km^2 &&\Rightarrow\ a_{\min}\simeq \mathbf{3{,}290{,}712.635}\ km\\ b_{\max}^2 &= (r_1+r_2)^2 \simeq \mathbf{10{,}985{,}116{,}495{,}353.055828}\ km^2 &&\Rightarrow\ b_{\max}=r_1+r_2\simeq \mathbf{3{,}314{,}380.258}\ km\\ b_{\min}^2 &= (r_1-r_2)^2 \simeq \mathbf{10{,}821{,}726{,}751{,}149.042710}\ km^2 &&\Rightarrow\ b_{\min}=r_1-r_2\simeq \mathbf{3{,}289{,}639.304}\ km \end{aligned} \quad (220)$$

Further, whenever $\{Cos[(\omega_1-\omega_2)t]=0\}$ (41) over a 2π cycle of $[\phi_0(t)=(\omega_1-\omega_2)t]$ (4), $[a^2(t)]$ (14) and $[b^2(t)]$ (15) have equal values, as for magnitudes of radius vectors of a circle, which is approximately the value in (221) based on the found values of $(r_1=r_{Moon})$ (209) and $(r_2=r_{Earth})$ (45), which is a value close to $(r_1=r_{Moon})$ (209).

$$\begin{aligned} Cos[(\omega_1-\omega_2)t]=0\ &\Rightarrow\ a^2(t)=b^2(t)=r_1^2+r_2^2\simeq \mathbf{10{,}903{,}421{,}623{,}251.049269}\ km^2\\ &\Rightarrow\ a(t)=b(t)\simeq \mathbf{3{,}302{,}032.953}\ km \end{aligned} \quad (221)$$

As a result of the analyses we have made this Article, we have also reached a number of conclusions for the actual Sun-Earth-Moon orbital motion topology in space, as described and summarized below.

- Even if we have made approximations in the analytical calculation, the result we have obtained for the radius $(r_1 = r_{Moon})$ (209) of the Moon in its individual circular orbit of revolution is important in comparing the order of magnitudes of the radii of individual circular orbits of revolution of Earth, Sun, and the Moon, as $(r_{Earth} \ll r_{Moon} \ll r_{Sun})$ (222), based on our findings of r_1 (209), r_2 (45), and r_{Sun} (46).

$$(r_2 = r_{Earth} \simeq \mathbf{12,370.477}\ km) \ll (r_1 = r_{Moon} \simeq \mathbf{3,302,009.781}\ km) \ll (r_{Sun} \simeq \mathbf{149,609,642.749}\ km) \quad (222)$$

- Earth-Moon distance[5,11] $\vec{\boldsymbol{d}}[\phi(t)]$ (51) consists of two main components. One component is based on $\vec{\boldsymbol{r}}_2[\phi(t)]$ (6) due to individual circular orbit of revolution of the Earth[12] in its yearly cycle of angular velocity $(\omega_2 = \omega_{Earth})$ (35), and another component is $\vec{\boldsymbol{d}}_{MoonOrbit}[\phi(t)]$ (51) that is due to Moon[11] Orbit[19], which in turn is based on $\vec{\boldsymbol{r}}_1[\phi(t)]$ (5) due to individual circular orbit of revolution of the Moon[11] in its lunar cycle of angular velocity $(\omega_1 = \omega_{Moon})$ (34), as well as vector distance $\vec{\ell}[\phi(t)]$ (8) between centers of individual circular orbits of revolution of Earth and Moon.
- The "individual circular orbit of revolution of the Moon with radius vector $\vec{\boldsymbol{r}}_1[\phi(t)]$ (5)" as a whole is seemingly revolving around the Earth with an instantanous vector radius of $\vec{\ell}[\phi(t)]$ (8) at each $\phi(t)$ (3), during every orbital period[10,11] T_{Moon} (31) of the Moon[11], or vice versa, which is compliant with observation[19], such that the Moon and the Earth move in a relative trajectory in which the Moon passes between the Sun and the Earth once every Lunar[11,33] Month and the Earth passes between the Sun and Moon once every Lunar[11,33] Month, as visualized in **Figure 3**, and the Moon phases[11,18] occur over every synodic period[10,11] $T_{(Moon-Earth)}$ (38) of the Moon[11], observed[30] as sketched in **Figure 4**, a fact supporting our findings and assertions in this Article about the relative motions of the Earth and the Moon. In this sense, the Earth and the Moon must be forming a twin or binary[20] system of bodies in space, which is an already observed[19] phenomenon, whose individual circular orbits of revolution we claim must also be revolving around each other with an instantaneous vector radius of $\vec{\ell}[\phi(t)]$ (8) at each $\phi(t)$ (3), leading to this resultant observation[19,20].
- The Moon[11] is observed to move in counterclockwise[35] direction as seen from a vantage point from above the North Pole[23,34] of the Earth throughout a lunar[11,33] month of period T_{Moon} (31). Therefore, based on our analysis above, the Moon[11] Orbit[19] vector $\vec{\boldsymbol{d}}_{MoonOrbit}[\phi(t)]$ (51) directed from Earth towards the Moon must be moving in counterclockwise[35] direction as seen from a vantage point from above the North Pole[23,34] of the Earth, in its plane of motion. Further, based on our topology analysis above as well as observation[19], in order to maintain its synchronous[19] rotation with Earth, the Moon[11]'s self-rotation must also be in counterclockwise[35] direction from a vantage point from above the North Pole[23,34] of the Earth.

- We reason that the self-rotation of the Moon[11] has the same period T_{Moon} (31) and angular velocity $(\omega_1 = \omega_{Moon})$ (34) as its individual circular orbit of revolution, based on the synchronous[19] self-rotation of the Moon with Earth.
- We claim that lunar[11] monthly libration[21,22] of the Moon observed from Earth is mainly due to Moon's individual circular orbit of revolution around a radius of $\vec{r}_1\left[\phi(t)\right]$ (5), and the more-than-half view[11] of lunar surface from Earth is apparently based on the relative view angle of the Earth-bound observers, based on *both* the direction of $\vec{\ell}\left[\phi(t)\right]$ (8) vector and position of the Moon in its individual circular orbit of revolution around a vector radius of $\vec{r}_1\left[\phi(t)\right]$ (5), on that date throughout a cycle.
- Additionally, libration[19,21,22] of the Moon perceived by an Earth-bound observer aligned with the North Pole[23,34] direction of the Earth is observed[21,22] seemingly clockwise[35] throughout a lunar[11,33] month of period T_{Moon} (31). Therefore, the Moon's individual circular orbit of revolution around radius $\vec{r}_1\left[\phi(t)\right]$ (5) is in clockwise[35] direction from a vantage point from above the North Pole[23,34] of the Earth.
- Based on the assertion in the **ARTICLE 3** "Analytical calculation of earth and sun orbital parameters from distance data", all moving bodies in the Universe must be obeying "Faraday's Law of Induction"[24] in their motions, i.e. each one is possibly moving in space in external magnetic fields with changing flux[25] through the area of their individual circle of revolution, which is the cause that triggers their self rotation in the reverse direction, as seen from a vantage point above their North Pole[23,26,34], where we always accept the North Pole[23,26,34] of a body to be the pole rotating in counterclockwise[35] direction according to "right hand rule"[27]. As seen from a vantage point from above the North Pole[23,34] of the Earth, the Moon[11]'s observed *clockwise*[35] motion in its individual circular orbit of revolution around vector radius $\vec{r}_1\left[\phi(t)\right]$ (5) and its *counterclockwise*[35] self-rotation synchronous[19] with Earth are in alignment with this assertion.
- The angular velocity of the Moon's circular orbit around its own center of revolution $(\omega_1 = \omega_{Moon})$ (34) determines the higher frequency distance[5,11] variation of Earth-Moon system, and the angular velocity of Earth's circular orbit around its own center of revolution $(\omega_2 = \omega_{Earth})$ (35) determines the frequency of Earth-Moon distance[5] envelope sinusoid.
- As that there are approximately 13.3687466 orbital[10,11] (33) cycles of a distance variation sinusoid in approximately 1 cycle of a distance envelope sinusoid for the Earth-Moon system, which is also compliant with the observation in **Figure 1** and the data listed in **Table 1**, we infer that once in every 3-year cycles (86) of the Earth, which is approximately 40 cycles (86) of the Moon, both the Moon (88) and Earth (89) moving around their own circles of revolution must be passing from the same aligned phase of a multiple of 2π (76) - (77) with n being a multiple of 40 (86) around the same date, after their initial phase alignment of $\left[\phi_0(t_0) = 0\right]$ (4) at time $(t = t_0 = 0)$ with $(n \simeq 0)$, in the configuration of **Figure 2**.

- Checking the data in **Table 1**, between the dates of January 10, 2017 and January 9, 2021, minimum Earth-Moon distance[5] is observed as $356,565$ km occurring on January 1, 2018, and maximum Earth-Moon distance[5] is observed as $406,692$ km occurring on March 24, 2020. Based on our analysis for the Earth-Moon configuration described in **Figure 2**, the assessment is made that around every 10 cycles of Moon (31), Earth has a phase change of $\left(\phi \simeq \frac{3\pi}{2}\right)$ (112) - (116) in its individual circular orbit of revolution, with a phase of a multiple of 6π on March 18, 2017 (93) and March 24, 2020 (92). More specifically, in its individual circular orbit of revolution, the Earth passes through phase $(\phi \simeq 0)$ (112) in its motion relative to Moon on March 18, 2017 (93), and the Earth passes through phase $\left(\phi \simeq \frac{3\pi}{2}\right)$ (113) in its motion relative to Moon on December 19, 2017 (105), and the Earth passes through phase $(\phi \simeq 3\pi)$ (114) in its motion relative to the Moon on September 20, 2018 (99), and Earth passes through phase $\left(\phi \simeq \frac{9\pi}{2}\right)$ (115) in its motion relative to the Moon on June 23, 2019 (111), and the Earth passes through phase $(\phi \simeq 6\pi)$ (116) in its motion relative to the Moon on March 24, 2020 (92). Looking at the dates of phases $(\phi \simeq 0)$ (112), $\left(\phi \simeq \frac{3\pi}{2}\right)$ (113), $(\phi \simeq 3\pi)$ (114), $\left(\phi \simeq \frac{9\pi}{2}\right)$ (115), and $(\phi \simeq 6\pi)$ (116) of the Earth in its individual circular orbit of revolution within a 3-year cycle with respect to the Moon in the configuration described in **Figure 2**, which is approximately 40 cycles (87) of the Moon, we observe that the dates of these $\left(\phi \simeq \frac{3\pi}{2}\right)$ (112) - (116) multiples of phase for the Earth-Moon system occur around equinoxes[14,16] and solstices[15,16] of the Earth with respect to the Sun during their yearly relative cycles.
- Along the lines of our analysis in (86) - (116), we conclude that based on (86), (3), and (34) - (35), around every 1 cycle of the Moon (31), the Earth has a phase change of $\left(\phi \simeq \frac{3\pi}{20}\right)$ (117) in its individual circular orbit of revolution.
- Based on (112) and (119), as well as (116) and (121), we infer that whenever the Earth and the Moon moving around their individual circular orbits of revolution pass from the same aligned phase of a multiple of 2π around the same date every 3-year cycles of the Earth, which is approximately every 40 cycles (87) of the Moon, after their initial phase alignment of $\left[\phi_0(t_0) = 0\right]$ (4) at time $(t = t_0 = 0)$, in the configuration described in **Figure 2**, the Earth passes through its Spring Equinox[14,16] with respect to the Sun around March 20 (120), one of the dates around which Solar and Lunar Eclipses[11,19] occur due to Sun-Earth-Moon motion topology.

- Whenever the phase of the Earth in its individual circular orbit of revolution is around $\left(\phi = \phi_{Earth-Moon} \simeq 3\pi = 180^{\circ}\right)$ (114), or an odd multiple of 3π in the more general sense, in its motion relative to Moon, the phase of the Earth in its individual circular orbit of revolution is around $\left(\phi_{Earth-Sun} \simeq 240^{\circ} = -120^{\circ}\right)$ (126) with respect to its motion relative to the Sun, and Earth passes through its Autumn Equinox[14,16] with respect to the Sun around September 20 (126) and September 22 (127), another one of the dates around which Solar and Lunar Eclipses[11,19] occur due to motion topology of Sun-Earth-Moon system.
- Whenever the phase of the Earth in its individual circular orbit of revolution is around $\left(\phi = \phi_{Earth-Moon} \simeq \frac{9\pi}{2} = 90^{\circ}\right)$ (115) in its motion relative to Moon, phase of Earth in its individual circular orbit of revolution is like $\left(\phi_{Earth-Sun} \simeq -\phi_{0,Earth-Sun} \simeq 151^{\circ}\right)$ (123) - (124) with respect to its motion relative to the Sun, and Earth passes through its Summer Solstice[15,16] with respect to the Sun. Further, whenever the phase of the Earth in its individual circular orbit of revolution is around $\left(\phi = \phi_{Earth-Moon} \simeq \frac{3\pi}{2} = -90^{\circ}\right)$ (113) in its motion relative to Moon, the phase of the Earth in its individual circular orbit of revolution is around $\left(\phi_{Earth-Sun} \simeq \pi - \phi_{0,Earth-Sun} \simeq -30^{\circ}\right)$ (129) - (130) with respect to its motion relative to the Sun, and Earth passes through its Winter Solstice[15,16] with respect to the Sun.
- With the orbital period[10,11] T_{Moon} (31) of the Moon[11] and Earth[12]'s yearly[13] orbital period[10,12] T_{Earth} (32), which is the same as the orbital period[10,12] of Sun, all consistently constant, and thus having constant values for Moon[11] and Earth[12]'s angular frequencies $\left(\omega_1 = \omega_{Moon}\right)$ (34) and $\left(\omega_2 = \omega_{Earth}\right)$ (35), respectively, we naturally expect a constant difference between the phase value of Earth relative to the Earth-Sun configuration and the phase value of Earth relative to the Earth-Moon configuration at any date, within Earth's individual circular orbit of revolution. Based on the configuration in **Figure 2**, whenever the phase of Earth in its individual circular orbit of revolution is around $\left(\phi = \phi_{Earth-Moon} \simeq 0^{\circ}\right)$ (112) with respect to its motion relative to the Moon, the phase of Earth in its individual circular orbit of revolution is around $\left(\phi_{Earth-Sun} \simeq 60^{\circ}\right)$ (119) - (121) with respect to its motion relative to the Sun. Therefore, the $\left(\phi_{Earth-Moon} \simeq 0^{\circ}\right)$ (112) phase of Earth in its Earth-Moon configuration of **Figure 2** corresponds to the phase $\left(\phi_{Earth-Sun} \simeq 60^{\circ}\right)$ (119) - (121) of Earth in its Earth-Sun configuration of **Figure 2**, revealing that the $\left(\phi = \phi_{Earth-Moon} = 0^{\circ}\right)$ (112) plane of Earth-Moon configuration and the $\left(\phi = \phi_{Earth-Sun} = 0^{\circ}\right)$ (118) plane Earth-Sun configuration are around 60° (212) apart on the Earth's individual plane of revolution. As a result of the observations (112) - (130), we have reached the overall conclusion that, in its individual circular orbit of revolution, the phase $\phi_{Earth-Sun}$ (131) of the Earth in its motion relative to the Earth- Sun configuration

at any date is around 60° (212) ahead of its phase $\phi_{Earth-Moon}$ (3) for its motion relative to the Earth-Moon configuration.

- Based on observation[19], the shape of Moon[11]'s Orbit[19] is pretty much fixed in its plane of motion over a lunar[11,19,33] month of period T_{Moon} (31), but the *orientation* of the Moon[11]'s Orbit[19] with respect to the Earth and Sun is *not* fixed in space, and moves over time. As the Moon[11]'s motion in its individual circular orbit of revolution around vector radius $\vec{r}_1\left[\phi(t)\right]$ (5) has an angular velocity of $\left(\omega_1 = \omega_{Moon}\right)$ (34), the observation[19] that the shape of Moon[11]'s Orbit[19] is pretty much fixed in its plane of motion over a lunar[11,19,33] month of period T_{Moon} (31) has lead us to determine, based on $\vec{d}_{MoonOrbit}\left[\phi(t)\right]$ (51), that the $\ell\left[\phi(t)\right]$ (9) magnitude cycle of $\vec{\ell}\left[\phi(t)\right]$ (8) over a lunar[11,19,33] month of period T_{Moon} (31) must also be pretty much fixed in its plane of motion, and must also be changing with an angular velocity of $\left(\omega_1 = \omega_{Moon}\right)$ (34). Apart from this lunar[11,19,33] monthly variation of Moon[11]'s motion in its individual circular orbit of revolution around vector radius $\vec{r}_1\left[\phi(t)\right]$ (5) and the lunar[11,19,33] monthly almost-fixed magnitude cycle of $\vec{\ell}\left[\phi(t)\right]$ (8), we conclude that any other variation in the orientation of Moon[11]'s Orbit[19] in space over different periods, i.e. with angular velocities other than $\left(\omega_1 = \omega_{Moon}\right)$ (34), to be due to *only* orientation variations in $\vec{\ell}\left[\phi(t)\right]$ (8), based on $\vec{d}_{MoonOrbit}\left[\phi(t)\right]$ (51).
- One orbital precession of the Moon[11]'s Orbit[19] is called the apsidal[32] precession, and is the rotation of the Moon[11]'s Orbit[19] within its orbital plane, i.e. the axes of its apparent elliptical orbit change direction. The Moon[11]'s major axis[7] – the longest diameter of the orbit, joining its nearest and farthest points, the perigee[36] and apogee[36], respectively – is observed[19] to make one revolution every 8.85 Earth[12] years, or 3,232.6054 days, as it rotates slowly in the same direction as the self-rotation of the Moon[11] itself, which is in *counterclockwise*[35] direction from a vantage point from above the North Pole[23,34] $\left(\hat{u}_{Earth\perp} = -\hat{z}\right)$ (49) of the Earth[12] in the configuration of **Figure 2**. Considering that the Moon[11] Orbit[19]'s major axis[7] revolves to come back to the same position in 8.85 Earth[12] years, or 3,232.6054 days, in which case the position of the perigee[36] and apogee[36] are reversed with respect to the so called ecliptic[29] plane, we have concluded that the complete period $T_{A,\vec{d}_{MoonOrbit}\left[\phi(t)\right]}$ (223) for the perigee[36] and apogee[36] to come back to the same position with respect to the so called ecliptic[29] plane, namely one full apsidal[32] precession cycle, must be double this time, that is 17.7 Earth[12] years, or 6,465.2108 days, with angular velocity $\omega_{A,\vec{d}_{MoonOrbit}\left[\phi(t)\right]}$ (224). As the Moon[11]'s motion in its individual circular orbit of revolution around the vector radius $\vec{r}_1\left[\phi(t)\right]$ (5) has an angular velocity of $\left(\omega_1 = \omega_{Moon}\right)$ (34), and as $\vec{\ell}\left[\phi(t)\right]$ (8) has an almost fixed magnitude cycle over a lunar[11,19,33] month of period T_{Moon} (31), we conclude that this observed apsidal[32] precession of $\vec{d}_{MoonOrbit}\left[\phi(t)\right]$ (51) with an angular velocity $\omega_{A,\vec{d}_{MoonOrbit}\left[\phi(t)\right]}$ (54) different

from $(\omega_1 = \omega_{Moon})$ (34) must be due to an orientation variation of $\vec{\ell}[\phi(t)]$ (8). As a result, we infer, to an approximation, that the variation in the orientation of $\vec{\ell}[\phi(t)]$ (8) cycle, is very small over a couple of consecutive lunar[11,19,33] months of period T_{Moon} (31), compared to the individual motion frequencies of the Earth and the Moon. Further, following along the lines of the assertion in the **ARTICLE 3** "Analytical calculation of earth and sun orbital parameters from distance data" that all moving bodies in the Universe must be obeying "Faraday's Law of Induction"[24] in their motions, we also assert that each revolution in space, which is in external magnetic fields with changing flux[25] through the area of this revolution, triggers a rotation in the reverse direction. Hence, as Moon[11] and Earth[12] revolve in *clockwise*[35] directions in their individual circular orbits of revolution, as seen from a vantage point from above North Pole[23,34] $(\hat{\boldsymbol{u}}_{Earth\perp} = -\hat{z})$ (49) of the Earth[12] in the configuration of **Figure 2**, with angular velocities $(\omega_1 = \omega_{Moon})$ (34) and $(\omega_2 = \omega_{Earth})$ (35), respectively, the observed *counterclockwise*[35] revolution of Moon[11] Orbit[19]'s major axis[7], or equivalently lunar[11,19,33] orbit magnitude cycle orientation, from a vantage point from above North Pole[23,34] $(\hat{\boldsymbol{u}}_{Earth\perp} = -\hat{z})$ (49) of the Earth[12] in the configuration of **Figure 2**, is also in alignment with the expectation that all bodies in the Universe must be obeying "Faraday's Law of Induction"[24] in their motions, and even extending the scope of our previous assertion such that the harmonic motion of the distance vector $\vec{\ell}(\phi)$ (8) between centers of individual circular orbits of revolutions of twin or binary[20] system of bodies in space are triggered to move in reverse direction to their revolution directions in their individual circular orbits, because they obey "Faraday's Law of Induction"[24] as macro-scale charged "system of particles" moving in external magnetic fields with changing flux[25] through the area of their "twin or binary[20] system" motion planes.

$$T_{A,\vec{d}_{MoonOrbit}[\phi(t)]} \simeq 17.7\ Earth\ years \simeq 6{,}465.2108\ days\ \left(\vec{\boldsymbol{d}}_{MoonOrbit}[\phi(t)]\ Apsidal\ Variation\ Period\right) \quad (223)$$

$$\omega_{A,\vec{d}_{MoonOrbit}[\phi(t)]} = \frac{2\pi}{T_{A,\vec{d}_{MoonOrbit}[\phi(t)]}} \simeq \mathbf{0.000971845389353675}\ radians/day$$
$$\left(\vec{\boldsymbol{d}}_{MoonOrbit}[\phi(t)]\ Apsidal\ Variation\ Angular\ Velocity\right) \quad (224)$$

- Another observation[19] is based on the nodes or points at which the Moon[11]'s Orbit[19] crosses the so called ecliptic[29] plane. The line of nodes, the intersection between the two respective planes, namely the orbital plane of the Moon[11] and the so called ecliptic[29] plane, has a retrograde motion for an observer on Earth[12], such that it rotates westward along the ecliptic[29], or equivalently in a *clockwise*[35] direction from a vantage point from above the North Pole[23,34] $(\hat{\boldsymbol{u}}_{Earth\perp} = -\hat{z})$ (49) of Earth[12] in the configuration of **Figure 2**, with a period of 18.6 years (230) or $-19^\circ.3549$ (229) per year. The draconic year[40] $T_{Draconic\ Year}$ (225) is the time taken for the Sun, as seen from Earth, to complete one revolution with respect to the same lunar[11] node, namely a point at which the Moon[11]'s Orbit[19] crosses the so called ecliptic[29] plane, and has an observed period of about 346.6 days (225). The draconic[40] year is associated with eclipses; Lunar and Solar Eclipses are observed[19] to occur when the nodes align with the Sun, roughly every 173.3 days. The Moon[11] crosses the same node every 27.2122 days (226), an interval called the Draconic[19,33] Month $T_{Draconic\ Month}$ (226), which we figure, according to our accepted Moon-Earth topology, must be the synodic[38] period $T_{(Moon-\vec{\ell}[\phi(t)])}$ (227) of the motion of the Moon[11] in its individual circular orbit of revolution around radius $\vec{r}_1[\phi(t)]$ (5) with an angular velocity of $(\omega_1 = \omega_{Moon})$ (34), with respect to the nodal orientation variation motion of $\vec{\ell}[\phi(t)]$ (8) with an angular velocity $\omega_{N,\vec{\ell}[\phi(t)]}$ (229). We reach this conclusion based on our analysis result above which states that, as the Moon[11]'s motion in its individual circular orbit of revolution around vector radius $\vec{r}_1[\phi(t)]$ (5) has an angular velocity of $(\omega_1 = \omega_{Moon})$ (34), any other motion variation of $\vec{d}_{MoonOrbit}[\phi(t)]$ (51) with an angular velocity different from $(\omega_1 = \omega_{Moon})$ (34), such as this observed rotation of line of nodes of the Moon[11]'s Orbit[19] with angular velocity $\omega_{N,\vec{\ell}[\phi(t)]}$ (229), must be entirely due to an orientation variation of $\vec{\ell}[\phi(t)]$ (8). This synodic[38] period $T_{(Moon-\vec{\ell}[\phi(t)])}$ (227) is a little less than the orbital period[10,11] T_{Moon} (31) of the Moon[11], from which we calculate the corresponding synodic[38] angular velocity $\omega_{(Moon-\vec{\ell}[\phi(t)])}$ (228). It is significantly worth noting that $\omega_{(Moon-\vec{\ell}[\phi(t)])}$ (228) leads to a phase difference of $-152^{\circ}.1$ (228) per year in the synodic[38] motion of the motion of the nodal orientation variation motion of $\vec{\ell}[\phi(t)]$ (8) with respect to the Moon[11] in its individual circular orbit of revolution around radius $\vec{r}_1[\phi(t)]$ (5), which is very close to the value $-151^{\circ}.11$ (50) of the constant phase difference $\phi_{0,Earth-Sun}$ (50) of the Sun in its individual circular orbit of revolution throughout a year relative to Earth's phase in its own individual circular orbit of revolution, based on our findings in the **ARTICLE 3** "Analytical calculation of earth and sun orbital parameters from distance data" We conclude that this points out an alignment of the nodal orientation variation motion of $\vec{\ell}[\phi(t)]$ (8) with the yearly motion of the Sun, also supported by the observation[11,19] that the Moon[11] orbits closer to the so called ecliptic[29] plane than the equatorial plane of Earth, which apparently must be due to orientation of $\vec{\ell}[\phi(t)]$ (8). Utilizing $(\omega_1 = \omega_{Moon})$ (34) and $\omega_{(Moon-\vec{\ell}[\phi(t)])}$ (228), the value of the angular velocity $\omega_{N,\vec{\ell}[\phi(t)]}$ (229) of nodal orientation variation motion of $\vec{\ell}[\phi(t)]$ (8) is calculated, which turns out to be $-19^{\circ}.3549$ (229), from which we calculate the value of the period $T_{N,\vec{\ell}[\phi(t)]}$ (230) of the nodal orientation variation motion of $\vec{\ell}[\phi(t)]$ (8), that happens to be 18.59 years (230), both consistent with the observation[19] that the line of nodes has a retrograde motion for an observer on Earth[12], such that it rotates westward along the ecliptic[29], or equivalently in *clockwise*[35] direction, with a period of 18.6 years (230) or $-19^{\circ}.3549$ (229) per year from a vantage point from above North

Pole[23,34] $(\hat{\boldsymbol{u}}_{Earth\perp} = -\hat{z})$ (49) of the Earth[12] in the configuration of **Figure 2**. This is a reassuring result that indicates we are on the right track in our Earth-Moon topology analysis. Subsequently, we infer that the synodic[38] motion of Earth[12] with respect to the nodal orientation variation of Moon[11] Orbit[19] $\vec{\boldsymbol{d}}_{MoonOrbit}\left[\phi(t)\right]$ (51), or equivalently the nodal orientation variation motion of $\vec{\ell}\left[\phi(t)\right]$ (8) based on our analysis above, has a nodal synodic[38] angular velocity $\omega_{\left(Earth-\vec{\ell}[\phi(t)]\right)}$ (231), which can be computed using the values of $(\omega_2 = \omega_{Earth})$ (35) and $\omega_{N,\vec{\ell}[\phi(t)]}$ (229), from which the nodal synodic[38] cycle period $T_{\left(Earth-\vec{\ell}[\phi(t)]\right)}$ (232) is also calculated, yielding a value of 346.6 days (225). This is also a result confirming our Earth-Moon topology analysis, such that the observed[19] draconic year[40] $T_{Draconic\ Year}$ (225) turns out to be the period $T_{\left(Earth-\vec{\ell}[\phi(t)]\right)}$ (232) of the synodic[38] motion of Earth[12] with respect to the nodal orientation variation motion of $\vec{\ell}\left[\phi(t)\right]$ (8) in Moon[11] Orbit[19] $\vec{\boldsymbol{d}}_{MoonOrbit}\left[\phi(t)\right]$ (51).

$$T_{Draconic\ Year} = 346.620075883\ days = 346\ days\ 14\ hours\ 52\ min.s\ 54\ seconds \quad (Draconic\ Year) \qquad (225)$$

$$T_{Draconic\ Month} \simeq 27.2122247\ days \qquad (Moon's\ Draconic\ Month) \qquad (226)$$

$$T_{\left(Moon-\vec{\ell}[\phi(t)]\right)} = \frac{2\pi}{\omega_{\left(Moon-\vec{\ell}[\phi(t)]\right)}} = T_{Draconic\ Month} \simeq \mathbf{27.2122247}\ days \quad \left(Moon-\vec{\ell}\left[\phi(t)\right]\ Synodic\ Period\right) \qquad (227)$$

$$\omega_{\left(Moon-\vec{\ell}[\phi(t)]\right)} = \frac{2\pi}{T_{\left(Moon-\vec{\ell}[\phi(t)]\right)}} = \omega_{Moon} - \omega_{N,\vec{\ell}[\phi(t)]} \quad \left(Moon-\vec{\ell}\left[\phi(t)\right]\ Synodic\ Angular\ Velocity\right) \qquad (228)$$

$$\simeq \mathbf{0.2308956866}\ radians/day \simeq \mathbf{2.65470974}\ radians/year \simeq \mathbf{152^\circ.1}/year$$

$$\omega_{N,\vec{\ell}[\phi(t)]} = \frac{2\pi}{T_{N,\vec{\ell}[\phi(t)]}} = \omega_{Moon} - \omega_{\left(Moon-\vec{\ell}[\phi(t)]\right)} \simeq \mathbf{-0.3378}\ rad.s/year \simeq \mathbf{-19^\circ.3549}/year \qquad (229)$$

$$\left(Angular\ Velocity\ of\ \vec{\ell}\left[\phi(t)\right]\ Nodal\ Orientation\ Variation\right)$$

$$T_{N,\vec{\ell}[\phi(t)]} = \frac{2\pi}{\left|\omega_{N,\vec{\ell}[\phi(t)]}\right|} \simeq \mathbf{18.599}\ years \quad \left(Nodal\ Orientation\ Variation\ Period\ of\ \vec{\ell}\left[\phi(t)\right]\right) \qquad (230)$$

$$\omega_{\left(Earth-\vec{\ell}[\phi(t)]\right)} = \frac{2\pi}{T_{\left(Earth-\vec{\ell}[\phi(t)]\right)}} = \omega_{Earth} - \omega_{N,\vec{\ell}[\phi(t)]} \simeq \mathbf{6.621}\ rad.s/year \simeq \mathbf{379^\circ.3549}/year \qquad (231)$$

$$\left(Earth-\vec{\ell}\left[\phi(t)\right]\ Synodic\ Angular\ Velocity\right)$$

$$T_{\left(Earth-\vec{\ell}[\phi(t)]\right)} = \frac{2\pi}{\omega_{\left(Earth-\vec{\ell}[\phi(t)]\right)}} = T_{Draconic\ Year} \simeq \mathbf{0.94898}\ years \simeq \mathbf{346.620}\ days \qquad (232)$$

$$\left(Earth-\vec{\ell}\left[\phi(t)\right]\ Synodic\ Period\right)$$

- Another significant outcome (233) we obtain is that, the period of nodal orientation variation cycle $T_{N,\vec{\ell}[\phi(t)]}$ (230) of $\vec{\ell}\left[\phi(t)\right]$ (8), multiplied by the synodic[38] cycle $T_{\left(Earth-\vec{\ell}[\phi(t)]\right)}$ (232) of Earth[12] per year with respect to the nodal orientation variation motion of $\vec{\ell}\left[\phi(t)\right]$ (8), yields a result which is about the same as the apsidal[32] variation period $T_{A,\vec{d}_{MoonOrbit}\left[\phi(t)\right]}$ (223) of $\vec{d}_{MoonOrbit}\left[\phi(t)\right]$ (51) mentioned above. The slight difference is most probably due to the effect of orientation of $\vec{r}_1\left[\phi(t)\right]$ (5) relative to $\vec{\ell}\left[\phi(t)\right]$ (8) at the end of a $\left[T_{N,\vec{\ell}[\phi(t)]}\times T_{\left(Earth-\vec{\ell}[\phi(t)]\right)}\right]$ (233) cycle, leading to a number of additional days for reaching a local maximum magnitude for $\vec{d}_{MoonOrbit}\left[\phi(t)\right]$ (51), in determining the apsidal[32] variation period $T_{A,\vec{d}_{MoonOrbit}\left[\phi(t)\right]}$ (223). As a result, we have concluded that the observed[19] apsidal[32] precession of lunar[11,19,33] orbit magnitude in *counterclockwise*[35] direction, and the observed[19] rotation of the line of nodes of the Moon[11] Orbit[19] in *clockwise*[35] direction, from a vantage point from above North Pole[23,34] $\left(\hat{u}_{Earth\perp}=-\hat{z}\right)$ (49) of the Earth[12] in the configuration of **Figure 2**, are both phenomena related to the same orientation variation motion of $\vec{\ell}\left[\phi(t)\right]$ (8) as part of $\vec{d}_{MoonOrbit}\left[\phi(t)\right]$ (51). Further, following along the lines of the assertion in the **ARTICLE 3** "Analytical calculation of earth and sun orbital parameters from distance data" that all moving bodies in the Universe must be obeying "Faraday's Law of Induction"[24] in their motions, we also assert that each revolution in space, which is in external magnetic fields with changing flux[25] through the area of this revolution, triggers a rotation in the reverse direction. As a result, we conclude that the observed[19] apsidal[32] precession of lunar[11,19,33] orbit magnitude in *counterclockwise*[35] direction, and the observed[19] rotation of the line of nodes of the Moon[11] Orbit[19] in *clockwise*[35] direction, from a vantage point from above North Pole[23,34] $\left(\hat{u}_{Earth\perp}=-\hat{z}\right)$ (49) of the Earth[12] in the configuration of **Figure 2**, are a motion pair in alignment with this assertion.

$$T_{N,\vec{\ell}[\phi(t)]}\times {}^{T_{\left(Earth-\vec{\ell}[\phi(t)]\right)}}\!\Big/_{year}\simeq \mathbf{17.651}\ years\simeq T_{A,\vec{d}_{MoonOrbit}\left[\phi(t)\right]} \tag{233}$$

- The observed[11,19] topology of relative angles between poles[26,23] and planes of Moon, Earth, and Sun are sketched in **Figure 5** to provide a visual idea. The Moon[11] is observed[11,19] to orbit closer to the so called ecliptic[29] plane than the equatorial plane of Earth. Based on our findings and assertions about the actual topology of motions of the Sun, Earth, and Moon in the **ARTICLE 1** "Points on two circles in space mimic an ellipse with respect to a point" and **ARTICLE 3** "Analytical calculation of earth and sun orbital parameters from distance data", as well as in this Article, we have made the following assertions:
  - The equatorial plane of the Earth is in fact the plane of individual circular orbit of revolution of Earth, which is the horizontal plane of $Circle_2$ in **Figure 2**;

- The Moon's observed plane of orbit around the Earth should be the plane of motion of $\vec{d}_{MoonOrbit}\left[\phi(t)\right]$ (51). However, based on our Moon[11] Orbit[19] analysis so far, and our assertion in the **ARTICLE 1** "Points on two circles in space mimic an ellipse with respect to a point" that the self-rotation axis and the normal of the individual orbit of revolution of every spatial body is the same, the angles made by the Moon[11]'s North Pole[26] vector determine angles made by the plane of individual circular orbit of revolution of the Moon[11] around $\vec{r}_1\left[\phi(t)\right]$ (5), and thus the angles said to be made by the Moon's observed[19] plane of orbit around the Earth should be the angles made by the plane of motion of the vector distance $\vec{\ell}\left[\phi(t)\right]$ (8), which is between the centers of individual circular orbits of revolution of the Earth and the Moon, such that the individual circular orbit of revolution of the Moon around $\vec{r}_1\left[\phi(t)\right]$ (5) in turn revolves as a whole around the Earth with a radius of $\vec{\ell}\left[\phi(t)\right]$ (8) at each $\phi(t)$ (3) through every orbital period[10,11] T_{Moon} (31) of the Moon[11], or vice versa;
- The so called observed ecliptic[29] plane of the Earth in its motion around the Sun is in fact not really a plane, but the approximate angles said to be observed with respect to the ecliptic[29] are angles observed with respect to the plane of the distance vector between the Earth and the Sun around the year, and in fact the so called observed ecliptic[29] plane can approximately be considered to be the plane of individual circular orbit of revolution of the Sun, as $(\mathbf{12{,}370.477}\ km \simeq r_2 \ll r_{Sun} \simeq \mathbf{149{,}609{,}642.749}\ km)$ (45) - (46).

In light of all this information, we make the following conclusions. The above information tells us that the observed axial tilt[11,19,28] angle of the Moon[11]'s North Pole[26] vector with respect to the so called observed ecliptic[29] plane, or lunar[11,19] obliquity to ecliptic[29], namely $\alpha_{Moon\measuredangle Ecliptic}$ (234), which is around $1^{\circ}.5424$, is in fact the observed axial tilt[11,28] angle of the Moon[11]'s North Pole[26] vector approximately with respect to the normal of the plane of individual circular orbit of revolution of the Sun, namely $\alpha_{Moon\measuredangle Sun\ Orbital\ Plane}$ (234). Further, the observed mean axial tilt[11,28] angle of the Moon[11]'s North Pole[26] vector with respect to the Moon's observed approximate plane of orbit around the Earth, or mean lunar[11,19] obliquity, namely $\alpha_{Moon\measuredangle Moon\ Orbit}$ (235), which is about $6^{\circ}.687$, is in fact the observed axial tilt[11,28] angle of the Moon[11]'s North Pole[26] vector with respect to the normal of the plane of motion of vector distance $\vec{\ell}\left[\phi(t)\right]$ (8), which is between centers of individual circular orbits of revolution of Earth and Moon, namely $\alpha_{Moon\measuredangle\vec{\ell}[\phi(t)]-plane}$ (235). Moreover, the maximum angle that the Moon[11]'s observed approximate plane of orbit around the Earth makes with respect to the so called ecliptic[29] plane, or mean lunar[11,19] orbital inclination, namely $\alpha_{Moon\ Orbit\measuredangle Ecliptic}$ (236), which is around $5^{\circ}.09'$ or equivalently $5^{\circ}.1446$, is in fact the observed angle that the plane of motion of the vector distance $\vec{\ell}\left[\phi(t)\right]$ (8), which is between centers of individual circular orbits of revolution of Earth and Moon, makes approximately with respect to the normal of the plane of individual circular orbit of revolution of the Sun, namely $\alpha_{\vec{\ell}[\phi(t)]-plane \measuredangle Sun\ Orbital\ Plane}$ (236). Additionally, the observed axial tilt[3,28] angle of the Earth[12], or Earth[12] obliquity, or equivalently the angle that the so called ecliptic[29] plane makes with respect to Earth[12]'s equatorial plane, namely $\alpha_{Ecliptic \measuredangle Earth\ Equator}$ (237), which is approximately $23^{\circ}.27'$ or equivalently $23^{\circ}.44$, that also sums up to 180° with β_{Sun} (47) found in the **ARTICLE 3** "Analytical calculation of earth and sun orbital parameters from distance data", is in fact the observed angle between individual circular orbits of revolution of the Earth and the Sun, namely $\alpha_{Sun\ Orbital\ Plane \measuredangle Earth\ Orbital\ Plane}$ (237).

$$\alpha_{Moon \measuredangle Ecliptic} \simeq \alpha_{Moon \measuredangle Sun\ Orbital\ Plane} \simeq 1^{\circ}.5424 \tag{234}$$

$$\alpha_{Moon \measuredangle Moon\ Orbit} = \alpha_{Moon \measuredangle \vec{\ell}[\phi(t)]-plane} \simeq 6^{\circ}.687 \tag{235}$$

$$\alpha_{Moon\ Orbit \measuredangle Ecliptic} \simeq \alpha_{\vec{\ell}[\phi(t)]-plane \measuredangle Sun\ Orbital\ Plane} \simeq 5^{\circ}.09' \simeq 5^{\circ}.1446 \simeq 6^{\circ}.687 - 1^{\circ}.5424 \tag{236}$$

$$\alpha_{Ecliptic \measuredangle Earth\ Equator} \simeq \alpha_{Sun\ Orbital\ Plane \measuredangle Earth\ Orbital\ Plane} \simeq 23^{\circ}.27' \simeq 23^{\circ}.44 \simeq 180^{\circ} - 156^{\circ}.56 = 180^{\circ} + \beta_{Sun} \tag{237}$$

- The observed axial tilt[11,28] angle of the Moon[11]'s North Pole[26] vector with respect to Earth[12]'s equatorial plane, namely $\alpha_{Moon \measuredangle Earth\ Equator}$ (238), which is said to be about 24°, is in fact the observed axial tilt[11,28] angle of Moon[11]'s North Pole[26] vector with respect to Earth's plane of individual circular orbit of revolution, namely $\alpha_{Moon \measuredangle Earth\ Orbital\ Plane}$ (238), which is what reveals the value of the inclination angle $(\beta = \beta_{Moon})$ (210) of the plane of the Moon's individual orbit $(Circle_1)$ with respect to the plane of the Earth's individual circular orbit of revolution $(Circle_2)$ in **Figure 2**, also based on our assertion in the **ARTICLE 1** "Points on two circles in space mimic an ellipse with respect to a point" that the self-rotation axis and the normal of the individual orbit of revolution of every spatial body is the same. The direction of Earth's North Pole[23,26] vector is $(\hat{\boldsymbol{u}}_{Earth\perp} = -\hat{z})$ (49) in the Earth-Sun the configuration of **Figure 2**, which is the same in the Earth-Moon configuration of **Figure 2**, as the plane of individual circular orbit of revolution of the Earth is chosen to be the horizontal $\hat{\boldsymbol{x}} - \hat{\boldsymbol{y}}$ plane in both the Earth-Moon topology and the Earth-Sun topology of the configuration in **Figure 2**. Based on our analysis above, as Moon[11]'s self-rotation must be in *counterclockwise*[35] direction from a vantage point from above the North Pole[23,34] $(\hat{\boldsymbol{u}}_{Earth\perp} = -\hat{z})$ (49) of the Earth in order to maintain its synchronous[19] rotation with Earth, we determine that the direction of Moon[11]'s North Pole[26] vector is $(\hat{\boldsymbol{u}}_{Moon\perp} = \hat{\boldsymbol{x}} Sin\,\beta - \hat{z} Cos\,\beta)$ (211) in the Earth-Moon configuration of **Figure 2**, with $\left(-\dfrac{\pi}{2} \leq \beta \leq \dfrac{\pi}{2}\right)$ (239). The fact that the Moon[11] is observed[11,19] to orbit close to the so called ecliptic[29] plane, or equivalently close to the plane of individual circular orbit of revolution of the Sun based on our analysis above, is what allows us to conclude that $(\beta = \beta_{Moon} \simeq +24^{\circ})$ (210) based on (238) and (237).

$$\alpha_{Moon \measuredangle Earth\ Equator} = \alpha_{Moon \measuredangle Earth\ Orbital\ Plane} \simeq 24^{\circ} \tag{238}$$

$$-\frac{\pi}{2} \le \beta \le \frac{\pi}{2} \tag{239}$$

- Every 18.6 years (230), the angle between the Moon[11]'s Orbit[19] and Earth[12]'s equator reaches a maximum $\alpha_{Moon\,Orbit\,\measuredangle\,Earth\,Equator,Max}$ (240) of $28^{\circ}.36'$ or equivalently $28^{\circ}.5846$, the sum of Earth[12]'s equatorial tilt $\alpha_{Ecliptic\,\measuredangle\,Earth\,Equator}$ (237), or equivalently $\alpha_{Sun\,Orbital\,Plane\,\measuredangle\,Earth\,Orbital\,Plane}$ (237), which is $23^{\circ}.27'$ or equivalently $23^{\circ}.44$, and the Moon[11]'s orbital inclination $\alpha_{Moon\,Orbit\,\measuredangle\,Ecliptic}$ (236) to the so called ecliptic[29], or equivalently $\alpha_{\vec{\ell}[\phi(t)]-plane\,\measuredangle\,Sun\,Orbital\,Plane}$ (236), which is $5^{\circ}.09'$ or equivalently $5^{\circ}.1446$. This is called a major lunar[11] standstill[11,41]. Conversely, 9.3 years later, the angle between the Moon[11]'s Orbit[19] and Earth[12]'s equator reaches a minimum $\alpha_{Moon\,Orbit\,\measuredangle\,Earth\,Equator,Min}$ (241) of $18^{\circ}.20'$ or equivalently $18^{\circ}.33$, approximately near $18^{\circ}.2954$, which is the Moon[11]'s orbital inclination $\alpha_{Moon\,Orbit\,\measuredangle\,Ecliptic}$ (236) to the so called ecliptic[29], or equivalently $\alpha_{\vec{\ell}[\phi(t)]-plane\,\measuredangle\,Sun\,Orbital\,Plane}$ (236), subtracted from the Earth[12]'s equatorial tilt $\alpha_{Ecliptic\,\measuredangle\,Earth\,Equator}$ (237), or equivalently $\alpha_{Sun\,Orbital\,Plane\,\measuredangle\,Earth\,Orbital\,Plane}$ (237). This is called a minor lunar[11] standstill[11,41], which last occurred in October 2015. This variation of the angle $\alpha_{Moon\,Orbit\,\measuredangle\,Earth\,Equator}$ (240) - (242) between the Moon[11]'s Orbit[19] and Earth[12]'s equator over a 18.6 year cycle is apparently related to the observed[19] rotation of the line of nodes of the Moon[11] Orbit[19] over a 18.6 year (230) cycle, which we had previously concluded to be related to orientation variation motion of $\vec{\ell}[\phi(t)]$ (8) as part of $\vec{d}_{MoonOrbit}[\phi(t)]$ (51), with a period $T_{N,\vec{\ell}[\phi(t)]}$ (230) of 18.6 years. Therefore, we conclude here that the angle $\alpha_{Moon\,Orbit\,\measuredangle\,Earth\,Equator}$ (240) - (242) observed[19] to be between the Moon[11]'s Orbit[19] and Earth[12]'s equator is equivalently the angle $\alpha_{\vec{\ell}[\phi(t)]-plane\,\measuredangle\,Earth\,Orbital\,Plane}$ (240) - (242) between the plane of motion of $\vec{\ell}[\phi(t)]$ (8), which is the vector between the centers of individual circular orbits of revolution of the Earth and the Moon, and Earth[12]'s equator.

$$\begin{aligned}\alpha_{Moon\,Orbit\,\measuredangle\,Earth\,Equator,Max} &= \alpha_{\vec{\ell}[\phi(t)]-plane\,\measuredangle\,Earth\,Orbital\,Plane,Max} \simeq 5^{\circ}.1446 + 23^{\circ}.44 \simeq 28^{\circ}.5846 \simeq 28^{\circ}.36' \\ &= \alpha_{\vec{\ell}[\phi(t)]-plane\,\measuredangle\,Sun\,Orbital\,Plane} + \alpha_{Sun\,Orbital\,Plane\,\measuredangle\,Earth\,Orbital\,Plane}\end{aligned} \tag{240}$$

$$\begin{aligned}\alpha_{Moon\,Orbit\,\measuredangle\,Earth\,Equator,Min} &= \alpha_{\vec{\ell}[\phi(t)]-plane\,\measuredangle\,Earth\,Orbital\,Plane,Min} \simeq -5^{\circ}.1446 + 23^{\circ}.44 \simeq 18^{\circ}.2954 \simeq 18^{\circ}.20' \\ &= -\alpha_{\vec{\ell}[\phi(t)]-plane\,\measuredangle\,Sun\,Orbital\,Plane} + \alpha_{Sun\,Orbital\,Plane\,\measuredangle\,Earth\,Orbital\,Plane}\end{aligned} \tag{241}$$

$$\alpha_{Moon\,Orbit\,\measuredangle\,Earth\,Equator,Min} \le \alpha_{Moon\,Orbit\,\measuredangle\,Earth\,Equator} \le \alpha_{Moon\,Orbit\,\measuredangle\,Earth\,Equator,Max} \quad (over\ 18.6\ years\ cycle) \tag{242}$$

- The resultant expected overall orientation variation motion of $\vec{\ell}[\phi(t)]$ (8) as part of $\vec{d}_{MoonOrbit}[\phi(t)]$ (51), with a period $T_{N,\vec{\ell}[\phi(t)]}$ (230) of 18.6 years, is depicted roughly with Sun-Earth Cartesian $(\hat{x},\hat{y},\hat{z})$ coordinates, in **Figure 6** at $(t=t_0)$ and

$(t = t_0 + 18.6\,years)$ corresponding to a major lunar[11] standstill[11,41], in **Figure 7** at $(t = t_0 + 4.65\,years)$ end of 1st Quarter Cycle, in **Figure 8** at $(t = t_0 + 9.3\,years)$ end of 1 Half Cycle at a minor lunar[11] standstill[11,41], and in **Figure 9** at $(t = t_0 + 13.95\,years)$ end of 3rd Quarter Cycle, not to scale and with the shape of lunar[11,19,33] monthly cycle of $\vec{\ell}[\phi(t)]$ (8) not necessarily accurate, just to enable visualization of the motion of the $\vec{\ell}[\phi(t)]$ (8) cycle, leading to the combined perception by an observer[19] on Earth[12], of Moon[11] Orbit[19]'s apsidal[32] precession, rotation of the line of nodes of Moon[11] Orbit[19] over a 18.6 year (230) cycle, and the variation of $\alpha_{Moon\,Orbit\,\measuredangle\,Earth\,Equator}$ (240) - (242) angle between the Moon[11]'s Orbit[19] and the Earth[12]'s equator, or equivalently $\alpha_{\vec{\ell}[\phi(t)]-plane\,\measuredangle\,Earth\,Orbital\,Plane}$ (240) - (242), over a 18.6 year (230) cycle.

- Our finding in the in the **ARTICLE 3** "Analytical calculation of earth and sun orbital parameters from distance data" that in its own cycle of individual revolution, the Sun is lagging with a phase difference of $(-\phi_{Earth-Sun} = \phi_{0,Earth-Sun} \simeq -151^\circ)$ (123) - (124) behind Earth in its own cycle of individual revolution, or equivalently, Earth in its own cycle of individual revolution is moving ahead of the Sun in its own cycle of individual revolution, by a phase difference of $(\phi_{Earth-Sun} = -\phi_{0,Earth-Sun} \simeq 151^\circ)$ (123) - (124), as well as our finding in this Article that in its individual circular orbit of revolution, the phase $\phi_{Earth-Sun}$ (131) of Earth in its motion relative to the Earth- Sun configuration at any date is around 60° ahead of its phase $\phi_{Earth-Moon}$ (3) in its motion relative to the Earth-Moon configuration, reveal and exhibit an alignment of angles in the Sun-Earth motion topology and Earth-Moon motion topology. This alignment is also discovered when it is found that $\omega_{(Moon-\vec{\ell}[\phi(t)])}$ (228) leads to a phase difference of $-152^\circ.1$ (228) per year in the synodic[38] motion of the nodal orientation variation motion of $\vec{\ell}[\phi(t)]$ (8) with respect to the motion of the Moon[11] in its individual circular orbit of revolution around radius $\vec{r}_1[\phi(t)]$ (5), which is very close to $-151^\circ.11$ (50) of the constant phase difference $\phi_{0,Earth-Sun}$ (50) of the Sun in its individual circular orbit of revolution throughout a year relative to Earth's phase in its own individual circular orbit of revolution, based on our findings in the **ARTICLE 3** "Analytical calculation of earth and sun orbital parameters from distance data", which points out an alignment of the nodal orientation variation motion of $\vec{\ell}[\phi(t)]$ (8) with the yearly motion of the Sun, also supported by the observation[11,19] that the Moon[11] orbits closer to the so called ecliptic[29] plane than the equatorial plane of Earth, which apparently must be due to orientation of $\vec{\ell}[\phi(t)]$ (8).
- Based on observation[19], the shape of Moon[11]'s Orbit[19] is pretty much fixed in its plane of motion over a lunar[11,19,33] month of period T_{Moon} (31), but the *orientation* of the Moon[11]'s Orbit[19] with respect to the Earth and Sun is *not* fixed in space, and moves over time, over periods much longer than T_{Moon} (31), such as $T_{N,\vec{\ell}[\phi(t)]}$ (230) that happen to be 18.59 years (230), which we have assessed to be due to the nodal orientation variation motion

of $\vec{\ell}[\phi(t)]$ (8). As the Moon[11]'s motion in its individual circular orbit of revolution around vector radius $\vec{r}_1[\phi(t)]$ (5) has an angular velocity of $(\omega_1 = \omega_{Moon})$ (34), the observation[19] that the shape of Moon[11]'s Orbit[19] is pretty much fixed in its plane of motion over a lunar[11,19,33] month of period T_{Moon} (31) has led us to determine, based on $\vec{d}_{MoonOrbit}[\phi(t)]$ (51), that $\ell[\phi(t)]$ (9) magnitude cycle of $\vec{\ell}[\phi(t)]$ (8) over a lunar[11,19,33] month of period T_{Moon} (31) must also be pretty much fixed in its plane of motion, and the direction of this almost constant $\ell[\phi(t)]$ (9) magnitude cycle changing with angular velocity of $(\omega_1 = \omega_{Moon})$ (34), or equivalently the orientation of $\vec{\ell}[\phi(t)]$ (8), must be varying very little over consecutive lunar[11,19,33] months of period T_{Moon} (31), as $\left(T_{N,\vec{\ell}[\phi(t)]} \gg T_{Moon}\right)$ (243). Therefore, for our purpose of analytically finding an approximate value for r_1 (7), for the dates in **Table 2** corresponding to yearly and local maximum Earth-Moon distances[5] around minor lunar[11] standstill[11,41] that has occurred[42] in October 2015, we have taken the variation between $\vec{\ell}\left(-\frac{126\pi}{40}\right)$ (142) on August 18, 2015, $\vec{\ell}(-3\pi)$ (145) on September 14, 2015, and $\vec{\ell}\left(-\frac{114\pi}{40}\right)$ (148) on October 11, 2015, to be sufficiently negligible (244), and we have thus taken the $\vec{\ell}[\phi(t)]$ (8) vector values to be approximately the same on these dates, as $\vec{\ell}_{2\pi}$ (245), whose found approximate Cartesian $(\hat{x},\hat{y},\hat{z})$ coordinate values have been expressed in (246) - (248) with an approximate magnitude of $\ell_{2\pi}$ (249), where the phase $(\omega_1 t)$ (2) of the Moon is a multiple of 2π as stated in (140), (143), and (146), respectively. Similarly, for the dates in **Table 2** corresponding to yearly and local minimum Earth-Moon distances[5] around the minor lunar[11] standstill[11,41] which has occurred[42] in October 2015, we have taken the variation between $\vec{\ell}\left(-\frac{123\pi}{40}\right)$ (151) on August 30, 2015 and $\vec{\ell}\left(-\frac{117\pi}{40}\right)$ (154) on September 28, 2015 to be sufficiently negligible (244), and we have thus taken the $\vec{\ell}[\phi(t)]$ (8) vector values to be approximately the same on these dates, as $\vec{\ell}_\pi$ (250), whose found approximate Cartesian $(\hat{x},\hat{y},\hat{z})$ coordinate values have been expressed in (251) - (253) with an approximate magnitude of ℓ_π (254), where the phase $(\omega_1 t)$ (2) of the Moon is an odd multiple of π as stated in (149) and (152), respectively.

$$T_{N,\vec{\ell}[\phi(t)]} \simeq 18.599\ years \quad \& \quad T_{Moon} \simeq 27.321661\ days \quad \Rightarrow \quad T_{N,\vec{\ell}[\phi(t)]} \gg T_{Moon} \tag{243}$$

$$T_{N,\vec{\ell}[\phi(t)]} \gg T_{Moon} \Rightarrow \vec{\ell}_{2\pi} \simeq \vec{\ell}\left(-\frac{126\pi}{40}\right) \simeq \vec{\ell}(-3\pi) \simeq \vec{\ell}\left(-\frac{114\pi}{40}\right) \ \& \ \vec{\ell}_\pi \simeq \vec{\ell}\left(-\frac{123\pi}{40}\right) \simeq \vec{\ell}\left(-\frac{117\pi}{40}\right) \tag{244}$$

$$\vec{\ell}_{2\pi} = \hat{x}\ell_{x,2\pi} + \hat{y}\ell_{y,2\pi} + \hat{z}\ell_{z,2\pi} \simeq \vec{\ell}\left(-\frac{126\pi}{40}\right) \simeq \hat{x}\ell_x\left(-\frac{126\pi}{40}\right) + \hat{y}\ell_y\left(-\frac{126\pi}{40}\right) + \hat{z}\ell_z\left(-\frac{126\pi}{40}\right)$$
$$\simeq \vec{\ell}(-3\pi) \simeq \hat{x}\ell_x(-3\pi) + \hat{y}\ell_y(-3\pi) + \hat{z}\ell_z(-3\pi) \quad (245)$$
$$\simeq \vec{\ell}\left(-\frac{114\pi}{40}\right) \simeq \hat{x}\ell_x\left(-\frac{114\pi}{40}\right) + \hat{y}\ell_y\left(-\frac{114\pi}{40}\right) + \hat{z}\ell_z\left(-\frac{114\pi}{40}\right)$$

$$\ell_{x,2\pi} \simeq \mathbf{-2,912,300.998}\ km \quad (246)$$

$$\ell_{y,2\pi} \simeq \mathbf{19,524.629}\ km \quad (247)$$

$$\ell_{z,2\pi} \simeq \mathbf{-954,159.024}\ km \quad (248)$$

$$\ell_{2\pi} = \sqrt{\ell_{x,2\pi}^{\ 2} + \ell_{y,2\pi}^{\ 2} + \ell_{z,2\pi}^{\ 2}} \simeq \mathbf{3,064,685.589}\ km \quad (249)$$

$$\vec{\ell}_{\pi} = \hat{x}\ell_{x,\pi} + \hat{y}\ell_{y,\pi} + \hat{z}\ell_{z,\pi} \simeq \vec{\ell}\left(-\frac{123\pi}{40}\right) \simeq \hat{x}\ell_x\left(-\frac{123\pi}{40}\right) + \hat{y}\ell_y\left(-\frac{123\pi}{40}\right) + \hat{z}\left(-\frac{123\pi}{40}\right)$$
$$\simeq \vec{\ell}\left(-\frac{117\pi}{40}\right) \simeq \hat{x}\ell_x\left(-\frac{117\pi}{40}\right) + \hat{y}\ell_y\left(-\frac{117\pi}{40}\right) + \hat{z}\left(-\frac{117\pi}{40}\right) \quad (250)$$

$$\ell_{x,\pi} \simeq \mathbf{5,553,153.005}\ km \quad (251)$$

$$\ell_{y,\pi} \simeq \mathbf{-87,481.846}\ km \quad (252)$$

$$\ell_{z,\pi} \simeq \mathbf{4,276,708.457}\ km \quad (253)$$

$$\ell_{\pi} = \sqrt{\ell_{x,\pi}^{\ 2} + \ell_{y,\pi}^{\ 2} + \ell_{z,\pi}^{\ 2}} \simeq \mathbf{7,009,664.513}\ km \quad (254)$$

- Although the results expressed in (245) - (254) are all approximate values obtained with analytical simplifications based on observation, they are sufficient to provide an idea about Earth-Moon topology. Looking at signs of the found Cartesian $(\hat{x}, \hat{y}, \hat{z})$ coordinate values of $\vec{\ell}_{2\pi}$ (245) and $\vec{\ell}_{\pi}$ (250) in the Earth-Moon system configuration of **Figure 2**, one directly notices that these vectors are somewhat in a reverse direction relative to the vector radius $\vec{r}_1$ (5) of the individual circular orbit of the Moon, as $\vec{r}_1$ (5) vector is in $(+\hat{x})$-direction in the configuration of **Figure 2** when $(\omega_1 t)$ (2) of the Moon[11,19] in its individual circular orbit of revolution is a multiple of 2π, and $\vec{r}_1$ (5) is in $(-\hat{x})$-direction in the configuration of **Figure 2** when $(\omega_1 t)$ (2) of the Moon[11,19] in its individual circular orbit of revolution is an odd multiple of π, based on our analysis. This finding is also very significant regarding the relative motion of $\vec{\ell}\left[\phi(t)\right]$ (8), namely the vector distance between the centers of individual circular orbits of revolution of the Earth and the Moon, and $\vec{r}_1\left[\phi(t)\right]$ (5), the vector radius of the Moon's individual circular orbit of revolution.
- This finding of $\ell_{2\pi}$ (198) and ℓ_{π} (199), even if an approximate values, also provides us the insight that the magnitude $\ell\left[\phi(t)\right]$ (9) of $\vec{\ell}\left[\phi(t)\right]$ (8), which is the vector between the centers of individual circular orbits of revolution of the Earth and the Moon, is on a

similar order of magnitude as $\left(r_1 = r_{Moon}\right)$ (209), the radius of individual circular orbit of revolution of the Moon.

FIGURES

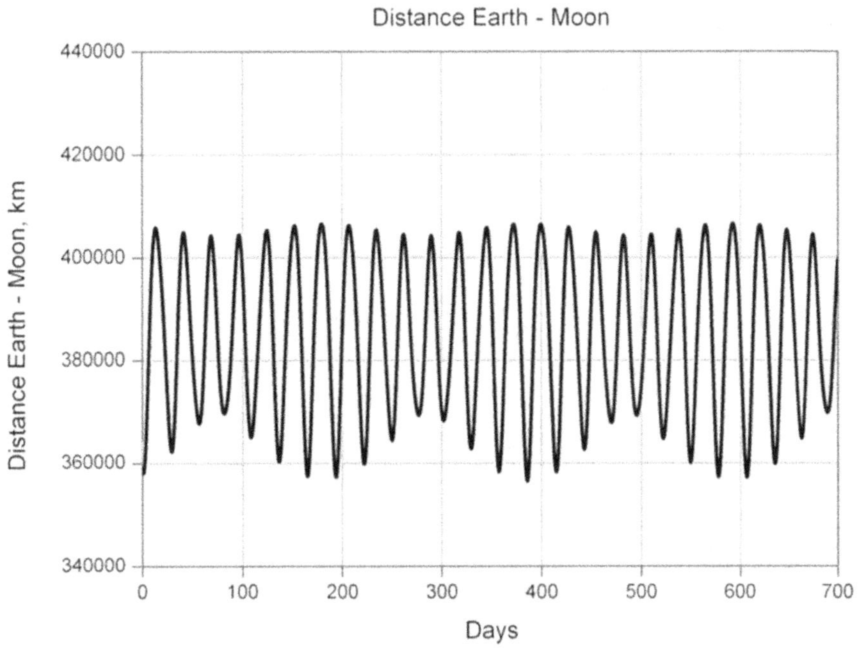

Figure 1 Earth-Moon distance[4,5] over 700 Days

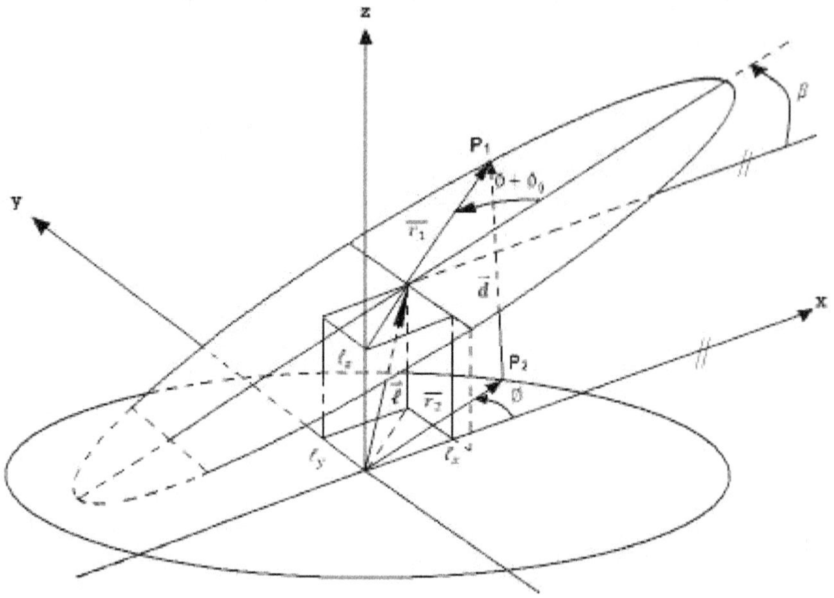

Figure 2 Earth-Moon Topology Moving Around Two Different Circles in Space

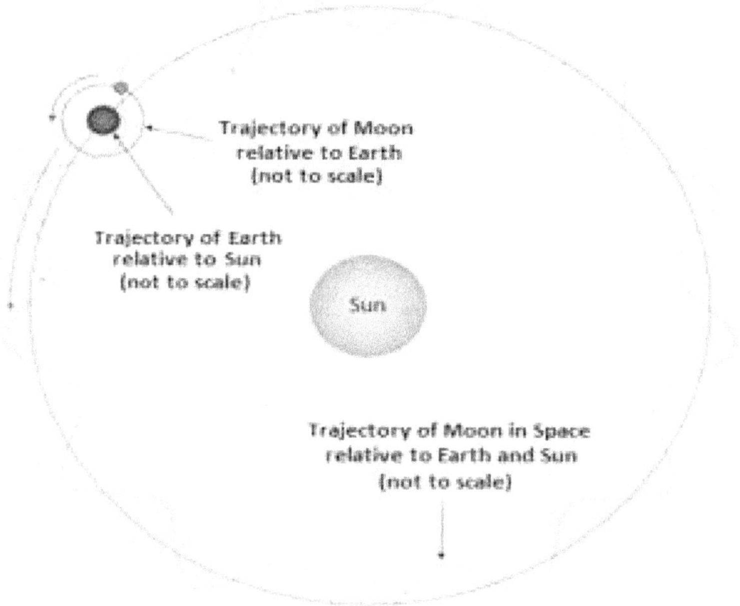

Figure 3 Observed[31] Earth-Moon Motion in Space Relative to the Sun over a Year

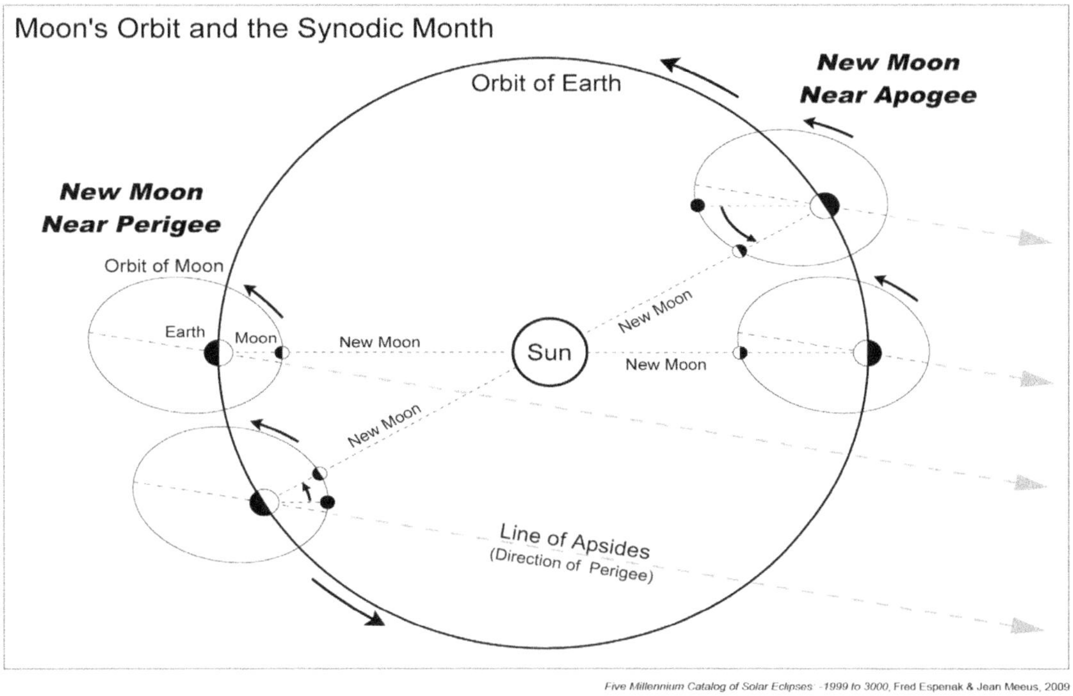

Figure 4 How Moon Phases are Observed[30] from Earth along Motion Relative to the Sun

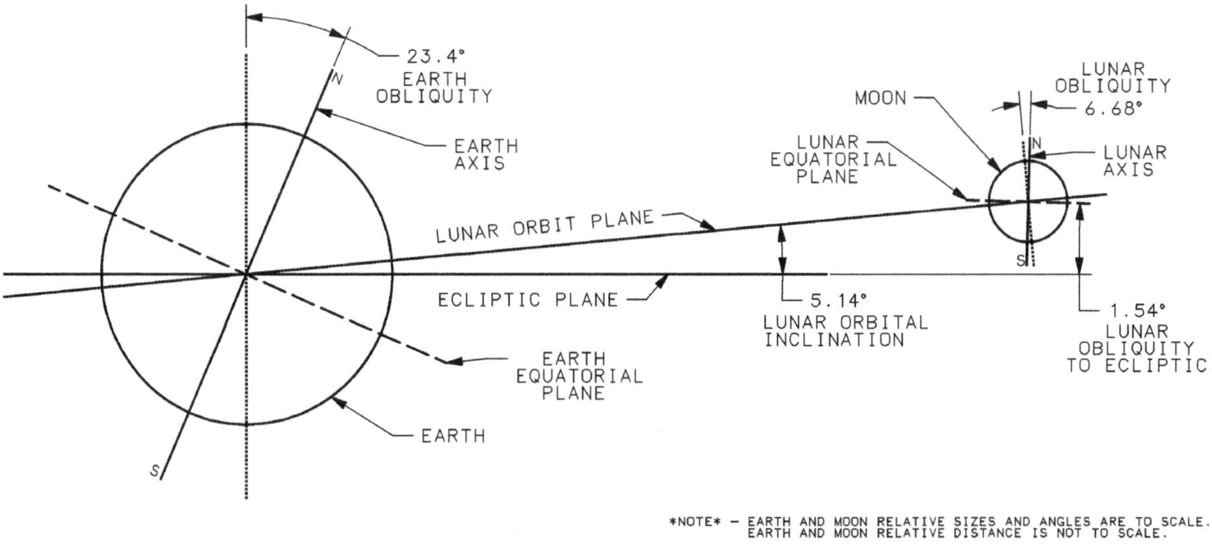

Figure 5 Relative Angles[19] of Poles and Lunar Orbit, Earth Orbit, and Solar Orbit

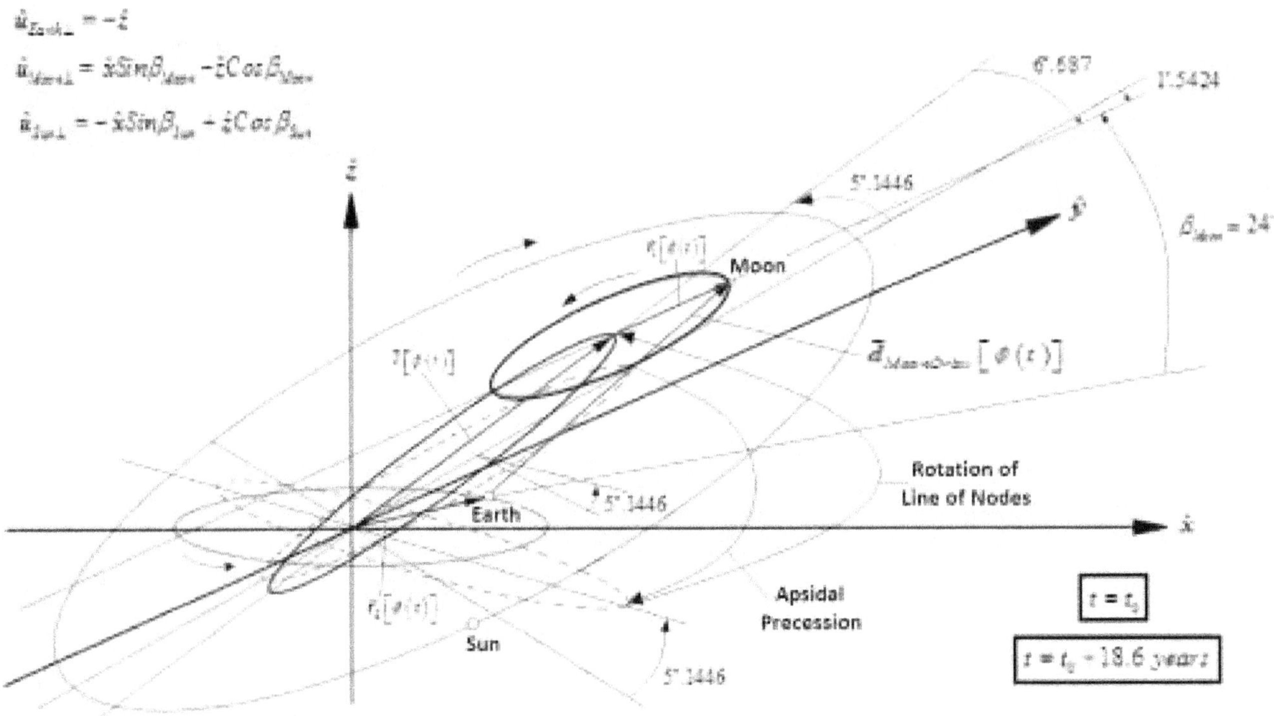

Figure 6 Orientation variation of $\vec{\ell}\left[\phi(t)\right]$ at $(t = t_0)$ and $(t = t_0 + 18.6\,years)$

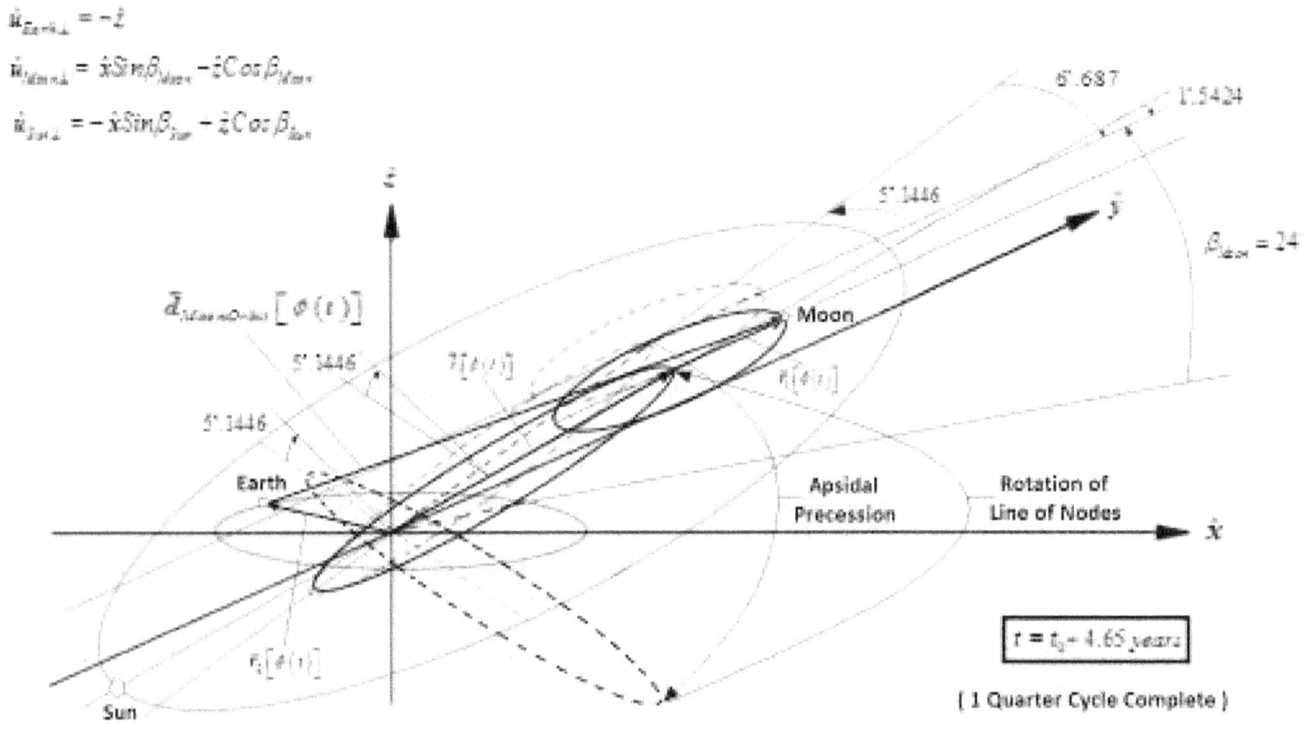

Figure 7 Orientation variation of $\vec{\ell}\left[\phi(t)\right]$ at $\left(t = t_0 + 4.65\,years\right)$ end of 1st Quarter Cycle

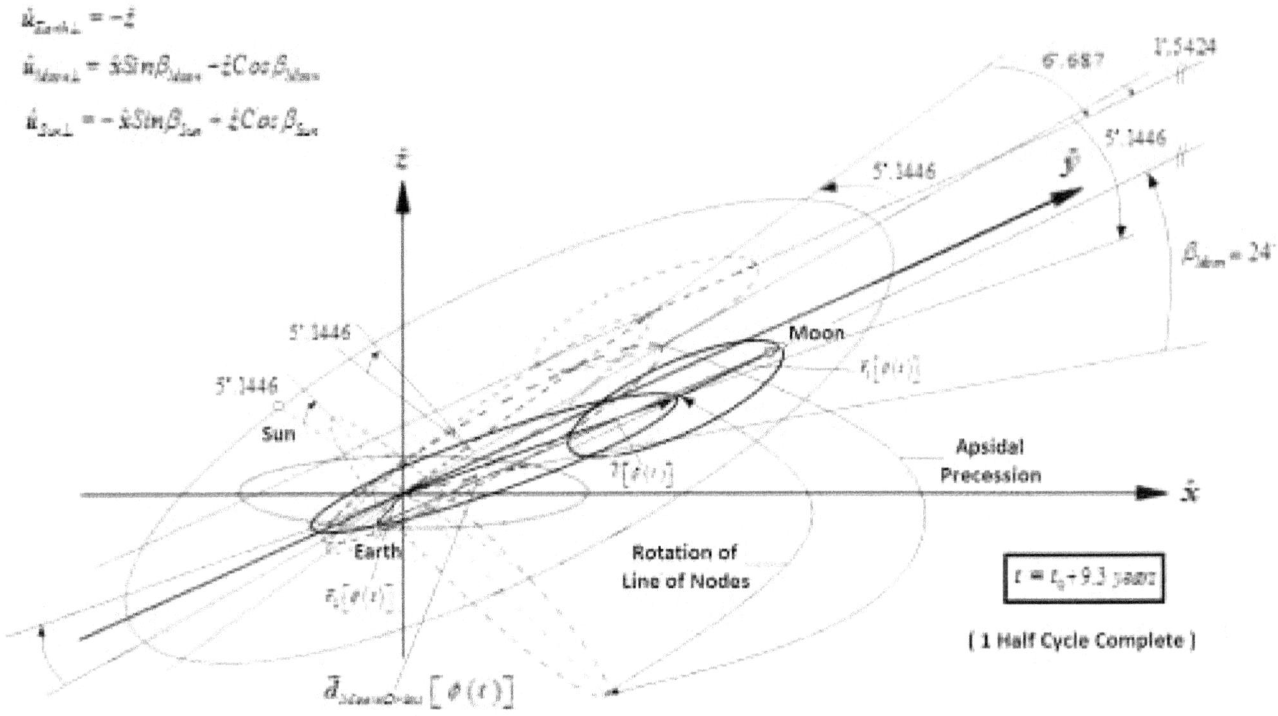

Figure 8 Orientation variation of $\vec{\ell}\left[\phi(t)\right]$ at $\left(t = t_0 + 9.3\,years\right)$ end of 1 Half Cycle

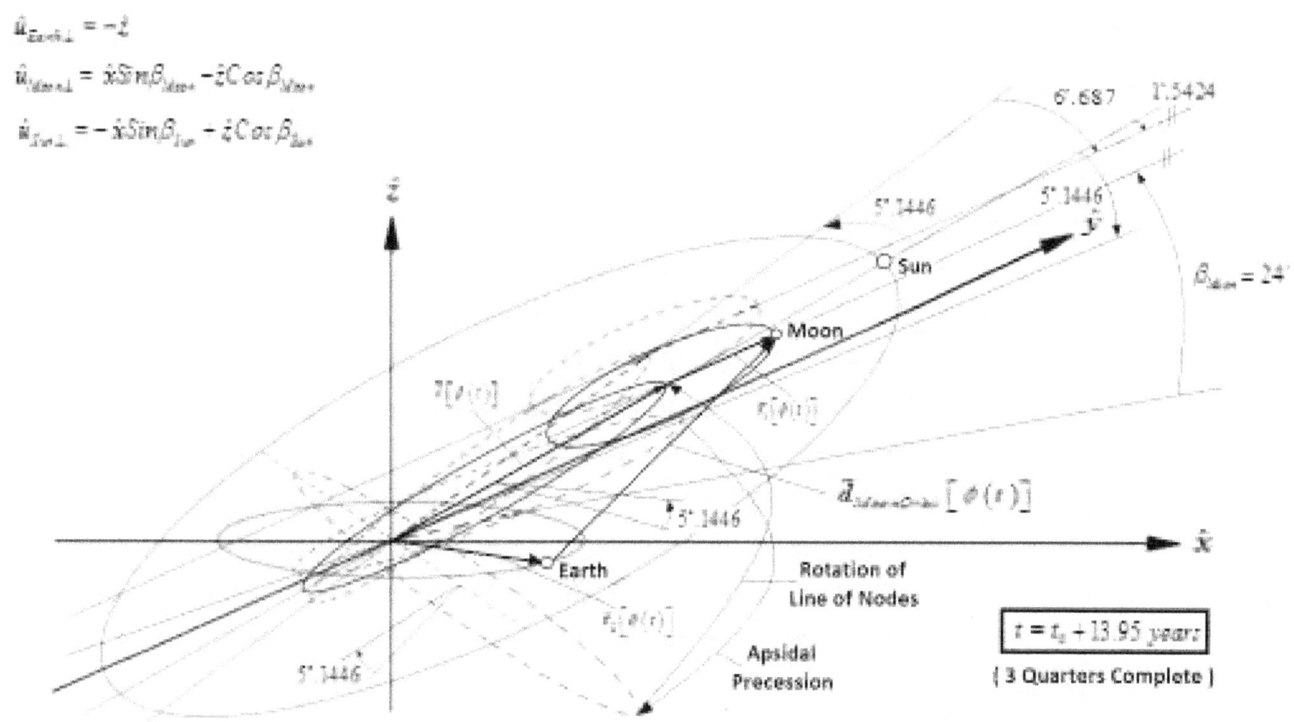

Figure 9 Orientation variation of $\vec{\ell}\left[\phi(t)\right]$ at $\left(t = t_0 + 13.95\,years\right)$ end of 3rd Quarter Cycle

REFERENCES

References in this Article can be any Physics and Calculus textbook, as the physics equations and mathematical identities used as a basis for the proof are all currently accepted theory in existing textbooks.

1. Halliday, D., Resnick, R., *Fundamentals of Physics (3rd Edition)*, John Wiley & Sons, 1988, ISBN 0-471-63735-1
2. Kepler's laws of planetary motion, Wikipedia, https://en.wikipedia.org/wiki/Kepler%27s_laws_of_planetary_motion, (2020)
3. Earth's orbit, Wikipedia, https://en.wikipedia.org/wiki/Earth%27s_orbit, (2020)
4. Lunar distance, Wikipedia, https://en.wikipedia.org/wiki/Lunar_distance_(astronomy), (2020)
5. Moon distances for UTC time zone, Time & Date, https://www.timeanddate.com/astronomy/moon/distance.html, (2020)
6. Ellipse, Wikipedia, https://en.wikipedia.org/wiki/Ellipse, (2020)
7. Semi-major and semi-minor axes, Wikipedia, https://en.wikipedia.org/wiki/Semi-major_and_semi-minor_axes, (2020)
8. Focus (geometry), Wikipedia, https://en.wikipedia.org/wiki/Focus_(geometry), (2020)
9. Dot product, Wikipedia, https://en.wikipedia.org/wiki/Dot_product, (2020)
10. Orbital period, Wikipedia, https://en.wikipedia.org/wiki/Orbital_period, (2020)

11. Moon, Wikipedia, https://en.wikipedia.org/wiki/Moon, (2020)
12. Earth, Wikipedia, https://en.wikipedia.org/wiki/Earth, (2020)
13. Year, Wikipedia, https://en.wikipedia.org/wiki/Year, (2020)
14. Equinox, Wikipedia, https://en.wikipedia.org/wiki/Equinox, (2020)
15. Solstice, Wikipedia, https://en.wikipedia.org/wiki/Solstice, (2020)
16. Equinox and solstice 2010-2019, Greenwich Mean Time, https://greenwichmeantime.com/longest-day/equinox-solstice-2010-2020/, (2020)
17. File: Tan, A. P., *Asli Pinar Tan Analysis Based on Earth-Sun distance (d) Landsat.xlsx*, https://www.dropbox.com/scl/fi/th5d3d5ur8d1mb60aitou/Asli-Pinar-Tan-Analysis-Based-on-Earth-Sun-distance-d-Landsat.xlsx?dl=0&rlkey=m3spxqxbxtnumrj2wvtg33l2v (2020)
18. Moon phases for UTC time zone, Time & Date, https://www.timeanddate.com/ moon/phases/timezone/utc, (2020)
19. Orbit of the Moon, Wikipedia, https://en.wikipedia.org/wiki/Orbit_of_the_Moon, (2020)
20. Double planet, Wikipedia, https://en.wikipedia.org/wiki/Double_planet, (2020)
21. Libration, Wikipedia, https://en.wikipedia.org/wiki/Libration, (2020)
22. File:Lunar libration with phase Oct2007 450px.gif, Wikimedia, (2020) https://upload.wikimedia.org/wikipedia/commons/c/c0/Lunar_libration_with_phase2.gif
23. North Pole, Wikipedia, https://en.wikipedia.org/wiki/North_Pole, (2020)
24. Faraday's law of induction, Wikipedia, https://en.wikipedia.org/wiki/Faraday%27s_law_of_induction, (2020)
25. Magnetic flux, Wikipedia, https://en.wikipedia.org/wiki/Magnetic_flux, (2020)
26. Geographical pole, Wikipedia, https://en.wikipedia.org/wiki/Geographical_pole, (2020)
27. Right hand rule, Wikipedia, https://en.wikipedia.org/wiki/Right-hand_rule, (2020)
28. Axial tilt, Wikipedia, https://en.wikipedia.org/wiki/Axial_tilt, (2020)
29. Ecliptic, Wikipedia, https://en.wikipedia.org/wiki/Ecliptic, (2020)
30. Moon's Orbit, https://www.tes.com/lessons/I6cXCV8KFsqOdg/the-moon-s-orbit, (2020)
31. How would I chart the path of the moon around the sun, https://www.quora.com/How-would-I-chart-the-path-of-the-moon-around-the-sun, (2020)
32. Apsidal precession, Wikipedia, https://en.wikipedia.org/wiki/Apsidal_precession, (2020)
33. Lunar month, Wikipedia, https://en.wikipedia.org/wiki/Lunar_month, (2020)
34. Celestial pole, Wikipedia, https://en.wikipedia.org/wiki/Celestial_pole, (2020)
35. Clockwise, Wikipedia, https://en.wikipedia.org/wiki/Clockwise, (2020)
36. Apsis, Wikipedia, https://en.wikipedia.org/wiki/Apsis, (2020)
37. Retrograde and prograde motion, Wikipedia, https://en.wikipedia.org/wiki/Retrograde_and_prograde_motion, (2020)
38. Sidereal vs. Synodic Motions, Astronomy Education at University of Nebraska-Lincoln, https://astro.unl.edu/naap/motion3/sidereal_synodic.html, (2020)
39. HORIZONS Web Interface, Jet Propulision Laboratory, California Institute of Technology, https://ssd.jpl.nasa.gov/horizons.cgi, (2020)
40. Year, Wikipedia, https://en.wikipedia.org/wiki/Year, (2020)
41. Lunar standstill, Wikipedia, https://en.wikipedia.org/wiki/Lunar_standstill, (2020)
42. Minor lunar standstill and harvest moon, EarthSky.org, https://earthsky.org/astronomy-essentials/minor-lunar-standstill-minimizes-harvest-and-hunters-moons, September 10, 2016
43. Astronomical unit, Wikipedia, https://en.wikipedia.org/wiki/Astronomical_unit, (2020)

ARTICLE 5

EVERYTHING IS A CIRCLE: A NEW MODEL FOR ORBITS OF BODIES IN THE UNIVERSE

Author: Aslı Pınar Tan[V]

SUMMARY

Based on measured astronomical position data of heavenly objects in the Solar System and other planetary systems, all bodies in space seem to move in some kind of elliptical motion with respect to each other. In this paper, our research results are summarized, where it is first mathematically demonstrated and proven that the "distance between points on any two different circles in three-dimensional space" is equivalent to the "distance of points on a vector ellipse from another fixed or moving point, as in two-dimensional space". Equivalently, it amounts to showing that bodies moving on two different circular orbits in space vector-wise mimic an elliptical path with respect to each other, even when they are moving with *different* and *changing* angular velocities with respect to their own centers of revolution. This mathematical revelation leads to far-reaching discoveries in physics, with a formulation of a fundamental new model of motion for bodies in the Universe, including the Sun, Planets, and Satellites in the Solar System and elsewhere, as well as at particle and sub-atomic level, enabling more insight into forces of nature. Based on this physical principle we have set forth based on the demonstrated mathematical analysis, the assertion is made that the Sun, the Earth, and the Moon must each be revolving in their individual circular orbits of revolution in space, as they exhibit almost fixed elliptic orbits relative to one another over time. Then, with this expectation, the individual orbital parameters of the Sun, the Earth, and the Moon are calculated based on observed Sun-Earth and Earth-Moon distance data, using an analytical method also developed as part of this research analysis to an approximation, revealing additional results along the way in alignment with observation, which support our initial assertion that the Sun, the Earth, and the Moon must each be revolving in individual circular orbits in actuality. Based on the same physical principle we have set forth based on the demonstrated mathematical analysis, the impact of particle motions in atomic and sub-atomic dimensions, which we assert must also be revolving in individual circular orbits of revolution, shall be analyzed in more detail in our subsequent publications.

ARTICLE

According to Kepler's 1st Law[1,2], "the orbit of a planet with respect to the Sun is an ellipse[5], with the Sun at one of the two foci", which is an empirical rule concluded through near estimation based on the observation of astronomical position data measured over time in the Solar System. The orbit of the Moon[3,19] with respect to the Earth is also distinctly elliptical, but with a varying eccentricity in a harmonic style as the Moon comes closer to and goes farther away from the Earth harmonically along a full cycle of this ellipse, again based on the observation of astronomical position data measured over time. In this research, it is initially mathematically demonstrated and proven that the "distance between points on any two different circles in three-dimensional space" is equivalent to the "distance of points on a vector ellipse from another fixed or moving point, similar to two-dimensional space". In other words, it is shown that moving

[V] *ASLI PINAR TAN*
Linkedin Website: *https://www.linkedin.com./in/apinartan*

points on two circles in space vector-wise mimic an elliptical path with respect to each other, whether they are moving with the *same* angular velocity, or *different* but *fixed* angular velocities, or even with *different* and *changing* angular velocities with respect to their own centers of revolution, virtually seeing each other as positioned at a instantaneously stationary point in space on their respective virtual ecliptic plane. The observation result by Very Large Baseline Array (VLBA) of antennas published[20] by National Radio Astronomy Observatory on August 4, 2020 about the orbits of a Saturn-sized planet and its small, cool star 35 light years from Earth is a proof of this assertion.

Consider a system of two circles in three-dimensional space, with the geometry of the system demonstrated as in **Figure 1** in Cartesian $(\hat{x},\hat{y},\hat{z})$ coordinates. Points $\mathbf{P}_1$ and $\mathbf{P}_2$ are defined on these two circles, respectively, phased apart by a constant or time(t)-dependent angle ϕ_0 (3), with the location of $\mathbf{P}_1$ phased at $\left[\phi_1(t)=\phi+\phi_0\right]$ (1) in its own circle, and the location of $\mathbf{P}_2$ phased at $\left[\phi_2(t)=\phi\right]$ (2) in its own circle, demonstrated as in **Figure 1** spatially in the most generic form, and expressed in the form of generic vector equations below in (1) - (8). When points $\mathbf{P}_1$ and $\mathbf{P}_2$ are moving as a function of time(t) with phases $\phi_1(t)$ (1) and $\phi_2(t)$ (2), respectively, with respect to the centers of their own circles, in the most general case they have a time(t) - varying phase difference $\phi_0(t)$ (3).

$$\phi_1(t)=\phi+\phi_0=\phi(t)+\phi_0(t)=\phi_2(t)+\phi_0(t) \qquad (Phase\ of\ \mathbf{P}_1) \qquad (1)$$

$$\phi_2(t)=\phi=\phi(t) \qquad (Phase\ of\ \mathbf{P}_2) \qquad (2)$$

$$\phi_0(t)=\phi_1(t)-\phi_2(t)=\phi_0 \qquad (Phase\ difference\ of\ \mathbf{P}_1\ and\ \mathbf{P}_2) \qquad (3)$$

At each $\left[\phi=\phi(t)\right]$ (2), the two circles have vector radii $\vec{r}_1$ (1) - (3) and $\vec{r}_2$ (4) - (5), with constant magnitudes r_1 (6) and r_2 (6), respectively, where the unit vector is $\hat{r}_1(\phi)$ (2) in the direction of $\vec{r}_1$ (1) - (3) and the unit vector is $\hat{r}_2(\phi)$ (5) in the direction of $\vec{r}_2$ (4) - (5). The scalar magnitude r_1 (6) of the radius vector $\vec{r}_1$ (1) - (3) is calculated as the square root of r_1^2 (6), which in turn is calculated in terms of the Dot Product[4] $\left[\vec{r}_1\cdot\vec{r}_1\right]$ (6) of the vector $\vec{r}_1$ (1) - (3) with itself. The scalar magnitude r_2 (6) of radius vector $\vec{r}_2$ (4) - (5) is calculated as square root of r_2^2 (6), which in turn is calculated in terms of the Dot Product[4] $\left[\vec{r}_2\cdot\vec{r}_2\right]$ (6) of the vector $\vec{r}_2$ (4) - (5) with itself. The centers of these two circles are displaced by a constant or variable vector $\vec{\ell}(\phi)$ (7) with magnitude $\ell(\phi)$ (8) at each phase ϕ (2). The scalar distance between the centers of the two circles at every phase ϕ (2), namely the magnitude $\ell(\phi)$ (8) of $\vec{\ell}(\phi)$ (7), is calculated as the square root of $\left[\ell^2(\phi)\right]$ (8), which in turn is calculated in terms of the Dot Product[4] $\left[\vec{\ell}(\phi)\cdot\vec{\ell}(\phi)\right]$ (8) of the vector $\vec{\ell}(\phi)$ (7) with itself.

Note that in all our operations, we take the "square of a vector" as the "Dot Product[4] of the vector with itself", which amounts to the scalar "square of the magnitude" for any vector.

More explicitly, in the Cartesian $(\hat{x},\hat{y},\hat{z})$ coordinate configuration of **Figure 1**, at each ϕ (2), location of $\mathbf{P}_1$ phased at $(\phi+\phi_0)$ (1) is defined by the vector $\left[\vec{r}_1(\phi)+\vec{\ell}(\phi)\right]$ based on (1) - (3) and (7), and the location of $\mathbf{P}_2$ phased at ϕ (2) is defined by the vector $\vec{r}_2(\phi)$ (4) - (5). Note that the inclination angle β (1) between the planes of these two circles is also taken as constant.

$$\vec{r}_1=\vec{r}_1(\phi+\phi_0)=\hat{x}\,r_1Cos(\phi+\phi_0)Cos\,\beta+\hat{y}\,r_1Sin(\phi+\phi_0)+\hat{z}\,r_1Cos(\phi+\phi_0)Sin\,\beta \quad (4)$$

$$\hat{r}_1(\phi)=\hat{x}\,Cos(\phi+\phi_0)Cos\,\beta+\hat{y}\,{}_1Sin(\phi+\phi_0)+\hat{z}\,Cos(\phi+\phi_0)Sin\,\beta \;\Rightarrow\; \vec{r}_1=\hat{r}_1(\phi)r_1 \quad (5)$$

$$\Rightarrow\; \vec{r}_1=\vec{r}_1\left[\phi(t)\right]=\left(\hat{x}\,r_1Cos\,\beta\,Cos\left[\phi_0(t)\right]+\hat{y}\,r_1Sin\left[\phi_0(t)\right]+\hat{z}\,r_1Sin\,\beta\,Cos\left[\phi_0(t)\right]\right)\,Cos\left[\phi(t)\right]+ \\ \left(-\hat{x}\,r_1Cos\,\beta\,Sin\left[\phi_0(t)\right]+\hat{y}\,r_1Cos\left[\phi_0(t)\right]-\hat{z}\,r_1Sin\,\beta\,Sin\left[\phi_0(t)\right]\right)Sin\left[\phi(t)\right] \quad (6)$$

$$\vec{r}_2=\vec{r}_2\left[\phi(t)\right]=\hat{x}\,r_2\,Cos\left[\phi(t)\right]+\hat{y}\,r_2\,Sin\left[\phi(t)\right] \quad (7)$$

$$\hat{r}_2(\phi)=\hat{x}\,Cos\,\phi+\hat{y}\,Sin\,\phi \;\Rightarrow\; \vec{r}_2=\hat{r}_2(\phi)r_2 \quad (8)$$

$$\vec{r}_1\cdot\vec{r}_1=r_1^2 \;;\; \left|\vec{r}_1\right|=r_1=\sqrt{\vec{r}_1\cdot\vec{r}_1} \;;\; \vec{r}_2\cdot\vec{r}_2=r_2^2 \;;\; \left|\vec{r}_2\right|=r_2=\sqrt{\vec{r}_2\cdot\vec{r}_2} \quad (9)$$

$$\vec{\ell}=\vec{\ell}\left[\phi(t)\right]=\hat{x}\,\ell_x\left[\phi(t)\right]+\hat{y}\,\ell_y\left[\phi(t)\right]+\hat{z}\,\ell_z\left[\phi(t)\right] \quad (10)$$

$$\left|\vec{\ell}(\phi)\right|=\ell(\phi)=\sqrt{\ell^2(\phi)}=\sqrt{\vec{\ell}(\phi)\cdot\vec{\ell}(\phi)} \;;\; \ell^2(\phi)=\vec{\ell}(\phi)\cdot\vec{\ell}(\phi)=\ell_x^2(\phi)+\ell_y^2(\phi)+\ell_z^2(\phi) \quad (11)$$

The vector distance $\vec{d}(\phi)$ (9) - (12) between any of these two points $\mathbf{P}_1$ phased at $(\phi+\phi_0)$ (1) and $\mathbf{P}_2$ phased at ϕ (2) on the two respective circles, and its magnitude $d(\phi)$ (13) which is also the scalar distance between $\mathbf{P}_1$ and $\mathbf{P}_2$, is as demonstrated in **Figure 2** and expressed in the following vector equations in (9) - (17), based on (1) - (8). It is worth noting that the scalar distance $d(\phi)$ (13) between $\mathbf{P}_1$ and $\mathbf{P}_2$ is calculated as the square root of $d^2(\phi)$ (14) - (16), which in turn is calculated in terms of Dot Product[4] $\left[\vec{d}(\phi)\cdot\vec{d}(\phi)\right]$ (14) - (16) of vector distance $\vec{d}(\phi)$ (9) - (12) with itself. Unit vector in the direction of $\vec{d}(\phi)$ (9) - (12) is $\hat{d}(\phi)$ (17).

$$\vec{d}(\phi)=\vec{r}_1(\phi+\phi_0)-\vec{r}_2(\phi)+\vec{\ell}(\phi)=\vec{r}_1-\vec{r}_2+\vec{\ell} \quad (12)$$

$$\Rightarrow\; \vec{d}(\phi)=\hat{x}\,[r_1Cos(\phi+\phi_0)Cos\,\beta-r_2\,Cos\,\phi+\ell_x(\phi)]+\hat{y}\,[r_1Sin(\phi+\phi_0)-r_2Sin\,\phi+\ell_y(\phi)] \\ +\hat{z}\,[r_1Cos(\phi+\phi_0)Sin\,\beta+\ell_z(\phi)] \quad (13)$$

$$\vec{d}\left[\phi(t)\right]=\left[\hat{x}\left(r_1Cos\,\beta\,Cos\left[\phi_0(t)\right]-r_2\right)+\hat{y}\,r_1Sin\left[\phi_0(t)\right]+\hat{z}\,r_1Sin\,\beta\,Cos\left[\phi_0(t)\right]\right]\,Cos\left[\phi(t)\right]+ \\ \left[-\hat{x}\,r_1Cos\,\beta\,Sin\left[\phi_0(t)\right]+\hat{y}\left(r_1Cos\left[\phi_0(t)\right]-r_2\right)-\hat{z}\,r_1Sin\,\beta\,Sin\left[\phi_0(t)\right]\right]Sin\left[\phi(t)\right]+ \\ \left[\hat{x}\,\ell_x\left[\phi(t)\right]+\hat{y}\,\ell_y\left[\phi(t)\right]+\hat{z}\,\ell_z\left[\phi(t)\right]\right] \quad (14)$$

$$\vec{d}(\phi)=\hat{x}d_x(\phi)+\hat{y}d_y(\phi)+\hat{z}d_z(\phi) \Rightarrow \begin{cases} d_x(\phi)=r_1 Cos(\phi+\phi_0)Cos\beta - r_2\, Cos\phi + \ell_x(\phi) \\ d_y(\phi)=r_1 Sin(\phi+\phi_0) - r_2 Sin\phi + \ell_y(\phi) \\ d_z(\phi)=r_1 Cos(\phi+\phi_0)Sin\beta + \ell_z(\phi) \end{cases} \quad (15)$$

$$d(\phi)=\left|\vec{d}(\phi)\right|=\sqrt{d^2(\phi)}=\sqrt{\vec{d}(\phi)\cdot\vec{d}(\phi)}=\sqrt{\left[d_x(\phi)\right]^2+\left[d_y(\phi)\right]^2+\left[d_z(\phi)\right]^2} \quad (16)$$

$$d^2(\phi)=\vec{d}(\phi)\cdot\vec{d}(\phi)=\left(\vec{r}_1-\vec{r}_2+\vec{\ell}\right)\cdot\left(\vec{r}_1-\vec{r}_2+\vec{\ell}\right)=\overline{\vec{r}_1\cdot\vec{r}_1}-2\overline{\vec{r}_1\cdot\vec{r}_2}+\overline{\vec{r}_2\cdot\vec{r}_2}+2\left(\vec{r}_1-\vec{r}_2\right)\cdot\vec{\ell}+\vec{\ell}\cdot\vec{\ell} \quad (17)$$

$$\begin{aligned} d^2(\phi)&=\vec{d}(\phi)\cdot\vec{d}(\phi) \\ &=[r_1 Cos(\phi+\phi_0)Cos\beta - r_2\, Cos\phi+\ell_x(\phi)]^2+[r_1 Sin(\phi+\phi_0)-r_2\, Sin\phi+\ell_y(\phi)]^2+ \\ &\quad [r_1 Cos(\phi+\phi_0)Sin\beta+\ell_z(\phi)]^2 \end{aligned} \quad (18)$$

$$d^2(\phi)=\vec{d}(\phi)\cdot\vec{d}(\phi)=\left[d_x(\phi)\right]^2+\left[d_y(\phi)\right]^2+\left[d_z(\phi)\right]^2 \quad (19)$$

$$\hat{d}(\phi)=\frac{\vec{d}(\phi)}{\left|\vec{d}(\phi)\right|}=\frac{\vec{d}(\phi)}{d(\phi)}=\hat{x}\frac{d_x(\phi)}{d(\phi)}+\hat{y}\frac{d_y(\phi)}{d(\phi)}+\hat{z}\frac{d_z(\phi)}{d(\phi)} \Rightarrow \vec{d}(\phi)=\hat{d}(\phi)\left|\vec{d}(\phi)\right|=\hat{d}(\phi)d(\phi) \quad (20)$$

The vector distance $\vec{d}(\phi)$ (11) at any value of the phase ϕ (2), between points $\mathbf{P}_1$ phased at $(\phi+\phi_0)$ (1) and $\mathbf{P}_2$ phased at ϕ (2) on the two respective circles, can be equivalently expressed as $\vec{d}(\phi)$ (18) in terms of virtual vectors $\vec{\mathbf{X}}(\phi)$ (19) and $\vec{\mathbf{Y}}(\phi)$ (20), defined utilizing $\vec{r}_1$ (1) - (3) and $\vec{r}_2$ (4) - (5). Based on the definition of $\vec{\mathbf{X}}(\phi)$ (19) from (18), (7) and (11), the virtual vector $\vec{a}$ (21) is also defined, and based on the definition of $\vec{\mathbf{Y}}(\phi)$ (20) from (18), (7) and (11), another virtual vector $\vec{b}$ (22) is also defined.

The magnitude $X(\phi)$ (19) of the virtual vector $\vec{\mathbf{X}}(\phi)$ (19) is calculated as the square root of $X^2(\phi)$ (19), which in turn is calculated in terms of the Dot Product[4] $\left[\vec{\mathbf{X}}(\phi)\cdot\vec{\mathbf{X}}(\phi)\right]$ (19) of the vector $\vec{\mathbf{X}}(\phi)$ (19) with itself. The magnitude $Y(\phi)$ (20) of the virtual vector $\vec{\mathbf{Y}}(\phi)$ (20) is calculated as the square root of $Y^2(\phi)$ (20), which in turn is calculated in terms of the Dot Product[4] $\left[\vec{\mathbf{Y}}(\phi)\cdot\vec{\mathbf{Y}}(\phi)\right]$ (20) of the vector $\vec{\mathbf{Y}}(\phi)$ (20) with itself. The magnitude a (23) of the virtual vector $\vec{a}$ (21) is calculated as the square root of a^2 (23), which in turn is calculated in terms of the Dot Product[4] $(\vec{a}\cdot\vec{a})$ (23) of the vector $\vec{a}$ (21) with itself. The magnitude b (24) of the virtual vector $\vec{b}$ (22) is calculated as the square root of b^2 (24), which in turn is calculated in terms of the Dot Product[4] $(\vec{b}\cdot\vec{b})$ (24) of the vector $\vec{b}$ (22) with itself. When the phase difference ϕ_0 (3) is constant for all values of ϕ (2), the magnitudes a (23) and b (24) are constant for all ϕ (2), and when phase difference ϕ_0 (3) is time(t)-dependent, i.e. if $\left[\phi_0=\phi_0(t)\right]$ (3), $a(t)$ (23)

and $b(t)$ (24) are also time(t)-dependent for different ϕ (2). Based on the definitions of $\vec{a}$ (21) and $\vec{b}$ (22), their Dot Product[4] $\left(\vec{a}\cdot\vec{b}\right)$ (25) is also defined.

$$\vec{d}(\phi)=\vec{\mathbf{X}}(\phi)+\vec{\mathbf{Y}}(\phi)+\vec{\ell}(\phi) \qquad where \qquad \vec{r}_1(\phi)-\vec{r}_2(\phi)=\vec{\mathbf{X}}(\phi)+\vec{\mathbf{Y}}(\phi)=\vec{a}\,Cos\,\phi+\vec{b}\,Sin\,\phi \quad (21)$$

$$\vec{\mathbf{X}}(\phi)=\vec{a}\,Cos\,\phi \quad ; \quad \vec{\mathbf{X}}(\phi)\cdot\vec{\mathbf{X}}(\phi)=X^2(\phi)=\vec{a}\cdot\vec{a}\;Cos^2\phi=a^2Cos^2\phi \quad ; \quad \left|\vec{\mathbf{X}}(\phi)\right|=X(\phi) \quad (22)$$

$$\vec{\mathbf{Y}}(\phi)=\vec{b}\,Sin\,\phi \quad ; \quad \vec{\mathbf{Y}}(\phi)\cdot\vec{\mathbf{Y}}(\phi)=Y^2(\phi)=\vec{b}\cdot\vec{b}\;Sin^2\,\phi=b^2Sin^2\,\phi \quad ; \quad \left|\vec{\mathbf{Y}}(\phi)\right|=Y(\phi) \quad (23)$$

$$\begin{aligned}\vec{a}=\vec{a}(t)&=\hat{x}\left\{r_1 Cos\,\beta\, Cos\left[\phi_0(t)\right]-r_2\right\}+\hat{y}\,r_1 Sin\left[\phi_0(t)\right]+\hat{z}\,r_1 Sin\,\beta\, Cos\left[\phi_0(t)\right]\\ \Rightarrow\quad \vec{a}&=\left[\vec{r}_1(\phi)-\vec{r}_2(\phi)\right](\phi=0)\end{aligned} \quad (24)$$

$$\begin{aligned}\vec{b}=\vec{b}(t)&=-\hat{x}\,r_1 Cos\,\beta\, Sin\left[\phi_0(t)\right]+\hat{y}\left\{r_1 Cos\left[\phi_0(t)\right]-r_2\right\}-\hat{z}\,r_1 Sin\,\beta\, Sin\left[\phi_0(t)\right]\\ \Rightarrow\quad \vec{b}&=\left[\vec{r}_1(\phi)-\vec{r}_2(\phi)\right]\left(\phi=\frac{\pi}{2}\right)\end{aligned} \quad (25)$$

$$\vec{a}(t)\cdot\vec{a}(t)=a^2(t)=r_1^2-2r_1\,r_2\,Cos\beta\,Cos\left[\phi_0(t)\right]+r_2^2 \quad ; \quad a=a(t)=\left|\vec{a}(t)\right|=\sqrt{\vec{a}(t)\cdot\vec{a}(t)} \quad (26)$$

$$\vec{b}(t)\cdot\vec{b}(t)=b^2(t)=r_1^2-2r_1\,r_2\,Cos\left[\phi_0(t)\right]+r_2^2 \quad ; \quad b=b(t)=\left|\vec{b}(t)\right|=\sqrt{\vec{b}(t)\cdot\vec{b}(t)} \quad (27)$$

$$\vec{a}\cdot\vec{b}=\vec{a}(t)\cdot\vec{b}(t)=r_1\,r_2\left(Cos\beta-1\right)Sin\left[\phi_0(t)\right] \quad (28)$$

Based on (18) - (25), square of the distance $d^2(\phi)$ (14) - (16) between the points $\mathbf{P}_1$ and $\mathbf{P}_2$ can also be expressed for all ϕ (2) also as in (31) - (34).

$$\left|\vec{d}(\phi)\right|^2=d^2(\phi)=\vec{d}(\phi)\cdot\vec{d}(\phi)=\left[\vec{\mathbf{X}}(\phi)+\vec{\mathbf{Y}}(\phi)+\vec{\ell}\right]\cdot\left[\vec{\mathbf{X}}(\phi)+\vec{\mathbf{Y}}(\phi)+\vec{\ell}\right] \;;\; \left|\vec{d}(\phi)\right|=d(\phi) \quad (29)$$

$$\Rightarrow\quad d^2(\phi)=\vec{\mathbf{X}}(\phi)\cdot\vec{\mathbf{X}}(\phi)+2\vec{\mathbf{X}}(\phi)\cdot\vec{\mathbf{Y}}(\phi)+\vec{\mathbf{Y}}(\phi)\cdot\vec{\mathbf{Y}}(\phi)+2\left[\vec{\mathbf{X}}(\phi)+\vec{\mathbf{Y}}(\phi)\right]\cdot\vec{\ell}+\vec{\ell}\cdot\vec{\ell} \quad (30)$$

$$\Rightarrow\quad d^2(\phi)=a^2Cos^2\phi+2\vec{a}\cdot\vec{b}\,Sin\phi Cos\,\phi+b^2Sin^2\phi+2\vec{a}\cdot\vec{\ell}\,Cos\,\phi+2\vec{b}\cdot\vec{\ell}\,Sin\,\phi+\ell^2 \quad (31)$$

$$d^2(\phi)=b^2+2r_1\,r_2\left(1-Cos\beta\right)Cos\phi\,Cos\left(\phi+\phi_0\right)+2\left(\vec{a}Cos\phi+\vec{b}Sin\phi\right)\cdot\vec{\ell}(\phi)+\ell^2(\phi) \quad (32)$$

According to definitions of vectors $\vec{\mathbf{X}}(\phi)$ (19), $\vec{\mathbf{Y}}(\phi)$ (20), $\vec{a}$ (21), and $\vec{b}$ (22), as described in (18) - (24), the relation in (26) is valid and holds for all ϕ (2) due to given trigonometric identity.

$$\frac{\vec{\mathbf{X}}(\phi)\cdot\vec{\mathbf{X}}(\phi)}{\vec{a}\cdot\vec{a}}+\frac{\vec{\mathbf{Y}}(\phi)\cdot\vec{\mathbf{Y}}(\phi)}{\vec{b}\cdot\vec{b}}=\frac{X^2(\phi)}{a^2}+\frac{Y^2(\phi)}{b^2}=Cos^2\phi+Sin^2\phi=1 \quad (33)$$

$$\Rightarrow\quad \boxed{\frac{\vec{\mathbf{X}}(\phi)\cdot\vec{\mathbf{X}}(\phi)}{\vec{a}\cdot\vec{a}}+\frac{\vec{\mathbf{Y}}(\phi)\cdot\vec{\mathbf{Y}}(\phi)}{\vec{b}\cdot\vec{b}}=1} \quad \left(Definition\ of\ Vector\ Ellipse\ in\ 3-Dimensions\right) \quad (34)$$

$$\Rightarrow\quad \boxed{\frac{X^2(\phi)}{a^2}+\frac{Y^2(\phi)}{b^2}=1} \quad \left(Definition\ of\ Scalar\ Ellipse\ in\ 2-Dimensions\right) \quad (35)$$

Therefore, the relation in (26) reveals the validity of (27) and (28) for the vector pair $\left[\vec{\mathbf{X}}(\phi),\vec{\mathbf{Y}}(\phi)\right]$ (19) - (20) and its magnitude pair $\left[X(\phi),Y(\phi)\right]$ (19) - (20), respectively. As (28) is the defining equation of an ellipse[5] in two dimensions, where a (23) is the semi-major[6] axis and b (24) is the semi-minor[6] axis of the ellipse[5] when $(a>b)$[5,6] (39) holds, and vice versa, with (28) reducing to the special case of a circle when $(a=b)$[5,6] (39) holds, we can claim that (27) indicates that the vector pair $\left[\vec{\mathbf{X}}(\phi),\vec{\mathbf{Y}}(\phi)\right]$ (19) - (20) defines points on a vector ellipse in three dimensions in the most general case. *In other words, the vector distance $\vec{\boldsymbol{d}}(\phi)$ (11) between two points $\mathbf{P}_1$ phased at $(\phi+\phi_0)$ (1) and $\mathbf{P}_2$ phased at ϕ (2) on two respective circles, whose centers are displaced by a constant or variable vector $\vec{\ell}(\phi)$ (7), can equivalently be mathematically expressed and interpreted as the distance $\vec{\boldsymbol{d}}(\phi)$ (18) of points on a virtual vector ellipse, whose locations with respect to a virtual origin at each ϕ (2) are determined by the sum of vector pair $\left[\vec{\mathbf{X}}(\phi),\vec{\mathbf{Y}}(\phi)\right]$ (19) - (20), from another fixed or moving point displaced from the same virtual origin of the ellipse by a constant or variable vector $\left[-\vec{\ell}(\phi)\right]$ (7), where $\vec{\boldsymbol{a}}$* (21) *and $\vec{\boldsymbol{b}}$* (22) *are the fixed or variable semi-major[6] and semi-minor[6] axis vectors of the vector ellipse based on (116) - (39). This result is mathematically valid even when the phase difference ϕ_0 (3) is a variable function of time(t), i.e. even if $\phi_0=\phi_0(t)$*. This revelation is the core and most significant finding of our research. Moreover, we have also mathematically introduced the concept of a "vector ellipse" (27) in three-dimensional space based on our analysis introduced along the lines of (1) - (39).

Based on basic trigonometry, $\left[(1-Cos\beta)\geq 0\right]$ (115) is always true. As a result, the sign of the difference (116) of $\left[a^2(t)\right]$ (23) and $\left[b^2(t)\right]$ (24), as well as the sign of the difference (38) of $a(t)$ (23) and $b(t)$ (24), is based on the sign of $Cos\left[\phi_0(t)\right]$ (116), thus the value of $\phi_0(t)$ (3) as described in (116) - (38). Therefore, depending on the values of the parameters (r_1,r_2,β) (1) - (6), and the value of $\phi_0(t)$ (3) at any time t, the expression in (39) holds regarding the instantaneous semi-major[6] and semi-minor[6] axes of the ellipse[5], or instantaneous radii $a(t)$ (23) and $b(t)$ (24) of the vector circle defining the relative motion of $\mathbf{P}_1$ and $\mathbf{P}_2$ whenever $a(t)=b(t)$.

$$-1\leq Cos\beta\leq 1 \quad\Rightarrow\quad (1-Cos\beta)\geq 0 \tag{36}$$

$$a^2(t)-b^2(t)=2\,r_1 r_2\,(1-Cos\beta)Cos\left[\phi_0(t)\right]\begin{cases}\geq 0 \;\; if \;\; Cos\left[\phi_0(t)\right]\geq 0 \Rightarrow -\dfrac{\pi}{2}\leq\phi_0(t)\leq\dfrac{\pi}{2}\\[2ex] \leq 0 \;\; if \;\; Cos\left[\phi_0(t)\right]\leq 0 \Rightarrow \dfrac{\pi}{2}\leq\phi_0(t)\leq\dfrac{3\pi}{2}\end{cases} \tag{37}$$

$$a^2(t)-b^2(t)=\left[a(t)-b(t)\right]\left[a(t)+b(t)\right] \quad \& \quad a(t),b(t),\left[a(t)+b(t)\right]\geq 0$$

$$\Rightarrow \left[a(t)-b(t)\right]\begin{cases}\geq 0 \;\; if \;\; Cos\left[\phi_0(t)\right]\geq 0 \Rightarrow -\dfrac{\pi}{2}\leq\phi_0(t)\leq\dfrac{\pi}{2} \\ \leq 0 \;\; if \;\; Cos\left[\phi_0(t)\right]\leq 0 \Rightarrow \dfrac{\pi}{2}\leq\phi_0(t)\leq\dfrac{3\pi}{2}\end{cases} \quad (38)$$

$$\begin{cases} a(t)>b(t) \;\Rightarrow\; \vec{a}(t) \text{ is semi-major axis } \& \; \vec{b}(t) \text{ is semi-minor axis of vector ellipse} \\ a(t)<b(t) \;\Rightarrow\; \vec{b}(t) \text{ is semi-major axis } \& \; \vec{a}(t) \text{ is semi-minor axis of vector ellipse} \\ a(t)=b(t) \;\Rightarrow\; \vec{a}(t) \text{ and } \vec{b}(t) \text{ are radii of vector circle} \end{cases} \quad (39)$$

Note that $\vec{a}$ (21) is the vector value of $\left[\vec{r}_1(\phi)-\vec{r}_2(\phi)\right]$ (18) when $(\phi=0)$, and $\vec{b}$ (22) is the vector value of $\left[\vec{r}_1(\phi)-\vec{r}_2(\phi)\right]$ (18) when $\left(\phi=\dfrac{\pi}{2}\right)$. Throughout a respective cycle of the phased points $\mathbf{P}_1$ and $\mathbf{P}_2$ moving around their own circles, namely points $\mathbf{P}_1$ phased at $(\phi+\phi_0)$ (1) and $\mathbf{P}_2$ phased at ϕ (2) on the two respective circles, the $\left[\vec{r}_1(\phi)-\vec{r}_2(\phi)\right]$ (18) vector has a vector value of $\left(\vec{a}\,Cos\,\phi+\vec{b}\,Sin\,\phi\right)$ (18) at each phase value of ϕ (2), and moves in the plane formed by the $\vec{a}$ (21) and $\vec{b}$ (22) vectors, namely the $\vec{a}-\vec{b}$ plane, which is a <u>variable</u> moving plane in three dimensions if ϕ_0 (3) is a variable function of time t, i.e. if $\phi_0=\phi_0(t)$, and a <u>fixed</u> plane otherwise.

Square of the instantaneous focal[5,7] distance c^2 (29) of the vector ellipse can be determined using (23) - (24), and the instantaneous eccentricity[5] e (30) of the vector ellipse can be found using (23) - (24) and (29).

$$c^2=c^2(t)=\left|a^2(t)-b^2(t)\right|=2\,r_1\,r_2\left(1-Cos\beta\right)\left|Cos\left[\phi_0(t)\right]\right| \qquad (Focal\ Distance\ Squared) \quad (40)$$

$$e=e(t)=\begin{cases}\dfrac{c(t)}{a(t)}=\sqrt{\dfrac{2\,r_1\,r_2\left(1-Cos\beta\right)\left|Cos\left[\phi_0(t)\right]\right|}{r_1^2-2r_1\,r_2\,Cos\beta\,Cos\left[\phi_0(t)\right]+r_2^2}} & if \;\; a(t)>b(t) \\ \dfrac{c(t)}{b(t)}=\sqrt{\dfrac{2\,r_1\,r_2\left(1-Cos\beta\right)\left|Cos\left[\phi_0(t)\right]\right|}{r_1^2-2r_1\,r_2 Cos\left[\phi_0(t)\right]+r_2^2}} & if \;\; a(t)<b(t)\end{cases} \qquad (Eccentricity) \quad (41)$$

Let us continue to analyze the features and potential physical meaning and implications of this virtual vector ellipse in three-dimensional space.

The major consequence in physics, of this analysis and mathematical results in (18) - (30) for moving points $\mathbf{P}_1$ and $\mathbf{P}_2$ on the two respective circles, is that "Particles or bodies moving around different circular orbits in space see themselves positioned on an elliptical path with respect to each other, with fixed or variable semi-major[6] and semi-minor[6] axes over time depending on the time dependencies of the angular velocities of the particles or bodies, where the particles or bodies instantaneously observe their counterpart virtually positioned at a fixed point in space, whose position is determined by the distance vector between the centers of the circles around which they revolve."

A special case of (1) - (3) is expressed in (61) - (63), when the moving points $\mathbf{P}_1$ and $\mathbf{P}_2$ are phased with ***fixed*** but ***different*** (60) angular velocities ω_1 (61) and ω_2 (62), respectively, with respect to the centers of their own circles of revolution, their phase difference being $\left[\phi_0(t_0)=\varphi_0\right]$ (63) at time $(t=t_0=0)$, the reference timestamp t_0 taken to be the point in time when $\mathbf{P_2}$ has a phase of $\left[\phi_2(t_0)=0\right]$ (62), in the configuration described in **Figure 2**.

$$\omega_1 \neq \omega_2 \qquad\qquad (\omega_1 \,\&\, \omega_2 \; constant) \tag{42}$$

$$\phi_1(t)=\phi+\phi_0=\phi(t)+\phi_0(t)=\phi_2(t)+\phi_0(t)=\omega_1 t+\varphi_0 \quad ; \quad \phi_1(t_0=0)=\varphi_0 \qquad (Phase\ of\ \mathbf{P}_1) \tag{43}$$

$$\phi_2(t)=\phi=\phi(t)=\omega_2 t \quad ; \quad \phi_2(t_0=0)=0 \qquad (Phase\ of\ \mathbf{P}_2) \tag{44}$$

$$\phi_0(t)=\phi_1(t)-\phi_2(t)=(\omega_1-\omega_2)t+\varphi_0 \quad ; \quad \phi_0(t_0=0)=\varphi_0 \quad ; \quad (Phase\ difference\ of\ \mathbf{P}_1\ and\ \mathbf{P}_2) \tag{45}$$

In this special case, at each $\phi(t)$ (62), the vector radii $\vec{r}_1$ (3) and $\vec{r}_2$ (4) of the two circles become $\vec{r}_1$ (64) and $\vec{r}_2$ (47), respectively, where constant or variable vector $\vec{\ell}$ (7), by which the centers of these two circles are displaced, can be expressed as $\vec{\ell}$ (48), with magnitude $\ell(\phi)$ (8). Hence, at each $\phi(t)$ (62), $\vec{\boldsymbol{d}}\left[\phi(t)\right]$ (11) can be expressed as $\vec{\boldsymbol{d}}\left[\phi(t)\right]$ (68), $d^2(\phi)$ (34) can be expressed as $d^2(\phi)$ (25), and elliptic coordinate vectors $\vec{\mathbf{X}}\left[\phi(t)\right]$ (19) and $\vec{\mathbf{Y}}\left[\phi(t)\right]$ (20) of the vector ellipse[5] become $\vec{\mathbf{X}}\left[\phi(t)\right]$ (53) and $\vec{\mathbf{Y}}\left[\phi(t)\right]$ (54), respectively, in which case the elliptic representation of vector distance $\vec{\boldsymbol{d}}\left[\phi(t)\right]$ (18) becomes $\vec{\boldsymbol{d}}\left[\phi(t)\right]$ (55). Note also that in this special case, at each $\phi(t)$ (62), the semi-major[6] and semi-minor[6] axis vectors $\vec{\boldsymbol{a}}(t)$ (69) and $\vec{\boldsymbol{b}}(t)$ (70) of the vector ellipse[5] are sinusoidal functions of time t, with $\phi_0(t)$ (63) replaced in (21) - (22). Throughout a respective cycle of points $\mathbf{P}_1$ and $\mathbf{P}_2$ moving around their own circles of revolution with ***fixed*** but ***different*** (60) angular velocities ω_1 (61) and ω_2 (62), respectively, the $(\vec{r}_1-\vec{r}_2)$ (18) vector has a vector value of $\left[\vec{\boldsymbol{a}}(t)Cos(\omega_2 t)+\vec{\boldsymbol{b}}(t)Sin(\omega_2 t)\right]$ (75) at each phase value of $\left[\phi(t)=\omega_2 t\right]$ (62), and moves in the plane formed by the $\vec{\boldsymbol{a}}(t)$ (69) and $\vec{\boldsymbol{b}}(t)$ (70) vectors, namely the $\vec{\boldsymbol{a}}(t)-\vec{\boldsymbol{b}}(t)$ plane, which is a <u>variable</u> moving plane in three dimensions for this case.

$$\vec{r}_1=\vec{r}_1\left[\phi(t)\right]=\begin{Bmatrix} \hat{\boldsymbol{x}}\, r_1 Cos\,\beta\, Cos\left[(\omega_1-\omega_2)t+\varphi_0\right]+\hat{\boldsymbol{y}}\, r_1 Sin\left[(\omega_1-\omega_2)t+\varphi_0\right]+ \\ \hat{\boldsymbol{z}}\, r_1 Sin\,\beta\, Cos\left[(\omega_1-\omega_2)t+\varphi_0\right] \end{Bmatrix} Cos(\omega_2 t)+$$
$$\begin{Bmatrix} -\hat{\boldsymbol{x}}\, r_1 Cos\,\beta\, Sin\left[(\omega_1-\omega_2)t+\varphi_0\right]+\hat{\boldsymbol{y}}\, r_1 Cos\left[(\omega_1-\omega_2)t+\varphi_0\right] \\ -\hat{\boldsymbol{z}}\, r_1 Sin\,\beta\, Sin\left[(\omega_1-\omega_2)t+\varphi_0\right] \end{Bmatrix} Sin(\omega_2 t) \tag{46}$$

$$\vec{r_2} = \vec{r}_2\left[\phi(t)\right] = \hat{\boldsymbol{x}}\, r_2\, Cos(\omega_2 t) + \hat{\boldsymbol{y}}\, r_2\, Sin(\omega_2 t) \tag{47}$$

$$\vec{\ell} = \vec{\ell}\left[\phi(t)\right] = \hat{\boldsymbol{x}}\,\ell_x(\omega_2 t) + \hat{\boldsymbol{y}}\,\ell_y(\omega_2 t) + \hat{\boldsymbol{z}}\,\ell_z(\omega_2 t) \tag{48}$$

$$\begin{aligned}\vec{\boldsymbol{d}}\left[\phi(t)\right] = &\left\{\begin{array}{l}\hat{\boldsymbol{x}}\left(r_1 Cos\,\beta\, Cos\left[(\omega_1-\omega_2)t+\varphi_0\right]-r_2\right)+\hat{\boldsymbol{y}}\, r_1 Sin\left[(\omega_1-\omega_2)t+\varphi_0\right]+\\ \hat{\boldsymbol{z}}\, r_1 Sin\,\beta\, Cos\left[(\omega_1-\omega_2)t+\varphi_0\right]\end{array}\right\} Cos(\omega_2 t)\\ &+\left\{\begin{array}{l}-\hat{\boldsymbol{x}}\, r_1 Cos\,\beta\, Sin\left[(\omega_1-\omega_2)t+\varphi_0\right]+\hat{\boldsymbol{y}}\left(r_1 Cos\left[(\omega_1-\omega_2)t+\varphi_0\right]-r_2\right)\\ -\hat{\boldsymbol{z}}\, r_1 Sin\,\beta\, Sin\left[(\omega_1-\omega_2)t+\varphi_0\right]\end{array}\right\} Sin(\omega_2 t)\\ &+\left\{\hat{\boldsymbol{x}}\,\ell_x(\omega_2 t)+\hat{\boldsymbol{y}}\,\ell_y(\omega_2 t)+\hat{\boldsymbol{z}}\,\ell_z(\omega_2 t)\right\}\end{aligned} \tag{49}$$

$$\begin{aligned}d^2(\phi) = d^2\left[\phi(t)\right] = \vec{\boldsymbol{d}}\left[\phi(t)\right]\cdot\vec{\boldsymbol{d}}\left[\phi(t)\right] = b^2(t) + 2r_1 r_2\left(1-Cos\beta\right)Cos(\omega_1 t)Cos(\omega_2 t)\\ +2\left\{\vec{\boldsymbol{a}}(t)Cos(\omega_2 t)+\vec{\boldsymbol{b}}(t)Sin(\omega_2 t)\right\}\cdot\vec{\ell}\left[\phi(t)\right]+\ell^2\left[\phi(t)\right]\end{aligned} \tag{50}$$

$$\begin{aligned}\vec{\boldsymbol{a}} = \vec{\boldsymbol{a}}(t) = \hat{\boldsymbol{x}}\left\{r_1 Cos\,\beta\, Cos\left[(\omega_1-\omega_2)t+\varphi_0\right]-r_2\right\}+\hat{\boldsymbol{y}}\, r_1 Sin\left[(\omega_1-\omega_2)t+\varphi_0\right]\\ +\hat{\boldsymbol{z}}\, r_1 Sin\,\beta\, Cos\left[(\omega_1-\omega_2)t+\varphi_0\right] \quad ; \quad a = a(t) = \left|\vec{\boldsymbol{a}}(t)\right| = \sqrt{\vec{\boldsymbol{a}}(t)\cdot\vec{\boldsymbol{a}}(t)} = \sqrt{a^2(t)}\end{aligned} \tag{51}$$

$$\begin{aligned}\vec{\boldsymbol{b}} = \vec{\boldsymbol{b}}(t) = -\hat{\boldsymbol{x}}\, r_1 Cos\,\beta\, Sin\left[(\omega_1-\omega_2)t+\varphi_0\right]+\hat{\boldsymbol{y}}\left\{r_1 Cos\left[(\omega_1-\omega_2)t+\varphi_0\right]-r_2\right\}\\ -\hat{\boldsymbol{z}}\, r_1 Sin\,\beta\, Sin\left[(\omega_1-\omega_2)t+\varphi_0\right] \quad ; \quad b = b(t) = \left|\vec{\boldsymbol{b}}(t)\right| = \sqrt{\vec{\boldsymbol{b}}(t)\cdot\vec{\boldsymbol{b}}(t)} = \sqrt{b^2(t)}\end{aligned} \tag{52}$$

$$\vec{\mathbf{X}}\left[\phi(t)\right] = \vec{\boldsymbol{a}}(t)Cos(\omega_2 t) \quad ; \quad \left|\vec{\mathbf{X}}\left[\phi(t)\right]\right| = X\left[\phi(t)\right] = \sqrt{\vec{\mathbf{X}}(\phi)\cdot\vec{\mathbf{X}}(\phi)} = \sqrt{\vec{\boldsymbol{a}}\cdot\vec{\boldsymbol{a}}\; Cos^2(\omega_2 t)} \tag{53}$$

$$\vec{\mathbf{Y}}\left[\phi(t)\right] = \vec{\boldsymbol{b}}(t)Sin(\omega_2 t) \quad ; \quad \left|\vec{\mathbf{Y}}\left[\phi(t)\right]\right| = Y\left[\phi(t)\right] = \sqrt{\vec{\mathbf{Y}}(\phi)\cdot\vec{\mathbf{Y}}(\phi)} = \sqrt{\vec{\boldsymbol{b}}\cdot\vec{\boldsymbol{b}}\; Sin^2(\omega_2 t)} \tag{54}$$

$$\vec{\boldsymbol{d}}\left[\phi(t)\right] = \vec{\mathbf{X}}\left[\phi(t)\right]+\vec{\mathbf{Y}}\left[\phi(t)\right]+\vec{\ell}\left[\phi(t)\right] = \vec{r}_1\left[\phi(t)\right]-\vec{r}_2\left[\phi(t)\right]+\vec{\ell}\left[\phi(t)\right] \tag{55}$$

$$\vec{r}_1-\vec{r}_2 = \vec{r}_1\left[\phi(t)\right]-\vec{r}_2\left[\phi(t)\right] = \vec{\mathbf{X}}\left[\phi(t)\right]+\vec{\mathbf{Y}}\left[\phi(t)\right] = \vec{\boldsymbol{a}}(t)Cos(\omega_2 t)+\vec{\boldsymbol{b}}(t)Sin(\omega_2 t) \tag{56}$$

Based on (68) and depending on the values of ω_1 (61), ω_2 (62), and angular frequencies in the variation of $\vec{\ell}\left[\phi(t)\right]$ (48), $d^2(\phi)$ (16) and therefore $d(\phi)$ (13) is expected to vary according to a sinusoid based on the <u>higher</u> of the angular frequencies ω_1 (61), ω_2 (62), $(\omega_1-\omega_2)$ (63), or angular frequencies in the variation of $\vec{\ell}\left[\phi(t)\right]$ (48), within a sinusoidal distance envelope varying according to the <u>smaller</u> of the angular frequencies ω_1 (61), ω_2 (62), $(\omega_1-\omega_2)$ (63), or angular frequencies in the variation of $\vec{\ell}\left[\phi(t)\right]$ (48). An example of such a d (25)-curve as a function of time t is demonstrated in **Figure 3**.

Note here that based on the astronomical position data measured[3,5] over time, the curve in **Figure 3** is in fact the variation of the Earth-Moon[3] distance over 700 days, which is an elliptical[5] motion curve with semi-major[6] and semi-minor[6] axis values varying in a harmonic style as the

Moon comes closer to and goes farther away from the Earth, giving an idea of the cyclic relative Earth-Moon[3] motion. As we know the orbital period[10,11] T_{Moon} (31) of the Moon[11] and the Earth[10]'s yearly[26] orbital period[10,10] T_{Earth} (32), we are able to calculate the values of the Moon[11] and the Earth[10]'s angular velocities ω_{Moon} (34) and ω_{Earth} (35), respectively, as well as their angular velocity difference $(\omega_{Moon} - \omega_{Earth})$ (37). We can also calculate the synodic period[10,11] $T_{(Moon-Earth)}$ (38) of the Moon[11] to an approximation using the relation in (37). Therefore, there are approximately 13.3687466 orbital[10,11] (33) cycles of the Moon[11] in one orbital[10,10] cycle of the Earth[10], and there are approximately 12.3687463 (39) synodic[10,11] cycles of the Moon[11] in one orbital[10,10] cycle of the Earth[10].

$$T_{Moon} = 27.321661\, days = 27\, days\ 7\, hours\ 43\, minutes\ 11.5\, seconds \quad (Moon's\ Orbital\ Period) \tag{57}$$

$$T_{Earth} = 365.256363004\, days = 365\, days\ 6\, hours\ 9\, minutes\ 9.76\, seconds\ (Earth's\ Orbital\ Period) \tag{58}$$

$$\frac{T_{Earth}}{T_{Moon}} = \frac{365.256363004\, days}{27.321661\, days} = 13.3687466 \tag{59}$$

$$\omega_{Moon} = \frac{2\pi}{T_{Moon}} = \mathbf{0.229970839151382}\ radians/day \quad (Moon's\ Angular\ Velocity) \tag{60}$$

$$\omega_{Earth} = \frac{2\pi}{T_{Earth}} = \mathbf{0.017202124161519}\ radians/day \quad (Earth's\ Angular\ Velocity) \tag{61}$$

$$\omega_{Moon} - \omega_{Earth} = \frac{2\pi}{T_{(\omega_{Moon} - \omega_{Earth})}} = \mathbf{0.212768714989863}\ radians/day$$
$$(Moon - Earth\ Angular\ Velocity\ Difference) \tag{62}$$

$$T_{(Moon-Earth)} = \frac{2\pi}{(\omega_{Moon} - \omega_{Earth})} = 29.530589\, days = 29\, days\ 12\, hours\ 44\, minutes\ 2.9\, seconds$$
$$(Moon's\ Synodic\ Period) \tag{63}$$

$$\frac{T_{Earth}}{T_{(Moon-Earth)}} = \frac{365.256363004\, days}{29.530589\, days} = 12.3687463 \tag{64}$$

Therefore, we conclude and assert here that it is equivalently possible for the Earth and the Moon to be each revolving in a circular motion with ***fixed*** but ***different*** angular velocities ω_{Earth} (35) and ω_{Moon} (34) with respect to the centers of their own individual circular orbits, respectively, where their orbital centers are displaced by an $\vec{\ell}\left[\phi(t)\right]$ (48) vector from each other. With this assumption, we have made analytical calculations and obtained orbital parameters of Earth and Moon to an approximation, whose results are presented in subsequent paragraphs of this paper.

Based on Earth-Moon distance[5] data provided in timeanddate.com website, as well as looking at **Figure 3**, we observe that there are a little more than 13 cycles of a sinusoid in approximately 1 cycle of an envelope sinusoid, for the Earth-Moon system. As a result, we reach the conclusion that mainly the $(\omega_1 t = \omega_{Moon} t)$ (61) component of $d^2\left[\phi(t)\right]$ (25) determines the frequency of the distance variation of Earth-Moon system within a distance envelope of angular velocity $(\omega_2 = \omega_{Earth})$ (62), and that the $\left[Cos(\omega_1 t)Cos(\omega_2 t)\right]$ term determines the maximum and minimum values of $d^2\left[\phi(t)\right]$ (25) in its sinusoidal cycles, mainly the $Cos(\omega_1 t)$ term, which must be at dates when the Moon's phase $(\omega_1 t = \omega_{Moon} t)$ (34) at time t is a multiple of π, corresponding to a global maximum or minimum Earth-Moon distance[5] when the Earth's phase $(\omega_2 t = \omega_{Earth} t)$ (35) at time t is close to a multiple of π at the same time. In other words, the angular velocity of the Moon's circular orbit around its own center of revolution ω_{Moon} (34) determines the higher frequency distance variation of the Earth-Moon system, and the angular velocity of Earth's circular orbit around its own center of revolution ω_{Earth} (35) determines the frequency of Earth-Moon distance[5] envelope sinusoid.

A more special case of (1) - (3), and in fact a more special case of (61) - (63), is expressed in (87) - (89), when the moving points $\mathbf{P}_1$ and $\mathbf{P}_2$ are moving with the ***same fixed*** angular velocity ω (86) - (88) with respect to the centers of their own circles of revolution, i.e. when $(\omega_1 = \omega_2 = \omega)$ (86), and are phased apart by a constant angle φ_0 (89) over all time t, the reference timestamp t_0 taken to be the point in time when $\mathbf{P_2}$ has a phase of $\left[\phi_2(t_0) = 0\right]$ (88), in the configuration described in **Figure 2**.

$$\omega_1 = \omega_2 = \omega \qquad (\omega\ constant) \tag{65}$$

$$\phi_1(t) = \phi + \phi_0 = \phi(t) + \phi_0(t) = \phi_2(t) + \phi_0(t) = \omega t + \varphi_0 \quad ; \quad \phi_1(t_0 = 0) = \varphi_0 \qquad (Phase\ of\ \mathbf{P}_1) \tag{66}$$

$$\phi_2(t) = \phi = \phi(t) = \omega t \quad ; \quad \phi_2(t_0 = 0) = 0 \qquad (Phase\ of\ \mathbf{P}_2) \tag{67}$$

$$\phi_0(t) = \phi_1(t) - \phi_2(t) = (\omega - \omega)t + \varphi_0 = \varphi_0 \quad ; \quad \varphi_0\ constant \quad ; \quad (Phase\ difference\ of\ \mathbf{P}_1\ and\ \mathbf{P}_2) \tag{68}$$

In this more special case, at each $\phi(t)$ (88), the vector radii $\vec{r}_1$ (3) and $\vec{r}_2$ (4) of the two circles become $\vec{r}_1$ (90) and $\vec{r}_2$ (70), respectively, where the constant or variable vector $\vec{\ell}$ (7), by which the centers of these two circles are displaced, can be expressed as $\vec{\ell}$ (71). Therefore, at each $\phi(t)$ (88), $\vec{\boldsymbol{d}}\left[\phi(t)\right]$ (11) can be expressed as $\vec{\boldsymbol{d}}\left[\phi(t)\right]$ (94), and the elliptic coordinate vectors $\vec{\mathbf{X}}\left[\phi(t)\right]$ (19) and $\vec{\mathbf{Y}}\left[\phi(t)\right]$ (20) of vector ellipse[5] become $\vec{\mathbf{X}}\left[\phi(t)\right]$ (75) and $\vec{\mathbf{Y}}\left[\phi(t)\right]$ (76), respectively, in which case the elliptic representation of vector distance $\vec{\boldsymbol{d}}\left[\phi(t)\right]$ (18) becomes $\vec{\boldsymbol{d}}\left[\phi(t)\right]$ (77). In this more special case, at each $\phi(t)$ (88), the semi-major[6] and semi-minor[6] axis vectors $\vec{\boldsymbol{a}}$ (95) and $\vec{\boldsymbol{b}}$ (96) of the vector ellipse[5] are <u>constant</u> over time t, with $\phi_0(t)$ (89) replaced in (21) - (22). Throughout a respective cycle of the points $\mathbf{P}_1$ and $\mathbf{P}_2$ moving with the ***same fixed*** angular velocity ω (86) around the centers of their own circles of revolution, the $(\vec{r}_1 - \vec{r}_2)$ (18) vector has a vector value of $\left[\vec{\boldsymbol{a}}(t)Cos(\omega t) + \vec{\boldsymbol{b}}(t)Sin(\omega t)\right]$ (101) at each phase value of $\left[\phi(t) = \omega t\right]$ (88), and moves in the plane formed by the constant $\vec{\boldsymbol{a}}$ (95) and $\vec{\boldsymbol{b}}$ (96)

vectors, namely the $\vec{a}-\vec{b}$ plane, which is a fixed plane in three dimensions for this case. Note that $\vec{a}$ (95) is the vector value of $\left[\vec{r}_1(\omega t)-\vec{r}_2(\omega t)\right]$ (101) when $(\omega t=2n\pi)$, and $\vec{b}$ (96) is the vector value of $\left[\vec{r}_1(\omega t)-\vec{r}_2(\omega t)\right]$ (101) when $\left(\omega t=2n\pi+\frac{\pi}{2}\right)$, where n is an integer.

$$\begin{aligned}\vec{r}_1=\vec{r}_1\left[\phi(t)\right]=&\{\ \hat{x}r_1 Cos\,\beta\, Cos\,\varphi_0+\hat{y}\,r_1 Sin\,\varphi_0+\hat{z}\,r_1 Sin\,\beta\, Cos\,\varphi_0\}\,Cos(\omega t)+\\&\{-\hat{x}\,r_1 Cos\,\beta\, Sin\,\varphi_0+\hat{y}\,r_1 Cos\,\varphi_0-\hat{z}\,r_1 Sin\,\beta\, Sin\,\varphi_0\}\ Sin(\omega t)\end{aligned} \tag{69}$$

$$\vec{r}_2=\vec{r}_2\left[\phi(t)\right]=\hat{x}\,r_2 Cos(\omega t)+\hat{y}\,r_2 Sin(\omega t) \tag{70}$$

$$\vec{\ell}=\vec{\ell}\left[\phi(t)\right]=\hat{x}\ell_x(\omega t)+\hat{y}\ell_y(\omega t)+\hat{z}\ell_z(\omega t) \tag{71}$$

$$\begin{aligned}\vec{d}\left[\phi(t)\right]=&\{\ \hat{x}(r_1 Cos\,\beta\, Cos\,\varphi_0-r_2)+\hat{y}\,r_1 Sin\,\varphi_0+\hat{z}\,r_1 Sin\,\beta\, Cos\,\varphi_0\}\,Cos(\omega t)\\&+\{-\hat{x}\,r_1 Cos\,\beta\, Sin\,\varphi_0+\hat{y}(r_1 Cos\,\varphi_0-r_2)-\hat{z}\,r_1 Sin\,\beta\, Sin\,\varphi_0\}\,Sin(\omega t)\\&+\{\hat{x}\ell_x(\omega t)+\hat{y}\ell_y(\omega t)+\hat{z}\ell_z(\omega t)\}\end{aligned} \tag{72}$$

$$\vec{a}(t)=\vec{a}=\hat{x}\{r_1 Cos\,\beta\, Cos\varphi_0-r_2\}+\hat{y}\,r_1 Sin\,\varphi_0+\hat{z}\,r_1 Sin\,\beta\, Cos\varphi_0 \tag{73}$$

$$\vec{b}(t)=\vec{b}=-\hat{x}\,r_1 Cos\,\beta\, Sin\,\varphi_0+\hat{y}\{r_1 Cos\varphi_0-r_2\}-\hat{z}\,r_1 Sin\,\beta\, Sin\,\varphi_0 \tag{74}$$

$$\vec{\mathbf{X}}\left[\phi(t)\right]=\vec{a}\,Cos(\omega t)\ \ ;\ \ \left|\vec{\mathbf{X}}\left[\phi(t)\right]\right|=X\left[\phi(t)\right]=\sqrt{\vec{\mathbf{X}}(\phi)\cdot\vec{\mathbf{X}}(\phi)}=\sqrt{\vec{a}\cdot\vec{a}\,Cos^2(\omega t)} \tag{75}$$

$$\vec{\mathbf{Y}}\left[\phi(t)\right]=\vec{b}\,Sin(\omega t)\ \ ;\ \ \left|\vec{\mathbf{Y}}\left[\phi(t)\right]\right|=Y\left[\phi(t)\right]=\sqrt{\vec{\mathbf{Y}}(\phi)\cdot\vec{\mathbf{Y}}(\phi)}=\sqrt{\vec{b}\cdot\vec{b}\,Sin^2(\omega t)} \tag{76}$$

$$\vec{d}\left[\phi(t)\right]=\vec{\mathbf{X}}\left[\phi(t)\right]+\vec{\mathbf{Y}}\left[\phi(t)\right]+\vec{\ell}\left[\phi(t)\right]=\vec{r}_1\left[\phi(t)\right]-\vec{r}_2\left[\phi(t)\right]+\vec{\ell}\left[\phi(t)\right] \tag{77}$$

$$\vec{r}_1-\vec{r}_2=\vec{r}_1\left[\phi(t)\right]-\vec{r}_2\left[\phi(t)\right]=\vec{\mathbf{X}}\left[\phi(t)\right]+\vec{\mathbf{Y}}\left[\phi(t)\right]=\vec{a}(t)Cos(\omega t)+\vec{b}(t)Sin(\omega t) \tag{78}$$

Based on (94), $d^2(\phi)$ (16) and therefore $d(\phi)$ (13) is expected to vary according to a sinusoid depending on the value of the angular frequency ω (86) - (88).

Based on Sun-Earth distance observation[8], this scenario matches that of the Sun and Earth couple, as the orbit of the Earth with respect to the Sun is a distinct fixed ellipse that completes its one cycle through the course of a year[26] of duration T_{Sun} (1) same as Earth[10]'s yearly[26] orbital period[10,10] T_{Earth} (32), with the Sun at a seemingly fixed point with respect to the Earth throughout this cycle. Therefore, it is highly possible that the Sun and the Earth are moving around their individual circular obits, with the ***same fixed*** angular velocity ω (86) - (88) with respect to the centers of their own circles of revolution, with their orbital centers displaced by an $\vec{\ell}(\phi)$ (71) vector from each other, where $(\omega=\omega_{Sun}=\omega_{Earth})$ (2) with ω_{Earth} (35). With this assumption, we have made analytical calculations and obtained the orbital parameters of the Sun and Earth to an approximation, based on yearly Sun-Earth distance data observed by NASA Landsat[3], and using an analytical method we have developed to calculate orbital parameters of two heavenly bodies moving with the *same* angular velocity in their own circular orbital revolutions, whose results are also presented in subsequent paragraphs of this paper.

$$T_{Sun} = T_{Earth} = 365.256363004\, days \qquad (Sun's\ Orbital\ Period) \tag{79}$$

$$\omega = \omega_{Sun} = \frac{2\pi}{T_{Sun}} = \omega_{Earth} = \mathbf{0.017202124161519}\ radians / day \quad (Sun's\ Angular\ Velocity) \tag{80}$$

Based on the analysis demonstrated above, one can reach the conclusion that the <u>cyclical distance-based observation</u> of the Sun and the planets in the Solar System, which has lead to the empirical Kepler's Law[1,2] stating that the "orbit of a planet with respect to the Sun is elliptical, with the Sun at one of the two foci", may equivalently be due to "the Sun, the Planets, and their Moons each revolving in a circular motion with ***fixed*** but ***different*** angular velocities with respect to the centers of their own individual circular orbits, where those orbital centers are displaced by different $\vec{\ell}(\phi)$ (7) vectors from each other."

This mathematical revelation would have very far-reaching consequences in physics. It would lead to the formulation of a fundamental new model of motion for spatial bodies in the Universe, including the Sun, the Planets, and the Satellites in the Solar System and elsewhere, as well as at particle and sub-atomic level, where "all bodies move at some angular velocity around their own circular orbits of revolution with different radii and centers of revolution in space", but "each appear to be moving in elliptical motion with respect to each other, where they see the other body located at a fixed or variable point in their respective virtual plane of motion". This subsequently would lead to more insight into forces in nature.

In the irregular case when centers of the two circles are displaced by an $\vec{\ell}\left[\phi(t)\right]$ (7) vector <u>variable</u> over time t, the relative elliptic behavior of the respective motions of points $\mathbf{P}_1$ and $\mathbf{P}_2$ on the two circles still holds, but the analysis of their relative motion just becomes more complicated, as they no longer see each other positioned at a fixed point in their respective plane of motion, but rather see the other moving according to the $\vec{\ell}\left[\phi(t)\right]$ (7) vector.

Observation also tells us that, apart from their relative motions with respect to each other, spatial bodies such as the Sun[9], the Earth[10,13], and the Moon[11] also rotate about their own axes[12,13], which determine their geographical poles[14], namely North Poles[15] and South Poles[16], along the normal to their equatorial[17] planes. Further, from a vantage point above the North Pole[15] of the Earth[10], the Earth appears to rotate[8] in a *counterclockwise*[16] direction about its axis, which is expected based on the "right hand rule"[19]. Based on these observations and our findings above as a result of our research, we make the broader assertion that "Spatial bodies must be revolving around individual circular orbits of revolution in space, which determine their individual equatorial[17] planes, whose normal and the direction of motion of the body around its individual circular orbit determine the axis of self-revolution according to 'right hand rule'[19]. The direction of self-rotation axis[12,13] of each spatial body is also determined according to 'right hand rule'[19] based on the direction of self-rotation of the body. This *self-revolution axis and the self-rotation axis*[12,13] *of each spatial body must be aligned along the normal of its equatorial*[17] *plane.*" Based on this assertion, we can also infer that *the axial tilt*[13] *of a body in its motion relative to another body must topologically be based on the angle between their planes of individual circular orbits of revolution in space.*

In fact, all moving bodies in space must obey "Faraday's Law of Induction"[13] in their motions, which states that "Electromotive Force (EMF)[14] around a closed path is equal to the negative of time rate of change of the Magnetic Flux[15] enclosed by the path, implying that charged particles moving in an external magnetic field with changing flux[15] move in a direction to reverse the change in that magnetic flux[15]. Therefore, we assert that all bodies in space possibly behave like macro-scale charged particles moving in external magnetic fields with changing flux[15] through the area of their individual circular orbit of revolution while moving in *clockwise*[16] direction (Revolution Orbit "West" to Revolution Orbit "East") from a vantage point above their North Pole[14,15], which we expect must be the main cause that triggers their self rotation in the reverse (*counterclockwise*[16]) direction (Self Rotation "West" to Self Rotation "East"), as seen from a vantage point above their North Pole[14,15], hence the two Easts and two Wests of bodies in the Universe, where we always accept the North Pole[14,15] of a body or revolution to be the pole rotating in *counterclockwise*[16] direction according to "right hand rule"[19]. This assertion is also supported by our research results for the Sun-Earth-Moon system, as described further below.

Based on our analysis demonstrated above, our calculation results regarding the Sun-Earth-Moon system, as well as further conclusions we have reached as part of our research, are presented below in this paper.

We have found out that the actual Sun-Earth-Moon topology in space can be represented roughly as in the configuration of **Figure 6**, not to scale, which in fact is the Sun-Earth configuration of **Figure 2** superimposed on the Earth-Moon configuration of **Figure 2**. Note that in the configuration of **Figure 6** with Cartesian $(\hat{\boldsymbol{x}}, \hat{\boldsymbol{y}}, \hat{\boldsymbol{z}})$ coordinates, the plane of Earth's individual circular orbit of revolution is taken to be the horizontal $\hat{\boldsymbol{x}} - \hat{\boldsymbol{y}}$ plane, and the direction of the individual circular orbit of revolution of Earth is taken to be *counterclockwise*[16] in $\left[+\phi(t)\right]$ (2) direction, looking down on from $\hat{\boldsymbol{z}}$-axis onto $\hat{\boldsymbol{x}} - \hat{\boldsymbol{y}}$ plane, which corresponds to *clockwise*[16] direction from a vantage point above the North Pole[15] of the Earth[10], as our calculations have led to the result that $\left(\hat{\boldsymbol{u}}_{Earth\perp} = -\hat{\boldsymbol{z}}\right)$ (49) is the unit vector in the direction of self-rotation axis[12,13] of the Earth[10] in this **Figure 6** configuration, which is also the direction of the North Pole[15] of the Earth[10], normal to the Earth's plane of revolution in its individual circular orbit, but reverse in direction to the unit vector $\hat{\boldsymbol{z}}$ defining its revolution axis according to 'right hand rule'[19]. This makes us further confirm that Earth's yearly revolution around its own circular orbit is in a *clockwise*[16] direction as seen from above the North Pole[15] of the Earth[10], opposite to its daily *counterclockwise*[16] self-rotation as seen from the same vantage point above the North Pole[15] of the Earth[10], is supporting our expectation above that the main cause triggering self-rotation of bodies in space in the reverse direction to their individual circular orbit revolution is most probably because they obey "Faraday's Law of Induction"[13] as macro-scale charged particles moving in external magnetic fields with changing flux[15] through the area of their individual circular orbit of revolution.

$$\hat{\boldsymbol{u}}_{Earth\perp} = -\hat{\boldsymbol{z}} \qquad \left(\textit{North Pole unit vector of Earth}\right) \tag{81}$$

The inclination angle of the plane of individual circular orbit of revolution of the Sun with respect to the plane of individual circular orbit revolution of the Earth is found as $\beta_{Sun-Earth}$ (47).

$$\beta_{Sun-Earth} \simeq \mathbf{-156^{\circ}.56} \simeq \mathbf{-2.732513127}\ \textit{radians}\left(\textit{angle between Sun \& Earth revolution planes}\right) \tag{82}$$

As an observer aligned with the North Pole[15] of the Earth sees, the sunspots on the Sun travel[10] from east to west, going across the Sun's disk in June and December, but their path dipping somewhat in March and September. This observation[10] implies that the Sun's self rotation is in *clockwise*[16] direction from a vantage point above the North Pole[15] of the Earth, meaning that the pole vector of the Sun that is angle-wise closer to the Earth's North Pole[15] should be the Sun's South Pole[14] vector, as the North Pole[14,15] is expected to be rotationg *counterclockwise*[16] based on the "right hand rule"[19]. Hence, we have found $\left(\hat{\boldsymbol{u}}_{Sun\perp} = -\hat{\boldsymbol{x}}Sin\beta_{Sun-Earth} + \hat{\boldsymbol{z}}Cos\beta_{Sun-Earth}\right)$ (48) to be the unit vector normal to the Sun's revolution in its individual circular orbit in the direction of its self-rotation axis[12,13], as is also the direction of the North Pole[14,15] of the Sun in the configuration of **Figure 6**, which subsequently indicates that the Sun's yearly revolution around its own circular orbit is in a *clockwise*[16] direction as seen from above the North Pole[14,15] of the Sun, reverse to its *counterclockwise*[16] self rotation direction as seen from the same vantage point above North Pole[14,15] of the Sun. This also ensures that the Sun also obeys "Faraday's Law of Induction"[13], supporting our assertion that the main cause triggering self-rotation of bodies in space in reverse direction to their individual circular orbit revolution is most probably because they obey "Faraday's Law of Induction"[13] as macro-scale charged particles moving in external magnetic fields with changing flux[15] through area of their individual circular orbit of revolution.

$$\hat{\boldsymbol{u}}_{Sun\perp} = -\hat{\boldsymbol{x}}Sin\beta_{Sun-Earth} + \hat{\boldsymbol{z}}Cos\beta_{Sun-Earth} \quad (North\ Pole\ unit\ vector\ of\ Sun) \qquad (83)$$

Our analysis has revealed that inclination angle of the plane of individual circular orbit of revolution of the Moon with respect to the plane of individual circular orbit revolution of Earth is $\beta_{Moon-Earth}$ (210), whereas the unit vector normal to plane of the Moon's individual circular orbit of revolution in the direction of its self-rotation axis[12], as is also the direction of North Pole[14,15] of the Moon in the configuration of **Figure 6**, is $\left(\hat{\boldsymbol{u}}_{Moon\perp} = \hat{\boldsymbol{x}}Sin\beta_{Moon-Earth} - \hat{\boldsymbol{z}}Cos\beta_{Moon-Earth}\right)$ (211).

$$\beta_{Moon-Earth} \simeq 24^{\circ} \quad (angle\ between\ Moon\ \&\ Earth\ revolution\ planes) \qquad (84)$$

$$\hat{\boldsymbol{u}}_{Moon\perp} = \hat{\boldsymbol{x}}Sin\beta_{Moon-Earth} - \hat{\boldsymbol{z}}Cos\beta_{Moon-Earth} \quad (North\ Pole\ unit\ vector\ of\ Moon) \qquad (85)$$

We have obtained the following radii magnitudes r_{Sun} (46), r_{Earth} (45), and r_{Moon} (89) for individual circular orbits of revolution of the Sun, the Earth, and the Moon, respectively, in terms of kilometers and astronomical units[4] (au) (3), based on yearly Sun-Earth distance data observed by NASA Landsat[3] as well as Earth-Moon distance[5] data provided in timeanddate.com website, and using an analytical method we have developed to calculate orbital parameters of two heavenly bodies, to some approximations.

$$1\ astronomical\ unit\ (au)\ =\ 149,597,870.700\ km\ (defined\ exactly) \qquad (86)$$

$$r_{Sun} = \mathbf{1.00007869128446}\ au = \mathbf{149,609,642.749}\ km \quad (radius\ of\ Sun's\ revolution orbit) \quad (87)$$

$$r_{Earth} = \mathbf{0.0000826915340950721}\ au = \mathbf{12,370.477}\ km \quad (radius\ of\ Earth's\ revolution\ orbit) \quad (88)$$

$$r_{Moon} \simeq \mathbf{3,302,009.781}\ km \simeq \mathbf{0.0220725720604161}\ au \quad (radius\ of\ Moon's\ revolution\ orbit) \quad (89)$$

Even if we have made approximations in the analytical calculation, the result we have obtained is important in comparing the order of magnitude of the radii of individual circular orbits of revolution of the Earth, the Sun, and the Moon, as $\left(r_{Earth} \ll r_{Moon} \ll r_{Sun}\right)$ (192), based on our findings of r_{Sun} (46), r_{Earth} (45), and r_{Moon} (89).

$$\left(r_{Earth} \simeq \mathbf{12,370.477}\ km\right) \ll \left(r_{Moon} \simeq \mathbf{3,302,009.781}\ km\right) \ll \left(r_{Sun} \simeq \mathbf{149,609,642.749}\ km\right) \quad (90)$$

We have also found the constant phase difference $\left[\phi_0(t) = \varphi_0\right]$ (89) between the Sun and the Earth in their individual orbits of revolution throughout a year to be $\left(\varphi_0 = \phi_{0,Earth-Sun}\right)$ (91), such that in its own cycle of individual revolution, the Sun is lagging with a fixed phase difference of $\left(\varphi_0 \simeq -\mathbf{151^\circ.11}\right)$ (91) behind Earth in its own cycle of individual revolution, or equivalently, in its own cycle of individual revolution, the Earth is moving ahead of the Sun in its own cycle of individual revolution by a phase difference of $\left(-\varphi_0 \simeq \mathbf{151^\circ.11}\right)$ (91). For the Earth-Moon system, the reference timestamp t_0 for $\left[\phi_0(t_0 = 0) = \varphi_0\right]$ (63) is taken to be a point in time when both of the moving points $\mathbf{P}_1$ (Moon) and $\mathbf{P}_2$ (Earth) are aligned in phase, such that $\left[\varphi_0 = \phi_{0,Earth-Moon}(t_0 = 0) = 0\right]$ (92). In light of this, considering the Sun-Earth-Moon configuration of **Figure 6**, we have discovered that in its individual circular orbit of revolution, phase $\left[\phi(t) = \phi_{Earth-Sun}\right]$ (88) of Earth in its motion relative to the Sun in the Sun-Earth configuration of **Figure 2** at any date is around 60° ahead of its phase $\left[\phi(t) = \phi_{Earth-Moon}\right]$ (62) in its motion relative to the Moon in Earth-Moon configuration of **Figure 2**, as expressed in (93).

$$\phi_{0,Earth-Sun} = -\mathbf{2.63736997198951}\ rad = -\mathbf{151^\circ.11} \quad \left(\textit{Phase difference for Sun - Earth system}\right) \quad (91)$$

$$\phi_{0,Earth-Moon}(t) = \left(\omega_{Moon} - \omega_{Earth}\right)t \quad ; \quad \phi_{0,Earth-Moon}(t_0 = 0) = \varphi_0 = 0 \quad \left(\textit{Phase difference for Earth - Moon system}\right) \quad (92)$$

$$\phi_{Earth-Sun} \simeq \phi_{Earth-Moon} + 60^\circ \qquad \left(\textit{at any date of the year}\right) \quad (93)$$

Phase values $\left[\phi(t) = \phi_{Earth-Sun}\right]$ (88), in the Sun-Earth configuration of **Figure 2**, of Earth in its individual circular orbit of revolution at any date over a yearly Sun-Earth cycle are calculated[30] and plotted in **Figure 14**. Phase values of Earth for equinoxes[9], solstices[17], and dates of minimum and maximum Sun-Earth distances, as well as dates of some significant phases in the Earth-Sun yearly cycle are listed in (94) - (103).

$$\phi_{Earth-Sun,Min\ Sun-Earth\ Distance} = -\mathbf{0.193236295254927}\ radians = -\mathbf{11^\circ.072} \qquad \left(\textit{January } 3\right) \quad (94)$$

$$\phi_{Earth-Sun} = \mathbf{0.00058673254949726}\ radians = \mathbf{0^\circ.034} \simeq 0 \qquad \left(\textit{January } 17\right) \quad (95)$$

$$\phi_{Earth-Sun,Spring\ Equinox} = \mathbf{1.02676771671574}\ radians = \mathbf{58^\circ.829} \qquad \left(\textit{March } 20\right) \quad (96)$$

$$\phi_{Earth-Sun} = \mathbf{1.57138305934439}\ radians = \mathbf{90^\circ.034} \simeq \frac{\pi}{2} \qquad \left(\textit{April } 18\right) \quad (97)$$

$$\phi_{Earth-Sun,Summer\ Solstice} = -\phi_{0,Earth-Sun} = \mathbf{2.63736997198951}\ rad = \mathbf{151^\circ.11} \qquad \left(\textit{June } 21\right) \quad (98)$$

$$\phi_{Earth-Sun,Max\ Sun-Earth\ Distance} = \mathbf{2.94835635833487}\ radians = \mathbf{168^\circ.928} \qquad \left(\textit{July } 5\right) \quad (99)$$

$$\phi_{Earth-Sun} = \mathbf{3.14217938613929}\ radians = \mathbf{180^\circ.034} \simeq \pi \qquad \left(\textit{July } 18\right) \quad (100)$$

$$\phi_{Earth-Sun,Autumn\ Equinox} = -\mathbf{2.05488176708229}\ radians = \mathbf{242^\circ.264} = -\mathbf{117^\circ.736} \qquad \left(\textit{September } 22\right) \quad (101)$$

$$\phi_{Earth-Sun} = \mathbf{-1.5702095942454}\, radians = \mathbf{-89^{\circ}.966} \simeq -\frac{\pi}{2} \qquad (October\ 17) \quad (102)$$

$$\phi_{Earth-Sun,Winter\ Solstice} = \pi - \phi_{0,Earth-Sun} = \mathbf{5.7789626255793}\, rad = \mathbf{331^{\circ}.11} = \mathbf{-28^{\circ}.89} \quad (December\ 21) \quad (103)$$

For the relative motion ellipse[5] of the Sun-Earth system, as $\left[Cos\left(\phi_{0,Earth-Sun}\right) < 0\right]$ (91), $\left[a(t) < b(t)\right]$ (23) - (24) based on (38) - (39), which is also confirmed by the values we have determined for $\left[a(t) = a_{Earth-Sun}\right]$ (104), making it the semi-minor[6] axis magnitude, and $\left[b(t) = b_{Earth-Sun}\right]$ (105), making it the semi-major[6] axis magnitude. Subsequently, their focal[5,7] distance $\left[c(t) = c_{Earth-Sun}\right]$ (107) and eccentricity[5] $\left[e(t) = e_{Earth-Sun}\right]$ (108) are also found based on (29) - (30), respectively. The fixed semi-minor[6] axis vector $\left[\vec{\boldsymbol{a}}(t) = \vec{\boldsymbol{a}}_{Earth-Sun}\right]$ (10) and the fixed semi-major[6] axis vector $\left[\vec{\boldsymbol{b}}(t) = \vec{\boldsymbol{b}}_{Earth-Sun}\right]$ (11) are also revealed in Sun-Earth configuration of **Figure 2**, based on (95) - (96), respectively as well as their Dot Product[4] $\left[\vec{\boldsymbol{a}}(t) \cdot \vec{\boldsymbol{b}}(t) = \vec{\boldsymbol{a}}_{Earth-Sun} \cdot \vec{\boldsymbol{b}}_{Earth-Sun}\right]$ (111) based on (25). The orbital parameter values in (104) - (111) for the Sun-Earth system in the elliptic (27) - (28) representation (77) - (101) of the vector distance $\left\{\vec{\boldsymbol{d}}\left[\phi(t)\right] = \vec{\boldsymbol{d}}_{Earth-Sun}\left[\phi(t)\right]\right\}$ (94) have all been determined using found values of $(r_1 = r_{Sun})$ (46), $(\beta = \beta_{Sun-Earth})$ (47), $(r_2 = r_{Earth})$ (45), $(\omega = \omega_{Sun} = \omega_{Earth})$ (2) with ω_{Earth} (35), and $(\varphi_0 = \phi_{0,Earth-Sun})$ (91), to an approximation.

$$a_{Earth-Sun} = \mathbf{1.00001226586661}\, au = \mathbf{149{,}599{,}705.647527}\, km \quad (Semi-minor\ axis\ magnitude\ of\ Sun-Earth\ system) \quad (104)$$

$$b_{Earth-Sun} = \mathbf{1.00015109267839}\, au = \mathbf{149{,}620{,}473.842965}\, km \quad (Semi-major\ axis\ magnitude\ of\ Sun-Earth\ system) \quad (105)$$

$$\left(b_{Earth-Sun} - a_{Earth-Sun}\right) = \mathbf{0.000138826811781456}\, au = \mathbf{20{,}768.195439}\, km \quad (106)$$

$$c_{Earth-Sun} = \mathbf{0.0166636221185271}\, au = \mathbf{2{,}492{,}842.4}\, km \quad (Focal\ distance\ of\ Sun-Earth\ system) \quad (107)$$

$$e_{Earth-Sun} = \frac{c_{Earh-Sun}}{b_{Earh-Sun}} = \mathbf{0.0166611047475858} \quad (Eccentricity\ of\ Sun-Earth\ system) \quad (108)$$

$$\vec{\boldsymbol{a}}_{Earth-Sun} = \hat{\boldsymbol{x}}\, 120{,}168{,}467.040\, km - \hat{\boldsymbol{y}}\, 72{,}280{,}843.623\, km + \hat{\boldsymbol{z}}\, 52{,}106{,}536.351\, km \quad (109)$$

$$\vec{\boldsymbol{b}}_{Earth-Sun} = -\hat{\boldsymbol{x}}\, 66{,}316{,}021.722\, km - \hat{\boldsymbol{y}}\, 131{,}002{,}922.948\, km - \hat{\boldsymbol{z}}\, 28{,}752{,}488.898\, km \quad (110)$$

$$\vec{\boldsymbol{a}}_{Earth-Sun} \cdot \vec{\boldsymbol{b}}_{Earth-Sun} = \mathbf{0.00007661063}\, au^2 \quad (111)$$

The difference (289) between semi-major[6] axis magnitude $b_{Earth-Sun}$ (105) and semi-minor[6] axis magnitude $a_{Earth-Sun}$ (104) of the Sun-Earth system is about **20,768.2** *km* (289), which is relatively small on the Sun-Earth distance scale (3). Therefore, the ellipse[5] formed by the relative motions of the Sun-Earth system reveals to be almost circular, as expected. The focal distance

$c_{Earth-Sun}$ (107) of the ellipse[5] formed by the relative Sun-Earth motion is about $1/60$ of an astronomical unit[4] (au) (3). We further understand from the found r_{Sun} (46) and r_{Earth} (45) values that it is the radius r_{Sun} (46) of the Sun's individual circular orbit of revolution which predominantly determines the distances between the Sun and the Earth over a yearly cycle, as well as the semi-major[6] axis magnitude $b_{Earth-Sun}$ (105) and the semi-minor[6] axis magnitude $a_{Earth-Sun}$ (104) of the ellipse[5] formed by the relative Sun-Earth motion.

The values of the time (t) -varying semi-major[6] and semi-minor[6] axis vectors $\left[\vec{\boldsymbol{a}}(t)=\vec{\boldsymbol{a}}_{Earth-Moon}(t)\right]$ (213) and $\left[\vec{\boldsymbol{b}}(t)=\vec{\boldsymbol{b}}_{Earth-Moon}(t)\right]$ (214) of the relative vector ellipse[5] of the Earth-Moon system based on (69) - (70), their magnitudes squared $\left[a^2(t)=a^2_{Earth-Moon}(t)\right]$ (216) and $\left[b^2(t)=b^2_{Earth-Moon}(t)\right]$ (217) based on (23) - (24) and (69) - (70), respectively, their Dot Product[4] $\left[\vec{\boldsymbol{a}}(t)\cdot\vec{\boldsymbol{b}}(t)=\vec{\boldsymbol{a}}_{Earth-Moon}(t)\cdot\vec{\boldsymbol{b}}_{Earth-Moon}(t)\right]$ (215) based on (25), with square of the instantaneous focal[5,7] distance $\left[c^2(t)=c^2_{Earth-Moon}(t)\right]$ (218) and instantaneous eccentricity[5] $\left[e(t)=e_{Earth-Moon}(t)\right]$ (219) of the vector ellipse[5], based on (29) - (30), respectively, in the elliptic (27) - (28) representation (55) - (75) of the vector distance $\left\{\vec{\boldsymbol{d}}\left[\phi(t)\right]=\vec{\boldsymbol{d}}_{Earth-Moon}\left[\phi(t)\right]\right\}$ (68), have been determined using the found values of $\left(r_1=r_{Moon}\right)$ (89), $\left(\beta=\beta_{Moon-Earth}\right)$ (210), $\left(r_2=r_{Earth}\right)$ (45), $\left(\omega_1=\omega_{Moon}\right)$ (34), $\left(\omega_2=\omega_{Earth}\right)$ (35), and $\left(\omega_1-\omega_2=\omega_{Moon}-\omega_{Earth}\right)$ (37), to an approximation, for the case of $\left[\phi_0(t)=\phi_{0,Earth-Moon}(t)\right]$ (63) with $\phi_{0,Earth-Moon}(t)$ (92).

$$\begin{aligned}\vec{\boldsymbol{a}}_{Earth-Moon}(t) \simeq\ & \hat{\boldsymbol{x}}\left[3{,}016{,}536.037\,Cos\left(0.212768714989863t\right)-12{,}370.477\right] km \\ & +\hat{\boldsymbol{y}}\,3{,}302{,}009.781\,Sin\left(0.212768714989863t\right) km \\ & +\hat{\boldsymbol{z}}\,1{,}343{,}048.374\,Cos\left(0.212768714989863t\right) km\end{aligned} \tag{112}$$

$$\begin{aligned}\vec{\boldsymbol{b}}_{Earth-Moon}(t) =\ & -\hat{\boldsymbol{x}}\,3{,}016{,}536.037\,Sin\left(0.212768714989863t\right) km \\ & +\hat{\boldsymbol{y}}\left[3{,}302{,}009.781\,Cos\left(0.212768714989863t\right)-12{,}370.477\right] km \\ & -\hat{\boldsymbol{z}}\,1{,}343{,}048.374\,Sin\left(0.212768714989863t\right) km\end{aligned} \tag{113}$$

$$\vec{\boldsymbol{a}}_{Earth-Moon}(t)\cdot\vec{\boldsymbol{b}}_{Earth-Moon}(t)=-3{,}531{,}446{,}390.263\,Sin\left(0.212768714989863t\right) km^2 \tag{114}$$

$$\begin{aligned}a^2_{Earth-Moon}(t)=\vec{\boldsymbol{a}}_{Earth-Moon}(t)\cdot\vec{\boldsymbol{a}}_{Earth-Moon}(t)=\ & 10{,}903{,}421{,}623{,}251.049\ km^2 \\ & -74{,}631{,}979{,}321.481\,Cos\left(0.212768714989863t\right) km^2\end{aligned} \tag{115}$$

$$\begin{aligned}b^2_{Earth-Moon}(t)=\vec{\boldsymbol{b}}_{Earth-Moon}(t)\cdot\vec{\boldsymbol{b}}_{Earth-Moon}(t)=\ & 10{,}903{,}421{,}623{,}251.049\ km^2 \\ & -81{,}694{,}872{,}102.007\,Cos\left(0.212768714989863t\right) km^2\end{aligned} \tag{116}$$

$$c^2_{Earth-Moon}(t) = \left|a^2_{Earth-Moon}(t) - b^2_{Earth-Moon}(t)\right| = 7{,}062{,}892{,}780.525\left|Cos(0.212768714989863t)\right| \, km^2 \tag{117}$$

$$(Focal\ Distance\ Squared)$$

$$e_{Earth-Moon}(t) = \begin{cases} \dfrac{c_{Earth-Moon}(t)}{a_{Earth-Moon}(t)} = \sqrt{\dfrac{7{,}062{,}892{,}780.525\left|Cos(0.212768714989863t)\right|}{10{,}903{,}421{,}623{,}251.049 - 74{,}631{,}979{,}321.481Cos(0.212768714989863t)}} \\ \qquad , \ if\ a_{Earth-Moon}(t) > b_{Earth-Moon}(t) \\ \dfrac{c_{Earth-Moon}(t)}{b_{Earth-Moon}(t)} = \sqrt{\dfrac{7{,}062{,}892{,}780.525\left|Cos(0.212768714989863t)\right|}{10{,}903{,}421{,}623{,}251.049 - 81{,}694{,}872{,}102.007Cos(0.212768714989863t)}} \\ \qquad , \ if\ a_{Earth-Moon}(t) < b_{Earth-Moon}(t) \end{cases} \tag{118}$$

$$(Eccentricity)$$

Note here that in (213) - (219), based on our analysis results, the time (t) must be measured in $[days]$ from a date that coincides with a $(t = t_0 = 0)$ (63) at the initial phase alignment $[\phi_{0,Earth-Moon}(t_0 = 0) = 0]$ (92) of Earth and Moon, which we found is when the Earth-Moon distance[5] is a maximum in the month of March once in every 3-year yearly[26] cycles of the Earth[10] of period[10,10] T_{Earth} (32), which also is approximately once every 40 orbital cycles of period[10,11] T_{Moon} (31) of the Moon[11], such as March 18, 2017 or March 24, 2020, when both the Moon and the Earth moving around their own circular orbits of revolution must be passing from the same aligned phase of a multiple of 2π around the same date. Based on our analysis results for the Earth-Moon system configuration described in **Figure 2**, the assessment is made that around every 10 cycles of the Moon (31), the Earth has a phase change of $\left[\Delta\phi(t) = \Delta\phi_{Earth-Moon} \simeq \frac{3\pi}{2}\right]$ (62) in its individual circular orbit of revolution, with a phase of a multiple of 6π on March 18, 2017 and March 24, 2020, and around every 1 Moon cycle (31), the Earth has a phase change of $\left[\Delta\phi(t) = \Delta\phi_{Earth-Moon} \simeq \frac{3\pi}{20}\right]$ (62) in its individual circular orbit of revolution. Between the dates of January 10, 2017 and January 9, 2021, in the Earth-Moon system configuration of **Figure 2**, we have assessed that Earth passes through phase $[\phi(t) = \phi_{Earth-Moon} \simeq 0]$ (62) in its motion relative to Moon on March 18, 2017, and Earth passes through phase $\left[\phi(t) = \phi_{Earth-Moon} \simeq \frac{3\pi}{2}\right]$ (62) in its motion relative to Moon on December 19, 2017, and Earth passes through phase $[\phi(t) = \phi_{Earth-Moon} \simeq 3\pi]$ (62) in its motion relative to the Moon on September 20, 2018, and Earth passes through phase $\left[\phi(t) = \phi_{Earth-Moon} \simeq \frac{9\pi}{2}\right]$ (62) in its motion relative to the Moon on June 23, 2019, and Earth passes through phase

$\left[\phi(t)=\phi_{Earth-Moon}\simeq 6\pi\right]$ (62) in its motion relative to the Moon on March 24, 2020. We observe that the dates of these $\left[\Delta\phi(t)=\Delta\phi_{Earth-Moon}\simeq\frac{3\pi}{2}\right]$ (62) multiples of phase for the Earth-Moon system occur around equinoxes[9,6] and solstices[17,6] of the Earth with respect to the Sun during their relative cycles.

In this Earth-Moon system configuration of **Figure 2**, after their initial phase alignment of $\left[\phi_{0,Earth-Moon}(t_0=0)=0\right]$ (92) at time $(t=t_0=0)$ (63):

- Whenever the Earth and the Moon moving around their individual circular orbits of revolution pass from the same aligned phase of a multiple of 2π around the same date every 3-year[26] cycles (32) of the Earth, which is approximately every 40 cycles (31) of the Moon, that is near dates when Earth has a phase of a multiple of $\left[\phi(t)=\phi_{Earth-Moon}\simeq 6\pi=0^{\circ}\right]$ (62), phase of Earth in its individual circular orbit of revolution is around $\left(\phi_{Earth-Sun}\simeq 60^{\circ}\right)$ (93) with respect to its motion relative to Sun, and Earth passes through its Spring Equinox[9,6] with respect to Sun, one of the dates around which Solar and Lunar Eclipses[11,19] occur due to Sun-Earth-Moon motion topology.
- Near dates when the Earth has a phase of an odd multiple of $\left[\phi(t)=\phi_{Earth-Moon}\simeq 3\pi=180^{\circ}\right]$ (62) in its motion relative to the Moon, the phase of Earth in its individual circular orbit of revolution is around $\left(\phi_{Earth-Sun}\simeq 240^{\circ}=-120^{\circ}\right)$ (93) with respect to its motion relative to the Sun, and Earth passes through its Autumn Equinox[9,6] with respect to the Sun, another one of the dates around which Solar and Lunar Eclipses[11,19] occur due to motion topology of Sun-Earth-Moon system.
- Whenever the phase of Earth in its individual circular orbit of revolution is around $\left[\phi(t)=\phi_{Earth-Moon}\simeq\frac{9\pi}{2}=90^{\circ}\right]$ (62) in its motion relative to the Moon, phase of Earth in its individual circular orbit of revolution is like $\left(\phi_{Earth-Sun}\simeq-\phi_{0,Earth-Sun}\simeq 151^{\circ}\right)$ (93) with respect to its motion relative to the Sun, and Earth passes through its Summer Solstice[17,6] with respect to the Sun.
- Whenever the phase of the Earth in its individual circular orbit of revolution is around $\left[\phi(t)=\phi_{Earth-Moon}\simeq\frac{3\pi}{2}=-90^{\circ}\right]$ (62) in its motion relative to the Moon, phase of Earth in its individual circular orbit of revolution is around $\left(\phi_{Earth-Sun}\simeq\pi-\phi_{0,Earth-Sun}\simeq-30^{\circ}\right)$ (93) with respect to its motion relative to the Sun, and Earth passes through its Winter Solstice[17,6] with respect to the Sun.

Further to (213) - (219), based on the found values of $(r_1=r_{Moon})$ (89), $(r_2=r_{Earth})$ (45), and $(\beta=\beta_{Moon-Earth})$ (210) to an approximation, the expected maximum and minimum values $(a_{\max,Earth-Moon}, a_{\min,Earth-Moon}, b_{\max,Earth-Moon}, b_{\min,Earth-Moon})$ (220) of the semi-major[6] and semi-minor[6] axis magnitudes and their squares are also found for the Earth-Moon system, over a 2π cycle of $\left[\phi_{0,Earth-Moon}(t)=(\omega_1-\omega_2)t=(\omega_{Moon}-\omega_{Earth})t\right]$ (92).

$$
\begin{aligned}
&a^2_{\max,Earth-Moon}=r_1^2+2r_1r_2Cos\beta+r_2^2\simeq\mathbf{10{,}978{,}053{,}602{,}572.530599}\ km^2\\
&\qquad\Rightarrow\ a_{\max,Earth-Moon}\simeq\mathbf{3{,}313{,}314.595}\ km\\
&a^2_{\min,Earth-Moon}=r_1^2-2r_1r_2Cos\beta+r_2^2\simeq\mathbf{10{,}828{,}789{,}643{,}929.567938}\ km^2\\
&\qquad\Rightarrow\ a_{\min,Earth-Moon}\simeq\mathbf{3{,}290{,}712.635}\ km\\
&b^2_{\max,Earth-Moon}=(r_1+r_2)^2\simeq\mathbf{10{,}985{,}116{,}495{,}353.055828}\ km^2\\
&\qquad\Rightarrow\ b_{\max,Earth-Moon}=r_1+r_2\simeq\mathbf{3{,}314{,}380.258}\ km\\
&b^2_{\min,Earth-Moon}=(r_1-r_2)^2\simeq\mathbf{10{,}821{,}726{,}751{,}149.042710}\ km^2\\
&\qquad\Rightarrow\ b_{\min,Earth-Moon}=r_1-r_2\simeq\mathbf{3{,}289{,}639.304}\ km
\end{aligned}
\tag{119}
$$

Moreover, over a 2π cycle of $\left[\phi_{0,Earth-Moon}(t)=(\omega_1-\omega_2)t=(\omega_{Moon}-\omega_{Earth})t\right]$ (92), whenever $\left\{Cos\left[(\omega_{Moon}-\omega_{Earth})t\right]=0\right\}$ (221), $\left[a^2_{Earth-Moon}(t)\right]$ (216) and $\left[b^2_{Earth-Moon}(t)\right]$ (217) have equal values based on (23) - (24), as for magnitudes of radius vectors of a circle, which is approximately the value in (221) based on found values of $(r_1=r_{Moon})$ (89) and $(r_2=r_{Earth})$ (45), which is a value almost the same as r_{Moon} (89).

$$
\begin{aligned}
&Cos\left[(\omega_{Moon}-\omega_{Earth})t\right]=0\\
&\qquad\Rightarrow\ a^2_{Earth-Moon}(t)=b^2_{Earth-Moon}(t)=r_1^2+r_2^2\simeq\mathbf{10{,}903{,}421{,}623{,}251.049}\ km^2\\
&\qquad\Rightarrow\ a_{Earth-Moon}(t)=b_{Earth-Moon}(t)\simeq\mathbf{3{,}302{,}032.953}\ km
\end{aligned}
\tag{120}
$$

We have calculated[30] the Cartesian $(\hat{x},\hat{y},\hat{z})$ coordinates of the $\left\{\vec{\ell}\left[\phi(t)\right]=\vec{\ell}_{Earth-Sun}(\phi)\right\}$ (71) vector between the centers of individual circular orbits of revolution of the Sun and the Earth, and its magnitude $\left[\ell(\phi)=\ell_{Earth-Sun}(\phi)\right]$ (8), in the Sun-Earth configuration of **Figure 2**, over a yearly[26] cycle. The magnitude $\left[\ell(\phi)=\ell_{Earth-Sun}(\phi)\right]$ (8), which is the scalar distance between centers of individual circular orbits of revolution of the Sun and the Earth, is plotted over a yearly[26] Sun-Earth cycle in **Figure 19**.

As a result, we have discovered that the vector distance $\left\{\vec{\ell}\left[\phi(t)\right]=\vec{\ell}_{Earth-Sun}(\phi)\right\}$ (71) and its magnitude $\left[\ell(\phi)=\ell_{Earth-Sun}(\phi)\right]$ (8) between the centers of individual circular orbits of revolution of the Sun and the Earth is *not constant* but rather *harmonically varying* over a year, and $\left\{\vec{\ell}\left[\phi(t)\right]=\vec{\ell}_{Earth-Sun}(\phi)\right\}$ (71) repeats itself vector-wise as well as magnitude-wise every quarter of a yearly[26] Sun-Earth cycle. The maximum and minimum values of the magnitude $\left[\ell(\phi)=\ell_{Earth-Sun}(\phi)\right]$ (8) of the vector distance $\left\{\vec{\ell}\left[\phi(t)\right]=\vec{\ell}_{Earth-Sun}(\phi)\right\}$ (71) over a yearly[26] Sun-Earth cycle are $\left[\ell_{Earth-Sun}(\phi)\right]_{Max}$ (121) and $\left[\ell_{Earth-Sun}(\phi)\right]_{Min}$ (122), respectively. The maximum distance $\left[\ell_{Earth-Sun}(\phi)\right]_{Max}$ (121) between centers of individual circular orbits of revolution of the Sun and the Earth occur on February 23, May 26, August 25, and November 24 over a yearly[26] Sun-Earth cycle. The minimum distance $\left[\ell_{Earth-Sun}(\phi)\right]_{Min}$ (122) between centers of individual circular orbits of revolution of the Sun and the Earth occur on January 1, April 2, July 2, and October 1 over a yearly[26] Sun-Earth cycle.

$$\left[\ell_{Earth-Sun}(\phi)\right]_{Max} = \mathbf{0.0168346184954446}\ au\ (23\ February, 26\ May, 25\ August, 24\ November) \quad (121)$$

$$\left[\ell_{Earth-Sun}(\phi)\right]_{Min} = \mathbf{0.0165381684638675}\ au \qquad (1\ January, 2\ April, 2\ July, 1\ October) \quad (122)$$

Our analysis has also revealed that throughout a year, $\left\{\vec{\ell}\left[\phi(t)\right] = \vec{\ell}_{Earth-Sun}(\phi)\right\}$ (71) must be varying in a narrow angle range, which is closer to the sector of Earth's self-revolution when the Earth moves between $\left[\phi(t) = \phi_{Earth-Sun} = \mathbf{163^\circ.621}\right]$ (88) and $\left[\phi(t) = \phi_{Earth-Sun} = \mathbf{169^\circ.860}\right]$ (88) in its individual circular orbit of revolution, in the Sun-Earth configuration of **Figure 2**, which approximately corresponds[30] to the sector of Earth's self-revolution between dates of July 2 and July 7 over a yearly[26] Sun-Earth cycle. Furthermore, the direction of the distance vector $\left\{\vec{\boldsymbol{d}}\left[\phi(t)\right] = \vec{\boldsymbol{d}}_{Earth-Sun}\left[\phi(t)\right]\right\}$ (94) directed from Earth to Sun is most aligned with the direction of $\left\{\vec{\ell}\left[\phi(t)\right] = \vec{\ell}_{Earth-Sun}(\phi)\right\}$ (71) on July 5, which is the date of maximum Sun-Earth distance over a year[26], and the direction of the distance vector $\left\{\vec{\boldsymbol{d}}\left[\phi(t)\right] = \vec{\boldsymbol{d}}_{Earth-Sun}\left[\phi(t)\right]\right\}$ (94) directed from Earth to Sun is most aligned with the reverse direction of $\left\{\vec{\ell}\left[\phi(t)\right] = \vec{\ell}_{Earth-Sun}(\phi)\right\}$ (71) on January 3, which is the date of minimum Sun-Earth distance over a year[26]. The main factor influencing the presence of $\left\{\vec{\ell}\left[\phi(t)\right] = \vec{\ell}_{Earth-Sun}(\phi)\right\}$ (71) vector, which is between centers of individual circular orbits of revolution of the Sun and the Earth, especially in this narrow direction range, may be the impact of the Milky Way[32]'s Galactic Center[33] located in the reverse direction, which is observed from Earth at the sector of the sky in the direction of constellation Sagittarius[34] in the same sector of the sky as the observed location of the Sun around Winter Solstice[17] on December 21. Note that as $\left(\beta_{Sun-Earth} \simeq \mathbf{-156^\circ.56}\right)$ (47), this sector of the sky where $\left\{\vec{\ell}\left[\phi(t)\right] = \vec{\ell}_{Earth-Sun}(\phi)\right\}$ (71) exists, when Earth moving in its individual circular orbit of revolution is between $\left[\phi(t) = \phi_{Earth-Sun} = \mathbf{163^\circ.621}\right]$ (88) and $\left[\phi(t) = \phi_{Earth-Sun} = \mathbf{169^\circ.860}\right]$ (88) in the Sun-Earth configuration of **Figure 2**, is also one of the two times over a yearly[26] Sun-Earth cycle when the motion of the Earth and the Sun seemingly cross each other in *reverse* direction, in their motion within their individual circular orbits of revolution, between the point when the phase $\phi_1(t)$ (61) of the Sun is $\left(\phi_{Earth-Sun,Summer\ Solstice} + \phi_{0,Earth-Sun} = 0\right)$ (98) at the Summer Solstice[17] on June 21, and the point when the phase $\phi(t)$ (62) of Earth is $\left(\phi_{Earth-Sun} = \mathbf{180^\circ.034} \simeq \pi\right)$ (100) on July 18. The other of the two times over a yearly[26] Sun-Earth cycle when the motion of Earth and Sun seemingly cross each other in *reverse* direction, in their motion within their individual circular orbits of revolution, is between the point when the phase $\phi_1(t)$ (61) of the Sun is

$\left(\phi_{Earth-Sun,Winter\ Solstice}+\phi_{0,Earth-Sun}=\pi\right)$ (103) at Winter Solstice[17] on December 21, and the point when he phase $\phi(t)$ (62) of the Earth is $\left(\phi_{Earth-Sun}=\mathbf{0^{\circ}.034}\simeq 0\right)$ (95) on January 17.

The magnitude $\left[\ell(\phi)=\ell_{Earth-Sun}(\phi)\right]$ (8) of vector distance $\left\{\vec{\ell}\left[\phi(t)\right]=\vec{\ell}_{Earth-Sun}(\phi)\right\}$ (71) over a yearly[26] Sun-Earth cycle exhibits[30] three distinct main oscillation frequencies, as can also be seen in **Figure 19**. One oscillation of the magnitude $\left[\ell(\phi)=\ell_{Earth-Sun}(\phi)\right]$ (8) of the distance between the centers of individual orbits of revolutions of the Earth and the Sun is about 12 times a year, which is apparently based on the impact of the relative Earth-Moon motion. The second oscillation of $\left[\ell(\phi)=\ell_{Earth-Sun}(\phi)\right]$ (8) is seemingly based on the daily self-rotation of the Earth around its own axis. The possible cause of third oscillation of $\left[\ell(\phi)=\ell_{Earth-Sun}(\phi)\right]$ (8), which is about 4 times over a Sun-Earth year, can be due to the impact of relative Sun-Mercury motion which has about four cycles over a Sun-Earth year[26]. It seems that the Sun and Mercury form a twin or binary[25] system of bodies in space, in which Mercury behaves as the Sun's satellite, similar to Earth-Moon system, and implies the possibility that the individual circular orbits of revolution of the Sun and Mercury may also be revolving around each other. Furthermore, our analysis also implies that it is very well possible that the relative motions of the other planets and their satellites, as well as other moving bodies in the Solar system also impact the relative motion of the Sun and Earth in terms of additional minor oscillations of $\left[\ell(\phi)=\ell_{Earth-Sun}(\phi)\right]$ (8) over longer periods than a Sun-Earth year[26], but their impact on the distance between the centers of individual circular orbits of revolution of the Sun and the Earth can be observed *only* over longer periods than a yearly[26] cycle.

It is worth noticing that ancient people did *not* select the first days of months in the Solar Calendar cycle based on Min-Max Sun-Earth distance dates (Jan 3 and July 5), or the dates of Solstices[17] (June 21 and December 21), or the Equinoxes[9] (March 20 and September 23), which slightly vary every year. Apparently, the selection of dates for the first days of months in the Solar Calendar by ancient people was not coincidental, but rather based on phase $\phi_{0,Earth-Sun}$ (91) between individual circular revolutions of the Earth and the Sun in their own orbits, and also the aligned cycles of the harmonic $\left[\ell(\phi)=\ell_{Earth-Sun}(\phi)\right]$ (8), the distance between centers of individual circular orbits of revolutions of the Sun and the Earth, which subsequently has resulted in different duration months of 28, 30, and 31 days. Our ancestors possibly based these date selections on accurate Sun-Earth distance observations and the intermediate harmonic variations in it based on phase $\phi_{0,Earth-Sun}$ (91) and $\left[\ell(\phi)=\ell_{Earth-Sun}(\phi)\right]$ (8), even if they were not aware of the existence of phase $\phi_{0,Earth-Sun}$ (91), or $\left\{\vec{\ell}\left[\phi(t)\right]=\vec{\ell}_{Earth-Sun}(\phi)\right\}$ (71) with magnitude $\left[\ell(\phi)=\ell_{Earth-Sun}(\phi)\right]$ (8).

We have defined Earth's North Pole[15] tilt[13] $\gamma_{Earth}(\phi)$ (91) as the angle between $\hat{\boldsymbol{u}}_{Earth\perp}$ (49) and distance vector $\left\{\vec{\boldsymbol{d}}\left[\phi(t)\right]=\vec{\boldsymbol{d}}_{Earth-Sun}\left[\phi(t)\right]\right\}$ (94) directed from Earth to Sun, as, and the Sun's North Pole[14,15] tilt[13,10,37] $\gamma_{Sun}(\phi)$ (90) as the angle between $\hat{\boldsymbol{u}}_{Sun\perp}$ (48) and distance vector

$\left\{-\vec{d}\left[\phi(t)\right]=-\vec{d}_{Earth-Sun}\left[\phi(t)\right]\right\}$ (94) directed from Sun to Earth, at each $\left[\phi(t)=\phi_{Earth-Sun}\right]$ (88), where $\left\{\vec{\ell}\left[\phi(t)\right]=\vec{\ell}_{Earth-Sun}(\phi)\right\}$ (71) is the vector distance between centers of individual circular orbits of revolution of the Sun and the Earth, and utilizing $(r_1=r_{Sun})$ (46), $(r_2=r_{Earth})$ (45), and $(\beta=\beta_{Sun-Earth})$ (47).

$$\begin{aligned}\vec{d}_{Earth-Sun}(\phi)\cdot\hat{u}_{Earth\perp}&=\left|\vec{d}_{Earth-Sun}(\phi)\right|Cos\left[\gamma_{Earth}(\phi)\right]\\&=-[r_{Sun}Cos\left(\phi_{Earth-Sun}+\phi_{0,Earth-Sun}\right)Sin\left(\beta_{Sun-Earth}\right)+\ell_{Earth-Sun,z}\left(\phi_{Earth-Sun}\right)]\end{aligned} \quad (123)$$

$$\begin{aligned}\left[-\vec{d}_{Earth-Sun}(\phi)\right]\cdot\hat{u}_{Sun\perp}&=\left|\vec{d}_{Earth-Sun}(\phi)\right|Cos\left[\gamma_{Sun}(\phi)\right]\\&=-r_{Earth}Cos\left(\phi_{Earth-Sun}\right)Sin\left(\beta_{Sun-Earth}\right)\\&\quad+\ell_{Earth-Sun,x}\left(\phi_{Earth-Sun}\right)Sin\left(\beta_{Sun-Earth}\right)-\ell_{Earth-Sun,z}\left(\phi_{Earth-Sun}\right)Cos\left(\beta_{Sun-Earth}\right)\end{aligned} \quad (124)$$

We have calculated[30] values of $\gamma_{Earth}(\phi)$ (91) and $\gamma_{Sun}(\phi)$ (90) for all dates over a year[26], i.e. at each $\left[\phi(t)=\phi_{Earth-Sun}\right]$ (88).

We have discovered about the Earth's North Pole[15] tilt[13] $\gamma_{Earth}(\phi)$ (91) that $\gamma_{Earth}(\phi)=90^\circ$ on March 22 and on September 20, $\gamma_{Earth}(\phi)=66^\circ.560=\left(90^\circ-23^\circ.440\right)$ at the Summer Solstice[17] on June 21 – 22, and $\gamma_{Earth}(\phi)=113^\circ.440=\left(90^\circ+23^\circ.440\right)$ at the Winter Solstice[17] on December 21.

We have discovered about Sun's North Pole[14,15] tilt[13,10,37] $\gamma_{Sun}(\phi)$ (90) that $\gamma_{Sun}(\phi)=90^\circ$ on January 2 around minimum Sun-Earth distance, on July 5 at maximum Sun-Earth distance, at the Summer Solstice[17] on June 20 – 21, and at the Winter Solstice[17] on December 22, in a yearly[26] cycle. We have also found that in a yearly[26] cycle, $\gamma_{Sun}(\phi)=90^\circ.411=\left(90^\circ+0^\circ.411\right)$ between the dates of March 26 – 30, $\gamma_{Sun}(\phi)=89^\circ.993=\left(90^\circ-0^\circ.007\right)$ between the dates of June 26 - 29, $\gamma_{Sun}(\phi)=90^\circ.408=\left(90^\circ+0^\circ.408\right)$ between the dates of September 26 - 30, and $\gamma_{Sun}(\phi)=89^\circ.996=\left(90^\circ-0^\circ.004\right)$ between the dates of December 27 - 29. We can thus make the conclusion that Sun's North Pole[14,15] vector is tilted[13] $\gamma_{Sun}(\phi)$ (90) around $\mathbf{90}^\circ$ with respect to the Sun-Earth distance vector $\left\{-\vec{d}\left[\phi(t)\right]=-\vec{d}_{Earth-Sun}\left[\phi(t)\right]\right\}$ (94) directed from Sun to Earth throughout a yearly[26] cycle, and it is exactly at $\mathbf{90}^\circ$ with respect to the Sun-Earth distance vector $\left\{-\vec{d}\left[\phi(t)\right]=-\vec{d}_{Earth-Sun}\left[\phi(t)\right]\right\}$ (94) around the maximum and minimum Sun-Earth distance, as well as at the Summer and Winter Solstices[17], consistent with the observation[10,11] that in June and December, as viewed from Earth, the Sun's North and South Poles[14,16] do not to tip toward or away from Earth. Further, we have found that it is tilted *most towards* Earth twice a year, once a few days after the Spring Equinox[9] and secondly a few days after the Autumn Equinox[9]. Additionally, it is tilted *most away* from Earth twice a year, once between Winter Solstice[17] on December 21 and minimum Sun-Earth distance date of January 3 in a yearly[26] cycle, namely between $\left(\phi_{Earth-Sun,Winter\ Solstice}=\pi-\phi_{0,Earth-Sun}=331^\circ.11=-28^\circ.89\right)$ (103) and

$\left(\phi_{Earth-Sun,Min\ Sun-Earth\ Distance} = -11^\circ.07\right)$ (94), secondly between Summer Solstice[17] on June 21 and the maximum Sun-Earth distance date of July 5 in a yearly[26] cycle, namely between $\left(\phi_{Earth-Sun,Summer\ Solstice} = -\phi_{0,Earth-Sun} = 151^\circ.11\right)$ (98) and $\left(\phi_{Earth-Sun,Max\ Sun-Earth\ Distance} = 168^\circ.93\right)$ (99), that is when the Sun and Earth are travelling in opposite directions towards each other, based on the topology in **Figure 2**.

At each $\left[\phi(t) = \phi_{Earth-Sun}\right]$ (88), we have defined $\alpha_{Sun,wobble}(\phi)$ (330) as the relative angle that the Sun's South Pole[14,16] unit vector $\left(-\hat{\boldsymbol{u}}_{Sun\perp}\right)$ (48) seemingly makes towards or away from the Earth, with respect to the line of sight of an observer on Earth aligned with the North Pole[15] unit vector $\hat{\boldsymbol{u}}_{Earth\perp}$ (49) of Earth, and have also defined $\alpha_{Sun,sideways}(\phi)$ (331) as the relative "sideways" swing angle that the Sun's South Pole[14,16] unit vector $\left(-\hat{\boldsymbol{u}}_{Sun\perp}\right)$ (48) seemingly makes with respect to the line of sight of an observer on Earth aligned with North Pole[15] vector $\hat{\boldsymbol{u}}_{Earth\perp}$ (49) of Earth. Note here that $\left[\hat{\boldsymbol{d}}(\phi) = \hat{\boldsymbol{d}}_{Earth-Sun}(\phi)\right]$ (17) is the unit vector in the direction of the distance vector $\left\{\vec{\boldsymbol{d}}\left[\phi(t)\right] = \vec{\boldsymbol{d}}_{Earth-Sun}\left[\phi(t)\right]\right\}$ (94) directed from Earth to Sun, and $\left[-\hat{\boldsymbol{d}}(\phi) = -\hat{\boldsymbol{d}}_{Earth-Sun}(\phi)\right]$ is the unit vector in the direction of distance vector $\left\{-\vec{\boldsymbol{d}}\left[\phi(t)\right] = -\vec{\boldsymbol{d}}_{Earth-Sun}\left[\phi(t)\right]\right\}$ (94) directed from Sun to Earth. Further, in (330) - (331), $\left[\hat{\boldsymbol{u}}_{Earth\perp}\cdot\hat{\boldsymbol{d}}_{Earth-Sun}(\phi)\right]\hat{\boldsymbol{d}}_{Earth-Sun}(\phi)$ is the vector component of $\hat{\boldsymbol{u}}_{Earth\perp}$ (49) parallel to $\left\{\vec{\boldsymbol{d}}\left[\phi(t)\right] = \vec{\boldsymbol{d}}_{Earth-Sun}\left[\phi(t)\right]\right\}$ (94), $\left\{\hat{\boldsymbol{u}}_{Earth\perp} - \left[\hat{\boldsymbol{u}}_{Earth\perp}\cdot\hat{\boldsymbol{d}}_{Earth-Sun}(\phi)\right]\hat{\boldsymbol{d}}_{Earth-Sun}(\phi)\right\}$ is the vector component of $\hat{\boldsymbol{u}}_{Earth\perp}$ (49) normal to $\left\{\vec{\boldsymbol{d}}\left[\phi(t)\right] = \vec{\boldsymbol{d}}_{Earth-Sun}\left[\phi(t)\right]\right\}$ (94), $\left(\left[-\hat{\boldsymbol{u}}_{Sun\perp}\right]\cdot\left[-\hat{\boldsymbol{d}}_{Earth-Sun}(\phi)\right]\right)\left[-\hat{\boldsymbol{d}}_{Earth-Sun}(\phi)\right]$ is vector component of $\left(-\hat{\boldsymbol{u}}_{Sun\perp}\right)$ (48) parallel to $\left\{-\vec{\boldsymbol{d}}\left[\phi(t)\right] = -\vec{\boldsymbol{d}}_{Earth-Sun}\left[\phi(t)\right]\right\}$ (94), and $\left\{-\hat{\boldsymbol{u}}_{Sun\perp} - \left(\left[-\hat{\boldsymbol{u}}_{Sun\perp}\right]\cdot\left[-\hat{\boldsymbol{d}}_{Earth-Sun}(\phi)\right]\right)\left[-\hat{\boldsymbol{d}}_{Earth-Sun}(\phi)\right]\right\}$ is the vector component of $\left(-\hat{\boldsymbol{u}}_{Sun\perp}\right)$ (48) normal to $\left\{-\vec{\boldsymbol{d}}\left[\phi(t)\right] = -\vec{\boldsymbol{d}}_{Earth-Sun}\left[\phi(t)\right]\right\}$ (94).

$$Tan\left[\alpha_{Sun,wobble}(\phi)\right] = \frac{\left[-\hat{\boldsymbol{u}}_{Sun\perp}\right]\cdot\left[-\hat{\boldsymbol{d}}_{Earth-Sun}(\phi)\right] + \hat{\boldsymbol{u}}_{Earth\perp}\cdot\hat{\boldsymbol{d}}_{Earth-Sun}(\phi)}{\left\{\begin{matrix}-\hat{\boldsymbol{u}}_{Sun\perp} \\ -\left(\left[-\hat{\boldsymbol{u}}_{Sun\perp}\right]\cdot\left[-\hat{\boldsymbol{d}}_{Earth-Sun}(\phi)\right]\right)\left[-\hat{\boldsymbol{d}}_{Earth-Sun}(\phi)\right]\end{matrix}\right\}\cdot\left\{\frac{\hat{\boldsymbol{u}}_{Earth\perp} - \left[\hat{\boldsymbol{u}}_{Earth\perp}\cdot\hat{\boldsymbol{d}}_{Earth-Sun}(\phi)\right]\hat{\boldsymbol{d}}_{Earth-Sun}(\phi)}{\left|\hat{\boldsymbol{u}}_{Earth\perp} - \left[\hat{\boldsymbol{u}}_{Earth\perp}\cdot\hat{\boldsymbol{d}}_{Earth-Sun}(\phi)\right]\hat{\boldsymbol{d}}_{Earth-Sun}(\phi)\right|}\right\}} \quad (125)$$

$$Cos\left[\alpha_{Sun,sideways}(\phi)\right]$$

$$=\frac{\begin{Bmatrix}-\hat{\boldsymbol{u}}_{Sun\perp}\\-\left(\left[-\hat{\boldsymbol{u}}_{Sun\perp}\right]\cdot\left[-\hat{\boldsymbol{d}}_{Earth-Sun}(\phi)\right]\right)\left[-\hat{\boldsymbol{d}}_{Earth-Sun}(\phi)\right]\end{Bmatrix}\cdot\begin{Bmatrix}\hat{\boldsymbol{u}}_{Earth\perp}\\-\left[\hat{\boldsymbol{u}}_{Earth\perp}\cdot\hat{\boldsymbol{d}}_{Earth-Sun}(\phi)\right]\hat{\boldsymbol{d}}_{Earth-Sun}(\phi)\end{Bmatrix}}{\begin{vmatrix}-\hat{\boldsymbol{u}}_{Sun\perp}\\-\left(\left[-\hat{\boldsymbol{u}}_{Sun\perp}\right]\cdot\left[-\hat{\boldsymbol{d}}_{Earth-Sun}(\phi)\right]\right)\left[-\hat{\boldsymbol{d}}_{Earth-Sun}(\phi)\right]\end{vmatrix}\begin{vmatrix}\hat{\boldsymbol{u}}_{Earth\perp}\\-\left[\hat{\boldsymbol{u}}_{Earth\perp}\cdot\hat{\boldsymbol{d}}_{Earth-Sun}(\phi)\right]\hat{\boldsymbol{d}}_{Earth-Sun}(\phi)\end{vmatrix}} \quad (126)$$

We have calculated[30] values of $\alpha_{Sun,wobble}(\phi)$ (125) and $\alpha_{Sun,sideways}(\phi)$ (126) for all dates over a year[22], i.e. at each $\left[\phi(t)=\phi_{Earth-Sun}\right]$ (67). Found values of $\alpha_{Sun,wobble}(\phi)$ (125) at some critical dates can be seen in (127), and found values of $\alpha_{Sun,sideways}(\phi)$ (126) at some critical dates are listed in (128).

$$\begin{aligned}
\alpha_{Sun,wobble}(\phi) &= \mathbf{-7^{\circ}.308} && (\mathit{March}\ 3)\\
\alpha_{Sun,wobble}(\phi) &= \mathbf{0^{\circ}.024} && (\mathit{March}\ 21)\\
\alpha_{Sun,wobble}(\phi) &= \mathbf{7^{\circ}.214} && (\mathit{April}\ 2)\\
\alpha_{Sun,wobble}(\phi) &= \mathbf{21^{\circ}.691} && (\mathit{June}\ 21)\\
\alpha_{Sun,wobble}(\phi) &= \mathbf{7^{\circ}.257} && (\mathit{September}\ 3)\\
\alpha_{Sun,wobble}(\phi) &= \mathbf{0^{\circ}.002} && (\mathit{September}\ 21)\\
\alpha_{Sun,wobble}(\phi) &= \mathbf{-7^{\circ}.262} && (\mathit{October}\ 5)\\
\alpha_{Sun,wobble}(\phi) &= \mathbf{-21^{\circ}.691} && (\mathit{December}\ 21)
\end{aligned} \quad (127)$$

$$\begin{aligned}
\alpha_{Sun,sideways}(\phi)_{Data\ Points\,[i=317-320]} &= \mathbf{23^{\circ}.439} && (\mathit{March}\ 21)\\
\alpha_{Sun,sideways}(\phi)_{Data\ Point\,[i=687]} &= \mathbf{0^{\circ}.092} && (\mathit{June}\ 21)\\
\alpha_{Sun,sideways}(\phi)_{Data\ Points\,[i=1056-1059]} &= \mathbf{23^{\circ}.439} && (\mathit{September}\ 21-22)\\
\alpha_{Sun,wobble}(\phi)_{Data\ Point\,[i=1418]} &= \mathbf{0^{\circ}.026} && (\mathit{December}\ 21)
\end{aligned} \quad (128)$$

Calculated $\alpha_{Sun,wobble}(\phi)$ (125) values demonstrate that $\alpha_{Sun,wobble}(\phi)=-7^{\circ}.308$ (127) on March 3 and $\alpha_{Sun,wobble}(\phi)=7^{\circ}.257$ (127) on September 3, reinforcing our analysis result that it must be the South Pole[14,16] of the Sun which is observed[27,37] to be tipping 7.25° towards the Earth around the first week of September and to be tipping 7.25° away from the Earth around the first week of March. Hence, we have discovered and mathematically demonstrated that this observed[27,37] $\alpha_{Sun,wobble}=7.25°$ (127) is *not* the maximum axial tilt $\gamma_{Sun}(\phi)$ (124) of the Sun's rotational pole[14] with respect to the virtual ecliptic[39] plane along a yearly Sun-Earth cycle, but rather the angle that the Sun's South Pole[14,16] seemingly makes towards or away from the Earth during first weeks of September and March, respectively. This alignment of observation[27,37] and our calculated results is another justification that our analysis is correct.

The variation of the calculated results for $\alpha_{Sun,sideways}(\phi)$ (126), including the values listed in (128), are consistent with the animation[38] of the sideways swing of the Sun's axis during the span of one year[22], as seen from the line of sight of an observer on Earth aligned with the North Pole[15] vector $\hat{\boldsymbol{u}}_{Earth\perp}$ (81) of Earth, $\alpha_{Sun,sideways}(\phi)$ (126) swinging from $23^{\circ}.439$ (128) to the right side of the observer on March 21, to $23^{\circ}.439$ (128) to the left side of the observer on September 21-22. This alignment of observation and our results is yet another justification of our analysis and calculations.

The results we have obtained for Earth's North Pole[15] tilt[13] $\gamma_{Earth}(\phi)$ (123), Sun's South Pole[14,16] "wobble" angle $\alpha_{Sun,wobble}(\phi)$ (125), and Sun's South Pole[14,16] relative "sideways" swing angle $\alpha_{Sun,sideways}(\phi)$ (126) all confirm that the Spring Equinox[31] must be around March 21-22 and the Autumn Equinox[31] must be around September 20-22.

Earth-Moon distance[11,19] $\left\{\vec{\boldsymbol{d}}\left[\phi(t)\right]=\vec{\boldsymbol{d}}_{Earth-Moon}\left[\phi(t)\right]\right\}$ (55) consists of two main components as expressed in $\vec{\boldsymbol{d}}_{Earth-Moon}\left[\phi(t)\right]$ (129). One component is based on $\left\{\vec{\boldsymbol{r}}_2\left[\phi(t)\right]=\vec{\boldsymbol{r}}_{Earth}\left[\phi(t)\right]\right\}$ (47) due to individual circular orbit of revolution of the Earth[10] in its yearly[22] cycle of angular velocity $(\omega_2=\omega_{Earth})$ (61), where $(\beta=\beta_{Moon-Earth})$ (84) and $(r_2=r_{Earth})$ (88). The other component of $\left\{\vec{\boldsymbol{d}}\left[\phi(t)\right]=\vec{\boldsymbol{d}}_{Earth-Moon}\left[\phi(t)\right]\right\}$ (55) is $\vec{\boldsymbol{d}}_{MoonOrbit}\left[\phi(t)\right]$ (130), which is due to Moon[11] Orbit[40]. $\vec{\boldsymbol{d}}_{MoonOrbit}\left[\phi(t)\right]$ (130) is in turn is based on two components, one component $\left\{\vec{\boldsymbol{r}}_1\left[\phi(t)\right]=\vec{\boldsymbol{r}}_{Moon}\left[\phi(t)\right]\right\}$ (46) due to individual circular orbit of revolution of the Moon[11] in its lunar cycle of angular velocity $(\omega_1=\omega_{Moon})$ (60), where $(\omega_1-\omega_2=\omega_{Moon}-\omega_{Earth})$ (62), $(r_1=r_{Moon})$ (89), and $\left[\phi_0(t)=\phi_{0,Earth-Moon}(t)\right]$ (45) with $\phi_{0,Earth-Moon}(t)$ (92), and another component consisting of the vector distance $\left\{\vec{\ell}\left[\phi(t)\right]=\vec{\ell}_{Earth-Moon}(\phi)\right\}$ (48) between centers of individual circular orbits of revolution of the Earth[10] and the Moon[11].

$$\vec{\boldsymbol{d}}_{Earth-Moon}\left[\phi(t)\right]=\vec{\boldsymbol{d}}_{MoonOrbit}\left[\phi(t)\right]-\vec{\boldsymbol{r}}_{Earth}\left[\phi(t)\right] \tag{129}$$

$$\vec{\boldsymbol{d}}_{MoonOrbit}\left[\phi(t)\right]=\vec{\boldsymbol{r}}_{Moon}\left[\phi(t)\right]+\vec{\ell}_{Earth-Moon}(\phi) \tag{130}$$

The "individual circular orbit of revolution of the Moon with radius vector $\left\{\vec{\boldsymbol{r}}_1\left[\phi(t)\right]=\vec{\boldsymbol{r}}_{Moon}\left[\phi(t)\right]\right\}$ (46)" as a whole is seemingly revolving around the Earth with an instantanous vector radius of $\left\{\vec{\ell}\left[\phi(t)\right]=\vec{\ell}_{Earth-Moon}(\phi)\right\}$ (48) at each $\left[\phi(t)=\phi_{Earth-Moon}\right]$ (44), during every orbital period[21,11] T_{Moon} (57) of the Moon[11], or vice versa, which is compliant with observation[40], such that the Moon and the Earth move in a relative trajectory in which the Moon passes between the Sun and the Earth once every Lunar[11,41] Month and the Earth passes between the Sun and Moon once every Lunar[11,41] Month, and the Moon phases[11,42] occur over every synodic period[21,11] $T_{(Moon-Earth)}$ (63) of the Moon[11], a fact supporting our findings and assertions about the relative motions of the Earth and the Moon. In this sense, the Earth and the Moon must be forming a twin or binary[33] system of bodies in space, which is an already observed[40] phenomenon, whose individual circular orbits of revolution we claim must also be revolving around each other with an instantaneous vector radius of $\left\{\vec{\ell}\left[\phi(t)\right]=\vec{\ell}_{Earth-Moon}(\phi)\right\}$ (48) at each $\left[\phi(t)=\phi_{Earth-Moon}\right]$ (44), leading to this resultant observation[40,33].

The Moon[11] is observed to move in *counterclockwise*[23] direction as seen from a vantage point from above the North Pole[15,45] of the Earth throughout a lunar[11,41] month of period T_{Moon} (57). Therefore, based on our analysis, the Moon[11] Orbit[40] vector $\vec{\boldsymbol{d}}_{MoonOrbit}\left[\phi(t)\right]$ (130) directed from Earth towards the Moon must be moving in *counterclockwise*[23] direction as seen from a vantage point from above the North Pole[15,45] of the Earth, in its plane of motion. Further, based on our topology analysis above as well as observation[40], in order to maintain its synchronous[40] rotation with Earth, the Moon[11]'s self-rotation must also be in *counterclockwise*[23] direction from a vantage point from above the North Pole[15,45] of the Earth.

We reason that the self-rotation of the Moon[11] has the same period T_{Moon} (57) and angular velocity $(\omega_1=\omega_{Moon})$ (60) as its individual circular orbit of revolution, based on the synchronous[40] self-rotation of the Moon with Earth.

We claim that lunar[11] monthly libration[43,44] of the Moon observed from Earth is must be due to Moon's individual circular orbit of revolution around a radius of $\left\{\vec{r}_1\left[\phi(t)\right]=\vec{\boldsymbol{r}}_{Moon}\left[\phi(t)\right]\right\}$ (46), and the more-than-half view[11] of lunar surface from Earth is apparently based on the relative view angle of the Earth-bound observers, based on *both* the direction of $\left\{\vec{\ell}\left[\phi(t)\right]=\vec{\ell}_{Earth-Moon}(\phi)\right\}$ (48) vector and position of the Moon in its individual circular orbit of revolution around vector radius of $\left\{\vec{r}_1\left[\phi(t)\right]=\vec{\boldsymbol{r}}_{Moon}\left[\phi(t)\right]\right\}$ (46), on that date throughout a cycle.

Additionally, libration[40,43,44] of the Moon perceived by an Earth-bound observer aligned with the North Pole[15,45] direction of the Earth is observed[43,44] seemingly *clockwise*[23] throughout a lunar[11,41] month of period T_{Moon} (57). Therefore, we reason that the Moon's individual circular orbit of revolution around radius $\left\{\vec{r}_1\left[\phi(t)\right]=\vec{\boldsymbol{r}}_{Moon}\left[\phi(t)\right]\right\}$ (46) must be in *clockwise*[23] direction from a vantage point from above the North Pole[15,45] of the Earth.

Based on the results of our research analysis, as seen from a vantage point from above the North Pole[15,45] of Earth, the Moon[11]'s observed *clockwise*[23] motion in its individual circular orbit of revolution around vector radius $\left\{\vec{r}_1\left[\phi(t)\right]=\vec{\boldsymbol{r}}_{Moon}\left[\phi(t)\right]\right\}$ (46) leading to libration[40,43,44], and its *counterclockwise*[23] self-rotation synchronous[40] with Earth ensure that the Moon[11] also obeys "Faraday's Law of Induction"[24], in alignment with our assertion that the main cause triggering self-rotation of bodies in space in reverse direction to their individual circular orbit revolution is most probably because they obey "Faraday's Law of Induction"[24] as macro-scale charged particles moving in external magnetic fields with changing flux[26] through area of their individual circular orbit of revolution, where we always accept the North Pole[14,15] of a body or revolution to be the pole rotating in *counterclockwise*[23] direction according to "right hand rule"[18].

Based on observation[40], the shape of Moon[11]'s Orbit[40] is pretty much fixed in its plane of motion over a lunar[11,40,41] month of period T_{Moon} (57), but the *orientation* of the Moon[11]'s Orbit[40] with respect to the Earth and Sun is *not* fixed in space, and moves over time. As the Moon[11]'s motion in its individual circular orbit of revolution around vector radius $\{\vec{r}_1[\phi(t)] = \vec{r}_{Moon}[\phi(t)]\}$ (46) has an angular velocity of $(\omega_1 = \omega_{Moon})$ (60), the observation[40] that the shape of Moon[11]'s Orbit[40] is pretty much fixed in its plane of motion over a lunar[11,40,41] month of period T_{Moon} (57) has lead us to determine, based on $\vec{d}_{MoonOrbit}[\phi(t)]$ (130), that the $[\ell(\phi) = \ell_{Earth-Moon}(\phi)]$ (11) magnitude cycle of $\{\vec{\ell}[\phi(t)] = \vec{\ell}_{Earth-Moon}(\phi)\}$ (48) over a lunar[11,40,41] month of period T_{Moon} (57) must also be pretty much fixed in its plane of motion, and must also be changing with an angular velocity of $(\omega_1 = \omega_{Moon})$ (60). Apart from this lunar[11,40,41] monthly variation of Moon[11]'s motion in its individual circular orbit of revolution around vector radius $\{\vec{r}_1[\phi(t)] = \vec{r}_{Moon}[\phi(t)]\}$ (46) and the lunar[11,40,41] monthly almost-fixed magnitude cycle of $[\ell(\phi) = \ell_{Earth-Moon}(\phi)]$ (11), we conclude that any other variation in the orientation of Moon[11]'s Orbit[40] in space over different periods, i.e. with angular velocities other than $(\omega_1 = \omega_{Moon})$ (60), to be due to *only* orientation variations in $\{\vec{\ell}[\phi(t)] = \vec{\ell}_{Earth-Moon}(\phi)\}$ (48), based on $\vec{d}_{MoonOrbit}[\phi(t)]$ (130).

One orbital precession of the Moon[11]'s Orbit[40] is called the apsidal[47] precession, and is the rotation of the Moon[11]'s Orbit[40] within its orbital plane, i.e. the axes of its apparent elliptical orbit change direction. The Moon[11]'s major axis[6] – the longest diameter of the orbit, joining its nearest and farthest points, the perigee[48] and apogee[48], respectively – is observed[40] to make one revolution every 8.85 Earth[10] years, or 3,232.6054 days, as it rotates slowly in the same direction as the self-rotation of the Moon[11] itself, which is in *counterclockwise*[23] direction from a vantage point from above the North Pole[15,45] $(\hat{u}_{Earth\perp} = -\hat{z})$ (81) of the Earth[10] in the configuration of **Figure 2**. Considering that the Moon[11] Orbit[40]'s major axis[6] revolves to come back to the same position in 8.85 Earth[10] years, or 3,232.6054 days, in which case the position of the perigee[48] and apogee[48] are reversed with respect to the so called ecliptic[39] plane, we have concluded that the complete period $T_{A,\vec{d}_{MoonOrbit}[\phi(t)]}$ (131) for the perigee[48] and apogee[48] to come back to the same position with respect to the so called ecliptic[39] plane, namely one full apsidal[47] precession cycle, must be double this time, that is 17.7 Earth[10] years, or 6,465.2108 days, with angular velocity $\omega_{A,\vec{d}_{MoonOrbit}[\phi(t)]}$ (132). As the Moon[11]'s motion in its individual circular orbit of revolution around the vector radius $\{\vec{r}_1[\phi(t)] = \vec{r}_{Moon}[\phi(t)]\}$ (46) has an angular velocity of $(\omega_1 = \omega_{Moon})$ (60), and as $\{\vec{\ell}[\phi(t)] = \vec{\ell}_{Earth-Moon}(\phi)\}$ (48) has an almost fixed magnitude cycle over a lunar[11,40,41] month of period T_{Moon} (57), we conclude that this observed apsidal[47] precession of $\vec{d}_{MoonOrbit}[\phi(t)]$ (129) with an angular velocity $\omega_{A,\vec{d}_{MoonOrbit}[\phi(t)]}$ (132) different from $(\omega_1 = \omega_{Moon})$ (60) must be due to an orientation variation of $\{\vec{\ell}[\phi(t)] = \vec{\ell}_{Earth-Moon}(\phi)\}$ (48). As a result, we infer, to an approximation, that the variation in the orientation of $\{\vec{\ell}[\phi(t)] = \vec{\ell}_{Earth-Moon}(\phi)\}$ (48) cycle, is very small over a couple of consecutive lunar[11,40,41] months of period T_{Moon} (57), compared to the individual motion frequencies of the Earth and the

Moon. Further, following along the lines of the assertion in the **ARTICLE 3** "Analytical calculation of earth and sun orbital parameters from distance data" that all moving bodies in the Universe must be obeying "Faraday's Law of Induction"[24] in their motions, we also assert that each revolution in space, which is in external magnetic fields with changing flux[26] through the area of this revolution, triggers a rotation in the reverse direction. Hence, we observe that as the Moon[11] and the Earth[10] revolve in *clockwise*[23] directions in their individual circular orbits of revolution, as seen from a vantage point from above North Pole[15,45] $\left(\hat{\boldsymbol{u}}_{Earth\perp} = -\hat{z}\right)$ (81) of the Earth[10] in the configuration of **Figure 2**, with angular velocities $\left(\omega_1 = \omega_{Moon}\right)$ (60) and $\left(\omega_2 = \omega_{Earth}\right)$ (61), respectively, the observed *counterclockwise*[23] revolution of Moon[11] Orbit[40]'s major axis[6], or equivalently lunar[11,40,41] orbit magnitude-cycle orientation, from a vantage point from above North Pole[15,45] $\left(\hat{\boldsymbol{u}}_{Earth\perp} = -\hat{z}\right)$ (81) of the Earth[10] in the configuration of **Figure 2**, is also in alignment our assertion that all bodies in the Universe must be obeying "Faraday's Law of Induction"[24] in their motions, and even extending the scope of our previous assertion such that the harmonic motion of the distance vector $\vec{\ell}(\phi)$ (10) between centers of individual circular orbits of revolutions of twin or binary[33] system of bodies in space are triggered to move in reverse direction to their revolution directions in their individual circular orbits, because they obey "Faraday's Law of Induction"[24] as macro-scale charged "system of particles" moving in external magnetic fields with changing flux[26] through the area of their "twin or binary[33] system" motion planes.

$$T_{A,\vec{\boldsymbol{d}}_{MoonOrbit}[\phi(t)]} \simeq 17.7\ Earth\ years \simeq 6{,}465.2108\ days\ \left(\vec{\boldsymbol{d}}_{MoonOrbit}\left[\phi(t)\right]\ Apsidal\ Variation\ Period\right) \quad (131)$$

$$\omega_{A,\vec{\boldsymbol{d}}_{MoonOrbit}[\phi(t)]} = \frac{2\pi}{T_{A,\vec{\boldsymbol{d}}_{MoonOrbit}[\phi(t)]}} \simeq \mathbf{0.000971845389353675}\ radians\,/\,day$$
$$\left(\vec{\boldsymbol{d}}_{MoonOrbit}\left[\phi(t)\right]\ Apsidal\ Variation\ Angular\ Velocity\right) \quad (132)$$

Another observation[40] is based on the nodes or points at which the Moon[11]'s Orbit[40] crosses the so called ecliptic[39] plane. The line of nodes, the intersection between the two respective planes, namely the orbital plane of the Moon[11] and the so called ecliptic[39] plane, has a retrograde motion for an observer on Earth[10], such that it rotates westward along the ecliptic[39], or equivalently in a *clockwise*[23] direction from a vantage point from above the North Pole[15,45] $\left(\hat{\boldsymbol{u}}_{Earth\perp} = -\hat{z}\right)$ (81) of Earth[10] in the configuration of **Figure 2**, with a period of 18.6 years (138) or $-19^{\circ}.3549$ (137) per year. The draconic year[22] $T_{Draconic\ Year}$ (133) is the time taken for the Sun, as seen from Earth, to complete one revolution with respect to the same lunar[11] node, namely a point at which the Moon[11]'s Orbit[40] crosses the so called ecliptic[39] plane, and has an observed period of about 346.6 days (133). The draconic year[22] is associated with eclipses; Lunar and Solar Eclipses are observed[40] to occur when the nodes align with the Sun, roughly every 173.3 days. The Moon[11] crosses the same node every 27.2122 days (134), an interval called the Draconic Month[40,41] $T_{Draconic\ Month}$ (134), which we figure, according to our accepted Moon-Earth topology, must be the synodic[49] period $T_{\left(Moon-\vec{\ell}_{Earth-Moon}[\phi(t)]\right)}$ (135) of the motion of the Moon[11] in its individual circular orbit of revolution around radius $\left\{\vec{r}_1\left[\phi(t)\right] = \vec{\boldsymbol{r}}_{Moon}\left[\phi(t)\right]\right\}$ (46) with angular velocity

$(\omega_1 = \omega_{Moon})$ (60), with respect to the nodal orientation variation motion of $\{\vec{\ell}[\phi(t)] = \vec{\ell}_{Earth-Moon}(\phi)\}$ (48) with an angular velocity $\omega_{N,\vec{\ell}_{Earth-Moon}[\phi(t)]}$ (137). We reach this conclusion based on our analysis result above which states that, as the Moon[11]'s motion in its individual circular orbit of revolution around vector radius $\{\vec{r}_1[\phi(t)] = \vec{r}_{Moon}[\phi(t)]\}$ (46) has an angular velocity of $(\omega_1 = \omega_{Moon})$ (60), any other motion variation of $\vec{d}_{MoonOrbit}[\phi(t)]$ (130) with an angular velocity different from $(\omega_1 = \omega_{Moon})$ (60), such as this observed rotation of line of nodes of the Moon[11]'s Orbit[40] with angular velocity $\omega_{N,\vec{\ell}_{Earth-Moon}[\phi(t)]}$ (137), must be entirely due to an orientation variation of $\{\vec{\ell}[\phi(t)] = \vec{\ell}_{Earth-Moon}(\phi)\}$ (48). This synodic[49] period $T_{(Moon-\vec{\ell}_{Earth-Moon}[\phi(t)])}$ (135) is a little less than Moon[11]'s orbital period[21,11] T_{Moon} (57), from which we calculate the corresponding synodic[49] angular velocity $\omega_{(Moon-\vec{\ell}_{Earth-Moon}[\phi(t)])}$ (136). It is significantly worth noting that $\omega_{(Moon-\vec{\ell}_{Earth-Moon}[\phi(t)])}$ (136) leads to a phase difference of $-152^{\circ}.1$ (136) per year in the synodic[49] motion of nodal orientation variation motion of $\{\vec{\ell}[\phi(t)] = \vec{\ell}_{Earth-Moon}(\phi)\}$ (48) with respect to the motion of the Moon[11] in its individual circular orbit of revolution around radius $\{\vec{r}_1[\phi(t)] = \vec{r}_{Moon}[\phi(t)]\}$ (46), which is very close to the value $-151^{\circ}.11$ (91) of the constant phase difference $\phi_{0,Earth-Sun}$ (91) of the Sun in its individual circular orbit of revolution throughout a year relative to Earth's phase in its own individual circular orbit of revolution, based on our findings. We conclude that this points out an alignment of the nodal orientation variation motion of $\{\vec{\ell}[\phi(t)] = \vec{\ell}_{Earth-Moon}(\phi)\}$ (48) with the yearly motion of Sun, also supported by the observation[11,40] that Moon[11] orbits closer to so called ecliptic[39] plane than the equatorial plane of Earth, which apparently must be due to orientation of $\{\vec{\ell}[\phi(t)] = \vec{\ell}_{Earth-Moon}(\phi)\}$ (48). Utilizing $(\omega_1 = \omega_{Moon})$ (60) and $\omega_{(Moon-\vec{\ell}_{Earth-Moon}[\phi(t)])}$ (136), the value of the angular velocity $\omega_{N,\vec{\ell}_{Earth-Moon}[\phi(t)]}$ (137) of nodal orientation variation motion of $\{\vec{\ell}[\phi(t)] = \vec{\ell}_{Earth-Moon}(\phi)\}$ (48) is calculated, which turns out to be $-19^{\circ}.3549$ (137), from which we calculate the value of the period $T_{N,\vec{\ell}_{Earth-Moon}[\phi(t)]}$ (138) of the nodal orientation variation motion of $\{\vec{\ell}[\phi(t)] = \vec{\ell}_{Earth-Moon}(\phi)\}$ (48), that happens to be 18.59 years (138), both consistent with the observation[40] that the line of nodes has a retrograde motion for an observer on Earth[10], such that it rotates westward along the ecliptic[39], or equivalently in *clockwise*[23] direction, with a period of 18.6 years (138) or $-19^{\circ}.3549$ (137) per year from a vantage point from above North Pole[15,45] $(\hat{u}_{Earth\perp} = -\hat{z})$ (81) of the Earth[10] in the configuration of **Figure 2**. This is a reassuring result that indicates we are on the right track in our Earth-Moon topology analysis. Subsequently, we infer that the synodic[49] motion of Earth[10] with respect to the nodal orientation variation of Moon[11] Orbit[40] $\vec{d}_{MoonOrbit}[\phi(t)]$ (130), or equivalently the nodal orientation variation motion of

$\{\vec{\ell}[\phi(t)] = \vec{\ell}_{Earth-Moon}(\phi)\}$ (48) based on our analysis above, has a nodal synodic[49] angular velocity $\omega_{(Earth-\vec{\ell}_{Earth-Moon}[\phi(t)])}$ (139), which can be computed using values of $(\omega_2 = \omega_{Earth})$ (61) and $\omega_{N,\vec{\ell}_{Earth-Moon}[\phi(t)]}$ (137), from which the nodal synodic[49] cycle period $T_{(Earth-\vec{\ell}_{Earth-Moon}[\phi(t)])}$ (140) is also calculated, yielding a value of 346.6 days (133). This is also a result confirming our Earth-Moon topology analysis, such that the observed[40] draconic year[22] $T_{Draconic\ Year}$ (133) turns out to be in fact the period $T_{(Earth-\vec{\ell}_{Earth-Moon}[\phi(t)])}$ (140) of the synodic[49] motion of Earth[10] with respect to the nodal orientation variation motion of $\{\vec{\ell}[\phi(t)] = \vec{\ell}_{Earth-Moon}(\phi)\}$ (48) in Moon[11] Orbit[40] $\vec{\boldsymbol{d}}_{MoonOrbit}[\phi(t)]$ (130).

$$T_{Draconic\ Year} = 346.620075883\ days = 346\ days\ 14\ hours\ 52\ min.s\ 54\ seconds \quad (Draconic\ Year) \quad (133)$$

$$T_{Draconic\ Month} \simeq 27.2122247\ days \qquad (Moon's\ Draconic\ Month) \qquad (134)$$

$$T_{(Moon-\vec{\ell}_{Earth-Moon}[\phi(t)])} = \frac{2\pi}{\omega_{(Moon-\vec{\ell}_{Earth-Moon}[\phi(t)])}} = T_{Draconic\ Month} \simeq \mathbf{27.2122247}\ days \qquad (135)$$
$$\left(Moon - \vec{\ell}_{Earth-Moon}[\phi(t)]\ Synodic\ Period\right)$$

$$\omega_{(Moon-\vec{\ell}_{Earth-Moon}[\phi(t)])} = \frac{2\pi}{T_{(Moon-\vec{\ell}_{Earth-Moon}[\phi(t)])}} = \omega_{Moon} - \omega_{N,\vec{\ell}_{Earth-Moon}[\phi(t)]}$$
$$\simeq \mathbf{0.2308956866}\ radians/day \simeq \mathbf{2.65470974}\ radians/year \simeq \mathbf{152^\circ.1}/year \quad (136)$$
$$\left(Moon - \vec{\ell}_{Earth-Moon}[\phi(t)]\ Synodic\ Angular\ Velocity\right)$$

$$\omega_{N,\vec{\ell}_{Earth-Moon}[\phi(t)]} = \frac{2\pi}{T_{N,\vec{\ell}_{Earth-Moon}[\phi(t)]}} = \omega_{Moon} - \omega_{(Moon-\vec{\ell}_{Earth-Moon}[\phi(t)])} \simeq \mathbf{-0.3378}\ rad.s/yr \simeq \mathbf{-19^\circ.3549}/yr \qquad (137)$$
$$\left(Angular\ Velocity\ of\ \vec{\ell}_{Earth-Moon}[\phi(t)]\ Nodal\ Orientation\ Variation\right)$$

$$T_{N,\vec{\ell}_{Earth-Moon}[\phi(t)]} = \frac{2\pi}{\left|\omega_{N,\vec{\ell}_{Earth-Moon}[\phi(t)]}\right|} \simeq \mathbf{18.599}\ years \qquad (138)$$
$$\left(Nodal\ Orientation\ Variation\ Period\ of\ \vec{\ell}_{Earth-Moon}[\phi(t)]\right)$$

$$\omega_{(Earth-\vec{\ell}_{Earth-Moon}[\phi(t)])} = \frac{2\pi}{T_{(Earth-\vec{\ell}_{Earth-Moon}[\phi(t)])}} = \omega_{Earth} - \omega_{N,\vec{\ell}_{Earth-Moon}[\phi(t)]} \simeq \mathbf{6.621}\ rad.s/yr \simeq \mathbf{379^\circ.3549}/yr \qquad (139)$$
$$\left(Earth - \vec{\ell}_{Earth-Moon}[\phi(t)]\ Synodic\ Angular\ Velocity\right)$$

$$T_{(Earth-\vec{\ell}_{Earth-Moon}[\phi(t)])} = \frac{2\pi}{\omega_{(Earth-\vec{\ell}_{Earth-Moon}[\phi(t)])}} = T_{Draconic\ Year} \simeq \mathbf{0.94898}\ years \simeq \mathbf{346.620}\ days \qquad (140)$$
$$\left(Earth - \vec{\ell}_{Earth-Moon}[\phi(t)]\ Synodic\ Period\right)$$

Another significant outcome (141) we obtain is that, the period of nodal orientation variation cycle $T_{N,\vec{\ell}_{Earth-Moon}[\phi(t)]}$ (138) of $\{\vec{\ell}[\phi(t)]=\vec{\ell}_{Earth-Moon}(\phi)\}$ (48), multiplied by the synodic[49] cycle $T_{(Earth-\vec{\ell}_{Earth-Moon}[\phi(t)])}$ (140) of Earth[10] per year with respect to the nodal orientation variation motion of $\{\vec{\ell}[\phi(t)]=\vec{\ell}_{Earth-Moon}(\phi)\}$ (48), yields a result which is about the same as the apsidal[47] variation period $T_{A,\vec{d}_{MoonOrbit}[\phi(t)]}$ (131) of $\vec{d}_{MoonOrbit}[\phi(t)]$ (130) mentioned above. The slight difference is most probably due to the effect of orientation of $\{\vec{r}_1[\phi(t)]=\vec{r}_{Moon}[\phi(t)]\}$ (46) relative to $\{\vec{\ell}[\phi(t)]=\vec{\ell}_{Earth-Moon}(\phi)\}$ (48) at the end of a $\left[T_{N,\vec{\ell}_{Earth-Moon}[\phi(t)]}\times T_{(Earth-\vec{\ell}_{Earth-Moon}[\phi(t)])}\right]$ (141) cycle, leading to a number of additional days for reaching a local maximum magnitude for $\vec{d}_{MoonOrbit}[\phi(t)]$ (130), in determining the apsidal[47] variation period $T_{A,\vec{d}_{MoonOrbit}[\phi(t)]}$ (131). As a result, we have concluded that the observed[40] apsidal[47] precession of lunar[11,40,41] orbit magnitude in *counterclockwise*[23] direction, and the observed[40] rotation of the line of nodes of the Moon[11] Orbit[40] in *clockwise*[23] direction, from a vantage point from above North Pole[15,45] $(\hat{u}_{Earth\perp}=-\hat{z})$ (81) of the Earth[10] in the configuration of **Figure 2**, are both phenomena related to the same orientation variation motion of $\{\vec{\ell}[\phi(t)]=\vec{\ell}_{Earth-Moon}(\phi)\}$ (48) as part of $\vec{d}_{MoonOrbit}[\phi(t)]$ (130). Further, following along the lines our assertion that all moving bodies in the Universe must be obeying "Faraday's Law of Induction"[24] in their motions, this conclusion leads us to extend our assertion such that each revolution in space, which is in external magnetic fields with changing flux[26] through the area of this revolution, triggers a rotation in the reverse direction. As a result, we conclude that the observed[40] apsidal[47] precession of lunar[11,40,41] orbit magnitude in *counterclockwise*[23] direction, and the observed[40] rotation of the line of nodes of the Moon[11] Orbit[40] in *clockwise*[23] direction, from a vantage point from above North Pole[15,45] $(\hat{u}_{Earth\perp}=-\hat{z})$ (81) of the Earth[10] in the configuration of **Figure 2**, are a motion pair in alignment with this assertion.

$$T_{N,\vec{\ell}_{Earth-Moon}[\phi(t)]}\times T_{(Earth-\vec{\ell}_{Earth-Moon}[\phi(t)])}\Big/year \simeq \mathbf{17.651}\ years \simeq T_{A,\vec{d}_{MoonOrbit}[\phi(t)]} \qquad (141)$$

The Moon[11] is observed[11,40] to orbit closer to the so called ecliptic[39] plane than the equatorial plane of Earth. Based on our findings and assertions about the actual topology of motions of the Sun, Earth, and Moon, we have made the following assertions:

- The equatorial plane of the Earth is in fact the plane of individual circular orbit of revolution of Earth, which is the horizontal plane in **Figure 2**;
- The Moon[11]'s observed plane of orbit around the Earth should be the plane of motion of $\vec{d}_{MoonOrbit}[\phi(t)]$ (130). However, based on our Moon[11] Orbit[40] analysis so far, and our assertion that the self-rotation axis and the normal of the individual orbit of revolution of every spatial body is the same, the angles made by the Moon[11]'s North Pole[14] vector determine angles made by the plane of individual circular orbit of revolution of the

Moon[11] around $\left\{\vec{r}_1\left[\phi(t)\right]=\vec{\boldsymbol{r}}_{Moon}\left[\phi(t)\right]\right\}$ (46), and thus the angles said to be made by the Moon's observed[40] plane of orbit around the Earth should be the angles made by the plane of motion of the vector distance $\left\{\vec{\ell}\left[\phi(t)\right]=\vec{\ell}_{Earth-Moon}(\phi)\right\}$ (48), which is between the centers of individual circular orbits of revolution of the Earth and the Moon, such that individual circular orbit of revolution of the Moon around $\left\{\vec{r}_1\left[\phi(t)\right]=\vec{\boldsymbol{r}}_{Moon}\left[\phi(t)\right]\right\}$ (46) in turn revolves as a whole around Earth with a radius of $\left\{\vec{\ell}\left[\phi(t)\right]=\vec{\ell}_{Earth-Moon}(\phi)\right\}$ (48) at each $\left[\phi(t)=\phi_{Earth-Moon}\right]$ (44) through every orbital period[21,11] T_{Moon} (57) of the Moon[11], or vice versa;

- The so called observed ecliptic[39] plane of the Earth in its motion around the Sun is in fact not really a plane, but the approximate angles said to be observed with respect to the ecliptic[39] are angles observed with respect to the plane of the distance vector between the Earth and the Sun around the year, and in fact the so called observed ecliptic[39] plane can approximately be considered to be the plane of individual circular orbit of revolution of the Sun, as $\left(\mathbf{12,370.477}\ km \simeq r_{Earth} \ll r_{Sun} \simeq \mathbf{149,609,642.749}\ km\right)$ (90).

In light of all this information, and based on the observed[11,40] topology of relative angles between poles[14,15] and planes of Moon, Earth, and Sun, we make the following conclusions. The above information tells us that the observed axial tilt[11,40,13] angle of the Moon[11]'s North Pole[14] vector with respect to the so called observed ecliptic[39] plane, or lunar[11,40] obliquity to ecliptic[39], namely $\alpha_{Moon \measuredangle Ecliptic}$ (142), which is around $1^{\circ}.5424$, is in fact the observed axial tilt[11,13] angle of the Moon[11]'s North Pole[14] vector approximately with respect to the normal of the plane of individual circular orbit of revolution of the Sun, namely $\alpha_{Moon \measuredangle Sun\ Orbital\ Plane}$ (142). Further, the observed mean axial tilt[11,13] angle of the Moon[11]'s North Pole[14] vector with respect to the Moon's observed approximate plane of orbit around the Earth, or mean lunar[11,40] obliquity, namely $\alpha_{Moon \measuredangle Moon\ Orbit}$ (143), which is about $6^{\circ}.687$, is in fact the observed axial tilt[11,13] angle of the Moon[11]'s North Pole[14] vector with respect to the normal of the plane of motion of vector distance $\left\{\vec{\ell}\left[\phi(t)\right]=\vec{\ell}_{Earth-Moon}(\phi)\right\}$ (48), which is between centers of individual circular orbits of revolution of Earth and Moon, namely $\alpha_{Moon \measuredangle \vec{\ell}_{Earth-Moon}\left[\phi(t)\right]-plane}$ (143). Moreover, the maximum angle that the Moon[11]'s observed approximate plane of orbit around Earth makes with respect to the so called ecliptic[39] plane, or mean lunar[11,40] orbital inclination, namely $\alpha_{Moon\ Orbit \measuredangle Ecliptic}$ (144), which is around $5^{\circ}.09'$ or equivalently $5^{\circ}.1446$, is in fact the observed angle that the plane of motion of the vector distance $\left\{\vec{\ell}\left[\phi(t)\right]=\vec{\ell}_{Earth-Moon}(\phi)\right\}$ (48), which is between centers of individual circular orbits of revolution of Earth and Moon, makes approximately with respect to the normal of the plane of individual circular orbit of revolution of the Sun, namely $\alpha_{\vec{\ell}_{Earth-Moon}\left[\phi(t)\right]-plane \measuredangle Sun\ Orbital\ Plane}$ (144). Additionally, the observed axial tilt[8,13] angle of the Earth[10], or Earth[10] obliquity, or equivalently the angle that the so called ecliptic[39] plane makes with respect to Earth[10]'s equatorial plane, namely $\alpha_{Ecliptic \measuredangle Earth\ Equator}$ (145), which is approximately $23^{\circ}.27'$ or equivalently $23^{\circ}.44$, that also sums up to 180° with $\beta_{Sun-Earth}$ (82), is in fact the observed angle between individual circular orbits of revolution of the Earth and the Sun, namely $\alpha_{Sun\ Orbital\ Plane \measuredangle\ Earth\ Orbital\ Plane}$ (145).

$$\alpha_{Moon \measuredangle\ Ecliptic} \simeq \alpha_{Moon \measuredangle\ Sun\ Orbital\ Plane} \simeq 1^{\circ}.5424 \qquad (142)$$

$$\alpha_{Moon \measuredangle\ Moon\ Orbit} = \alpha_{Moon \measuredangle\ \vec{\ell}_{Earth-Moon}[\phi(t)]-plane} \simeq 6^{\circ}.687 \qquad (143)$$

$$\alpha_{Moon\ Orbit \measuredangle\ Ecliptic} \simeq \alpha_{\vec{\ell}_{Earth-Moon}[\phi(t)]-plane \measuredangle\ Sun\ Orbital\ Plane} \simeq 5^{\circ}.09' \simeq 5^{\circ}.1446 \simeq 6^{\circ}.687 - 1^{\circ}.5424 \qquad (144)$$

$$\begin{aligned}\alpha_{Ecliptic \measuredangle\ Earth\ Equator} \simeq \alpha_{Sun\ Orbital\ Plane \measuredangle\ Earth\ Orbital\ Plane} &\simeq 23^{\circ}.27' \simeq 23^{\circ}.44 \\ &\simeq 180^{\circ} - 156^{\circ}.56 = 180^{\circ} + \beta_{Sun-Earth}\end{aligned} \qquad (145)$$

The observed axial tilt[11,13] angle of the Moon[11]'s North Pole[14] vector with respect to Earth[10]'s equatorial plane, namely $\alpha_{Moon \measuredangle\ Earth\ Equator}$ (146), which is said to be about 24°, is in fact the observed axial tilt[11,13] angle of Moon[11]'s North Pole[14] vector with respect to Earth's plane of individual circular orbit of revolution, namely $\alpha_{Moon \measuredangle\ Earth\ Orbital\ Plane}$ (146), which is what reveals the value of the inclination angle $\beta_{Moon-Earth}$ (84) of the plane of the Moon's individual orbit with respect to the plane of the Earth's individual circular orbit of revolution in **Figure 2**, also based on our assertion that the self-rotation axis and the normal of the individual orbit of revolution of every spatial body is the same, and also taking into consideration the fact that the Moon[11] is observed[11,40] to orbit close to the so called ecliptic[39] plane, or equivalently close to the plane of individual circular orbit of revolution of the Sun based on our analysis above.

$$\alpha_{Moon \measuredangle\ Earth\ Equator} = \alpha_{Moon \measuredangle\ Earth\ Orbital\ Plane} \simeq 24^{\circ} \simeq \beta_{Moon-Earth} \qquad (146)$$

Every 18.6 years (138), the angle between the Moon[11]'s Orbit[40] and Earth[10]'s equator reaches a maximum $\alpha_{Moon\ Orbit \measuredangle\ Earth\ Equator, Max}$ (147) of $28^{\circ}.36'$ or equivalently $28^{\circ}.5846$, the sum of Earth[10]'s equatorial tilt $\alpha_{Ecliptic \measuredangle\ Earth\ Equator}$ (145), or equivalently $\alpha_{Sun\ Orbital\ Plane \measuredangle\ Earth\ Orbital\ Plane}$ (145), which is $23^{\circ}.27'$ or equivalently $23^{\circ}.44$, and the Moon[11]'s orbital inclination $\alpha_{Moon\ Orbit \measuredangle\ Ecliptic}$ (144) to the so called ecliptic[39], or equivalently $\alpha_{\vec{\ell}_{Earth-Moon}[\phi(t)]-plane \measuredangle\ Sun\ Orbital\ Plane}$ (144), which is $5^{\circ}.09'$ or equivalently $5^{\circ}.1446$. This is called a major lunar[11] standstill[11,50]. Conversely, 9.3 years later, the angle between Moon[11]'s Orbit[40] and Earth[10]'s equator reaches a minimum $\alpha_{Moon\ Orbit \measuredangle\ Earth\ Equator, Min}$ (148) of $18^{\circ}.20'$ or equivalently $18^{\circ}.33$, approximately near $18^{\circ}.2954$, which is the Moon[11]'s orbital inclination $\alpha_{Moon\ Orbit \measuredangle\ Ecliptic}$ (144) to the so called ecliptic[39], or equivalently $\alpha_{\vec{\ell}_{Earth-Moon}[\phi(t)]-plane \measuredangle\ Sun\ Orbital\ Plane}$ (144), subtracted from Earth[10]'s equatorial tilt $\alpha_{Ecliptic \measuredangle\ Earth\ Equator}$ (145), or equivalently $\alpha_{Sun\ Orbital\ Plane \measuredangle\ Earth\ Orbital\ Plane}$ (145). This is called a minor lunar[11] standstill[11,50], which last occurred in October 2015. This variation of the angle $\alpha_{Moon\ Orbit \measuredangle\ Earth\ Equator}$ (147) - (149) between the Moon[11]'s Orbit[40] and Earth[10]'s equator over a 18.6 year cycle is apparently related to the observed[40] rotation of the line of nodes of the Moon[11] Orbit[40] over a 18.6 year (138) cycle, which we had previously concluded to be related to orientation variation motion of $\{\vec{\ell}[\phi(t)] = \vec{\ell}_{Earth-Moon}(\phi)\}$ (48) as part of $\vec{\boldsymbol{d}}_{MoonOrbit}[\phi(t)]$ (130), with a period $T_{N,\vec{\ell}_{Earth-Moon}[\phi(t)]}$ (138) of 18.6 years. Therefore, we conclude here that the angle

$\alpha_{Moon\ Orbit\ \measuredangle\ Earth\ Equator}$ (147) - (149) observed[40] to be between the Moon[11]'s Orbit[40] and Earth[10]'s equator is equivalently the angle $\alpha_{\vec{\ell}_{Earth-Moon}[\phi(t)]-plane\ \measuredangle\ Earth\ Orbital\ Plane}$ (147) - (149) between the plane of motion of the vector $\{\vec{\ell}[\phi(t)] = \vec{\ell}_{Earth-Moon}(\phi)\}$ (48), which is between centers of individual circular orbits of revolution of the Earth and the Moon, and Earth[10]'s equator.

$$\begin{aligned}\alpha_{Moon\ Orbit\ \measuredangle\ Earth\ Equator,Max} &= \alpha_{\vec{\ell}_{Earth-Moon}[\phi(t)]-plane\ \measuredangle\ Earth\ Orbital\ Plane,Max} \simeq 5^{\circ}.1446 + 23^{\circ}.44 \simeq 28^{\circ}.5846 \simeq 28^{\circ}.36' \\ &= \alpha_{\vec{\ell}_{Earth-Moon}[\phi(t)]-plane\ \measuredangle\ Sun\ Orbital\ Plane} + \alpha_{Sun\ Orbital\ Plane\ \measuredangle\ Earth\ Orbital\ Plane}\end{aligned} \tag{147}$$

$$\begin{aligned}\alpha_{Moon\ Orbit\ \measuredangle\ Earth\ Equator,Min} &= \alpha_{\vec{\ell}_{Earth-Moon}[\phi(t)]-plane\ \measuredangle\ Earth\ Orbital\ Plane,Min} \simeq -5^{\circ}.1446 + 23^{\circ}.44 \simeq 18^{\circ}.2954 \simeq 18^{\circ}.20' \\ &= -\alpha_{\vec{\ell}_{Earth-Moon}[\phi(t)]-plane\ \measuredangle\ Sun\ Orbital\ Plane} + \alpha_{Sun\ Orbital\ Plane\ \measuredangle\ Earth\ Orbital\ Plane}\end{aligned} \tag{148}$$

$$\alpha_{Moon\ Orbit\ \measuredangle\ Earth\ Equator,Min} \leq \alpha_{Moon\ Orbit\ \measuredangle\ Earth\ Equator} \leq \alpha_{Moon\ Orbit\ \measuredangle\ Earth\ Equator,Max} \quad (over\ 18.6\ years\ cycle) \tag{149}$$

The resultant expected overall orientation variation motion of $\{\vec{\ell}[\phi(t)] = \vec{\ell}_{Earth-Moon}(\phi)\}$ (48) as part of $\vec{\boldsymbol{d}}_{MoonOrbit}[\phi(t)]$ (130), with a period $T_{N,\vec{\ell}_{Earth-Moon}[\phi(t)]}$ (138) of 18.6 years, is depicted roughly with Sun-Earth Cartesian $(\hat{\boldsymbol{x}}, \hat{\boldsymbol{y}}, \hat{\boldsymbol{z}})$ coordinates, in **Figure 6** at $(t = t_0)$ and $(t = t_0 + 18.6\,years)$ corresponding to a major lunar[11] standstill[11,50], in **Figure 7** at $(t = t_0 + 4.65\,years)$ end of 1st Quarter Cycle, in **Figure 8** at $(t = t_0 + 9.3\,years)$ end of 1 Half Cycle at a minor lunar[11] standstill[11,50], and in **Figure 9** at $(t = t_0 + 13.95\,years)$ end of 3rd Quarter Cycle, not to scale and with the shape of the lunar[11,40,41] monthly cycle of $\{\vec{\ell}[\phi(t)] = \vec{\ell}_{Earth-Moon}(\phi)\}$ (48) not necessarily accurate, just to enable visualization of the motion of the $\{\vec{\ell}[\phi(t)] = \vec{\ell}_{Earth-Moon}(\phi)\}$ (48) cycle, leading to the combined perception by an observer[40] on Earth[10], of Moon[11] Orbit[40]'s apsidal[47] precession, rotation of the line of nodes of Moon[11] Orbit[40] over a 18.6 year (138) cycle, and variation of $\alpha_{Moon\ Orbit\ \measuredangle\ Earth\ Equator}$ (147) - (149) angle between Moon[11]'s Orbit[40] and the Earth[10]'s equator, or equivalently $\alpha_{\vec{\ell}_{Earth-Moon}[\phi(t)]-plane\ \measuredangle\ Earth\ Orbital\ Plane}$ (147) - (149), over a 18.6 year (138) cycle.

Based on the observation[40] that the shape of Moon[11]'s Orbit[40] is pretty much fixed in its plane of motion over a lunar[11,40,41] month of period T_{Moon} (57), but the *orientation* of the Moon[11]'s Orbit[40] with respect to the Earth and Sun is *not* fixed in space, and moves over periods much longer than T_{Moon} (57), such as $T_{N,\vec{\ell}_{Earth-Moon}[\phi(t)]}$ (138) where $\left(T_{N,\vec{\ell}_{Earth-Moon}[\phi(t)]} \gg T_{Moon}\right)$ (150), we have assessed that $[\ell(\phi) = \ell_{Earth-Moon}(\phi)]$ (11) magnitude cycle of $\{\vec{\ell}[\phi(t)] = \vec{\ell}_{Earth-Moon}(\phi)\}$ (48) over a lunar[11,40,41] month of period T_{Moon} (57) must also be pretty much fixed in its plane of motion, and the direction cycle of this $\{\vec{\ell}[\phi(t)] = \vec{\ell}_{Earth-Moon}(\phi)\}$ (48) vector changing with angular velocity of $(\omega_1 = \omega_{Moon})$ (60), or equivalently the orientation of $\{\vec{\ell}[\phi(t)] = \vec{\ell}_{Earth-Moon}(\phi)\}$ (48), must be varying very little over consecutive lunar[11,40,41] months of period T_{Moon} (57), also taking into consideration our definition of $\vec{d}_{MoonOrbit}\left[\phi(t)\right]$ (130). Based on this assessment, for our purpose of analytically finding an approximate value for $\left\{\vec{r}_1\left[\phi(t)\right]=\vec{r}_{Moon}\left[\phi(t)\right]\right\}$ (46) and obtaining a better idea about $\left\{\vec{\ell}\left[\phi(t)\right]=\vec{\ell}_{Earth-Moon}(\phi)\right\}$ (48), we have made the assumption in (151) for the dates in **Table 1** corresponding to yearly and local maximum and minimum Earth-Moon distances[19] around the minor lunar[11] standstill[11,50] that has occurred[51] in October 2015, such that we have taken $\vec{\ell}_{Earth-Moon}\left(-\frac{126\pi}{40}\right)$ (152) on August 18, 2015, $\vec{\ell}_{Earth-Moon}(-3\pi)$ (152) on September 14, 2015, and $\vec{\ell}_{Earth-Moon}\left(-\frac{114\pi}{40}\right)$ (152) on October 11, 2015 to be approximately the same as $\vec{\ell}_{2\pi}$ (152), when the phase $\left[\phi_1(t)=\omega_1 t=\omega_{Moon}t\right]$ (43) of the Moon[11,40] in its individual circular orbit of revolution is a multiple of 2π based on ω_{Moon} (60), and similarly we have taken $\vec{\ell}\left(-\frac{123\pi}{40}\right)$ (153) on August 30, 2015 and $\vec{\ell}\left(-\frac{117\pi}{40}\right)$ (153) on September 28, 2015 to be approximately the same as $\vec{\ell}_{\pi}$ (153), when the phase $\left[\phi_1(t)=\omega_1 t=\omega_{Moon}t\right]$ (43) of the Moon[11,40] in its individual circular orbit of revolution is an odd multiple of π based on ω_{Moon} (60).

Table 1 Min-Max Earth-Moon distances[19] in kilometers (km) around 2015 lunar[11] standstill[11,50]

Earth-Moon Distance Data					
Date	Time (UTC)	Earth-Moon Distance [km]	Approximate Earth Phase $(\phi=\omega_{Earth}t)$ [radians]	Approximate Moon Phase $(\omega_{Moon}t)$ [radians]	Notes
18.08.2015	02:32	405.848	-3.150π	-42π	MAX (Local)
30.08.2015	15:21	358.290	-3.075π	-41π	MIN (Local)
14.09.2015	11:27	**406.464**	-3.000π	-40π	**MAX (Yearly)**
28.09.2015	01:45	**356.877**	-2.925π	-39π	**MIN (Yearly)**
11.10.2015	13:17	406.388	-2.850π	-38π	MAX (Local)

$$T_{N,\vec{\ell}[\phi(t)]}\simeq 18.599\,years \quad \& \quad T_{Moon}\simeq 27.321661\,days \quad \Rightarrow \quad T_{N,\vec{\ell}[\phi(t)]}\gg T_{Moon} \tag{150}$$

$$T_{N,\vec{\ell}[\phi(t)]}\gg T_{Moon}\Rightarrow \vec{\ell}_{2\pi}\simeq\vec{\ell}\left(-\frac{126\pi}{40}\right)\simeq\vec{\ell}(-3\pi)\simeq\vec{\ell}\left(-\frac{114\pi}{40}\right)\&\,\vec{\ell}_{\pi}\simeq\vec{\ell}\left(-\frac{123\pi}{40}\right)\simeq\vec{\ell}\left(-\frac{117\pi}{40}\right) \tag{151}$$

$$\bar{\ell}_{2\pi} = \hat{\boldsymbol{x}}\ell_{x,2\pi} + \hat{\boldsymbol{y}}\ell_{y,2\pi} + \hat{\boldsymbol{z}}\ell_{z,2\pi} \simeq \bar{\ell}\left(-\frac{126\pi}{40}\right) \simeq \hat{\boldsymbol{x}}\ell_x\left(-\frac{126\pi}{40}\right) + \hat{\boldsymbol{y}}\ell_y\left(-\frac{126\pi}{40}\right) + \hat{\boldsymbol{z}}\ell_z\left(-\frac{126\pi}{40}\right)$$
$$\simeq \bar{\ell}(-3\pi) \simeq \hat{\boldsymbol{x}}\ell_x(-3\pi) + \hat{\boldsymbol{y}}\ell_y(-3\pi) + \hat{\boldsymbol{z}}\ell_z(-3\pi) \quad (152)$$
$$\simeq \bar{\ell}\left(-\frac{114\pi}{40}\right) \simeq \hat{\boldsymbol{x}}\ell_x\left(-\frac{114\pi}{40}\right) + \hat{\boldsymbol{y}}\ell_y\left(-\frac{114\pi}{40}\right) + \hat{\boldsymbol{z}}\ell_z\left(-\frac{114\pi}{40}\right)$$

$$\bar{\ell}_{\pi} = \hat{\boldsymbol{x}}\ell_{x,\pi} + \hat{\boldsymbol{y}}\ell_{y,\pi} + \hat{\boldsymbol{z}}\ell_{z,\pi} \simeq \bar{\ell}\left(-\frac{123\pi}{40}\right) \simeq \hat{\boldsymbol{x}}\ell_x\left(-\frac{123\pi}{40}\right) + \hat{\boldsymbol{y}}\ell_y\left(-\frac{123\pi}{40}\right) + \hat{\boldsymbol{z}}\left(-\frac{123\pi}{40}\right)$$
$$\simeq \bar{\ell}\left(-\frac{117\pi}{40}\right) \simeq \hat{\boldsymbol{x}}\ell_x\left(-\frac{117\pi}{40}\right) + \hat{\boldsymbol{y}}\ell_y\left(-\frac{117\pi}{40}\right) + \hat{\boldsymbol{z}}\left(-\frac{117\pi}{40}\right) \quad (153)$$

We have thus determined the approximate Cartesian $(\hat{x}, \hat{y}, \hat{z})$ coordinate values of $\bar{\ell}_{2\pi}$ (152) as expressed in (154) - (156), with an approximate magnitude of $\ell_{2\pi}$ (157), and the approximate Cartesian $(\hat{x}, \hat{y}, \hat{z})$ coordinate values of $\bar{\ell}_{\pi}$ (153) as expressed in (158) - (160), with an approximate magnitude of ℓ_{π} (161), in the Earth-Moon configuration of **Figure 2**.

$$\ell_{x,2\pi} \simeq \mathbf{-2,912,300.998}\ km \quad (154)$$

$$\ell_{y,2\pi} \simeq \mathbf{19,524.629}\ km \quad (155)$$

$$\ell_{z,2\pi} \simeq \mathbf{-954,159.024}\ km \quad (156)$$

$$\ell_{2\pi} = \sqrt{\ell_{x,2\pi}^{\ 2} + \ell_{y,2\pi}^{\ 2} + \ell_{z,2\pi}^{\ 2}} \simeq \mathbf{3,064,685.589}\ km \quad (157)$$

$$\ell_{x,\pi} \simeq \mathbf{5,553,153.005}\ km \quad (158)$$

$$\ell_{y,\pi} \simeq \mathbf{-87,481.846}\ km \quad (159)$$

$$\ell_{z,\pi} \simeq \mathbf{4,276,708.457}\ km \quad (160)$$

$$\ell_{\pi} = \sqrt{\ell_{x,\pi}^{\ 2} + \ell_{y,\pi}^{\ 2} + \ell_{z,\pi}^{\ 2}} \simeq \mathbf{7,009,664.513}\ km \quad (161)$$

Although the results expressed for $\bar{\ell}_{2\pi}$ (152) and $\bar{\ell}_{\pi}$ (153) in (154) - (161) are all approximate values obtained with analytical simplifications based on observation, they are sufficient to provide an idea about Earth-Moon motion topology. Looking at the signs of the found Cartesian $(\hat{x}, \hat{y}, \hat{z})$ coordinate values of $\bar{\ell}_{2\pi}$ (152) and $\bar{\ell}_{\pi}$ (153) in the Earth-Moon system configuration of **Figure 2**, one directly notices that these vectors are somewhat in a reverse direction relative to the vector radius $\{\bar{r}_1[\phi(t)] = \bar{r}_{Moon}[\phi(t)]\}$ (46) of the individual circular orbit of the Moon, as $\{\bar{r}_1[\phi(t)] = \bar{r}_{Moon}[\phi(t)]\}$ (46) vector is in $(+\hat{x})$-direction in the configuration of **Figure 2** when the phase $[\phi_1(t) = \omega_1 t = \omega_{Moon} t]$ (43) of the Moon[11,40] in its individual circular orbit of revolution is a multiple of 2π based on ω_{Moon} (60), and $\{\bar{r}_1[\phi(t)] = \bar{r}_{Moon}[\phi(t)]\}$ (46) is in $(-\hat{x})$-direction in the configuration of **Figure 2** when the phase $[\phi_1(t) = \omega_1 t = \omega_{Moon} t]$ (43) of the Moon[11,40] in its individual circular orbit of revolution is an odd multiple of π based on ω_{Moon} (60), based on our analysis. This finding is also very significant regarding the relative motion of $\{\vec{\ell}[\phi(t)] = \vec{\ell}_{Earth-Moon}(\phi)\}$ (48), namely the vector distance between the centers of individual circular orbits of revolution of the Earth and the Moon, and $\{\vec{r}_1[\phi(t)] = \vec{\boldsymbol{r}}_{Moon}[\phi(t)]\}$ (46), the vector radius of the Moon's individual circular orbit of revolution.

This finding of $\vec{\ell}_{2\pi}$ (152) and $\vec{\ell}_{\pi}$ (153), even if an approximate values, also provides us the insight that the magnitude $[\ell(\phi) = \ell_{Earth-Moon}(\phi)]$ (11) of $\{\vec{\ell}[\phi(t)] = \vec{\ell}_{Earth-Moon}(\phi)\}$ (48), which is the vector between the centers of individual circular orbits of revolution of the Earth and the Moon, is on a similar order of magnitude as r_{Moon} (89), the radius of individual circular orbit of revolution of the Moon.

The most significant and interesting consequence of these orbital parameter values found for the Sun, the Earth, and the Moon is, the obtained results indicate that the Earth is *not* revolving around the Sun, and the Moon is not directly revolving around the Earth, but in fact the Sun is revolving around a larger individual circular orbit which encompasses the individual circular orbits of the Earth and the Moon with an inclination, whose centers of revolutions are shifted compared to the center of revolution of the Sun, the center of revolution of the Moon being in continuous harmonic motion relative to other bodies. This revelation raises the need to reevaluate existing theories about the formation, expected motional behavior, and topologies of solar systems observed in the Universe in general as well as our Solar System, bringing new vision to the formation and motion structure of the bodies in the Universe. We can also conclude that planets do *not* necessarily revolve around stars, and Earth is *not* revolving around the Sun.

All the obtained results strongly support our expectation that the Sun and the Earth may in reality be revolving around their individual circular orbits with the *same* angular velocity, demonstrating their relative elliptical orbital behavior, just as well as the Moon possibly revolving around its individual circular orbit with a *different* but *fixed* angular velocity. As a result, we make the more generalized assertion for the motion of spatial bodies in the Universe, including the Sun, the Planets, and the Satellites in the Solar System and elsewhere, as well as at particle and sub-atomic level, that "possibly all bodies move at some angular velocity around their own circular orbits of revolution with different radii and centers of revolution in space", but "each appear to be moving in elliptical motion with respect to each other, where they see the other body located at a fixed or variable point in their respective virtual plane of motion."

FIGURES

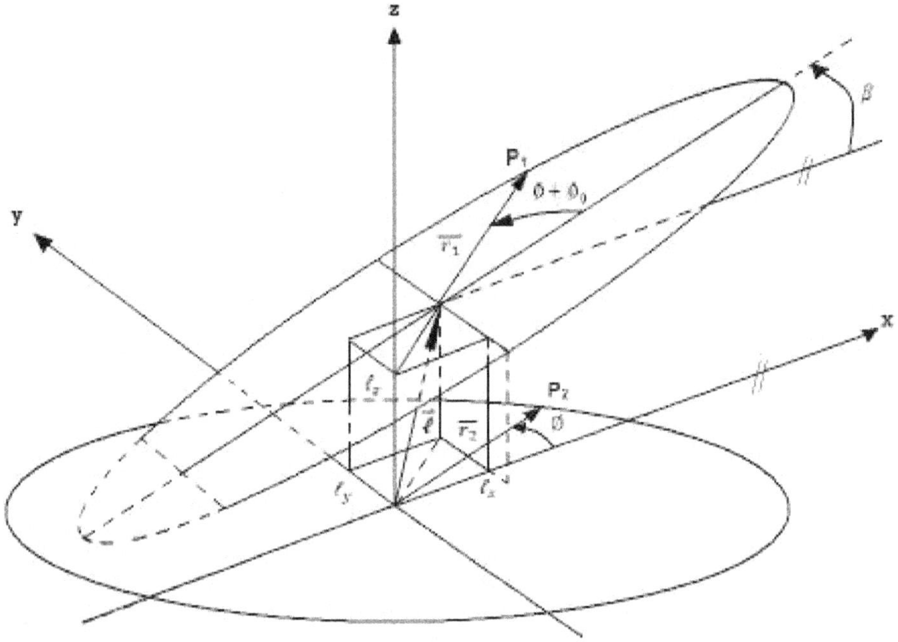

Figure 1 Points $\mathbf{P}_1$ and $\mathbf{P}_2$ on Two Circles in Space

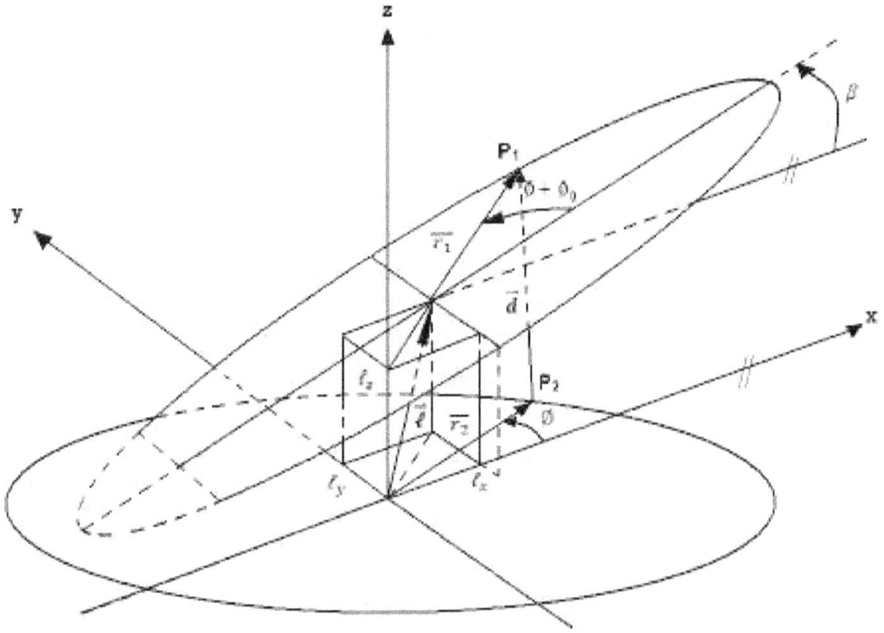

Figure 2 Distance Between Points $\mathbf{P}_1$ and $\mathbf{P}_2$ on Two Different Circles in Space

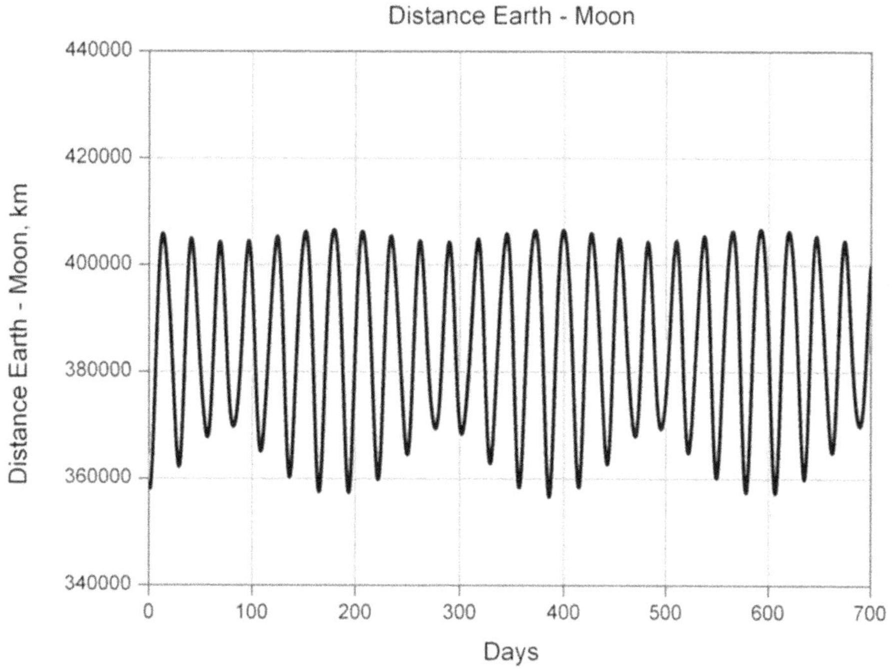

Figure 3 A Distance d Curve Between Moving Points $\mathbf{P}_1$ and $\mathbf{P}_2$ as a Function of Time t (Observed Earth-Moon distance[3,19] over 700 Days)

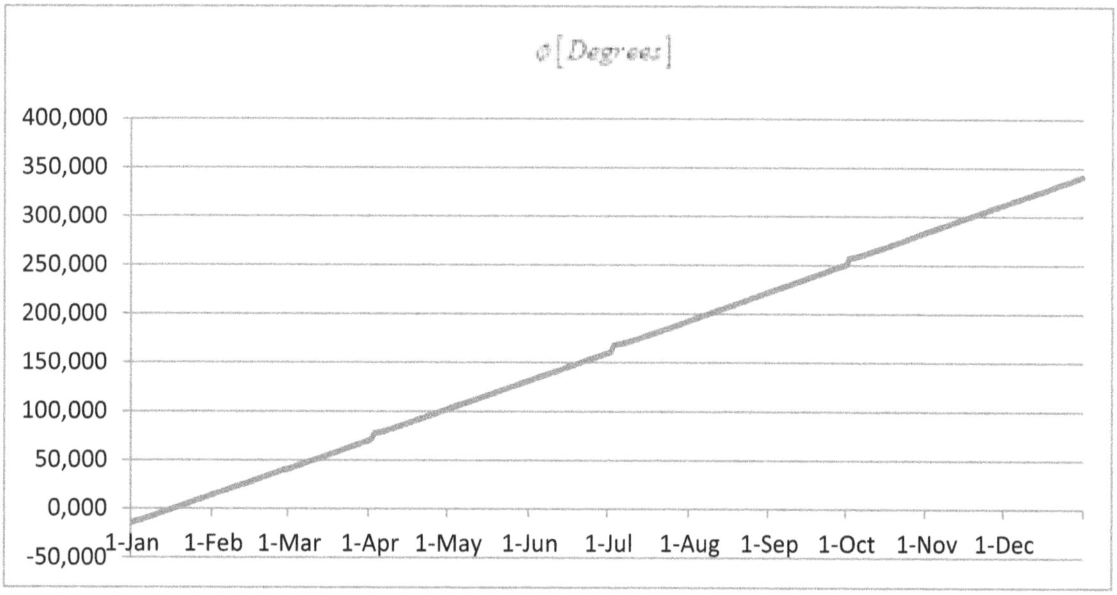

Figure 4 ϕ in $[Degrees]$ against dates for the Earth-Sun system over a year

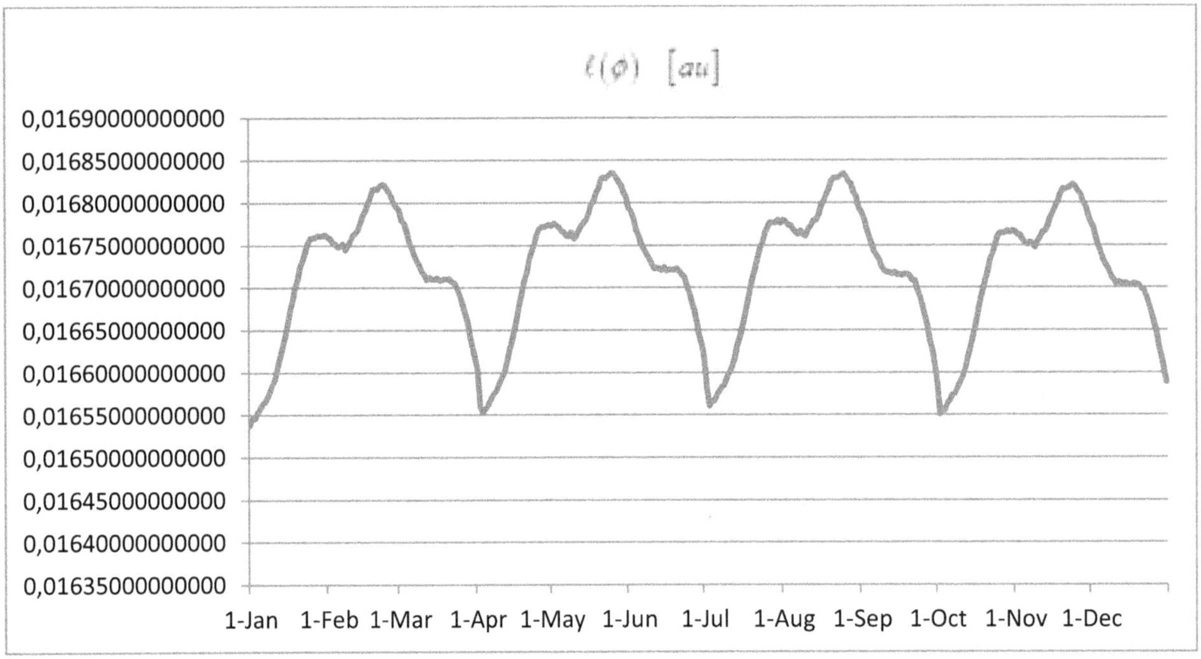

Figure 5 $\ell(\phi)$ in $[au]$ for the Sun-Earth system over a year

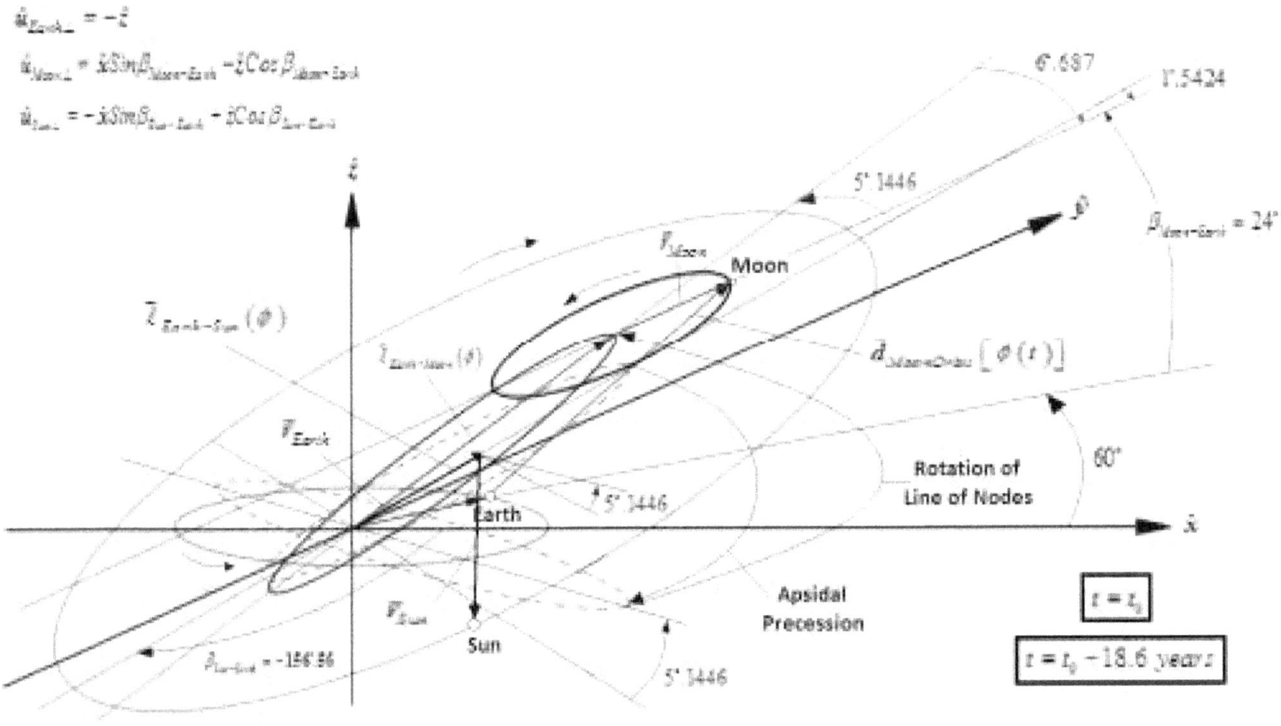

Figure 6 Orientation variation of $\bar{\ell}\left[\phi(t)\right]$ at $(t=t_0)$ and $(t=t_0+18.6\,years)$

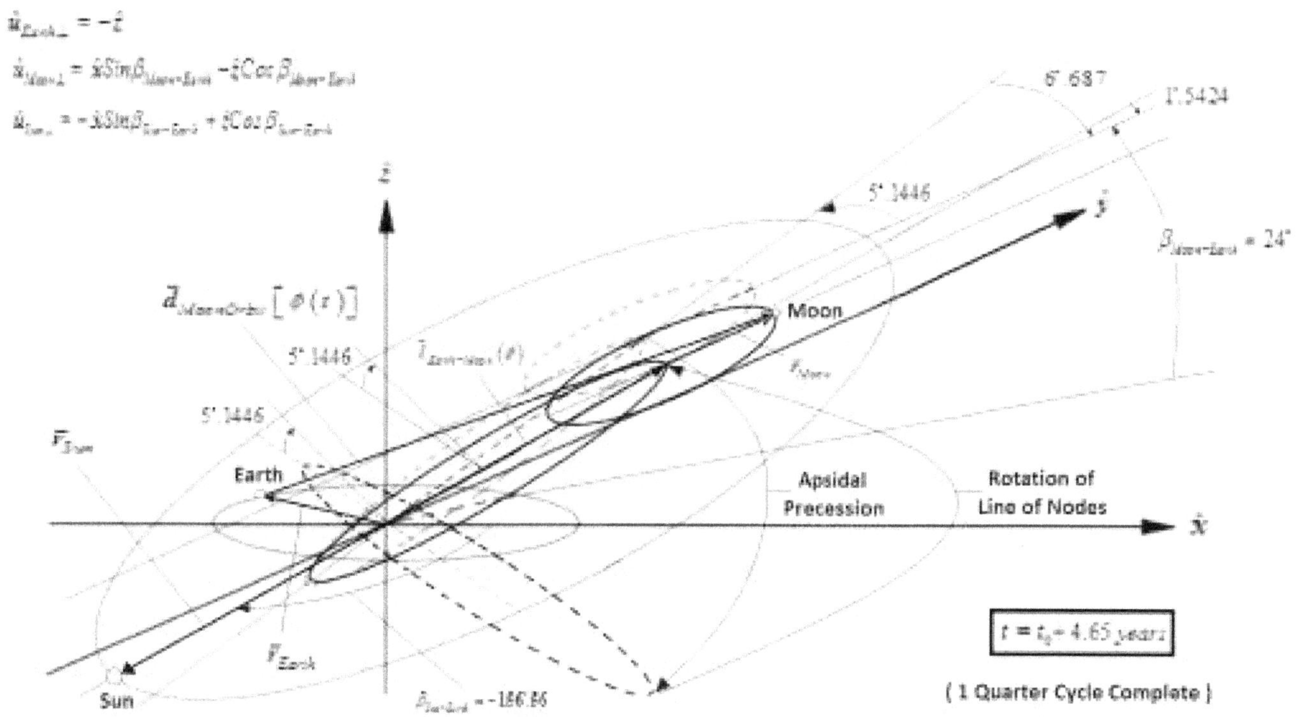

Figure 7 Orientation variation of $\vec{\ell}\left[\phi(t)\right]$ at $\left(t = t_0 + 4.65\,years\right)$ end of 1st Quarter Cycle

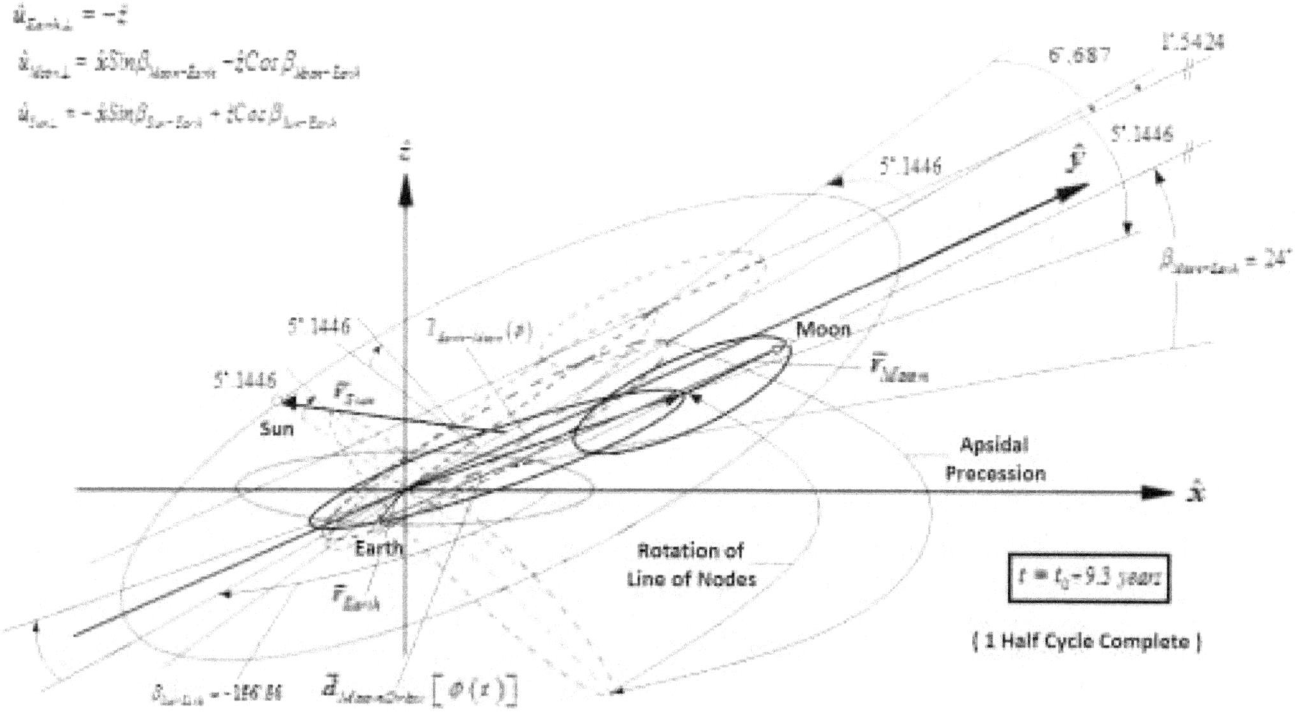

Figure 8 Orientation variation of $\vec{\ell}\left[\phi(t)\right]$ at $\left(t = t_0 + 9.3\,years\right)$ end of 1 Half Cycle

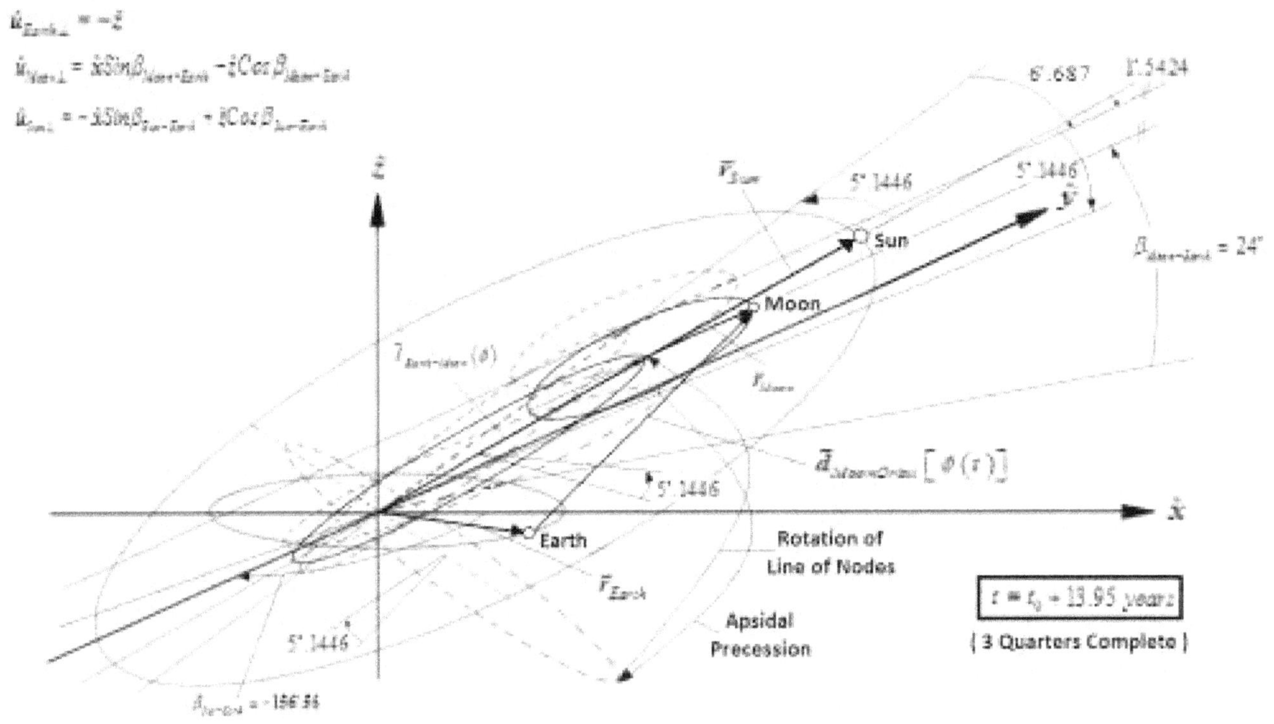

Figure 9 Orientation variation of $\bar{\ell}\left[\phi(t)\right]$ at $\left(t = t_0 + 13.95\, years\right)$ end of 3rd Quarter Cycle

REFERENCES

References in this Article can be any Physics, Electromagnetics, and Calculus textbook, as the physics equations and mathematical identities used as a basis for the proof are all currently accepted theory in existing textbooks.

1. Halliday, D., Resnick, R., *Fundamentals of Physics (3rd Edition)*, John Wiley & Sons, 1988, ISBN 0-471-63735-1
2. Kepler's laws of planetary motion, Wikipedia, https://en.wikipedia.org/wiki/Kepler%27s_laws_of_planetary_motion, (2020)
3. Lunar distance, Wikipedia, https://en.wikipedia.org/wiki/Lunar_distance_(astronomy), (2020)
4. Dot product, Wikipedia, https://en.wikipedia.org/wiki/Dot_product, (2020)
5. Ellipse, Wikipedia, http://en.wikipedia.org/wiki/Ellipse, (2020)
6. Semi-major and semi-minor axes, Wikipedia, http://en.wikipedia.org/wiki/Semi-major_and_semi-minor_axes, (2020)
7. Focus (geometry), Wikipedia, http://en.wikipedia.org/wiki/Focus_(geometry), (2020)
8. Earth's orbit, Wikipedia, https://en.wikipedia.org/wiki/Earth%27s_orbit, (2020)
9. Sun, Wikipedia, https://en.wikipedia.org/wiki/Sun, (2020)
10. Earth, Wikipedia, https://en.wikipedia.org/wiki/Earth, (2020)

11. Moon, Wikipedia, https://en.wikipedia.org/wiki/Moon, (2020)
12. Rotation around a fixed axis, Wikipedia, http://en.wikipedia.org/wiki/Rotation_around_a_fixed_axis, (2020)
13. Axial tilt, Wikipedia, https://en.wikipedia.org/wiki/Axial_tilt, (2020)
14. Geographical pole, Wikipedia, https://en.wikipedia.org/wiki/Geographical_pole, (2020)
15. North Pole, Wikipedia, https://en.wikipedia.org/wiki/North_Pole, (2020)
16. South Pole, Wikipedia, https://en.wikipedia.org/wiki/South_Pole, (2020)
17. Equator, Wikipedia, https://en.wikipedia.org/wiki/Equator, (2020)
18. Right hand rule, Wikipedia, https://en.wikipedia.org/wiki/Right-hand_rule, (2020)
19. Moon distances for UTC time zone, Time & Date, https://www.timeanddate.com/astronomy/moon/distance.html, (2020)
20. VLBA Finds Planet Orbiting Small Cool Star, National Radio Astronomy Observatory, https://public.nrao.edu/news/vlba-finds-planet/, August 4, 2020
21. Orbital period, Wikipedia, https://en.wikipedia.org/wiki/Orbital_period, (2020)
22. Year, Wikipedia, https://en.wikipedia.org/wiki/Year, (2020)
23. Clockwise, Wikipedia, https://en.wikipedia.org/wiki/Clockwise, (2020)
24. Faraday's law of induction, Wikipedia, https://en.wikipedia.org/wiki/Faraday%27s_law_of_induction, (2020)
25. Electromotive force, Wikipedia, https://en.wikipedia.org/wiki/Electromotive_force, (2020)
26. Magnetic flux, Wikipedia, https://en.wikipedia.org/wiki/Magnetic_flux, (2020)
27. June 2006: The Tilt of the Sun's Axis, Bruce McClure's Astronomy Page, http://www.idialstars.com/fipl.htm, (2020)
28. Astronomical unit, Wikipedia, https://en.wikipedia.org/wiki/Astronomical_unit, (2020)
29. NASA *Landsat* Observation Data, http://landsathandbook.gsfc.nasa.gov/excel_docs/d.xls
30. File: Tan, A. P., *Asli Pinar Tan Analysis Based on Earth-Sun distance (d) Landsat.xlsx*, https://www.dropbox.com/scl/fi/th5d3d5ur8d1mb60aitou/Asli-Pinar-Tan-Analysis-Based-on-Earth-Sun-distance-d-Landsat.xlsx?dl=0&rlkey=m3spxqxbxtnumrj2wvtg33l2v (2020)
31. Equinox, Wikipedia, https://en.wikipedia.org/wiki/Equinox, (2020)
32. Solstice, Wikipedia, https://en.wikipedia.org/wiki/Solstice, (2020)
33. Double planet, Wikipedia, https://en.wikipedia.org/wiki/Double_planet, (2020)
34. Milky Way, Wikipedia, https://en.wikipedia.org/wiki/Milky_Way, (2020)
35. Galactic Center, Wikipedia, https://en.wikipedia.org/wiki/Galactic_Center, (2020)
36. Sagittarius (constellation), Wikipedia, https://en.wikipedia.org/wiki/Sagittarius_(constellation), (2020)
37. Zell, Holly, *Solar Rotation Varies by Latitude* (August 7, 2017), NASA, https://www.nasa.gov/mssion_pages/sunearth/science/solar-rotation.html, (2020)
38. The Sun's Tilt Thrughout The Year, YouTube, https://youtu.be/j44q2xvNePQ, @astroguyz , July 2, 2015
39. Ecliptic, Wikipedia, https://en.wikipedia.org/wiki/Ecliptic, (2020)
40. Orbit of the Moon, Wikipedia, https://en.wikipedia.org/wiki/Orbit_of_the_Moon, (2020)
41. Lunar month, Wikipedia, https://en.wikipedia.org/wiki/Lunar_month, (2020)
42. Moon phases for UTC time zone, Time & Date, https://www.timeanddate.com/ moon/phases/timezone/utc, (2020)
43. Libration, Wikipedia, https://en.wikipedia.org/wiki/Libration, (2020)

44. File:Lunar libration with phase Oct2007 450px.gif, Wikimedia, (2020) https://upload.wikimedia.org/wikipedia/commons/c/c0/Lunar_libration_with_phase2.gif
45. Celestial pole, Wikipedia, https://en.wikipedia.org/wiki/Celestial_pole, (2020)
46. Equinox and solstice 2010-2019, Greenwich Mean Time, https://greenwichmeantime.com/longest-day/equinox-solstice-2010-2020/, (2020)
47. Apsidal precession, Wikipedia, https://en.wikipedia.org/wiki/Apsidal_precession, (2020)
48. Apsis, Wikipedia, https://en.wikipedia.org/wiki/Apsis, (2020)
49. Sidereal vs. Synodic Motions, Astronomy Education at University of Nebraska-Lincoln, https://astro.unl.edu/naap/motion3/sidereal_synodic.html, (2020)
50. Lunar standstill, Wikipedia, https://en.wikipedia.org/wiki/Lunar_standstill, (2020)
51. Minor lunar standstill and harvest moon, EarthSky.org, https://earthsky.org/astronomy-essentials/minor-lunar-standstill-minimizes-harvest-and-hunters-moons, September 10, 2016

ARTICLE 6 (TÜRKÇE)

HER ŞEY BİR DAİREDİR: YENİ BİR EVRENSEL YÖRÜNGE MODELİ

Yazar: Aslı Pınar Tan[VI]

ÖZET

Güneş Sistemi ve diğer gezegen sistemlerindeki gök cisimlerinin ölçülen astronomik pozisyon verilerine dayanarak, uzaydaki bütün cisimler birbirlerine göre bir çeşit eliptik yörüngede hareket eder gibi görünmektedirler. Kepler'in 1[inci] Kanunu'na göre, "bir gezegenin Güneş'e göre yörüngesi, Güneş'in iki odağından birinde bulunduğu bir elips şeklindedir". Ay'ın Dünya'ya göre yörüngesi de belirgin bir elips şeklindedir, fakat bu elipsin tam bir döngüsü boyunca Ay'ın Dünya'ya harmonik bir şekilde yaklaşıp uzaklaştığı değişken bir eksantrikliği vardır. Bu makalede, öncelikle matematiksel olarak "üç boyutlu uzayda iki ayrı daire çevresinde bulunan noktalar arası uzaklığın", "iki boyutlu uzayda olduğu gibi, bir vektörel elips çevresinde bulunan noktaların başka bir sabit veya hareketli noktaya uzaklığına" eşdeğer olduğunu gösterip ispatladığımız araştırma sonuçlarımız özetlenmektedir. Yapılan şey, uzayda iki ayrı dairesel yörüngede hareket eden gök cisminin, kendi devrim merkezlerine göre aynı açısal hızda, ya da farklı fakat sabit açısal hızlarda, hatta farklı ve değişen açısal hızlarda bile hareket etseler, vektörel olarak sanki birbirlerine göre eliptik bir güzergahta hareket ediyormuş gibi davrandıklarını ve sanal olarak birbirlerini uzayda göreceli ekliptik düzlemlerinde anlık olarak sabit bir noktada konumlanmış gibi gördüklerini göstermeye eşdeğerdir. Bu matematiksel bulgu, Güneş, gezegenler ve Güneş Sistemi ile başka yerlerdeki uydular, ayrıca parçacık ve atom altı seviyesi de dahil olmak üzere evrendeki cisimlerin hareketlerine ilişkin yeni bir temel modelin formülasyonu ile doğadaki kuvvetlere ilişkin daha çok içgörü sağlayarak, fizikte çok geniş kapsamlı keşiflere yol açma potansiyeline sahiptir. Gösterilen matematiksel analize dayanarak, zaman içerisinde birbirlerine göre neredeyse sabit eliptik yörüngeler sergiledikleri için, bu araştırma sonucunda Güneş, Dünya ve Ay'ın her birinin uzayda kendine ait dairesel devrim yörüngesinde dönüyor oldukları iddiasında bulunulmaktadır. Bu beklenti ile, gözlemlenen Dünya ile Güneş ve Dünya ile Ay aresı mesafe verilerine dayanarak, aynı zamanda yaklaşık olarak bu araştırma kapsamında geliştirilen analitik yöntemler kullanılarak, Güneş, Dünya ve Ay'ın şahsi yörüngesel parametreleri hesaplanmıştır. Bu hesaplama ve analiz süreci, gözlemle uyuşan ek sonuçlar da ortaya çıkartmıştır, ki bu da Güneş, Dünya ve Ay'ın gerçekte kendi dairesel yörüngelerinde döndükleri iddiamızı desteklemektedir.

MAKALE

Üç boyutlu uzayda, sistemin geometrisinin Kartezyen $(\hat{x}, \hat{y}, \hat{z})$ koordinatlarında **Figür 1**'deki şekilde gösterildiği ve tanımlayıcı vektörlerin (4) ve (9) arası genel vektör denklemleri ile ifade edildiği iki daireli bir system düşünün. Bu iki daire üzerinde, aralarındaki faz farkı zamana(t) dayalı ϕ_0 (3) açısı olan $\mathbf{P}_1$ ve $\mathbf{P}_2$ noktaları tanımlanmıştır. Kendi dairesi çevresinde $[\phi_1(t) = \phi + \phi_0]$ (1) faz açısında bulunan $\mathbf{P}_1$'in konumu, (4) ile (5) ve (8)'deki denklemlere

---

VI *ASLI PINAR TAN*
Linkedin Web Sitesi: *https://www.linkedin.com./in/apinartan*

dayalı olarak $\left[\vec{r}_1(\phi)+\vec{\ell}(\phi)\right]$ vektörüyle tanımlıdır ve kendi dairesi çevresinde $\left[\phi_2(t)=\phi\right]$ (2) faz açısında bulunan $\mathbf{P}_2$ 'nin konumu $\vec{r}_2(\phi)$ (6) vektörüyle tanımlıdır.

$$\phi_1(t)=\phi+\phi_0=\phi(t)+\phi_0(t)=\phi_2(t)+\phi_0(t) \qquad (\textit{Phase of } \mathbf{P}_1) \tag{1}$$

$$\phi_2(t)=\phi=\phi(t) \qquad (\textit{Phase of } \mathbf{P}_2) \tag{2}$$

$$\phi_0(t)=\phi_1(t)-\phi_2(t)=\phi_0 \qquad (\textit{Phase difference of } \mathbf{P}_1 \textit{ and } \mathbf{P}_2) \tag{3}$$

Her ϕ (2) faz açısında, sabit büyüklükleri r_1 (7) ve r_2 (7) olacak şekilde iki dairenin vektörel yarıçapları sırasıyla $\vec{r}_1$ (4) - (5) ve $\vec{r}_2$ (6)'dir ve iki dairenin merkezleri, büyüklüğü aynı zamanda iki dairenin merkezi arasındaki skaler mesafe $\ell(\phi)$ (9) olan, sabit veya değişken bir $\vec{\ell}(\phi)$ (8) vektörü ile ayrılmıştır.

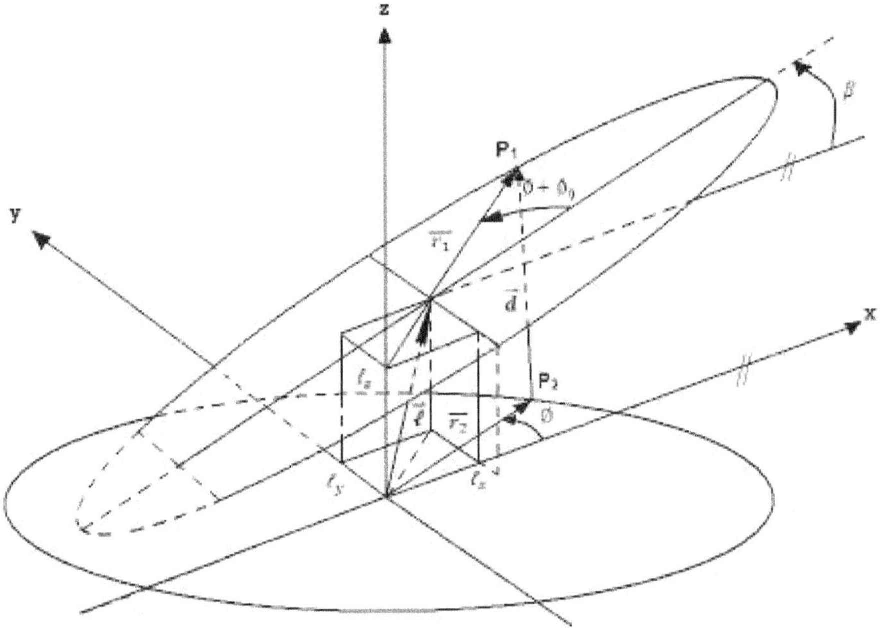

Figür 1 Uzayda İki Farklı Daire Üzerindeki $\mathbf{P}_1$ ve $\mathbf{P}_2$ Noktaları Arasındaki Mesafe

Bundan sonraki bütün işlemlerde, "bir vektörün karesi" olarak, herhangi bir vektör için skaler "büyüklüğün karesi"ne karşılık gelen "vektörün kendisi ile nokta çarpımı"nı aldığımıza dikkat ediniz.

$$\vec{r}_1=\vec{r}_1(\phi+\phi_0)=\hat{x}\,r_1 Cos(\phi+\phi_0)Cos\,\beta+\hat{y}\,r_1 Sin(\phi+\phi_0)+\hat{z}\,r_1 Cos(\phi+\phi_0)Sin\,\beta \tag{4}$$

$$\begin{aligned}\vec{r}_1=\vec{r}_1\left[\phi(t)\right]=&\left(\hat{x}\,r_1 Cos\,\beta\,Cos\left[\phi_0(t)\right]+\hat{y}\,r_1 Sin\left[\phi_0(t)\right]+\hat{z}\,r_1 Sin\,\beta\,Cos\left[\phi_0(t)\right]\right)\,Cos\left[\phi(t)\right]+\\&\left(-\hat{x}\,r_1 Cos\,\beta\,Sin\left[\phi_0(t)\right]+\hat{y}\,r_1 Cos\left[\phi_0(t)\right]-\hat{z}\,r_1 Sin\,\beta\,Sin\left[\phi_0(t)\right]\right)Sin\left[\phi(t)\right]\end{aligned} \tag{5}$$

$$\vec{r}_2=\vec{r}_2\left[\phi(t)\right]=\hat{x}\,r_2\,Cos\left[\phi(t)\right]+\hat{y}\,r_2\,Sin\left[\phi(t)\right] \tag{6}$$

$$r_1=\left|\vec{r}_1\right|=\sqrt{\vec{r}_1\cdot\vec{r}_1}=\sqrt{r_1^{\,2}} \quad ; \quad r_2=\left|\vec{r}_2\right|=\sqrt{\vec{r}_2\cdot\vec{r}_2}=\sqrt{r_2^{\,2}} \tag{7}$$

$$\vec{\ell}=\vec{\ell}\left[\phi(t)\right]=\hat{\boldsymbol{x}}\,\ell_x\left[\phi(t)\right]+\hat{\boldsymbol{y}}\,\ell_y\left[\phi(t)\right]+\hat{\boldsymbol{z}}\,\ell_z\left[\phi(t)\right] \qquad (8)$$

$$\left|\vec{\ell}(\phi)\right|=\ell(\phi)=\sqrt{\ell^2(\phi)}=\sqrt{\vec{\ell}(\phi)\cdot\vec{\ell}(\phi)} \quad ; \quad \ell^2(\phi)=\vec{\ell}(\phi)\cdot\vec{\ell}(\phi)=\ell_x^{\,2}(\phi)+\ell_y^{\,2}(\phi)+\ell_z^{\,2}(\phi) \qquad (9)$$

Herhangi bir $\mathbf{P}_2$ noktasından $\mathbf{P}_1$'e vektörel mesafe $\vec{\boldsymbol{d}}(\phi)$ (10) - (12) şeklinde ifade edilmektedir, ki büyüklüğü olan $d(\phi)$ (13) aynı zamanda iki ilgili daire üzerindeki $\mathbf{P}_1$ ve $\mathbf{P}_2$ arasındaki skaler mesafedir.

$$\vec{\boldsymbol{d}}(\phi)=\vec{\boldsymbol{r}}_1(\phi+\phi_0)-\vec{\boldsymbol{r}}_2(\phi)+\vec{\ell}(\phi)=\vec{\boldsymbol{r}}_1-\vec{\boldsymbol{r}}_2+\vec{\ell}=\vec{\boldsymbol{a}}Cos\,\phi+\vec{\boldsymbol{b}}Sin\phi+\vec{\ell}(\phi)=\vec{\mathbf{X}}(\phi)+\vec{\mathbf{Y}}(\phi)+\vec{\ell}(\phi) \qquad (10)$$

$$\begin{aligned}\vec{\boldsymbol{d}}(\phi)=\hat{\boldsymbol{x}}\ &[r_1Cos(\phi+\phi_0)Cos\,\beta-r_2\,Cos\,\phi+\ell_x(\phi)]+\hat{\boldsymbol{y}}\ [r_1Sin(\phi+\phi_0)-r_2Sin\phi+\ell_y(\phi)]\\ &+\hat{\boldsymbol{z}}\ [r_1Cos(\phi+\phi_0)Sin\,\beta+\ell_z(\phi)]\end{aligned} \qquad (11)$$

$$\begin{aligned}\vec{\boldsymbol{d}}\left[\phi(t)\right]=&\left[\hat{\boldsymbol{x}}\left(r_1Cos\,\beta\,Cos\left[\phi_0(t)\right]-r_2\right)+\hat{\boldsymbol{y}}\,r_1Sin\left[\phi_0(t)\right]+\hat{\boldsymbol{z}}\,r_1Sin\,\beta\,Cos\left[\phi_0(t)\right]\right]Cos\left[\phi(t)\right]+\\ &\left[-\hat{\boldsymbol{x}}\,r_1Cos\,\beta\,Sin\left[\phi_0(t)\right]+\hat{\boldsymbol{y}}\left(r_1Cos\left[\phi_0(t)\right]-r_2\right)-\hat{\boldsymbol{z}}\,r_1Sin\,\beta\,Sin\left[\phi_0(t)\right]\right]Sin\left[\phi(t)\right]+\\ &\left[\hat{\boldsymbol{x}}\,\ell_x\left[\phi(t)\right]+\hat{\boldsymbol{y}}\,\ell_y\left[\phi(t)\right]+\hat{\boldsymbol{z}}\,\ell_z\left[\phi(t)\right]\right]\end{aligned} \qquad (12)$$

$$d(\phi)=\left|\vec{\boldsymbol{d}}(\phi)\right|=\sqrt{d^2(\phi)}=\sqrt{\vec{\boldsymbol{d}}(\phi)\cdot\vec{\boldsymbol{d}}(\phi)} \qquad (13)$$

Herhangi bir faz açısı ϕ (2)'deki vektörel mesafe olan $\vec{\boldsymbol{d}}(\phi)$ (12), tanımlamış olduğumuz ve büyüklükleri $X(\phi)$ (14) ve $Y(\phi)$ (15) olan sanal vektörler $\vec{\mathbf{X}}(\phi)$ (14) ve $\vec{\mathbf{Y}}(\phi)$ (15) cinsinden, aynı zamanda tanımlamış olduğumuz ve büyüklükleri a (18) ve b (19) olan sanal vektörler $\vec{\boldsymbol{a}}$ (16) ve $\vec{\boldsymbol{b}}$ (17) cinsinden, eşdeğer olarak $\vec{\boldsymbol{d}}(\phi)$ (10)'daki gibi ifade edilebilir. Faz farkı ϕ_0 (3) bütün ϕ (2) için sabit olduğunda a (18) ve b (19) de sabittir, ama eğer $\left[\phi_0=\phi_0(t)\right]$ (3) zamana (t) bağımlı ise a (18) ve b (19) de öyledir.

$$\vec{\mathbf{X}}(\phi)=\vec{\boldsymbol{a}}\,Cos\,\phi \quad ; \quad \vec{\mathbf{X}}(\phi)\cdot\vec{\mathbf{X}}(\phi)=X^2(\phi)=\vec{\boldsymbol{a}}\cdot\vec{\boldsymbol{a}}\ Cos^2\phi=a^2Cos^2\phi \quad ; \quad \left|\vec{\mathbf{X}}(\phi)\right|=X(\phi) \qquad (14)$$

$$\vec{\mathbf{Y}}(\phi)=\vec{\boldsymbol{b}}\ Sin\phi \quad ; \quad \vec{\mathbf{Y}}(\phi)\cdot\vec{\mathbf{Y}}(\phi)=Y^2(\phi)=\vec{\boldsymbol{b}}\cdot\vec{\boldsymbol{b}}\ Sin^2\,\phi=b^2Sin^2\,\phi \quad ; \quad \left|\vec{\mathbf{Y}}(\phi)\right|=Y(\phi) \qquad (15)$$

$$\vec{\boldsymbol{a}}=\vec{\boldsymbol{a}}(t)=\hat{\boldsymbol{x}}\left\{r_1Cos\,\beta\,Cos\left[\phi_0(t)\right]-r_2\right\}+\hat{\boldsymbol{y}}\,r_1Sin\left[\phi_0(t)\right]+\hat{\boldsymbol{z}}\,r_1Sin\,\beta\,Cos\left[\phi_0(t)\right]=\left(\vec{\boldsymbol{r}}_1-\vec{\boldsymbol{r}}_2\right)(\phi=0) \qquad (16)$$

$$\vec{\boldsymbol{b}}=\vec{\boldsymbol{b}}(t)=-\hat{\boldsymbol{x}}\,r_1Cos\,\beta\,Sin\left[\phi_0(t)\right]+\hat{\boldsymbol{y}}\left\{r_1Cos\left[\phi_0(t)\right]-r_2\right\}-\hat{\boldsymbol{z}}\,r_1Sin\,\beta\,Sin\left[\phi_0(t)\right]=\left(\vec{\boldsymbol{r}}_1-\vec{\boldsymbol{r}}_2\right)\left(\phi=\frac{\pi}{2}\right) \qquad (17)$$

$$\vec{\boldsymbol{a}}(t)\cdot\vec{\boldsymbol{a}}(t)=a^2(t)=r_1^{\,2}-2\,r_1\,r_2\,Cos\beta\,Cos\left[\phi_0(t)\right]+r_2^{\,2} \quad ; \quad a=a(t)=\left|\vec{\boldsymbol{a}}(t)\right|=\sqrt{\vec{\boldsymbol{a}}(t)\cdot\vec{\boldsymbol{a}}(t)} \qquad (18)$$

$$\vec{\boldsymbol{b}}(t)\cdot\vec{\boldsymbol{b}}(t)=b^2(t)=r_1^{\,2}-2\,r_1\,r_2\,Cos\left[\phi_0(t)\right]+r_2^{\,2} \quad ; \quad b=b(t)=\left|\vec{\boldsymbol{b}}(t)\right|=\sqrt{\vec{\boldsymbol{b}}(t)\cdot\vec{\boldsymbol{b}}(t)} \qquad (19)$$

$\vec{\mathbf{X}}(\phi)$ (14), $\vec{\mathbf{Y}}(\phi)$ (15), $\vec{\boldsymbol{a}}$ (16), ve $\vec{\boldsymbol{b}}$ (17) vektörlerinin tanımlamalarına dayanarak ve (20) denklemindeki trigonometrik özdeşlik nedeniyle, (20) denklemindeki bağıntı doğrudur ve her ϕ (2) için geçerlidir. Bu nedenle, (20) denklemindeki bağıntı, $\left[\vec{\mathbf{X}}(\phi),\vec{\mathbf{Y}}(\phi)\right]$ (14) - (15) vektör çifti için (22) denklemindeki bağıntının geçerli olduğunu ve $\left[X(\phi), Y(\phi)\right]$ (14) - (15) büyüklük çifti için (21) denklemindeki bağıntının geçerli olduğunu ortaya çıkarmaktadır.

$$\frac{\vec{\mathbf{X}}(\phi)\cdot\vec{\mathbf{X}}(\phi)}{\vec{a}\cdot\vec{a}}+\frac{\vec{\mathbf{Y}}(\phi)\cdot\vec{\mathbf{Y}}(\phi)}{\vec{b}\cdot\vec{b}}=\frac{X^2(\phi)}{a^2}+\frac{Y^2(\phi)}{b^2}=Cos^2\phi+Sin^2\phi=1 \qquad (20)$$

$$\boxed{\frac{X^2(\phi)}{a^2}+\frac{Y^2(\phi)}{b^2}=1} \qquad (2-\mathit{Boyutta\ Skaler\ Elips\ Tanımı}) \qquad (21)$$

$$\boxed{\frac{\vec{\mathbf{X}}(\phi)\cdot\vec{\mathbf{X}}(\phi)}{\vec{a}\cdot\vec{a}}+\frac{\vec{\mathbf{Y}}(\phi)\cdot\vec{\mathbf{Y}}(\phi)}{\vec{b}\cdot\vec{b}}=1} \qquad (3-\mathit{Boyutta\ Vektörel\ Elips\ Tanımı}) \qquad (22)$$

$(a>b)$ (23) olduğunda a (18) yarı büyük eksen ve b (19) yarı küçük eksen iken diğer durumda tam tersi olacak şekilde (21) denklemi iki boyutta bir elipsi tanımlayan denklem olduğu için, $(a=b)$ (23) olduğunda ise (21) denklemi daire denklemi özel durumuna indirgendiği için, (22) denkleminin, $\left[\vec{\mathbf{X}}(\phi), \vec{\mathbf{Y}}(\phi)\right]$ (14) - (15) vektör çiftinin en genel anlamda üç boyutta vektörel bir elips üzerindeki noktaları tanımladığını gösterdiğini, $\vec{a}$ (16) ve $\vec{b}$ (17)'nin de onun yarı büyük eksen ve yarı küçük eksen vektörleri olduğunu iddia etmekteyiz. Başka bir deyişle, merkezleri sabit veya değişken bir $\vec{\ell}(\phi)$ (8) vektörü ile birbirinden ayrılmış olan iki daire çevresinde hareket eden iki $\mathbf{P}_1$ ve $\mathbf{P}_2$ noktası arasındaki vektörel mesafe olan $\vec{d}(\phi)$ (12), matematiksel olarak eşdeğer şekilde, her bir ϕ (2)'de sanal bir sıfır noktasına göre konumlarının $\left[\vec{\mathbf{X}}(\phi), \vec{\mathbf{Y}}(\phi)\right]$ (14) - (15) vektör çiftinin toplamı ile belirlendiği sanal bir vektörel elips üzerindeki noktaların, aynı sanal sıfır noktasından sabit veya değişken $\left[-\vec{\ell}(\phi)\right]$ (8) vektörü ile ayrılan bir noktaya göre olan $\vec{d}(\phi)$ (10) mesafesi olarak ifade edilip yorumlanabilir, ki burada $\vec{a}$ (16) ve $\vec{b}$ (17) bu vektörel elipsin sabit veya değişken yarı büyük eksen ve yarı küçük eksen vektörleridir.Bu sonuç, $\left[\phi_0=\phi_0(t)\right]$ (3) faz farkının zamana(t) göre değişken bir fonksiyon olması durumunda bile matematiksel olarak geçerlidir. Ortaya çıkan bu bulgu, araştırmamızın özündeki en önemli sonuçtur. Daha da ötesi, (22) denklemi ile matematiksel olarak bir "vektörel elips" kavarmını sunmuş bulunmaktayız. Araştıramamızın sonuçları hakkında daha fazla detay kitabımızda[1] bulunabilir.

$$\begin{cases} a(t)>b(t) & \Rightarrow \quad \vec{a}(t)\ \mathit{vektörel\ elipsin\ yarı\ büyük\ ekseni\ ve}\ \vec{b}(t)\ \mathit{yarı\ küçük\ ekseni} \\ a(t)<b(t) & \Rightarrow \quad \vec{b}(t)\ \mathit{vektörel\ elipsin\ yarı\ büyük\ ekseni\ ve}\ \vec{a}(t)\ \mathit{yarı\ küçük\ ekseni} \\ a(t)=b(t) & \Rightarrow \quad \vec{a}(t)\ \mathit{ve}\ \vec{b}(t)\ \mathit{bir\ vektörel\ dairenin\ yarıçapları} \end{cases} \qquad (23)$$

Vektörel elipsin anlık odak mesafesi c (24), (18) - (19)'daki denklemler kullanılarak belirlenebilir ve vektörel elipsin anlık eksantrikliği e (25) de (18) - (19) ve (24)'deki denklemler kullanılarak bulunabilir.

$$c=c(t)=\sqrt{c^2(t)}=\sqrt{\left|a^2(t)-b^2(t)\right|}=\sqrt{2\,r_1 r_2\,(1-Cos\beta)\left|Cos\left[\phi_0(t)\right]\right|} \qquad (\mathit{Odak\ Mesafesi}) \quad (24)$$

$$e=e(t)=\begin{cases} \dfrac{c(t)}{a(t)}=\sqrt{\dfrac{2\,r_1 r_2 (1-Cos\beta)\left|Cos[\phi_0(t)]\right|}{r_1^2-2r_1 r_2\,Cos\beta\,Cos[\phi_0(t)]+r_2^2}} & ,\quad a(t)>b(t)\quad ise \\ \dfrac{c(t)}{b(t)}=\sqrt{\dfrac{2\,r_1 r_2 (1-Cos\beta)\left|Cos[\phi_0(t)]\right|}{r_1^2-2r_1 r_2 Cos[\phi_0(t)]+r_2^2}} & ,\quad a(t)<b(t)\quad ise \end{cases} \qquad (Eksantriklik)\ (25)$$

İlgili iki daire üzerinde hareket eden $\mathbf{P}_1$ ve $\mathbf{P}_2$ noktaları için (1)'den (25)'e kadar olan bu analiz ve matematiksel sonuçların fizikte yol açtığı ana gelişme, "Uzayda farklı dairesel yörüngeler çevresinde hareket eden parçacık veya cisimlerin kendilerini, birbirlerine göre eliptik güzergahlarda konumlanmış gördüğü, parçacık veya cisimlerin açısal hızlarının zamana bağımlılıklarına göre bu eliptik yörüngelerin sabit veya zamana göre değişken yarı büyük eksen ve yarı küçük eksenlere sahip olduğu, parçacık veya cisimlerin karşı taraftakini de anlık olarak uzayda sabit bir noktada sanal olarak konumlanmış gözlemlediği, bu konumun da etrafında döndükleri şahsi dairesel yörüngelerinin merkezleri arasındaki mesafe vektörü tarafından belirlendiği"dir.

Bu matematiksel bulgu, fizikte çok geniş kapsamlı keşiflere yol açma potansiyeline sahiptir. Bu, Güneş, gezegenler ve Güneş Sistemi ile başka yerlerdeki uydular, ayrıca parçacık ve atom altı seviyesi de dahil olmak üzere evrendeki cisimlerin hareketlerine ilişkin, "bütün cisimlerin uzayda, farklı yarıçaplara ve devrim merkezlerine sahip kendilerine ait dairesel yörüngeler çevresinde bir açısal hızla hareket ettiği", fakat "her birinin birbirlerine göre sanki eliptik bir yörüngede hareket ediyormuş gibi göründüğü, diğer cismi de sanal karşılıklı hareket düzlemlerinde sabit veya değişken bir noktada konumlanmış gibi gördüğü" yeni bir temel modelin formülasyonuna yol açacaktır. Dolayısıyla bu da doğadaki kuvvetlere ilişkin daha çok içgörü sağlayacaktır.

Gösterilmiş olan matematiksel analize dayanarak, zaman içerisinde birbirlerine göre belirgin eliptik yörüngeler sergilemelerinden dolayı, Güneş, Dünya ve Ay'ın her birinin uzayda kendi şahsi dairesel yörüngelerinde döndüğü iddiasında bulunuyoruz. Bu beklentiyle, gözlemlenen Güneş ile Dünya arası ve Dünya ile Ay arası mesafe verilerine dayanarak ve aynı zamanda bu araştırma kapsamında geliştirilen analitik yöntemleri yaklaşık olarak kullanarak, (26)'dan (57)'ye kadar olan deklemlerde sunulduğu şekilde Güneş, Dünya ve Ay'ın yörüngesel parametreleri hesaplanmıştır. Analizimiz, gözlemle uyuşan ve böylelikle Güneş, Dünya ve Ay'ın gerçekte şahsi dairesel yörüngelerde döndüğü iddiamızı destekleyen ek sonuçlar da ortaya koymuştur, ki bunlar da kitabımızda[1] bulunabilir.

Gözlemlere ve araştırmamız sonucundaki bulgularımıza dayanarak, daha geniş kapsamlı olarak şunu keşfetmiş bulunuyor ve iddia ediyoruz: "Uzayda hareket eden cisimler, aynı zamanda kendi şahsi ekvator düzlemlerini de belirleyen devrim düzlemleri içerisinde, şahsi dairesel devrim yörüngeleri etrafında dönüyor olmalılar. Kendi kendine devrim eksenini de 'sağ el kuralı'na göre cismin devrim düzleminin normali ve şahsi dairesel yörüngesi etrafında dönüş yönü belirler. Uzayda her cismin kendi etrafında dönüş ekseninin yönü de 'sağ el kuralı'na göre cismin kendi etrafında dönüş ekseninin yönüne göre belirlenir. Uzayda hareket eden her cismin kendi kendine devrim ekseni ve kendi etrafında dönüş ekseni, ekvator düzleminin normali ile hizalı olmalıdır. Uzayda hareket eden her cismin 'Faraday'ın İndüksiyon Kanunu'na uyması gerektiği için, uzaydaki bütün cisimlerin kuzey kutuplarının bakış açısından saat yönünde (Devrim Yörüngesi 'Batı'sından Devrim Yörüngesi 'Doğu'suna doğru) dönerken, muhtemelen şahsi dairesel devrim yörüngelerinin alanından geçen değişken akıya sahip harici manyetik alanlarda hareket eden makro ölçekli yüklü parçacıklar gibi davrandığını, bunun da kuzey kutuplarının bakış açısından ters (saat yönünün tersine) yönde (Kendi Etrafında Dönüş 'Batı'sından Kendi Etrafında Dönüş 'Doğu'suna doğru) kendi etrafında dönüş hareketini tetikleyen asıl etken olmasını beklediğimizi, dolayısıyla evrendeki cisimlerin bu bağlamda iki batısı ve iki doğusu olduğunu iddia ediyoruz, ki bir cisim veya devrimin kuzey kutbunun, her zaman için 'sağ el kuralı'na göre cismin veya devrimin saat yönünün tersine dönen kutbu olduğunu kabul ediyoruz. Bu iddiaya dayalı olarak, aynı zamanda, bir cismin başka bir cisme göre olan hareketindeki eksen eğikliğinin, topolojik olarak uzaydaki şahsi dairesel devrim yörüngelerinin düzlemleri arasındaki yörünge eğiklik açısına dayalı olması gerektiğini de çıkartıyoruz." Bu iddia, detayları kitabımızda[1] bulunabilecek olan, Güneş-Dünya-Ay sistemine dair araştırma sonuçlarımız tarafından da desteklenmektedir.

Uzaydaki gerçek Güneş-Dünya-Ay topolojisinin, ölçekli olmamak kaydıyla kabaca **Figür 2**'deki konfigürasyonla temsil edilebileceğini bulmuş durumdayız, ki aslında bu da **Figür 1**'in Güneş-Dünya konfigürasyonu ile **Figür 1**'in Dünya-Ay konfigürasyonunun üst üste bindirilmiş halidir. **Figür 2**'deki Kartezyen $\left(\hat{x},\hat{y},\hat{z}\right)$ koordinatlı konfigürasyonda dikkat ediniz ki, Dünya'nın şahsi dairesel devrim yörüngesinin düzlemi yatay $\hat{x}-\hat{y}$ düzlemi olarak alınmış ve $\hat{z}$ − ekseninden aşağı $\hat{x}-\hat{y}$ düzlemine doğru bakarken Dünya'nın şahsi dairesel devrim yörüngesinin yönü de $\left[+\phi(t)\right]$ (2) doğrultusunda saat yönünün tersine olarak alınmıştır, ki bu da Dünya'nın kuzey kutbu üzerinden bir bakış açısından saat yönüne karşılık gelmektedir, çünkü hesaplamalarımız **Figür 2**'deki konfigürasyonda, aynı zamanda Dünya'nın kuzey kutbu yönü olan, Dünya'nın kendi etrafında dönüş ekseni doğrultusudaki birim vektörün $\left(\hat{\boldsymbol{u}}_{Dünya\perp}=-\hat{z}\right)$ (26) olduğunu ortaya çıkarmıştır.

$$\hat{\boldsymbol{u}}_{Dünya\perp}=\hat{\boldsymbol{u}}_{Earth\perp}=-\hat{z}\qquad \left(Dünya\text{'}nın\ Kuzey\ Kutbu\ Birim\ Vektörü\right)\qquad (26)$$

Güneş'in şahsi dairesel devrim yörüngesinin düzleminin, Dünya'nın şahsi dairesel devrim yörüngesinin düzlemine göre eğiklik açısı $\beta_{Güneş-Dünya}$ (27) olarak bulunmuştur.

$$\beta_{Güneş-Dünya}=\beta_{Sun-Earth}\simeq\mathbf{-156^{\circ}.56}\simeq\mathbf{-2.732513127}\ radyan$$
$$\left(Güneş\ ve\ Dünya\ devrim\ düzlemleri\ arasındaki\ açı\right)\qquad (27)$$

Gözleme dayalı olan analizimiz aynı zamanda, Güneş'in açısal olarak Dünya'nın kuzey kutbuna yakın olan kutup vektörünün, Güneş'in güney kutbu vektörü olması gerektiği sonucuna yol açmıştır. Buna müteakip, **Figür 2** konfigürasyonunda Güneş'in kuzey kutbunun yönünü, aynı zamanda Güneş'in kendi etrafında dönme ekseni doğrultusundaki birim vektörü $\hat{\boldsymbol{u}}_{Güneş\perp}$ (28) olarak bulmuş olduk. Dahası, Güneş'ten Dünya'ya doğru olan mesafe vektörü $\left[-\vec{\boldsymbol{d}}(\phi)\right]$ (10)'ye göre Güneş'in eksen eğikliği de hesaplamalarımıza göre yıl boyunca 90° civarındadır ve ayrıca yıl boyunca Dünya'dan gözlemlendiği şekilde Güneş'in yalpalama açısı ve yana doğru sallanma açısını da detaylı olarak kitabımızda[1] hesaplamış bulunmaktayız.

$$\hat{\boldsymbol{u}}_{Güneş\perp}=\hat{\boldsymbol{u}}_{Sun\perp}=-\hat{x}Sin\beta_{Sun-Earth}+\hat{z}Cos\beta_{Sun-Earth}\qquad \left(Güneş\text{'}in\ Kuzey\ Kutbu\ Birim\ Vektörü\right)\ (28)$$

Analizimiz, Ay'ın şahsi dairesel devrim yörüngesinin düzleminin, Dünya'nın şahsi dairesel devrim yörüngesinin düzlemine göre eğiklik açısının $\beta_{Ay-Dünya}$ (29) olduğunu, buna karşılık

Figür 2'deki konfigürasyonda, Ay'ın kuzey kutbunun yönü de olan Ay'ın kendi etrafında dönme ekseni doğrultusundaki birim vektörünün $\hat{\boldsymbol{u}}_{Ay\perp}$ (30) olduğunu ortaya çıkarmıştır.

$$\beta_{Ay-Dünya} = \beta_{Moon-Earth} \simeq 24^{\circ} \qquad \left(Ay\ ve\ Dünya\ devrim\ düzlemleri\ arasındaki\ açı\right) \quad (29)$$

$$\hat{\boldsymbol{u}}_{Ay\perp} = \hat{\boldsymbol{u}}_{Moon\perp} = \hat{x}Sin\beta_{Moon-Earth} - \hat{z}Cos\beta_{Moon-Earth} \quad \left(Ay'ın\ Kuzey\ Kutbu\ Birim\ Vektörü\right) \quad (30)$$

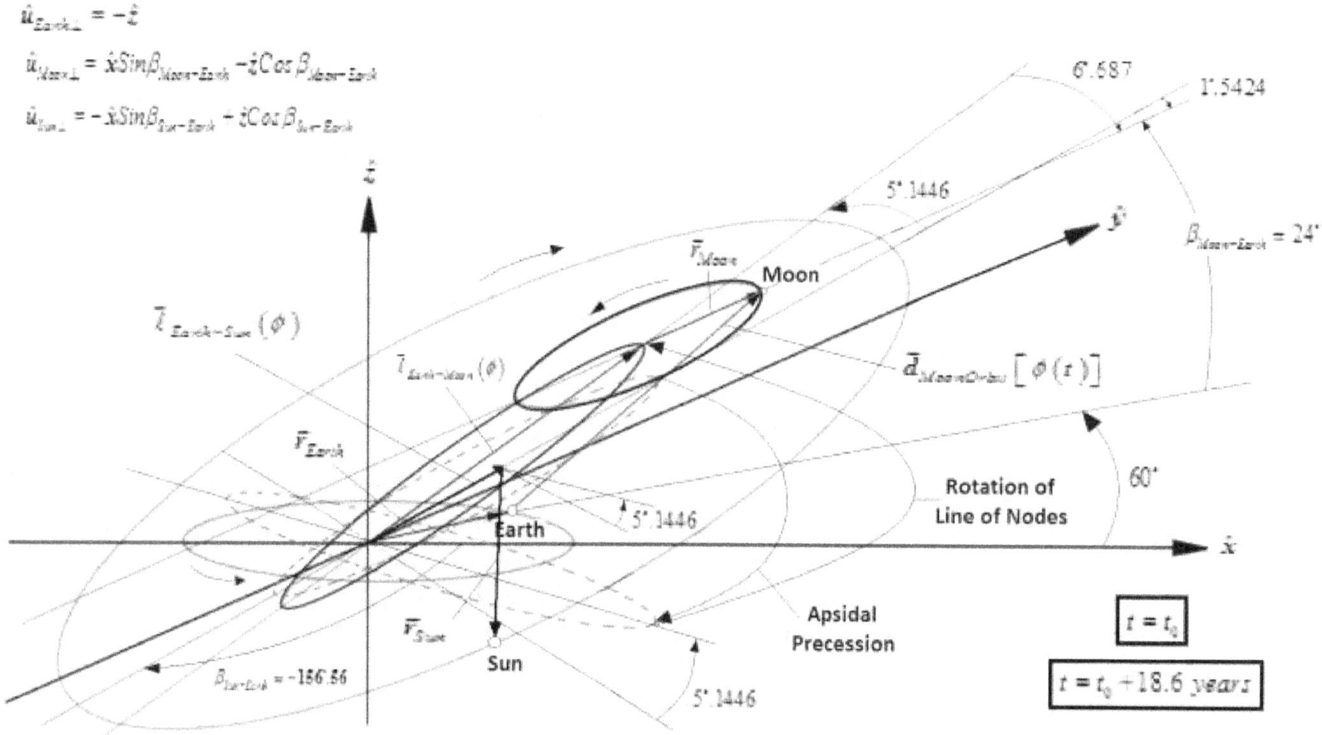

Figür 2 Dairesel Yörünge Modeline Göre Güneş-Dünya-Ay Konfigürasyonu

Yıl boyunca Güneş ile Dünya'nın kendi şahsi devrim yörüngelerindeki sabit faz farklarının $\left(\phi_{0,Dünya-Güneş} \simeq \mathbf{-151^{\circ}.11}\right)$ (31) olduğunu bulmuş bulunmaktayız, öyle ki kendi şahsi devrim yörüngesinde Dünya, kendi şahsi devrim döngüsündeki Güneş'ten **151°.11** (31) faz farkı ile önde gitmektedir. **Figür 2**'deki Güneş-Dünya-Ay konfigürasyonunda, **Figür 1**'in Güneş-Dünya konfigürasyonunda Dünya'nın Güneş'e göre göreceli hareketinde Dünya'nın kendi şahsi devrim yörüngesindeki faz açısı olan $\left[\phi(t) = \phi_{Dünya-Güneş}\right]$ (32)'in, **Figür 1**'in Dünya-Ay konfigürasyonunda Dünya'nın Ay'a göre göreceli hareketinde Dünya'nın kendi şahsi devrim yörüngesindeki faz açısı olan $\left[\phi(t) = \phi_{Dünya-Ay}\right]$ (32)'a göre her tarihte yaklaşık 60° ilerisinde olduğunu keşfetmiş bulunmaktayız.

$$\phi_{0,Dünya-Güneş} = \phi_{0,Earth-Sun} = \mathbf{-2.63736997198951}\ rad = \mathbf{-151^{\circ}.11} \quad (31)$$
$$\left(Güneş - Dünya\ Sistemi\ için\ Faz\ Farkı\right)$$

$$\phi_{Dünya-Güneş} = \phi_{Earth-Sun} \simeq \phi_{Earth-Moon} + 60^{\circ} = \phi_{Dünya-Ay} + 60^{\circ} \qquad \left(yıl\ boyunca\ her\ tarihte\right) \quad (32)$$

Güneş, Dünya ve Ay'ın kendi şahsi devrim yörüngeleri için, gözlemlenen yıllık Güneş ile Dünya arası mesafe verilerine ve Dünya ile Ay arası mesafe verilerine dayanarak ve gök cisimlerinin yörüngesel parametrelerini hesaplamak için geliştirmiş olduğumuz analitik bir yöntemi kullanarak, kilometre (km) ve astonomik birim (au) cinsinden yaklaşık olarak aşağıdaki $r_{Güneş}$ (33), $r_{Dünya}$ (34) ve r_{Ay} (35) yarıçap büyüklüklerini bulmuş bulunmaktayız. Analitik hesaplama ile yaklaşık değerler elde etmiş olsak da, varmış olduğumuz sonuç, Güneş, Dünya ve Ay'ın kendi şahsi devrim yörüngelerinin yarıçaplarının büyüklük ölçeklerini karşılaştırmak açısından önemlidir, çünkü $\left(r_{Dünya} \ll r_{Ay} \ll r_{Güneş}\right)$ (36)'tir.

$$r_{Güneş} = r_{Sun} = \mathbf{1.00007869128446}\ au = \mathbf{149,609,642.749}\ km \quad \left(Güneş'in\ devrim\ yörüngesinin\ yarıçapı\right) \tag{33}$$

$$r_{Dünya} = r_{Earth} = \mathbf{0.0000826915340950721}\ au = \mathbf{12,370.477}\ km \quad \left(Dünya'nın\ devrim\ yörüngesinin\ yarıçapı\right) \tag{34}$$

$$r_{Ay} = r_{Moon} \simeq \mathbf{3,302,009.781}\ km \simeq \mathbf{0.0220725720604161}\ au \quad \left(Ay'ın\ devrim\ yörüngesinin\ yarıçapı\right) \tag{35}$$

$$\left(r_{Dünya} \simeq \mathbf{12,370.477}\ km\right) \ll \left(r_{Ay} \simeq \mathbf{3,302,009.781}\ km\right) \ll \left(r_{Güneş} \simeq \mathbf{149,609,642.749}\ km\right) \tag{36}$$

Güneş, Dünya ve Ay için bulunan yörüngesel parametre değerlerine dayanarak en ilginç ve kayda değer çıkarım, elde edilen sonuçlar Dünya'nın Güneş etrafında dönmediğine ve Ay'ın da direk olarak Dünya etrafında dönmediğine işaret ediyor, ki aslında sonuçlar Güneş'in bir yörüngesel eğiklik ile Dünya ve Ay'ın şahsi dairesel yörüngelerini de içinde kapsayan daha büyük bir şahsi dairesel yörünge etrafında döndüğünü, her birinin devrim merkezinin de birbirininkine göre kayık olduğunu gösteriyor. Ay'ın devrim merkezi, diğer cisimlerin devrim merkezlerine göre devamlı olarak harmonik hareket içerisinde, esasen de Dünya'nın devrim merkezine göre yaklaşık olarak 18.6 yıllık bir döngü içerisinde, ki bu da daha detaylı olarak kitabımızda[1] analiz edilmiş bulunmaktadır. Dolayısıyla, gezegenlerin yıldızlar etrafında dönmesinin şart olmadığı ve Dünya'nın Güneş etrafında dönmediği sonucuna varmış bulunmaktayız. Bu bulgular, yeni bir vizyon getirerek, bizim Güneş Sistemimiz olduğu kadar genel olarak evrende gözlemlenen güneş sistemlerinin oluşumu, beklenen hareketsel davranışı ve topolojileri üzerine var olan teorileri yeniden değerlendirme ihtiyacını doğuruyor.

Figür 1'in Güneş-Dünya konfigürasyonunda, Dünya'nın ekinokslarda, gün dönümlerinde ve maksimum ile minimum Dünya ile Güneş arası mesafe tarihlerindeki $\left[\phi(t) = \phi_{Dünya-Güneş}\right]$ (2) faz açısı değerleri, aynı zamanda da Dünya-Güneş yıllık döngüsünde bazı önemli faz açısı değerlerinin tarihleri, (37)'den (46)'ya kadar listelenmiştir.

$$\phi_{Dünya-Güneş,Min\ Güneş-Dünya\ Mesafesi} = -\mathbf{0.193236295254927}\ radyan = -\mathbf{11^{\circ}.072} \quad \left(3\ Ocak\right) \tag{37}$$

$$\phi_{Dünya-Güneş} = \mathbf{0.00058673254949726}\ radyan = \mathbf{0^{\circ}.034} \simeq 0 \quad \left(17\ Ocak\right) \tag{38}$$

$$\phi_{Dünya-Güneş,İlkbahar\ Ekinoksu} = \mathbf{1.02676771671574}\ radyan = \mathbf{58^{\circ}.829} \quad \left(20\ Mart\right) \tag{39}$$

$$\phi_{Dünya-Güneş} = \mathbf{1.57138305934439}\ radyan = \mathbf{90^{\circ}.034} \simeq \frac{\pi}{2} \quad \left(18\ Nisan\right) \tag{40}$$

$$\phi_{Dünya-Güneş,Yaz\ Gün\ Dönümü} = -\phi_{0,Dünya-Güneş} = \mathbf{2.63736997198951}\ radyan = \mathbf{151^{\circ}.11} \quad (21\ Haziran) \quad (41)$$

$$\phi_{Dünya-Güneş,Max\ Güneş-Dünya\ Mesafesi} = \mathbf{2.94835635833487}\ radyan = \mathbf{168^{\circ}.928} \quad (5\ Temmuz) \quad (42)$$

$$\phi_{Dünya-Güneş} = \mathbf{3.14217938613929}\ radyan = \mathbf{180^{\circ}.034} \simeq \pi \quad (18\ Temmuz) \quad (43)$$

$$\phi_{Dünya-Güneş,Sonbahar\ Ekinoksu} = \mathbf{-2.05488176708229}\ radyan = \mathbf{242^{\circ}.264} = \mathbf{-117^{\circ}.736} \quad (22\ Eylül) \quad (44)$$

$$\phi_{Dünya-Güneş} = \mathbf{-1.5702095942454}\ radyan = \mathbf{-89^{\circ}.966} \simeq -\frac{\pi}{2} \quad (17\ Ekim) \quad (45)$$

$$\phi_{Dünya-Güneş,Kış\ Gün\ Dönümü} = \pi - \phi_{0,Dünya-Güneş} = \mathbf{5.7789626255793}\ radyan = \mathbf{-28^{\circ}.89} \quad (21\ Aralık) \quad (46)$$

Figür 1'in Güneş-Dünya konfigürasyonunda aynı zamanda, Güneş-Dünya sisteminin göreceli hareket elipsi için, sabit yarı küçük eksen ve yarı büyük eksen vektörleri $\left[\vec{a}(t) = \vec{a}_{Dünya-Güneş}\right]$ (47) ve $\left[\vec{b}(t) = \vec{b}_{Dünya-Güneş}\right]$ (48)'yi, ayrıca onların sabit büyüklükleri olan $\left[a(t) = a_{Dünya-Güneş}\right]$ (49) ve $\left[b(t) = b_{Dünya-Güneş}\right]$ (50) ile odak mesafesi $\left[c(t) = c_{Dünya-Güneş}\right]$ (51) ve eksantriklik $\left[e(t) = e_{Dünya-Güneş}\right]$ (52) değerlerini belirlemiş bulunmaktayız.

$$\vec{a}_{Dünya-Güneş} = \hat{x}\,120{,}168{,}467.040\ km - \hat{y}\,72{,}280{,}843.623\ km + \hat{z}\ 52{,}106{,}536.351\ km \quad (47)$$

$$\vec{b}_{Dünya-Güneş} = -\hat{x}\ 66{,}316{,}021.722\ km - \hat{y}\,131{,}002{,}922.948\ km - \hat{z}\,28{,}752{,}488.898\ km \quad (48)$$

$$a_{Dünya-Güneş} = \mathbf{1.00001226586661}\ au = \mathbf{149{,}599{,}705.647527}\ km \quad (Güneş - Dünya\ yarı\ küçük\ eksen\ büyüklüğü) \quad (49)$$

$$b_{Dünya-Güneş} = \mathbf{1.00015109267839}\ au = \mathbf{149{,}620{,}473.842965}\ km \quad (Güneş - Dünya\ yarı\ büyük\ eksen\ büyüklüğü) \quad (50)$$

$$c_{Dünya-Güneş} = \mathbf{0.0166636221185271}\ au = \mathbf{2{,}492{,}842.4}\ km \quad (Güneş - Dünya\ sisteminin\ odak\ mesafesi) \quad (51)$$

$$e_{Dünya-Güneş} = \frac{c_{Dünya-Güneş}}{b_{Dünya-Güneş}} = \mathbf{0.0166611047475858} \quad (Güneş - Dünya\ sisteminin\ eksantrikliği) \quad (52)$$

Dünya ile Güneş'in şahsi dairesel devrim yörüngelerinin merkezleri arasındaki mesafe vektörü $\left\{\vec{\ell}\left[\phi(t)\right] = \vec{\ell}_{Dünya-Güneş}(\phi)\right\}$ (8)'in büyüklüğü olan $\left[\ell(\phi) = \ell_{Dünya-Güneş}(\phi)\right]$ (9), yıllık Güneş-Dünya döngüsü boyunca üç ana belirgin salınım frekansı sergilemekedir. Bir salınım yılda yaklaşık 12 keredir, ki görünüşe göre göreceli Dünya-Ay hareketinin etkisine dayalıdır. İkinci salınım, göründüğü kadarıyla Dünya'nın kendi ekseni etrafında günlük dönüşüne dayalıdır. Bir Güneş-Dünya yılı boyunca yaklaşık 4 kere olan üçüncü salınımın muhtemel sebebi, bir Güneş-Dünya yılı boyunca yaklaşık dört döngüsü olan göreceli Güneş-Merkür hareketinin etkisi olabilir. Güneş ve Merkür'ün uzayda, Dünya-Ay sisteminde olduğu gibi Merkür'ün Güneş'in uydusu gibi davrandığı, bir ikiz ya da ikili gök cismi sistemi oluşturmuş olması mümkündür ve bu da aynı zamanda, Merkür'ün yörünge parametreleri hesaplandığı zaman teyit edilmek üzere, Güneş ve Merkür'ün şahsi dairesel devrim yörüngelerinin birbiri etrafında dönüyor olabilmesi olasılığını doğurmaktadır. Analizimiz aynı zamanda, başka gezegen ve uyduların ve hatta Güneş Sistemi içerisindeki başka hareket eden cisimlerin göreceli hareketlerinin, Güneş ve Dünya'nın göreceli hareketini $\left[\ell(\phi)=\ell_{Dünya-Güneş}(\phi)\right]$ (9)'nin ek küçük salınımları cinsinden etkiliyor olabilme olasılığı olduğu sonucunu da çıkarmaktadır, ki bunlar sadece bir Güneş-Dünya yılından daha uzun dönemlerde gözlemlenebilir.

Yıllık bir Güneş-Dünya döngüsü boyunca, Güneş ve Dünya'nın şahsi dairesel devrim yörüngelerinin merkezleri arasındaki maksimum mesafe olan $\left[\ell_{Dünya-Güneş}(\phi)\right]_{Max}$ (53) 23 Şubat, 26 Mayıs, 25 Ağustos ve 24 Kasım'da, Güneş ve Dünya'nın şahsi dairesel devrim yörüngelerinin merkezleri arasındaki minimum mesafe olan $\left[\ell_{Dünya-Güneş}(\phi)\right]_{Min}$ (54) ise 1 Ocak, 2 Nisan, 2 Temmuz ve 1 Ekim'de gerçekleşmektedir.

$$\left[\ell_{Dünya-Güneş}(\phi)\right]_{Max} = \mathbf{0.0168346184954446}\ au\ (23\ Şubat, 26\ Mayıs, 25\ Ağustos, 24\ Kasım) \quad (53)$$

$$\left[\ell_{Dünya-Güneş}(\phi)\right]_{Min} = \mathbf{0.0165381684638675}\ au \quad (1\ Ocak, 2\ Nisan, 2\ Temmuz, 1\ Ekim) \quad (54)$$

Dünya-Ay sisteminin göreceli vektörel elipsi için, zamana(t) göre değişken yarı küçük eksen ve yarı büyük eksen vektörleri $\left[\vec{a}(t)=\vec{a}_{Dünya-Ay}(t)\right]$ (55) ve $\left[\vec{b}(t)=\vec{b}_{Dünya-Ay}(t)\right]$ (56) ile anlık odak mesafesi olan $\left[c(t)=c_{Dünya-Ay}(t)\right]$ (57)'ın değerlerini de yaklaşık olarak belirlemiş bulunmaktayız.

$$\begin{aligned}\vec{a}_{Dünya-Ay}(t) \simeq\ & \hat{x}\left[3{,}016{,}536.037\,Cos(0.212768714989863t)-12{,}370.477\right] km \\ & +\hat{y}\,3{,}302{,}009.781\,Sin(0.212768714989863t)\ km \\ & +\hat{z}\,1{,}343{,}048.374\,Cos(0.212768714989863t)\ km\end{aligned} \quad (55)$$

$$\begin{aligned}\vec{b}_{Dünya-Ay}(t) =\ & -\hat{x}\,3{,}016{,}536.037\,Sin(0.212768714989863t)\ km \\ & +\hat{y}\left[3{,}302{,}009.781\,Cos(0.212768714989863t)-12{,}370.477\right] km \\ & -\hat{z}\,1{,}343{,}048.374\,Sin(0.212768714989863t)\ km\end{aligned} \quad (56)$$

$$c_{Dünya-Ay}(t) = 7{,}062{,}892{,}780.525\left|Cos(0.212768714989863t)\right|\ km^2 \quad (Odak\ Mesafesi) \quad (57)$$

Burada (55) - (57) arası denklemlerde dikkat ediniz ki, analizimize dayanarak zaman(t), Dünya'nın her 3 yıllık döngüsünde bir, 18 Mart 2017 ve 24 Mart 2020 gibi Mart ayında gerçekleşen, Dünya'dan Ay'a mesafe maksimumuna denk gelen bir tarihten [*günler*] cinsinden ölçülmelidir.

Eksen eğiklikleri, yalpalama açıları, librasyon ve Dünya ile Ay'ın devrim merkezleri arasındaki mesafe vektörünün oryantasyon varyasyon döngüleri gibi araştırma sonuçlarımız üzerine daha fazla detay kitabımızda[1] bulunabilir.

REFERANSLAR

Bu makale, aşağıdaki kitapta detaylı bir şekilde yayınlanmış olan araştırma sonuçlarının bir özetidir.

1. Tan, A.P., *Everything is a Circle: A New Model For Orbits of Bodies in the Universe*, Amazon Kindle Publishing, 2020, ISBN & EAN 979-8-56-757466-9

www.ingramcontent.com/pod-product-compliance
Lightning Source LLC
Chambersburg PA
CBHW081424220526
45466CB00008B/2264
* 9 7 9 8 5 6 7 5 7 4 6 6 9 *